W0230075

ANALOG AND DIGITAL ELECTRONICS

ANALOG AND DIGITAL ELECTRONICS

For B.Sc. (Physics Honours), B.Sc. (Electronics Pass & Honours) and B.Tech. (Electronics & Communication)

Taraprasad Chattopadhyay

M.Sc., Ph.D., SMIEEE (USA), FIETE
Professor, Department of Physics,
Visva-Bharati (a central University),
Santiniketan, West Bengal, India

CBS

CBS PUBLISHERS & DISTRIBUTORS PVT. LTD.
New Delhi • Bangalore • Pune • Cochin • Chennai (India)

ISBN : 978-81-239-1626-2

First Edition : 2008
Reprint : 2010

Copyright © 2008 by Taraprasad Chattopadhyay

All rights reserved. No part of this book may be reproduced or transmitted in any form or by any means, electronic or mechanical, including photocopying, recording, or any information storage and retrieval system without permission, in writing, from the publisher.

Published by Satish Kumar Jain and produced by V.K. Jain for
CBS Publishers & Distributors Pvt. Ltd.,
CBS Plaza, 4819/XI Prahlad Street, 24 Ansari Road, Daryaganj,
New Delhi - 110002, India.
e-mail: cbspubs@vsnl.com, cbspubs@airtelmail.in, delhi@cbspd.com
Website: www.cbspd.com

Branches:
- *Bangalore:* Seema House, 2975, 17th Cross, K.R. Road,
  Bansankari 2nd Stage, Bangalore - 560070
  Fax: 080-26771680 • e-mail: bangalore@cbspd.com
- *Pune:* Shaan Brahmha Complex, 631/632, Basement, Appa Balwant
  Chowk, Budhwar Peth, Next to Ratan Talkies, Pune - 411002
  Fax: 020-24464059 • e-mail: pune@cbspd.com
- *Cochin:* 36/14, Kalluvilakam, Lissie Hospital Road,
  Cochin - 682018, Kerala • e-mail: cochin@cbspd.com
- *Chennai:* 20, West Park Road, Shenoy Nagar,
  Chennai - 600030 • e-mail: chennai@cbspd.com

Printed at :
J.S. Offset Printers, Delhi

Preface

This book entitled, **"Analog and Digital Electronics"** is intended to serve as a text book of undergraduate students pursuing a first course in Electronics in colleges, Universities and in Technological institutions. This book is designed to cater to the needs of students studying B.Sc. Honours degree course in Physics, B.Sc. in Electronic Science, undergraduate course in Electronics in Engineering and Technological institutions. The special feature of this book is that it contains full derivations of all mathematical formulae discussed in connection with the theory of devices and circuits. The book has been written in simple language so that it becomes comprehensible for the readers. To go through this book what is required is a knowledge of algebra, trigonometry, determinants and Bessel functions. Every chapter of this book contains solved problems which will be of immense help for the students. Moreover, there are exercise problems and sample questions appended to each chapter. To make it convenient for the students, answers to exercise problems and in some cases hints to the exercise problems are also provided.

This book has sixteen chapters. The first chapter presents network theory which includes network theorems, network transformations and a RC notch filter. A study of this chapter will enable the students to follow the calculations in subsequent chapters. Although vacuum tubes may be said to be obsolete today, a brief introduction of vacuum tubes has been given in chapter 2. This will help the students to understand the history of development of electronic devices. Chapter 3 discusses the basic concepts of semiconductor physics which is essential for understanding the device physics. Chapter 4 and chapter 5 describe, respectively, two-terminal and three-terminal solid-state electronic devices – their theory and principles of operation. Although diac is a two terminal switching device, it has been placed in chapter 5 in order to keep a continuity in realizing the operation of a triac. Chapter 6 and chapter 7 present small signal bipolar junction transistor amplifiers, and power and tuned amplifiers respectively. Detailed analyses are given in these chapters. Chapter 8 describes feedback in amplifiers and chapter 9 presents the theory of transistorized sinusoidal and nonsinusoidal oscillators. Chapter 10 and 11 contain, respectively, detailed discussions on rectifiers, and on modulation, detection, radar and television. Chapter 12 presents radiowave propagation and chapter 13 describes operational amplifiers and its applications. Chapter 14 gives a detailed description of the number system related with digital electronics. Chapter 15 presents Boolean algebra and combinational logic while chapter 16 explains sequential logic.

Now, it is my turn to acknowledge the assistance and encouragement that I have received in course of writing this book. I must acknowledge the help of my Ph.D. students, Dr. Bipul Sarkar,

Dr. Sudipta Chattopadhyay, Sri Anuj Kumar Saw and Sri Sujit Das in typing the manuscript. Particularly, Dr. Bipul Sarkar has devoted much time in typing a considerable portion of the manuscript. I must acknowledge the inspiration of my wife, Dr. Madhumita Bhattacharya, Lecturer in Physics, Guskara College, Burdwan, West Bengal, in writing this book. I was also inspired by my daughter, Amrita Chattopadhyay and my son , Anish Chattopadhyay, while writing the book. I am also grateful to my mother for her encouragement.

I sincerely hope that my effort will be successful if the readers of this book find it useful for them.

February , 2008   **Taraprasad Chattopadhyay**
Santiniketan.

Contents

$$\boxed{1}$$

Network Theory

1.1 DEFINITIONS OF TERMS ASSOCIATED WITH NETWORK THEORY

- Circuit elements like resistance, inductance, capacitance, energy sources etc. are called circuit components.
- A group of circuit elements connected in series and through which the same current flows forms a branch.
- Node is a junction of two or more branches in a network.
- Loop is a closed path in an electrical circuit formed by a group of branches such that if any branch is removed the path is no longer closed. Loop is also called as mesh.
- Linear element is an element which has a linear current-voltage relationship. Resistance (R), inductance (L), capacitance (C) are linear elements since,

$$V_R = RI, \ V_L = L\frac{dI}{dt} \ \text{ and } \ V_C = \frac{1}{C}\int I\,dt$$

  where V is the voltage drop across the element, indicated by subscripts, and produced by a current I. Here, R is a constant and differentiation and integration are linear operations.
- Voltage source is a source of energy which delivers an output voltage across its terminals that is independent of the current drawn from it. The internal resistance of an ideal voltage source should be zero.
- Current source is a source of energy which delivers an output current that is independent of the output voltage existing across the terminals of the source. The internal resistance of an ideal current source is infinite.

The basic laws which make the analysis of an electrical circuit possible are Kirchhoff's laws. There are two laws of Kirchhoff — Kirchhoff's voltage law (KVL) and Kirchhoff's current law (KCL).

KVL : The algebraic sum of the potential drops across the passive circuit elements in any loop (or mesh) is equal to the algebraic sum of the emfs acting in that loop.
    The law leads to the loop or mesh analysis of a network.
Mathematically, $\Sigma \mathcal{E} = \Sigma IZ$; where $\mathcal{E}$ = emf, I = current through impedance Z. The summation Σ extends over all sources and all elements of the loop.

KCL : The algebraic sum of the currents meeting at any node or junction is zero.
    Mathematically, $\Sigma I = 0$. This law leads to the node or junction analysis of the network.

1.2 LOOP (OR MESH) ANALYSIS

Let us assume the currents flowing through different loops of the network. There are three meshes marked (1), (2), (3) of the network. shown in Fig. 1.1.

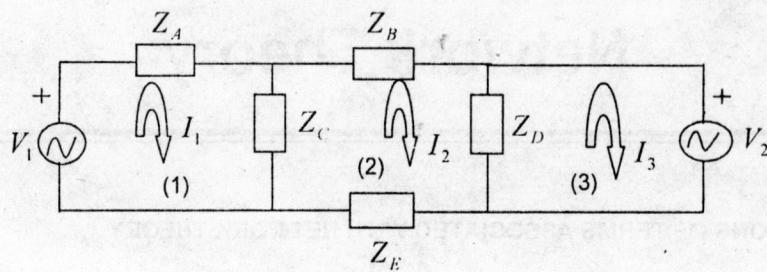

Fig. 1.1

Loop (1): $\quad I_1 Z_C - I_2 Z_C + I_1 Z_A = V_1$ $\qquad\qquad$ ...(1.1)

Loop (2): $\quad (I_2 - I_3)Z_D + I_2 Z_B + (I_2 - I_1)Z_C + I_2 Z_E = 0$ $\qquad$ ...(1.2)

Loop (3): $\quad (I_2 - I_3)Z_D = V_2$ $\qquad\qquad$ ...(1.3)

Rearranging,

$$(Z_A + Z_C)I_1 \qquad\quad - Z_C I_2 \qquad\qquad\qquad 0 \qquad = V_1 \qquad ...(1.4)$$

$$- Z_C I_1 \quad + (Z_B + Z_C + Z_D + Z_E)I_2 \quad - Z_D I_3 \; = 0 \qquad ...(1.5)$$

$$Z_D I_2 \qquad\qquad\qquad\quad - Z_D I_3 \; = V_2 \qquad ...(1.6)$$

These equations can be solved by Crammer's rule. Let the determinant

$$D = \begin{vmatrix} (Z_A + Z_C) & - Z_C & 0 \\ - Z_C & (Z_B + Z_C + Z_D + Z_E) & - Z_D \\ 0 & Z_D & - Z_D \end{vmatrix} \qquad ...(1.7)$$

Solutions for I_1, I_2 and I_3 can be obtained by replacing the first, second and third columns of D in (1.7) by the right hand side members in (1.4), (1.5) and (1.6) respectively, and then dividing the determinant by D. Thus,

$$I_1 = \frac{1}{D} \begin{vmatrix} V_1 & - Z_C & 0 \\ 0 & (Z_B + Z_C + Z_D + Z_E) & - Z_D \\ V_2 & Z_D & - Z_D \end{vmatrix} \qquad ...(1.8)$$

Similarly, for I_2 and I_3 we can get solutions.

If 'n' be the number of nodes and 'r' be the number of branches in a network, then r–(n–1) = r–n+1 equations are required in loop analysis. In Fig.1.1, n = 4, r = 6. So, 6 – 4 + 1 = 3 equations are required.

1.3 NODE (OR JUNCTION) ANALYSIS

Here, some junction is chosen as reference. We assume different branch currents I_1, I_2, I_3 and I_4 . The reference node (ground) is 0.

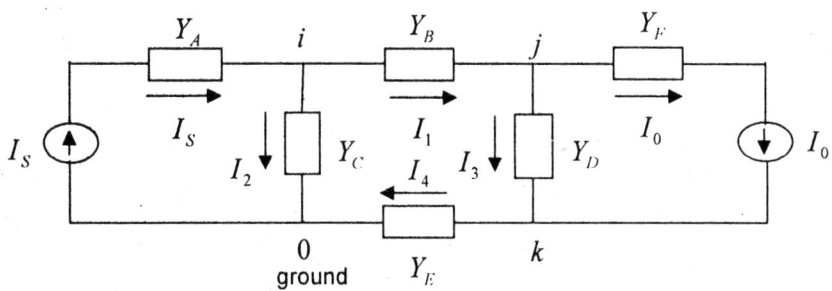

Fig. 1.2

There are 4 nodes in Fig. 1.2 which exist at a junction of 3 or more elements. We adopt the convention that current entering a node is positive and the current leaving the node as negative. Y_A, Y_B, Y_C, Y_D, Y_S and Y_F are admittances of the respective elements shown in Fig. 1.2.

Applying KCL in

node i: $\quad I_s - I_1 - I_2 = 0$

or, $\quad\quad I_s - (V_i - V_j)Y_B - V_iY_C = 0$ $\hspace{3cm}$ ...(1.9)

node j: $\quad I_1 - I_0 - I_3 = 0$

or, $\quad\quad (V_i - V_j)Y_B - I_0 - (V_j - V_k)Y_D = 0$ $\hspace{2cm}$ ...(1.10)

node k: $\quad I_0 + I_3 - I_4 = 0$

or, $\quad\quad I_0 + (V_j - V_k)Y_D - V_kY_E = 0$ $\hspace{3cm}$ ...(1.11)

Here, V_i, V_j, and V_k are voltages of i, j, and k nodes measured with respect to the ground.

Rearranging,

$$-(Y_B + Y_C)V_i \quad + Y_BV_j \quad\quad\quad +0 \quad\quad = -I_s \quad\quad ...(1.12)$$
$$Y_BV_i \quad\quad -(Y_B + Y_D)V_j \quad +Y_DV_k \quad\quad = I_0 \quad\quad ...(1.13)$$
$$0 \quad\quad +Y_DV_j \quad\quad -(Y_D + Y_E)V_k = -I_0 \quad\quad ...(1.14)$$

Let,

$$D = \begin{vmatrix} -(Y_B + Y_C) & Y_B & 0 \\ Y_B & -(Y_B + Y_D) & Y_D \\ 0 & Y_D & -(Y_D + Y_E) \end{vmatrix} \quad\quad ...(1.15)$$

As before, the solutions for V_i, V_j and V_k can be obtained as

$$V_i = \frac{1}{D} \begin{vmatrix} -I_s & Y_B & 0 \\ I_0 & -(Y_B + Y_D) & Y_D \\ -I_0 & Y_D & -(Y_D + Y_E) \end{vmatrix} \qquad \ldots(1.16)$$

$$V_j = \frac{1}{D} \begin{vmatrix} -(Y_B + Y_C) & -I_s & 0 \\ Y_B & I_0 & Y_D \\ 0 & -I_0 & -(Y_D + Y_E) \end{vmatrix} \qquad \ldots(1.17)$$

and

$$V_k = \frac{1}{D} \begin{vmatrix} -(Y_B + Y_C) & Y_B & -I_s \\ Y_B & -(Y_B + Y_D) & I_0 \\ 0 & Y_D & -I_0 \end{vmatrix} \qquad \ldots(1.18)$$

If 'n' be the number of nodes then (n–1) equations are required in node or junction analysis method. Here, n = 4 and so (n–1) = 3 equations are required. Where current sources are present in a network, or, node voltages with respect to a specific node is desired, node analysis will be preferred.

1.4 T AND Π NETWORKS

Analysis of complicated networks is difficult. However, so far as the performances of these complicated networks are concerned they can be reduced to some simple forms which will be equivalent in electrical performance. A network can act as an equivalent to an original complicated network if it produces the same voltage and current at the input and output terminals when operated in a similar condition. The equivalent network is not identical in internal structure with the original network but it yields identical values of voltages and currents at the terminals.

. There are four electrical quantities of a linear passive network to be measured. These are input voltage and input current, and output voltage and output current. These voltages and currents are, in general, phasors. A phasor has magnitude as well as phase angle. So, there are 8 quantities –4 magnitudes and 4 phase angles to be established by the equivalent network. If the load Z_l is given, we get a relation

$$V_0 = I_0 Z_l \qquad \ldots (1.19)$$

where V_0 and I_0 are output voltage and current respectively. Once I_0 is known V_0 is automatically known by (1.19). So, one magnitude and one phase angle can be deleted from our requirement. We are now left with 6 quantities which are to be established by the equivalent network. Since,

$$Z = |Z| e^{j\varphi} \qquad \ldots(1.20)$$

where φ is the phase angle of impedance Z, we require a minimum of 3 impedances to establish the equivalent network.

There are two possible structures which may be constructed using these 3 impedances. These are T or Y network and Π or Δ network.

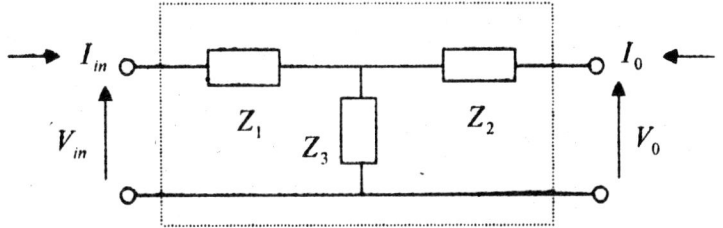

Fig.1.3. T-equivalent structure

For equivalence, the open and short-circuit measurements must be identical. For the equivalent T-section as shown in Fig. 1.3, we have

$$Z_{in0} = Z_1 + Z_3 \qquad \qquad \qquad \dots (1.21)$$

$$Z_{ins} = Z_1 + \frac{Z_2 Z_3}{Z_2 + Z_3} \qquad \qquad \qquad \dots (1.22)$$

$$Z_{00} = Z_2 + Z_3 \qquad \qquad \qquad \dots (1.23)$$

$$Z_{0s} = Z_2 + \frac{Z_1 Z_3}{Z_1 + Z_3} \qquad \qquad \qquad \dots (1.24)$$

Here, Z_{in0} and Z_{ins} are the open and short circuit input impedances respectively while Z_{00} and Z_{0s} are the open and short circuit output impedances respectively.

Since, there are 3 unknowns Z_1, Z_2, and Z_3, only 3 of the four equations (1.21)-(1.24) are needed. Now,

$$Z_{in0} - Z_{ins} = \frac{Z_3^2}{Z_2 + Z_3} = \frac{Z_3^2}{Z_{00}} \qquad \qquad \qquad \dots (1.25)$$

Then, $Z_3 = \pm [Z_{00}(Z_{in0} - Z_{ins})]^{1/2} \qquad \qquad \qquad \dots (1.26)$

Hence, $Z_1 = Z_{in0} - Z_3 \qquad \qquad \qquad \dots (1.27)$

and $Z_2 = Z_{00} - Z_3 \qquad \qquad \qquad \dots (1.28)$

The $\pm$ signs in Z_3 gives rise to two different T-equivalent networks. They are equivalent except for an ambiguity of $180°$ in the output phase angle. This ambiguity cannot be resolved without any information regarding the input-output phase relation of the original complicated network.

Now, let us consider the equivalent Π-section. This is shown in Fig. 1.4

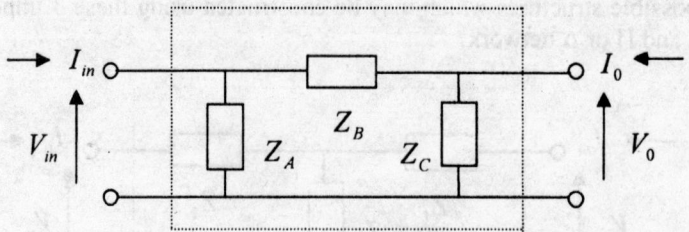

Fig. 1.4. Π-equivalent structure

For the equivalent Π-section, we have

$$Z_{in0} = \frac{Z_A(Z_B + Z_C)}{Z_A + Z_B + Z_C} \qquad \qquad \dots (1.29)$$

$$Z_{ins} = \frac{Z_A Z_B}{Z_A + Z_B} \qquad \qquad \dots (1.30)$$

$$Z_{00} = \frac{(Z_A + Z_B)Z_C}{Z_A + Z_B + Z_C} \qquad \qquad \dots (1.31)$$

Multiplying (1.30) and (1.31), we get

$$Z_{ins} Z_{00} = \frac{Z_A Z_B Z_C}{Z_A + Z_B + Z_C} \qquad \qquad \dots (1.32)$$

Subtracting (1.30) from (1.29), we obtain

$$Z_{in0} - Z_{ins} = \frac{Z_A^2 Z_C}{(Z_A + Z_B + Z_C)(Z_A + Z_B)} \qquad \qquad \dots (1.33)$$

Multiplying (1.31) and (1.33), we get

$$Z_{00}(Z_{in0} - Z_{ins}) = \frac{Z_A^2 Z_C^2}{(Z_A + Z_B + Z_C)^2} \qquad \qquad \dots (1.34)$$

Then, $\qquad \dfrac{Z_A Z_C}{Z_A + Z_B + Z_C} = \pm \sqrt{Z_{00}(Z_{in0} - Z_{ins})} \qquad \qquad \dots (1.35)$

Dividing (1.32) by (1.35), we get

$$Z_B = \pm \frac{Z_{ins} Z_{00}}{\sqrt{Z_{00}(Z_{in0} - Z_{ins})}} \qquad \qquad \dots (1.36)$$

From (1.30), we can write

$$\frac{1}{Z_A} + \frac{1}{Z_B} = \frac{1}{Z_{ins}} \qquad \qquad \dots (1.37)$$

or, $\dfrac{1}{Z_A} = \dfrac{1}{Z_{ins}} - \dfrac{1}{Z_B}$

$= \dfrac{1}{Z_{ins}} - \dfrac{\sqrt{Z_{00}(Z_{in0} - Z_{ins})}}{Z_{ins}Z_{00}}$, taking the positive sign before the radical in (1.36)

$= \dfrac{Z_{00} - \sqrt{Z_{00}(Z_{in0} - Z_{ins})}}{Z_{ins}Z_{00}} \cdot$

Thus, $Z_A = \dfrac{Z_{ins}Z_{00}}{Z_{00} - \sqrt{Z_{00}(Z_{in0} - Z_{ins})}}$ ... (1.38)

To find Z_C we proceed as follows. Equation (1.31) can be written as

$\dfrac{1}{Z_C} + \dfrac{1}{Z_A + Z_B} = \dfrac{1}{Z_{00}}$

or, $\dfrac{1}{Z_C} = \dfrac{1}{Z_{00}} - \dfrac{1}{Z_A + Z_B}$

$= \dfrac{1}{Z_{00}} - \dfrac{\sqrt{Z_{00}(Z_{in0} - Z_{ins})}}{Z_{ins}Z_{00}^2}(Z_{00} - \sqrt{Z_{00}(Z_{in0} - Z_{ins})})$

$= \dfrac{1}{Z_{00}} - \dfrac{\sqrt{Z_{in0} - Z_{ins}}}{Z_{ins}Z_{00}}(\sqrt{Z_{00}} - \sqrt{Z_{in0} - Z_{ins}})$

$= \dfrac{Z_{in0} - \sqrt{Z_{00}(Z_{in0} - Z_{ins})}}{Z_{ins}Z_{00}}$

Hence, $Z_C = \dfrac{Z_{ins}Z_{00}}{Z_{in0} - \sqrt{Z_{00}(Z_{in0} - Z_{ins})}}$ ... (1.39)

The ± sign in (1.36) gives rise to two possible solutions as obtained in the case of equivalent T- network. Now, Z is a function of signal frequency ω, in general. Thus, equivalence in terms of Z is all right at all frequencies. But, equivalence in terms of circuit elements R, L, C does hold good at a single frequency. As ω changes, $Z(\omega)$ also changes. But, the functional dependence of $Z(\omega)$ for the complicated circuit and the equivalent circuit are not the same for which the same equivalent circuit cannot hold good at all frequencies.

In conclusion, it is possible to find an equivalent T or a Π network at a single frequency for any passive electrical network which is linear and bilateral in character. (In bilateral network the magnitude of the current flowing through an element remains unchanged if the polarity of the supply is reversed.)

1.5 T-Π AND Π-T CONVERSIONS

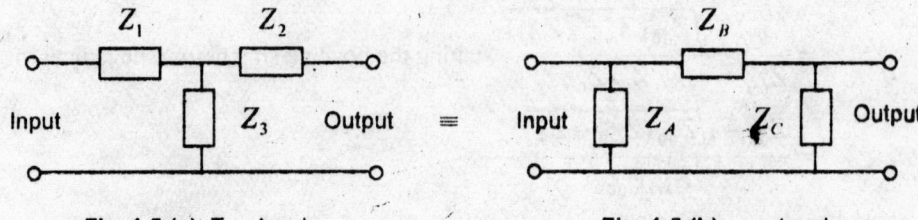

Fig. 1.5.(a) T-network $\qquad$ Fig. 1.5.(b) π-network

For T-network:

$$Z_{in0} = Z_1 + Z_3 \qquad \ldots (1.40a)$$

$$Z_{00} = Z_2 + Z_3 \qquad \ldots (1.40b)$$

$$Z_{ins} = Z_1 + \frac{Z_2 Z_3}{Z_2 + Z_3} \qquad \ldots (1.40c)$$

For Π-network:

$$Z_{in0} = \frac{Z_A(Z_B + Z_C)}{Z_A + Z_B + Z_C} \qquad \ldots (1.41a)$$

$$Z_{00} = \frac{Z_C(Z_A + Z_B)}{Z_A + Z_B + Z_C} \qquad \ldots (1.41b)$$

$$Z_{ins} = \frac{Z_A Z_B}{Z_A + Z_B} \qquad \ldots (1.41c)$$

Here, Z_{in0} stands for the input impedance measured with output open circuited and Z_{00} is the output impedance measured with input open circuited. Z_{ins} stands for input impedance measured with the output short circuited.

For equivalence of T and Π networks, we must have

$$Z_1 + Z_3 = \frac{Z_A(Z_B + Z_C)}{Z_A + Z_B + Z_C} \qquad \ldots (1.42)$$

$$Z_2 + Z_3 = \frac{Z_C(Z_A + Z_B)}{Z_A + Z_B + Z_C} \qquad \ldots (1.43)$$

$$Z_1 + \frac{Z_2 Z_3}{Z_2 + Z_3} = \frac{Z_A Z_B}{Z_A + Z_B} \qquad \ldots (1.44)$$

Subtracting (1.44) from (1.42), we get

$$Z_3 - \frac{Z_2 Z_3}{Z_2 + Z_3} = Z_A \left[\frac{Z_B + Z_C}{Z_A + Z_B + Z_C} - \frac{Z_B}{Z_A + Z_B} \right]$$

or, $\quad \dfrac{Z_3^2}{Z_2 + Z_3} = Z_A \left[\dfrac{Z_A Z_B + Z_B^2 + Z_A Z_C + Z_B Z_C - Z_A Z_B - Z_B^2 - Z_B Z_C}{(Z_A + Z_B + Z_C)(Z_A + Z_B)} \right]$

or, $\quad Z_3^2 = \dfrac{Z_A^2 Z_C}{(Z_A + Z_B + Z_C)(Z_A + Z_B)} \cdot \dfrac{Z_C(Z_A + Z_B)}{(Z_A + Z_B + Z_C)} \quad$ using (1.43)

$$= \dfrac{Z_A^2 Z_C^2}{(Z_A + Z_B + Z_C)^2}$$

Thus, $\quad Z_3 = \dfrac{Z_A Z_C}{Z_A + Z_B + Z_C} \qquad\qquad\qquad$ ... (1.45)

Substituting for Z_3 from (1.45) in equation (1.43) one can obtain

$$Z_2 = \dfrac{Z_C(Z_A + Z_B)}{Z_A + Z_B + Z_C} - \dfrac{Z_A Z_C}{Z_A + Z_B + Z_C}$$

$$= \dfrac{Z_B Z_C}{Z_A + Z_B + Z_C} \qquad\qquad\qquad\qquad \text{... (1.46)}$$

Similarly, substituting for Z_3 from (1.45) in equation (1.42) we get,

$$Z_1 = \dfrac{Z_A(Z_B + Z_C)}{Z_A + Z_B + Z_C} - \dfrac{Z_A Z_C}{Z_A + Z_B + Z_C}$$

$$= \dfrac{Z_A Z_B}{Z_A + Z_B + Z_C} \qquad\qquad\qquad\qquad \text{... (1.47)}$$

Thus, if the Π-network is given we can find Z_3, Z_2 and Z_1 from equations (1.45)-(1.47) and hence the equivalent T-network can be found out.

Now, to convert a given T-network into an equivalent Π-network we proceed as follows: From equation (1.44) we can write

$$\dfrac{Z_1 Z_2 + Z_1 Z_3 + Z_2 Z_3}{Z_2 + Z_3} = \dfrac{Z_A Z_B}{Z_A + Z_B}$$

$\therefore \quad Z_1 Z_2 + Z_1 Z_3 + Z_2 Z_3 = \dfrac{Z_A Z_B}{Z_A + Z_B} \cdot \dfrac{Z_C(Z_A + Z_B)}{Z_A + Z_B + Z_C}, \quad$ using (1.43)

$$= \dfrac{Z_A Z_B Z_C}{Z_A + Z_B + Z_C} \qquad\qquad\qquad \text{... (1.48)}$$

$$= Z_B \cdot Z_3 , \qquad\qquad\qquad \text{using (1.45)}$$

$\therefore \quad Z_B = \dfrac{Z_1 Z_2 + Z_1 Z_3 + Z_2 Z_3}{Z_3} \qquad\qquad\qquad \text{... (1.49)}$

Using (1.46) on the right hand side of equation (1.48) it is possible to get

$$Z_1 Z_2 + Z_1 Z_3 + Z_2 Z_3 = Z_A \cdot Z_2$$

$$\therefore \quad Z_A = \frac{Z_1 Z_2 + Z_1 Z_3 + Z_2 Z_3}{Z_2} \qquad \dots (1.50)$$

Similarly, use of equation (1.47) on the right hand side of equation (1.48) leads to

$$Z_1 Z_2 + Z_1 Z_3 + Z_2 Z_3 = Z_C \cdot Z_1$$

$$\therefore \quad Z_C = \frac{Z_1 Z_2 + Z_1 Z_3 + Z_2 Z_3}{Z_1} \qquad \dots (1.51)$$

Equations (1.49)-(1.51) indicate that once the T-network is given (i.e., Z_1, Z_2 and Z_3 given) we can construct the equivalent Π-network by calculating Z_A, Z_B and Z_C.

1.6 SUPERPOSITION THEOREM

Statement: In any network having electrical energy sources and linear, bilateral impedances, the current flowing through any element can be obtained by taking the vector sum of currents which flow in the same element due to individual energy sources.

Proof:

Any linear and bilateral network can be represented by its T-equivalent structure. So, in Fig. 1.6(a) we consider a T-network formed by impedances Z_1, Z_2 and Z_3. V_1 and V_2 are sources of emf while I_1 and I_2 are currents flowing through loops (1) and (2).

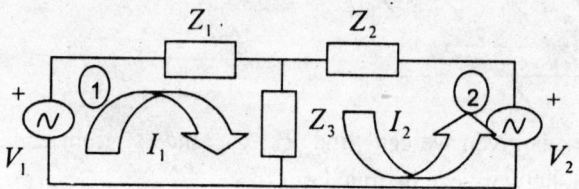

Fig. 1.6.(a)

Applying KVL to loops (1) and (2) respectively, we get

Loop-1: $\quad (I_1 + I_2)Z_3 + I_1 Z_1 = V_1 \qquad \dots (1.52)$

Loop-2: $\quad (I_1 + I_2)Z_3 + I_2 Z_2 = V_2 \qquad \dots (1.53)$

Rearranging (1.52) and (1.53), we obtain

$$I_1(Z_1 + Z_2) + I_2 Z_3 = V_1 \qquad \dots (1.54)$$

$$I_1 Z_3 + I_2(Z_2 + Z_3) = V_2 \qquad \dots (1.55)$$

Determinant formed by the coefficients of I_1 and I_2 is

$$D = \begin{vmatrix} (Z_1 + Z_3) & Z_3 \\ Z_3 & (Z_2 + Z_3) \end{vmatrix} = Z_1 Z_2 + Z_1 Z_3 + Z_2 Z_3$$

$$\therefore \qquad I_1 = \frac{1}{D}\begin{vmatrix} V_1 & Z_3 \\ V_2 & (Z_2 + Z_3) \end{vmatrix} = \frac{(Z_2 + Z_3)V_1 - Z_3 V_2}{Z_1 Z_2 + Z_1 Z_3 + Z_2 Z_3} \qquad \ldots (1.56)$$

$$I_2 = \frac{1}{D}\begin{vmatrix} (Z_1 + Z_3) & V_1 \\ Z_3 & V_2 \end{vmatrix} = \frac{-Z_3 V_1 + (Z_1 + Z_3)V_2}{Z_1 Z_2 + Z_1 Z_3 + Z_2 Z_3} \qquad \ldots (1.57)$$

Now, let us calculate the currents in loops (1) and (2) with one source of e.m.f. eliminated. First, we remove V_2 from Fig.1.6(a). The circuit now looks like the following as shown in Fig.1.6(b).

The loop equations can be written from (1.54) and (1.55) setting $V_2 = 0$ and modifying I_1 and I_2 by I_1' and I_2' respectively.

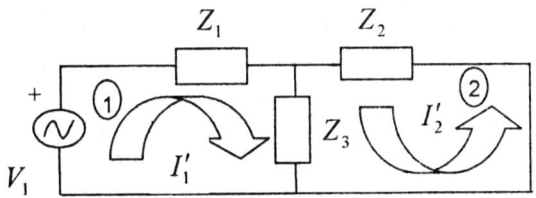

Fig. 1.6.(b)

Loop-1: $\qquad I_1'(Z_1 + Z_3) + I_2' Z_3 = V_1$ $\qquad \ldots (1.58)$

Loop-2: $\qquad I_1' Z_3 + I_2'(Z_2 + Z_3) = 0$ $\qquad \ldots (1.59)$

The determinant, $D = Z_1 Z_2 + Z_1 Z_3 + Z_2 Z_3$

which is the same as above. Now,

$$I_1' = \frac{1}{D}\begin{vmatrix} V_1 & Z_3 \\ 0 & Z_2 + Z_3 \end{vmatrix} = \frac{V_1(Z_2 + Z_3)}{Z_1 Z_2 + Z_1 Z_3 + Z_2 Z_3} \qquad \ldots (1.60)$$

and $\qquad I_2' = \frac{1}{D}\begin{vmatrix} Z_1 + Z_3 & V_1 \\ Z_3 & 0 \end{vmatrix} = \frac{-Z_3 V_1}{Z_1 Z_2 + Z_1 Z_3 + Z_2 Z_3} \qquad \ldots (1.61)$

Now, consider the following circuit with $V_1 = 0$ as shown in Fig.1.6(c).

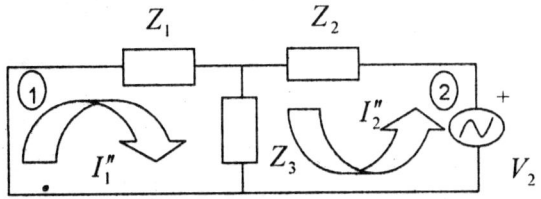

Fig. 1.6.(c)

The loop equations are obtained by setting $V_1 = 0$ in (1.54) and (1.55) and replacing I_1 and I_2 by I_1'' and I_2'' respectively as

Loop-1: $\qquad I_1''(Z_1 + Z_3) + I_2'' Z_3 = 0$ ... (1.62)

Loop-2: $\qquad I_1'' Z_3 + I_2''(Z_2 + Z_3) = V_2$ .... (1.63)

I_1'' and I_2'' are currents flowing in loops (1) and (2) respectively in the circuit of Fig. 1.6(c).
The determinant $D = Z_1 Z_2 + Z_1 Z_3 + Z_2 Z_3$ is the same as before.

Now, $I_1'' = \dfrac{1}{D} \begin{vmatrix} 0 & Z_3 \\ V_2 & (Z_2 + Z_3) \end{vmatrix} = \dfrac{-Z_3 V_2}{Z_1 Z_2 + Z_1 Z_3 + Z_2 Z_3}$ ... (1.64)

$\qquad I_2'' = \dfrac{1}{D} \begin{vmatrix} (Z_1 + Z_3) & 0 \\ Z_3 & V_2 \end{vmatrix} = \dfrac{(Z_1 + Z_3) V_2}{Z_1 Z_2 + Z_1 Z_3 + Z_2 Z_3}$ ... (1.65)

Now, $I_1' + I_1'' = I_1$ ... (1.66a)

and $I_2' + I_2'' = I_2$ ... (1.66b)

Hence, the theorem is proved.

1.7 THEVENIN'S THEOREM

This theorem gives the voltage-source equivalent representation of a circuit.

Statement: Any two-terminal linear network formed by impedances and electrical energy sources can be replaced by an equivalent circuit. The equivalent circuit consists of a voltage source (V_{eq}) in series with an impedance Z_{eq}. The value of V_{eq} is the open circuit voltage between the terminals of the network and Z_{eq} is the impedance measured between the terminals with all sources removed but retaining their internal impedances.

Proof : Let us consider an active network having the source of e.m.f. V_S which is connected to the load impedance Z_L through a passive network. This is shown in Fig. 1.7(a). A passive network is one which contains no energy source.

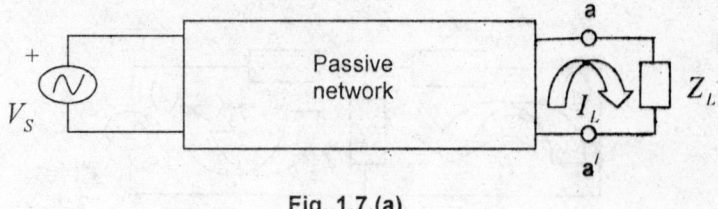

Fig. 1.7.(a)

The passive network can be transformed into its T-equivalent network as follows.

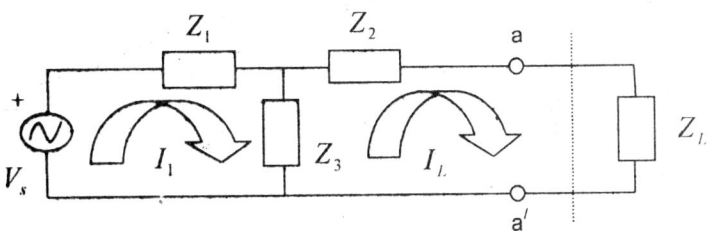

Fig. 1.7.(b)

The loop equations from Fig. 1.7(b) can be written as

Loop-1: $Z_3(I_1 - I_L) + I_1 Z_1 = V_s$ ... (1.67a)

Loop-2: $Z_3(I_1 - I_L) - Z_2 I_L - I_L Z_L = 0$ ... (1.67b)

Rearranging (1.67a) and (1.67b) we can write,

$$I_1(Z_1 + Z_3) - Z_3 I_L = V_s \qquad\qquad\qquad ... (1.68)$$

$$I_1 Z_3 - (Z_2 + Z_3 + Z_L)I_L = 0 \qquad\qquad ... (1.69)$$

The determinant formed by the coefficients of I_1 and I_L is

$$D = \begin{vmatrix} (Z_1 + Z_3) & -Z_3 \\ Z_3 & -(Z_2 + Z_3 + Z_L) \end{vmatrix} = -(Z_1 + Z_3)(Z_2 + Z_3 + Z_L) + Z_3^2 \qquad ... (1.70)$$

Then, $I_1 = \dfrac{1}{D}\begin{vmatrix} V_S & -Z_3 \\ 0 & -(Z_2 + Z_3 + Z_L) \end{vmatrix} = \dfrac{-(Z_2 + Z_3 + Z_L)V_S}{D}$ ... (1.71)

and $I_L = \dfrac{1}{D}\begin{vmatrix} (Z_1 + Z_3) & V_S \\ Z_3 & 0 \end{vmatrix} = \dfrac{-Z_3 V_S}{D}$

$$= \dfrac{Z_3 V_S}{(Z_1 + Z_3)(Z_2 + Z_3 + Z_L) - Z_3^2}$$

$$= \dfrac{V_S \left(\dfrac{Z_3}{(Z_1 + Z_3)}\right)}{Z_2 + Z_L + \dfrac{Z_1 Z_3}{Z_1 + Z_3}} \qquad\qquad ... (1.72)$$

Equation (1.72) gives the load current flowing in the original network. The open-circuit voltage between the terminals (a,a') in Fig. 1.7(b) is,

$$V_{eq} = \dfrac{Z_3}{Z_1 + Z_3} V_s \qquad\qquad ... (1.73)$$

Again, Z_{eq} = impedance measured between the terminals (a,a') in Fig. 1.7(b) with $V_s = 0$

$$Z_2 + (Z_1 \| Z_3) = Z_2 + \frac{Z_1 Z_3}{Z_1 + Z_3} \qquad \qquad \dots (1.74)$$

Equation (1.72) can be written as,

$$I_L = \frac{V_{eq}}{Z_{eq} + Z_L} \qquad \qquad \dots (1.75)$$

Equation (1.75) indicates an equivalent circuit of the following form shown in Fig.1.7(c):

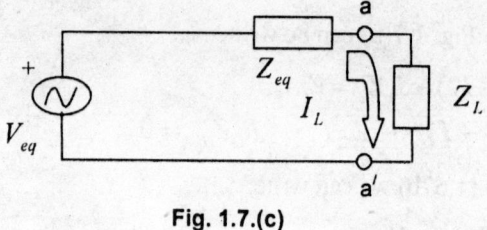

Fig. 1.7.(c)

This proves the theorem.

1.8 NORTON'S THEOREM

This theorem gives the current-source equivalent representation of a circuit.

Statement: Any two-terminal linear network formed by electrical energy sources and impedances can be replaced by an equivalent circuit. The equivalent circuit consists of a current source (I_{eq}) in parallel with an admittance Y_{eq}. The value of I_{eq} is the short-circuit current between the network terminals and Y_{eq} is the admittance measured between the terminals with all sources removed but retaining their internal admittances.

Proof : The original network is shown in Fig. 1.7(a). This network can be converted into its Thevenin's equivalent which is shown in Fig.1.7(c). The load current in Thevenin's equivalent circuit of Fig.1.7(c) is given by,

$$I_L = \frac{V_{eq}}{Z_{eq} + Z_L} = \frac{V_{eq}}{\dfrac{1}{Y_{eq}} + \dfrac{1}{Y_L}}$$

where $Y_{eq} = \dfrac{1}{Z_{eq}}$ and $Y_L = \dfrac{1}{Z_L}$ are the admittances.

$$I_L = \frac{V_{eq} Y_{eq} Y_L}{Y_{eq} + Y_L} = V_{eq} Y_{eq} \left(\frac{Y_L}{Y_{eq} + Y_L} \right) \qquad \qquad (1.76)$$

Consider the Norton's equivalent circuit as shown in Fig.1.8.

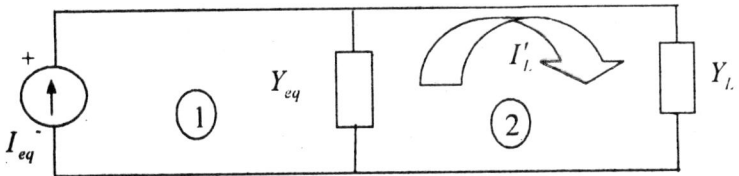

Fig.1.8. Norton's equivalent circuit.

In the Norton's equivalent circuit of Fig.1.8, the current source I_{eq} supplies current to the parallel combination of Y_{eq} and Y_L. Now,

$$\frac{I'_L}{I_{eq} - I'_L} = \frac{Y_L}{Y_{eq}}$$

or,

$$\frac{I'_L}{I_{eq}} = \frac{Y_L}{Y_L + Y_{eq}}$$

$\therefore$

$$I'_L = I_{eq}\left(\frac{Y_L}{Y_L + Y_{eq}}\right) \qquad \ldots(1.77)$$

The current I'_L and I_{eq} are shown in Fig.1.8. Let us compare equations (1.76) and (1.77). The load current (I'_L) in the Norton equivalent circuit of Fig.1.8 will be equal to the load current (I_L) of Thevenin's equivalent circuit in Fig.1.7(c) if and only if

$$I_{eq} = V_{eq} Y_{eq} \qquad \ldots(1.78)$$

and Y_{eq} be the same as in eqn. (1.76). Under this condition, Norton's equivalent circuit of Fig.1.8 will be equivalent to the Thevenin's equivalent circuit and hence to the original circuit. Let us now see what eqn.(1.78) tells us. It is seen that

$$I_{eq} = V_{eq} Y_{eq} = \frac{V_{eq}}{Z_{eq}} \qquad \ldots(1.79)$$

is the short circuit current in Thevenin equivalent circuit of Fig. 1.7(c) and $Y_{eq} = \dfrac{1}{Z_{eq}}$ is the admittance measured between the terminals (aa') of the Thevenin's equivalent circuit of Fig. 1.7(c) after removing the voltage source V_{eq}. Hence, the theorem is proved.

1.9 MAXIMUM POWER TRANSFER THEOREM

Statement: A linear network containing energy sources and impedances will deliver maximum power to the load if the load impedance is equal to the complex conjugate of the internal impedance of the network measured looking back into the terminals of the network.

Proof: The linear network containing energy sources and impedances can be represented by its Thevenin equivalent. The resultant equivalent circuit will contain an equivalent voltage source

E_s in series with an impedance Z_s. Let $Z_L = R_L + jX_L$ be the load impedance where R_L is the resistive part and X_L is the reactive part of the load. Also, let $Z_s = R_s + jX_s$ where R_s is the resistive part and X_s is the reactive part of the source impedance Z_s. The equivalent circuit has the following form:

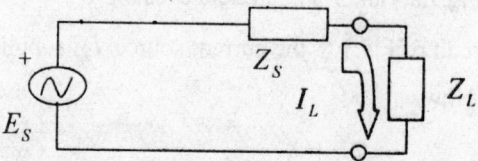

Fig. 1.9. Thevenin's equivalent circuit.

The load current,
$$I_L = \frac{E_s}{Z_s + Z_L}$$

$$= \frac{E_s}{(R_s + R_L) + j(X_s + X_L)} \qquad \ldots (1.80)$$

Power is delivered to the resistive load R_L which consumes power. The power developed is

$$P_L = |I_L|^2 R_L = \frac{E_s^2 R_L}{(R_s + R_L)^2 + (X_s + X_L)^2} \qquad \ldots (1.81)$$

There are two variables in the expression for P_L which are X_L and R_L. To maximize P_L with respect to X_L we set $\dfrac{\partial P_L}{\partial X_L} = 0$. This gives the condition $X_s + X_L = 0$

i.e., $\quad X_L = -X_s$ $\qquad \ldots (1.82)$

To confirm that P_L is maximum for $X_L = -X_s$, we can find $\dfrac{\partial^2 P_L}{\partial X_L^2}$ which turns to be negative.

Substituting $X_L = -X_s$ in (1.81), we get

$$P_L = \frac{E_s^2 R_L}{(R_s + R_L)^2} \qquad \ldots (1.83)$$

Let us now maximize P_L with respect to R_L. So, we set $\dfrac{\partial P_L}{\partial R_L} = 0$. This condition gives

$$R_L = R_s \qquad \ldots (1.84)$$

Further $\dfrac{\partial^2 P_L}{\partial R_L^2}$ turns out to be negative which confirms the maximum condition. Hence,

$$Z_L = R_L + jX_L = R_s - jX_s = Z_s^*.$$

Hence, the theorem is proved.

1.10 NOTCH FILTER

This filter consists of two parallel and symmetrical T networks, one placed above the other. The circuit is shown on Fig. 1.10.

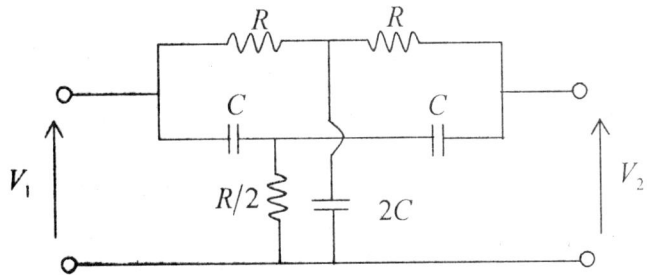

Fig.1.10. Circuit of twin-T notch filter

We convert the two T-networks $(R, R, 2C)$ and $\left(C, C, \dfrac{R}{2}\right)$ into their equivalent π-networks.

The first circuit is shown in Fig. 1.11.

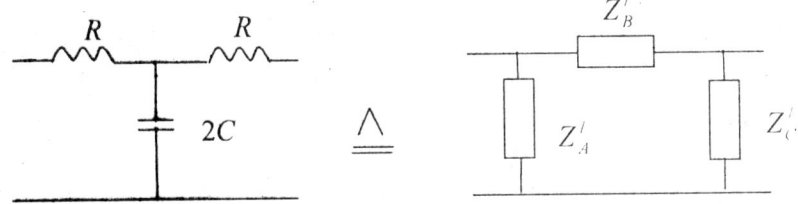

Fig. 1.11. T-Πconversion of (R, R, $2C$) network.

Now,

$$Z'_A = \frac{R.R + R.\dfrac{1}{j2\omega C} + \dfrac{R}{j2\omega C}}{R} = R_1 + \frac{1}{j\omega C} \qquad \ldots (1.85)$$

$$Z'_B = \frac{R.R + R.\dfrac{1}{j2\omega C} + \dfrac{R}{j2\omega C}}{\dfrac{1}{j2\omega C}} = 2R + j2\omega C R^2 \qquad \ldots (1.86)$$

$$Z'_C = \frac{R.R + R.\dfrac{1}{j2\omega C} + \dfrac{R}{j2\omega C}}{R} = R + \frac{1}{j\omega C} \qquad \ldots (1.87)$$

Now, we convert $\left(C, C, \dfrac{R}{2}\right)$. T-network into its equivalent Π-network. This is shown in Fig. 1.12.

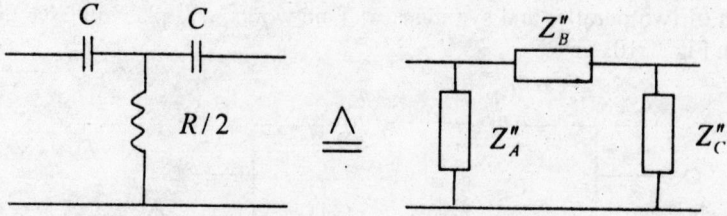

Fig.1.12. T to Π conversion of $(C, C, R/2)$ network.

Now,

$$Z''_A = \frac{\left(\dfrac{1}{j\omega C}\right)^2 + \dfrac{R}{j2\omega C} + \dfrac{R}{j2\omega C}}{\dfrac{1}{j\omega C}} = R + \frac{1}{j\omega C} \qquad \qquad \dots (1.88)$$

$$Z''_B = \frac{\left(\dfrac{1}{j\omega C}\right)^2 + \dfrac{R}{j2\omega C} + \dfrac{R}{j2\omega C}}{\dfrac{R}{2}} = \frac{2}{j\omega C} + \frac{2}{R}\left(\frac{1}{j\omega C}\right)^2 \qquad \dots (1.89)$$

$$Z''_C = \frac{\left(\dfrac{1}{j\omega C}\right)^2 + \dfrac{R}{j2\omega C} + \dfrac{R}{j2\omega C}}{\dfrac{1}{j\omega C}} = R + \frac{1}{j\omega C} \qquad \qquad \dots (1.90)$$

These two Π-networks $\left(Z'_A, Z'_B, Z'_C\right)$ and $\left(Z''_A, Z''_B, Z''_C\right)$ are placed one over the other in parallel. Finally, we get one Π-network $\left(Z_A, Z_B, Z_C\right)$. Then,

$$Z_A = Z'_A \| Z''_A = \frac{Z'_A Z''_A}{Z'_A + Z''_A} = \frac{\left(R + \dfrac{1}{j\omega C}\right)\left(R + \dfrac{1}{j\omega C}\right)}{2R + \dfrac{1}{j\omega}\left(\dfrac{1}{C} + \dfrac{1}{C}\right)}$$

$$= \frac{R^2 - \dfrac{1}{\omega^2 C^2} - j\dfrac{2R}{\omega C}}{2R + \dfrac{2}{j\omega C}} \qquad \qquad \dots (1.91)$$

$$Z_B = Z_B' \| Z_B'' = \frac{Z_B' Z_B''}{Z_B' + Z_B''} = \frac{\left(2R + j2\omega CR^2\right)\left(1 + \dfrac{1}{j\omega CR}\right) \cdot \dfrac{2}{j\omega C}}{2R + j2\omega CR^2 + \dfrac{2}{j\omega C} - \dfrac{2}{\omega^2 C^2 R}}$$

$$= \frac{\dfrac{4R}{j\omega C} + 4R^2 - \dfrac{4}{\omega^2 C^2} - j\dfrac{4R}{\omega C}}{2R - \dfrac{2}{\omega^2 C^2 R} + j\left(2\omega CR^2 - \dfrac{2}{\omega C}\right)}$$

$$= \frac{4R^2\left(1 - \dfrac{1}{\omega^2 C^2 R^2}\right) - j\dfrac{8R}{\omega C}}{2R\left(1 - \dfrac{1}{\omega^2 C^2 R^2}\right) + j\dfrac{2}{\omega C}\left(\omega^2 C^2 R^2 - 1\right)} \qquad \ldots (1.92)$$

and $\quad Z_C = Z_C' \| Z_C'' = \dfrac{Z_C' Z_C''}{Z_C' + Z_C''}$

$$= \frac{R^2 - \dfrac{1}{\omega^2 C^2} - j\dfrac{2R}{\omega C}}{2R + \dfrac{2}{j\omega C}} \qquad \ldots (1.93)$$

Since $\quad Z_C' = Z_A'$ and $Z_C'' = Z_A''$,

we get $Z_C = Z_A$.

To simplify we put $\omega_0^2 = \dfrac{1}{C^2 R^2}$. Then,

$$Z_A = \frac{R^2 - R^2 \dfrac{\omega_0^2}{\omega^2} - j\dfrac{2R}{\omega C}}{2R + \dfrac{2}{j\omega C}}$$

$$= \frac{R^2\left(1 - \dfrac{\omega_0^2}{\omega^2}\right) - j\dfrac{2R}{\omega C}}{2R + \dfrac{2}{j\omega C}} \qquad \ldots (1.94)$$

$$= Z_C$$

$$Z_B = \cfrac{4R^2\left(1 - \cfrac{\omega_0^2}{\omega^2}\right) - j\cfrac{8R}{\omega C}}{2R\left(1 - \cfrac{\omega_0^2}{\omega^2}\right) + j\cfrac{2}{\omega C}\left(\cfrac{\omega^2}{\omega_0^2} - 1\right)}$$

$$= \cfrac{2R\left(1 - \cfrac{\omega_0^2}{\omega^2}\right) - j\cfrac{4}{\omega C}}{\left(1 - \cfrac{\omega_0^2}{\omega^2}\right) - j\cfrac{1}{\omega CR}\left(1 - \cfrac{\omega^2}{\omega_0^2}\right)} \qquad \dots (1.95)$$

The final Π-equivalent circuit of the notch filter is given in Fig. 1.13.

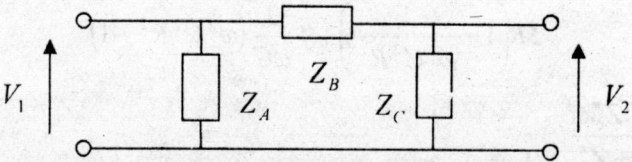

Fig. 1.13. Final Π-equivalent circuit of the notch filter.

Now, $\quad V_2 = \dfrac{V_1 Z_C}{Z_B + Z_C}$

$\therefore \qquad \dfrac{V_2}{V_1} = \dfrac{1}{1 + \dfrac{Z_B}{Z_C}} \qquad \dots (1.96)$

Now, $\quad \dfrac{Z_B}{Z_C} = \cfrac{4\left(1 - j\cfrac{\omega_0}{\omega}\right)}{1 - \cfrac{\omega_0^2}{\omega^2} - j\left(\cfrac{\omega_0}{\omega} - \cfrac{\omega}{\omega_0}\right)}$

Simplifying (1.96) we get

$$\dfrac{V_2}{V_1} = \cfrac{1 - \cfrac{\omega_0^2}{\omega^2} - j\left(\cfrac{\omega_0}{\omega} - \cfrac{\omega}{\omega_0}\right)}{5 - \cfrac{\omega_0^2}{\omega^2} - j\left(5\cfrac{\omega_0}{\omega} - \cfrac{\omega}{\omega_0}\right)}$$

$$= \left|\dfrac{V_2}{V_1}\right| < \varphi$$

$$\therefore \quad \left|\frac{V_2}{V_1}\right| = \left[\frac{\left(1-\frac{\omega_0^2}{\omega^2}\right)^2 + \left(\frac{\omega_0}{\omega}-\frac{\omega}{\omega_0}\right)^2}{\left(5-\frac{\omega_0^2}{\omega^2}\right)^2 + \left(5\frac{\omega_0}{\omega}-\frac{\omega}{\omega_0}\right)^2}\right]^{\frac{1}{2}} \qquad \dots (1.97)$$

Now, $\left|\dfrac{V_2}{V_1}\right| = 0$ when $\omega = \omega_0$. So, $\omega_0 = \dfrac{1}{RC}$ is the notch frequency. This signal of frequency ω_0 is completely cut off by this filter. The typical frequency response of this notch filter is shown in Fig. 1.14. When, $\omega \to 0$, $\left|\dfrac{V_2}{V_1}\right| \to 1$. Also, when $\omega \to \infty$, $\left|\dfrac{V_2}{V_1}\right| \to 1$.

Now,

$$\angle\varphi = \tan^{-1}\left(\frac{\omega}{\omega_0}\right) + \tan^{-1}\left(\frac{5\frac{\omega_0}{\omega}-\frac{\omega}{\omega_0}}{5-\frac{\omega_0^2}{\omega^2}}\right)$$

As $\omega \to 0$, $\angle\varphi \to 0$ and as $\omega \to \infty$, $\angle\varphi \to 0$.

When $\omega = \omega_0$, $\angle\varphi = \pi/2$.

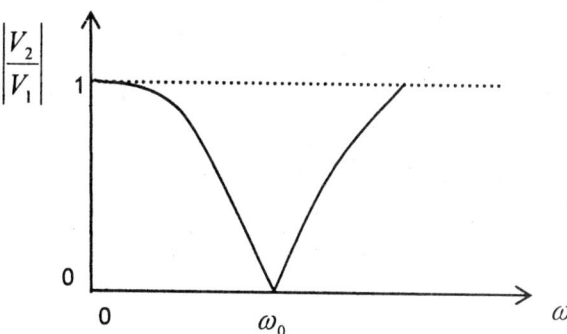

Fig. 1.14. Typical frequency response of the notch filter.

PHASE MEASUREMENT

To measure the phase difference between the input and output voltages of the notch filter, connect the output of the notch filter to the y-input of a CRO and connect the input to x-input of the CRO. Then, press the X-Y operation key of the CRO when an ellipse is displayed on the screen. By measuring y-intercept and y-maximum of the ellipse we can find the phase difference φ.

Let $V_1 = V_{01} \sin \omega t$ ... (1.98)

and $V_2 = V_{02} \sin(\omega t + \varphi)$ ... (1.99)

be the input and output voltages of the notch filter. ω is the angular frequency of the input signal, V_{01} and V_{02} are voltage amplitudes. The displacements of the spot of the CRO in x and y directions are proportional to the applied voltages.

Then, $\quad D_X = kV_1 = kV_{01} \sin \omega t$ $\qquad$ ... (1.100)

and $\quad D_y = kV_2 = kV_{02} \sin(\omega t + \varphi)$ $\qquad$ ... (1.101)

where k is the constant of proportionality and D_x, D_y are displacements. From equation (1.101), we write

$$\frac{D_y}{kV_{02}} = \sin(\omega t + \varphi)$$

$$= \sin \omega t \cos \varphi + \cos \omega t \sin \varphi$$

$$= \frac{D_x}{kV_{01}} \cos \varphi + \sqrt{1 - \left(\frac{D_x}{kV_{01}}\right)^2} \sin \varphi$$

or, $\quad \left[\frac{D_y}{kV_{02}} - \frac{D_x}{kV_{01}} \cos \varphi\right]^2 = \left[1 - \left(\frac{D_x}{kV_{01}}\right)^2\right] \sin^2 \varphi$

or, $\quad \dfrac{D_y^2}{(kV_{02})^2} + \left(\dfrac{D_x}{kV_{01}}\right)^2 - \dfrac{2D_x D_y}{k^2 V_{01} V_{02}} \cos \varphi = \sin^2 \varphi$ $\qquad$ ... (1.102)

Put $D_y = Y$, $D_x = X$, $kV_{01} = a$ and $kV_{02} = b$. Then, equation (1.102) reduces to

$$\frac{Y^2}{b^2} + \frac{X^2}{a^2} - \frac{2XY}{ab} \cos \varphi = \sin^2 \varphi \qquad \text{... (1.103)}$$

To find Y-intercept (Y_{int}), we put $X = 0$ in equation (1.103). Then,

$$Y_{int} = b \sin \varphi \qquad \text{... (1.104)}$$

To find the Y-maximum of the ellipse, we put $\dfrac{dY}{dX} = 0$ in equation (1.103). Differentiating (1.103) with respect to X we get

$$\frac{2Y}{b^2} \frac{dY}{dX} + \frac{2X}{a^2} - \frac{2}{ab} \cos \varphi \left[X \frac{dY}{dX} + Y\right] = 0$$

with $\dfrac{dY}{dX} = 0$, we get

$$\frac{X}{a} = \frac{Y}{b} \cos \varphi \qquad \text{... (1.105)}$$

Now, using equation (1.105) we get from equation (1.103)

$$\frac{Y_{max}^2}{b^2} + \frac{Y_{max}^2}{b^2}\cos^2\varphi - \frac{2Y_{max}^2}{b^2}\cos^2\varphi = \sin^2\varphi$$

$$\therefore \quad \frac{Y_{max}^2}{b^2}\left(1 - \cos^2\varphi\right) = \sin^2\varphi$$

$$\therefore \quad Y_{max} = b \qquad\qquad\qquad\qquad \ldots (1.106)$$

From equation (1.104), we write

$$\sin\varphi = \frac{Y_{int}}{b} = \frac{Y_{int}}{Y_{max}}$$

$$\therefore \quad \varphi = \sin^{-1}\left(\frac{Y_{int}}{Y_{max}}\right) \qquad\qquad\qquad \ldots (1.107)$$

So, measuring Y_{int} and Y_{max} from the ellipse on the screen of the CRO, we can measure φ.

SOLVED PROBLEMS

1.1 Apply Thevenin's theorem to find the galvanometer current under the unbalanced condition of the wheatstone bridge as shown in Fig. 1.15 (a).

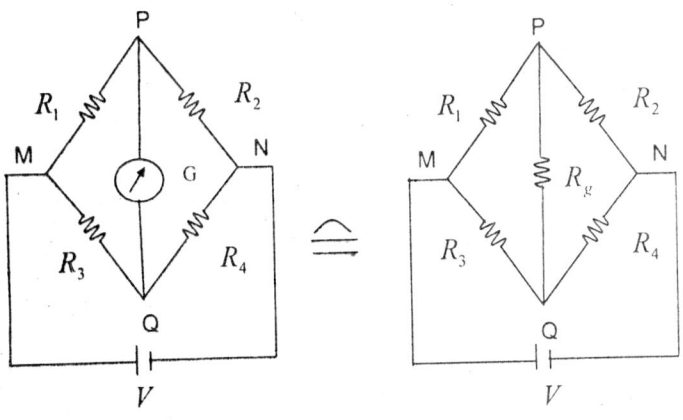

Fig. 1.15.(a)

Solution

Thevenin's equivalent circuit is as follows :

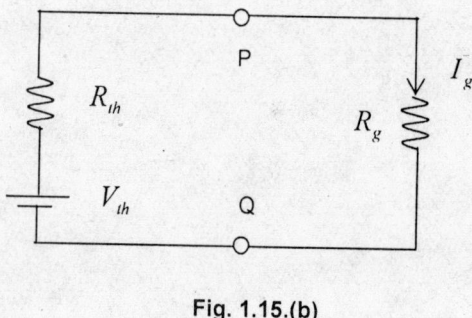

Fig. 1.15.(b)

We have to find the values the equivalent voltage V_{th} and the series resistance R_{th} in the Thevenin's equivalent circuit. Now,

V_{th} = open circuit voltage between P and Q

$$= V_P - V_Q.$$

The corresponding circuit is

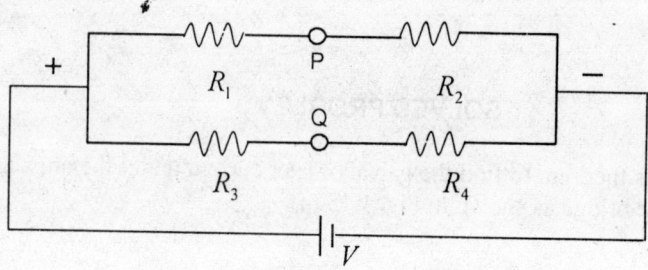

Fig. 1.15.(c)

Potential at P is $V_P = \dfrac{V R_2}{R_1 + R_2}$

Potential at Q is $V_Q = \dfrac{V R_4}{R_3 + R_4}$

$\therefore \qquad V_P - V_Q = V\left[\dfrac{R_2}{R_1 + R_2} - \dfrac{R_4}{R_3 + R_4}\right]$ ... (1.108)

Now, we have to find the Thevenin resistance, R_{th}. For this we remove the source V and put a short circuit here. The corresponding circuit now looks like the following.

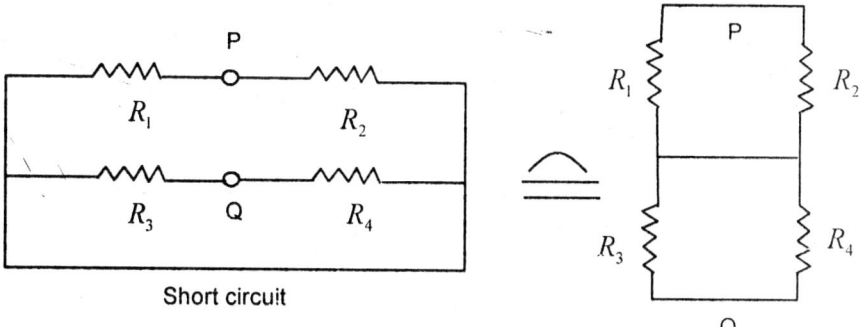

Fig. 1.15.(d)

$\therefore$ R_{th} = resistance measured between P and Q

$$= (R_1 \parallel R_2) + (R_3 \parallel R_4)$$

$$= \frac{R_1 R_2}{R_1 + R_2} + \frac{R_3 R_4}{R_3 + R_4}.$$

Now, from Fig. 1.15(b), the galvanometer current

$$I_g = \frac{V_{th}}{R_{th} + R_g}$$

$$= V \frac{\left(\dfrac{R_2}{R_1 + R_2} - \dfrac{R_4}{R_3 + R_4} \right)}{\dfrac{R_1 R_2}{R_1 + R_2} + \dfrac{R_3 R_4}{R_3 + R_4} + R_g} \qquad \qquad \dots (1.109)$$

Under balanced condition of the bridge, $I_g = 0$.

$$\therefore \qquad \frac{R_2}{R_1 + R_2} - \frac{R_4}{R_3 + R_4} = 0$$

or, $$\frac{R_1 + R_2}{R_2} = \frac{R_3 + R_4}{R_4}$$

$$\therefore \qquad \frac{R_1}{R_2} = \frac{R_3}{R_4} \qquad \qquad \dots (1.110)$$

This is the balanced condition of the bridge.

1.2   Apply Thevennin's theorem to find the current in the XY branch of the following circuit. Also, find the same current by applying superposition theorem alone.

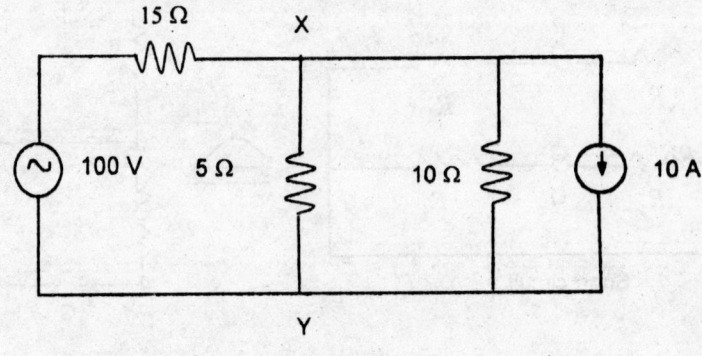

Fig. 1.16.(a)

Solution

The given circuit can be rearranged as

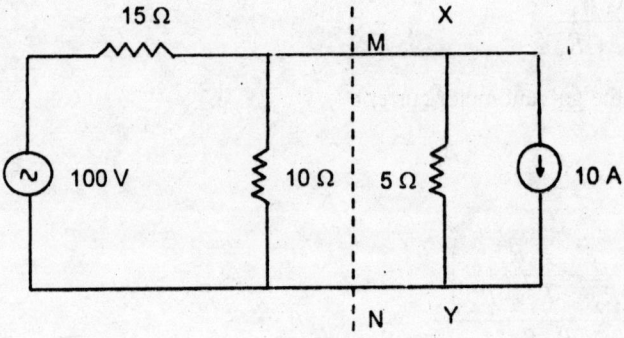

Fig. 1.16.(b)

Since $10\,\Omega$ and $5\,\Omega$ resistances are in parallel, their positions are interchanged. Applying Thevenin's theorem at the MN line of the circuit the left hand potion of the MN line can be replaced by a voltage source V_{MN} in series with a resistance R_{MN}.

By Thevenin's theorem,

$$V_{MN} = \text{open–circuit voltage at MN}$$

$$= \frac{100}{15+10}.10 = \frac{1000}{25} = 40 \text{ V}$$

R_{MN} = resistance measured at MN with the voltage source eliminated

$$= 15\,\Omega \| 10\,\Omega \quad = \frac{15.10}{15+10} = 6\,\Omega. \text{ The given circuit will now appear as}$$

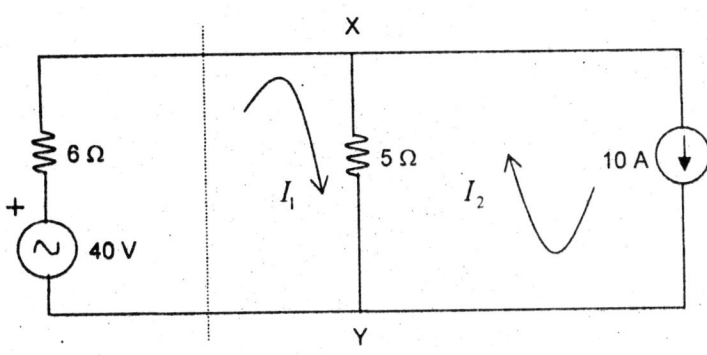

Fig. 1.16.(c)

We have to find the current through XY branch. Apply superposition theorem. Remove 10 A current source and make it open circuit. Then, the current I through XY due to 40 V source is

$$I_1 = \frac{40}{(6+5)} = \frac{40}{11} \text{ A flowing downwards.}$$

Now, remove 40 V source and put a short circuit in its place. The current I_2 flowing through XY due to 10 A source is

$$I_2 = \frac{(10 \times 6)}{(6+5)} = \frac{60}{11} \text{ A}$$

This current flows upwards.

$\therefore$ Resulting current through XY flowing upwards is

$$I_2 - I_1 = \frac{60}{11} - \frac{40}{11} = \frac{20}{11} \text{ A.}$$

Second method
We now solve this problem by superposition theorem alone. Firstly, remove 10 A current source and find current flowing through XY branch due to 100 V source alone. After removal of 10 A source, it becomes on open circuit. The corresponding circuit is

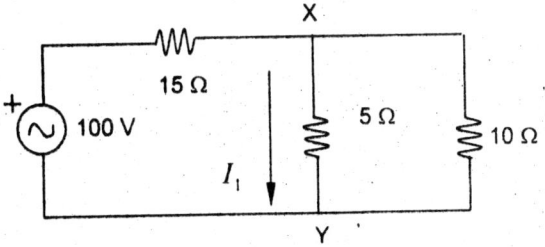

Fig. 1.16.(d)

Voltage across XY is $V_{XY} = \dfrac{100\left(\dfrac{5.10}{5+10}\right)}{15+\dfrac{5.10}{5+10}} = \dfrac{100.\dfrac{50}{15}}{15+\dfrac{50}{15}} = \dfrac{100 \times 50}{275} = \dfrac{200}{11}$ volt.

Current I_1 through XY is $I_1 = \dfrac{V_{XY}}{5.} = \dfrac{200}{11} \times \dfrac{1}{5} = \dfrac{40}{11}$ A.

Now, remove the 100 V source and put a short circuit in its place. Then, find the current I_2 flowing through XY branch due to 10 A source alone. The circuit will like the following :

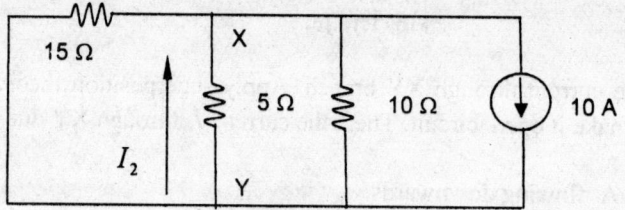

Fig. 1.16.(e)

Let $R = 15\,\Omega \parallel 10\,\Omega \parallel 5\,\Omega$

$\therefore \qquad \dfrac{1}{R} = \dfrac{1}{15} + \dfrac{1}{10} + \dfrac{1}{5} = \dfrac{11}{30}$ \qquad\qquad $\therefore R = \dfrac{30}{11}\,\Omega$

$\therefore \qquad V_{XY} = 10R = 10 \times \dfrac{30}{11} = \dfrac{300}{11}$ volt

$\therefore \qquad I_2 = \dfrac{V_{XY}}{5} = \dfrac{300}{11} \times \dfrac{1}{5} = \dfrac{60}{11}$ A

The resultant current through XY is

$I_2 - I_1 = \dfrac{60}{11} - \dfrac{40}{11} = \dfrac{20}{11}$ Amp. flowing upwards.

1.3 Apply superposition theorem and find the current through the $40\,\Omega$ resistance in the following circuit.

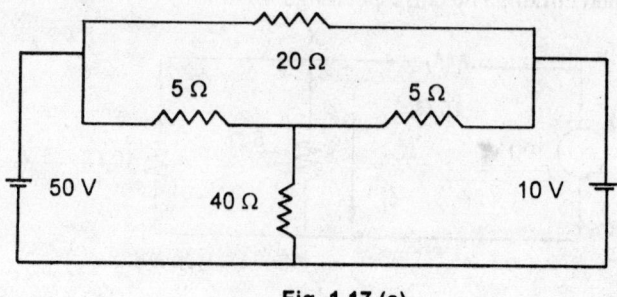

Fig. 1.17.(a)

Solution

Remove 10 V source and put a short circuit in its place. The resultant circuit looks like this.

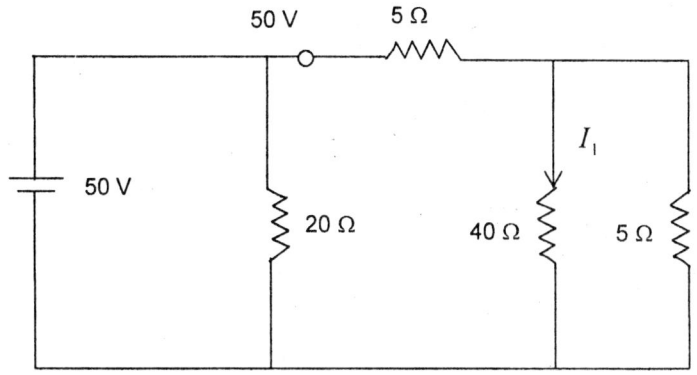

Fig.1.17.(b)

50 V appears across $5\,\Omega$ in series with $(40\,\Omega \parallel 5\,\Omega)$ combination.
Potential difference across $40\,\Omega$ resistance is

$$V_1 = \frac{50\left(\dfrac{40.5}{40+5}\right)}{5+\dfrac{40.5}{40+5}} = \frac{50 \times 200}{5 \times 45 + 200} = \frac{50 \times 200}{425} \text{ volt}$$

Current I_1 through $40\,\Omega$ due to 50 V source alone is

$$I_1 = \frac{V_1}{40} = \frac{50 \times 200}{425} \cdot \frac{1}{40} = \frac{250}{425} = \frac{10}{17} \text{ A}$$

Now, remove 50 V source and place a short circuit in its place. The resultant circuit looks like this.

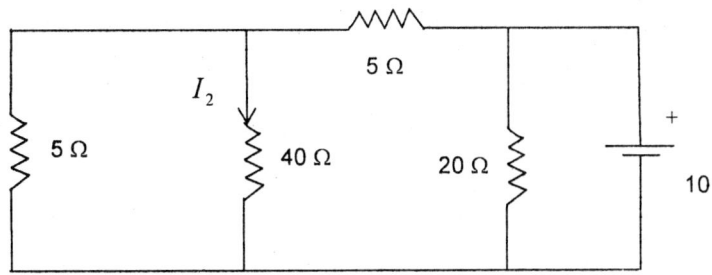

Fig. 1.17.(c)

Here, 10 V appears across $5\,\Omega$ in series with $(40\,\Omega \parallel 5\,\Omega)$ combination.
Potential difference across $40\,\Omega$ resistance is

$$V_2 = \frac{10\left(\dfrac{40 \times 5}{40 + 5}\right)}{5 + \dfrac{40 \times 5}{40 + 5}} = \frac{10 \times 200}{5 \times 45 + 200} = \frac{2000}{425} \text{ volts}$$

Current I_2 through $40\,\Omega$ resistance due to 10 V source alone is

$$I_2 = \frac{V_2}{40} = \frac{2000}{425} \times \frac{1}{40} = \frac{50}{425} = \frac{2}{17} \text{ A.}$$

Resultant current through $40\,\Omega$ resistance is

$$I_1 + I_2 = \frac{10}{17} + \frac{2}{17} = \frac{12}{17} \text{ A.}$$

Both I_1 and I_2 flow in the same direction.

1.4 Apply Norton's theorem to find the current through the XY branch of the following circuit.

[B.U.]

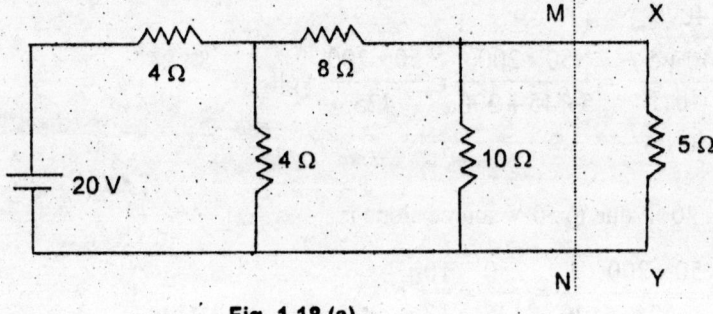

Fig. 1.18.(a)

Solution
We apply Norton's theorem at the point MN of the linear network on the left hand side of the line MN. The Norton equivalent circuit now looks a

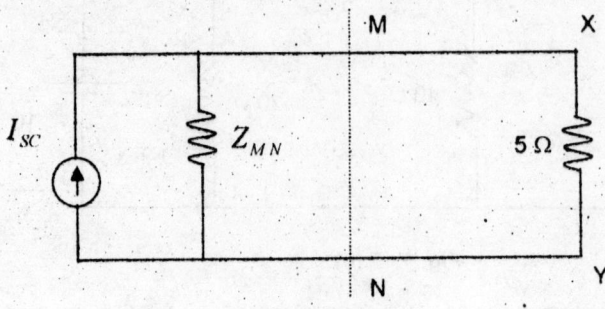

Fig. 1.18.(b)

We have to find the short circuit current (I_{sc}) through MN and the impedance Z_{MN}. The corresponding admittance is $Y_{MN} = 1/Z_{MN}$. Then 10 Ω resistance will be short circuited and 8 Ω appears in parallel with 4 Ω. The short circuit current

$$I_{sc} = \frac{20}{4 + \left(\dfrac{4.8}{4+8}\right)}\left(\frac{4}{8+4}\right) = \frac{20}{4+\dfrac{32}{12}}(1/3) = \frac{20}{4+\dfrac{8}{3}} = 1 \text{ Amp.}$$

The corresponding circuit is

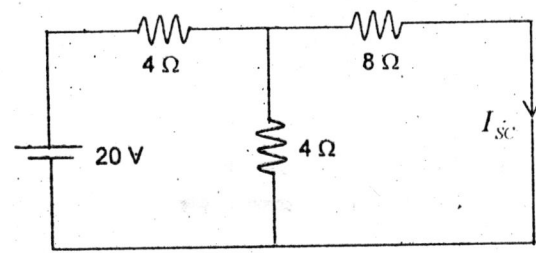

Fig. 1.18.(c)

Now remove 20 V source and put a short circuit here. The impedance measured at MN is

$$Z_{MN} = \frac{10 \times (8+2)}{10 + (8+2)} = \frac{100}{20} = 5\,\Omega$$

Two 4 Ω resistors in parallel produce a 2 Ω equivalent.
The corresponding circuit is

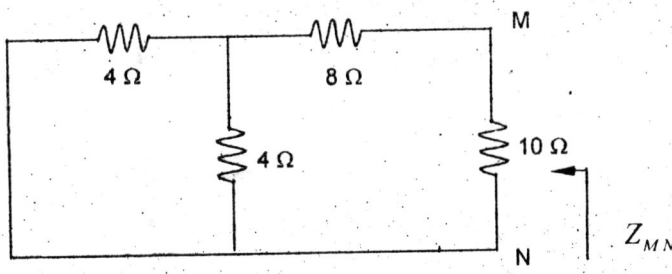

Fig. 1.18.(d)

The current through XY branch is

$$I_{XY} = I_{sc} \times \frac{5}{(5+5)} = \left(\frac{1}{2}\right)I_{sc} = \frac{1}{2} = 0.5 \text{ Amp.}$$

1.5 Find the voltage drop across the resistor R shown in the following circuit with the help of superposition theorem. Verify the result using Kirchhoff's law.   [B.U.]

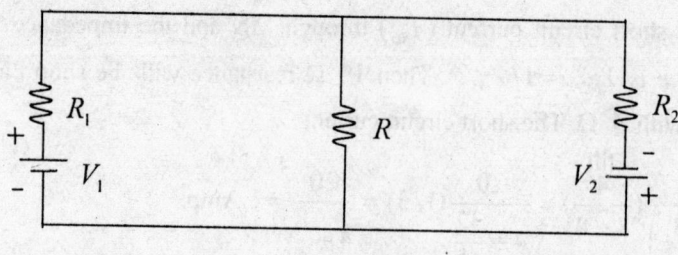

Fig. 1.19.(a)

Solution

Firstly, remove V_2 and put a short circuit in its place. The corresponding circuit looks as

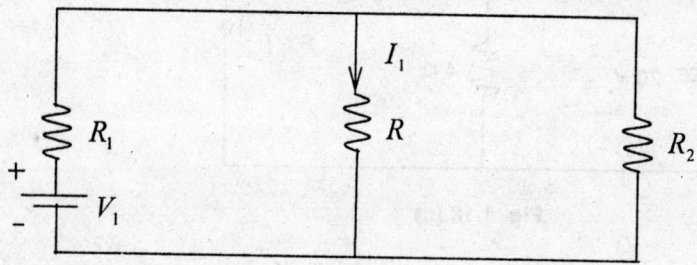

Fig. 1.19(b).

Current through R in this circuit is

$$I_1 = \frac{V_1}{R_1 + \dfrac{R R_2}{R + R_2}} \cdot \frac{R_2}{R + R_2}$$

$$= \frac{V_1 R_2}{R_1 (R + R_2) + R R_2} = \frac{V_1 R_2}{R_1 R_2 + R(R_1 + R_2)}$$

Now, remove V_1 and put a short circuit in its place. The corresponding circuit looks as

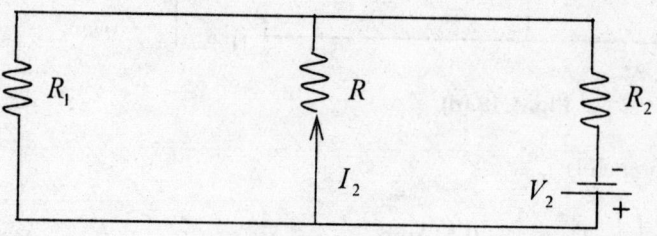

Fig. 1.19.(c)

The current I_2 through R is

$$I_2 = \frac{V_2}{R_2 + \dfrac{R R_1}{R + R_1}} \cdot \frac{R_1}{R + R_1}$$

$$= \frac{V_2 R_1}{R_2(R + R_1) + R R_1} = \frac{V_2 R_1.}{R_1 R_2 + R(R_1 + R_2)}$$

The resultant current through R by superposition theorem is $(I_1 - I_2)$ flowing downwards and the required voltage drop is

$$V_R = R(I_1 - I_2) = \frac{R[V_1 R_2 - V_2 R_1]}{R_1 R_2 + R(R_1 + R_2)}$$

Solution by Kirchhoff's law :

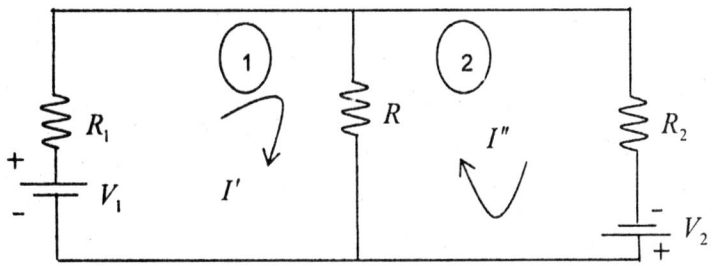

Fig. 1.19.(d)

Assume loop current I' and I'' flowing in two loops. Then, from the loop (1) we get by applying Kirchoff's law

$$V_1 = (I' - I'')R + R_1 I' \qquad \qquad \dots (1.111)$$

From loop (2) we get

$$-V_2 = (I' - I'')R - R_2 I'' \qquad \qquad \dots (1.112)$$

Rearranging (1.111) and (1.112) we write

$$(R + R_1)I' - RI'' = V_1 \qquad \qquad \dots (1.113)$$

$$RI' - (R + R_2)I'' = -V_2 \qquad \qquad \dots (1.114)$$

$$\therefore \quad I' = \frac{\begin{vmatrix} V_1 & -R \\ -V_2 & -(R + R_2) \end{vmatrix}}{\begin{vmatrix} (R + R_1) & -R \\ R & -(R + R_2) \end{vmatrix}}$$

$$= \frac{-(R + R_2)V_1 - RV_2}{-(R + R_1)(R + R_2) + R^2}$$

$$= \frac{(R+R_2)V_1 + RV_2}{R(R_1+R_2)+R_1 R_2} \qquad \qquad \dots (1.115)$$

Similarly, $I'' = \dfrac{\begin{vmatrix} (R+R_1) & V_1 \\ R & -V_2 \end{vmatrix}}{\begin{vmatrix} (R+R_1) & -R \\ R & -(R+R_2) \end{vmatrix}}$

$$= \frac{(R+R_1)V_2 + RV_1}{R(R_1+R_2)+R_1 R_2} \qquad \qquad \dots (1.116)$$

$\therefore$ Required voltage droop $V_R = R(I' - I'')$

$$= \left[\frac{(R+R_2)V_1 + RV_2 - (R+R_1)V_2 - RV_1}{R(R_1+R_2)+R_1 R_2} \right] R$$

$$= \frac{R(R_2 V_1 - R_1 V_2)}{R(R_1+R_2)+R_1 R_2}$$

The result is verified.

PROBLEMS

1.1 Find the voltage developed across the PQ branch of the following circuit using Thevenin's theorem. Calculate the same voltage using superposition theorem only.

[Ans. 3 volt]

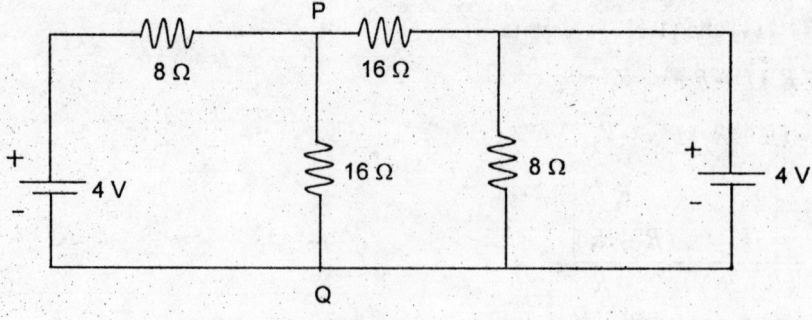

Fig. 1.20

1.2 Apply superposition theorem to find the current through the 12 Ω resistance (MN branch) in the following circuit.

[Ans. 0]

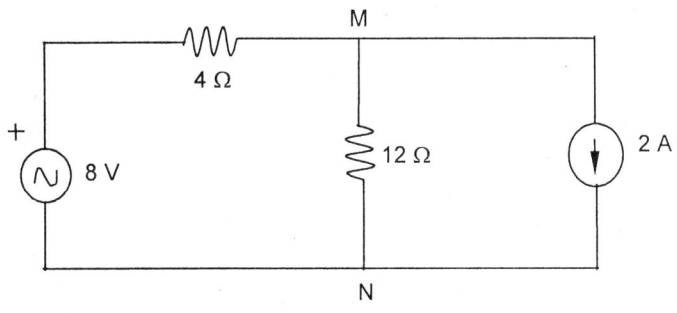

Fig. 1.21

1.3   Apply Norton's theorem and calculate the current flowing through the 3 Ω resistance in the following circuit. Perform the same calculation using superposition theorem alone.

[**Ans.** 2/9 A]

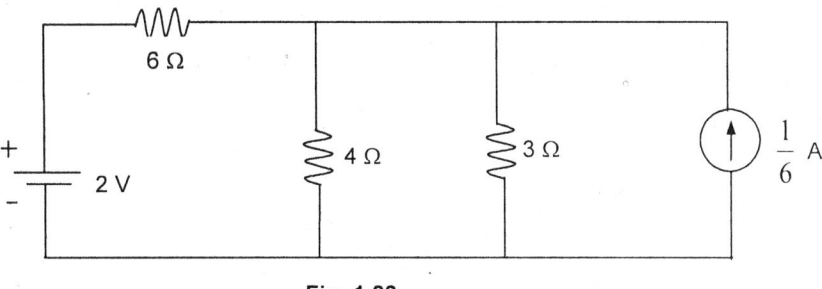

Fig. 1.22

1.4   Find the current passing through 8 Ω resistance in the following circuit by applying Norton's theorem. Verify your calculation using superposition theorem alone.

[**Ans.** 2/7 A]

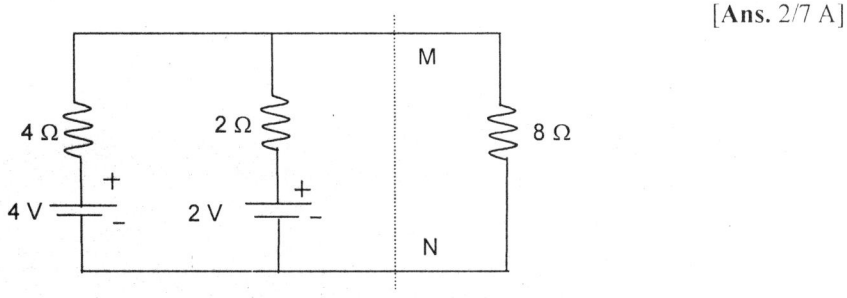

Fig. 1.23.

Hints : To find Norton's equivalent current source find short–circuit currents produced by each source through the short MN. Add the two currents to get Norton's current. To find Norton admittance remove two voltage sources and get a parallel combination of 2 Ω an 4 Ω resistances.

QUESTIONS

1.1 What do you mean by a 'linear network'? define a 'voltage source' and a 'current source'. Is a p–n junction diode a linear device? Is an LCR series resonant circuit a linear circuit? Explain your answer.

1.2 State Thevenin's theorem in network theory. Prove this theorem.

1.3 Give a statement of Norton's theorem. Establish the theorem from fundamental principles.

1.4 What is superposition theorem? Give a proof of the superposition theorem.

1.5 Write down maximum power transfer theorem. Prove this theorem from fundamental principles.

1.6 A T-network has impedances Z_1, Z_2 in the series arms and an impedance Z_3 in the shunt arm. Convert this T-network into its equivalent Π-network.

1.7 A Π-network has a impedance Z_B in the series arm and two impedances Z_A and Z_C in the shunt arms. Convert this Πnetwork into its equivalent T-network.

1.8 A twin-T parallel network formed by $(R, R, 2C)$ and $(C, C, R/2)$ components acts as a notch filter. Calculate the notch frequency. Describe a method for the measurement of phase difference between the input and output voltages of this notch filter.

2

Physics of Vacuum Tubes

2.1 EMISSION OF ELECTRONS FROM A METAL

In order that an electron can be emitted from a metal, some energy must be applied to it. There are four different methods of increasing the energy of the electron in a metal. One method is to increase the temperature of the metal by heating. When temperature of the metal is increased, the velocity of the electrons normal to the metal surface increases. Eventually, a condition is reached when the electrons can overcome the potential barrier existing on the surface of the metal. The emission of electrons by increasing the temperature of the metal is known as thermionic emission. The other three processes of electron emission are photoelectric emission, field emission and secondary emission. Field emission is also known as cold cathode emission.

In photoelectric emission, the additional energy required for the emission of electrons is supplied by light. The light quanta are called photons having an energy $h\upsilon$, where h is Planck's constant and υ is the frequency of light. For liberation of electrons, the light must have a frequency greater than that of a critical frequency (υ_0).

In field emission, a strong electric field is applied at the metal surface. This lowers the potential barrier existing on the metal surface. The electrons having an energy greater than the Fermi energy (E_f) can escape from the metal.

In secondary emission, the method of applying additional energy to the electrons in a metal is by the impact of high energy electrons or ions. The high energy electrons or ions impinge on the metal and impart their energy to the electrons on the metal surface. The metallic electrons can then escape from the metal. This type of electron emission is found in vacuum tubes like tetrode and pentode.

2.2 METALLIC WORK FUNCTION

In a metal, all energy levels below a certain level are completely filled with electrons and all energy levels above this level are completely empty at $0°$ K. This particular energy level is called the Fermi level of the metal.

The maximum energy of the electrons in a metal at $0°$ K is E_f, the Fermi energy level. Hence, if E_B be the height of the potential barrier on the metal surface, the minimum energy required by an electron on the Fermi surface at $0°$ K for emission from the metal is $E_w = E_B - E_f$, which is known as the work function of the metal.

The work function of a metal is defined as the minimum amount of energy which must be given to the highest energy electron on the Fermi surface at the absolute zero of temperature in order that this electron can just escape from the metal.

2.3 RICHARDSON – DUSHMAN EQUATION

The temperature dependence of the thermionic emission from a metal is described by Richardson-Dushman equation. According to this equation, the thermionic current density from a metal surface at temperature $T°$ K is given by

$$J = \alpha T^2 e^{-\frac{\phi}{kT}} \qquad \qquad \dots (2.1)$$

where $\alpha = \dfrac{4\pi qmk^2}{h^3}$ and ϕ is the work function of the metal in energy unit. q is the magnitude of electronic charge and m is the electron mass and k is the Boltzmann constant. From (2.1) we can write

$$\frac{J}{T^2} = \alpha e^{-\frac{\phi}{kT}}$$

or, $\quad \ln(\dfrac{J}{T^2}) = \ln\alpha - \dfrac{\phi}{kT} \qquad \qquad \dots (2.2)$

Plotting $\ln(\dfrac{J}{T^2})$ against $\dfrac{1}{T}$ we get a straight line. This gives an idea of experimental verification of Richardson–Dushman equation.

DERIVATION

The free electrons in a metal form an electron gas which obeys Fermi-Dirac statistics. We consider a rectangular Cartesian coordinate system on the surface of the metal such that the x-axis is perpendicular to the metal surface. Let E_B be the height of the barrier existing on the metal surface. The electrons which have x-component of velocity greater than or equal to v_{x0} will escape from the metal. So,

$$\frac{1}{2}mv_{x0}^2 = E_B \qquad \qquad \dots (2.3)$$

where m = electron mass.

Let us first calculate the electron density of the metal. For this, we have to calculate the density of states. Let N be the number of electrons in the electron gas which occupies a volume V in the coordinate space. For N electrons, there will be $3N$ coordinates and $3N$ momenta. So, we consider a $6N$ dimensional space known as phase space or Γ (gamma) space. The electrons having momenta in the range p to $p+dp$ occupies a volume $4\pi p^2 dp$ in the momentum space. The total phase volume at the disposal of these electrons in $4\pi p^2\, dp.V$.

The volume of a unit phase cell is h^3 where h =Planck's constant. This is because although classically the side length of a phase cell can be taken down to zero, quantum mechanically we can not have an accuracy better than h in the simultaneous measurement of x and p_x. The uncertainty relation is $\Delta x \Delta p_x \sim \hbar$. So, h is taken as the smallest side length of a unit cell. The number of phase cells in the momentum range p to $p+dp$ per unit volume is $\dfrac{4\pi p^2 dp}{h^3}$.

Again, electrons are spin $1/2$ particles. So, these states (cells) will have $(2 \times \frac{1}{2} + 1) = 2$ fold degeneracy. Then, the actual number of states per unit volume is $N(p)dp = \dfrac{2.4\pi p^2 dp}{h^3}$. This is the density of states. Now, $p = mv$ where v = electron velocity

$\therefore \qquad dp = m\,dv$.

In terms of velocity, we write $N(p)dp$ as

$$N(v)\,dv = 2.4\pi \left(\frac{m}{h}\right)^3 v^2 dv \qquad\qquad \ldots (2.4)$$

But, $4\pi v^2\,dv$ represents a volume element in the velocity space. In terms of velocity components, this volume element is given by $dv_x dv_y dv_z$. Then,

$$N(v)\,dv = 2\left(\frac{m}{h}\right)^3 dv_x\,dv_y\,dv_z \qquad\qquad \ldots (2.5)$$

This is the expression for density of states in the velocity space. Now, we proceed to calculate the thermionic current density. The probability of occupation of an electron energy state E is given by F-D statistics as

$$f(E) = \frac{1}{1 + e^{(E-E_f)/kT}} \qquad\qquad \ldots (2.6)$$

where E = energy of an electron

$$= \frac{1}{2}m\left(v_x^2 + v_y^2 + v_z^2\right) \qquad\qquad \ldots (2.7)$$

and E_f = fermi energy.

When the metal is heated, only those electrons will escape from the metal whose velocity components satisfy the conditions

$$v_{xo} \le v_x \le \infty, \; -\infty \le v_y \le \infty, \; -\infty \le v_z \le \infty.$$

The number of electrons per unit volume in the velocity range v_x to $v_x + dv_x$, v_y to $v_y + dv_y$ and v_z to $v_z + dv_z$ is

$$n = N(v)\,dv\,f(E)$$

$$= 2\left(\frac{m}{h}\right)^3 dv_x\,dv_y\,dv_z\,f(E) \qquad\qquad \ldots (2.8)$$

The number of electrons emitted per unit area per unit time from the metal surface is

$$n' = 2\left(\frac{m}{h}\right)^3 \int\limits_{v_{xo}}^{\infty} v_x\, dv_x \int\limits_{-\infty}^{\infty} dv_y \int\limits_{-\infty}^{\infty} dv_z\, \frac{1}{1 + e^{(E - E_f)/kT}}$$

$$\approx 2\left(\frac{m}{h}\right)^3 \int\limits_{v_{xo}}^{\infty} \int\limits_{-\infty}^{\infty} \int\limits_{-\infty}^{\infty} v_x\, dv_x\, dv_y\, dv_z\, e^{-(E - E_f)/kT}$$

since $(E_B - E_f) \sim 20\, kT$ for electron emission and $e^{(E - E_f)/kT} \gg 1$.

The thermionic current density is given by

$$J_S = qn' \qquad\qquad \text{where } q = \text{magnitude of electronic charge}$$

$$= 2q\left(\frac{m}{h}\right)^3 \int\limits_{v_{xo}}^{\infty} v x e^{-\frac{1}{2}\frac{mv_x^2}{kT}}\, dv_x \int\limits_{-\infty}^{\infty} e^{-\frac{mv_y^2}{2kT}}\, dv_y \int\limits_{-\infty}^{\infty} e^{-\frac{mv_z^2}{2kT}}\, dv_z \,.\, e^{E_f/kT}$$

$$= 2q\left(\frac{m}{h}\right)^3 e^{E_f/kT} \cdot \frac{kT}{m} e^{-\frac{mv_{xo}^2}{2kT}} \sqrt{\frac{\pi\, 2kT}{m}} \cdot \sqrt{\frac{\pi\, 2kT}{m}}$$

$$= \frac{4\pi q m}{h^3}(kT)^2\, e^{-(E_B - E_f)/kT}$$

$$= \alpha\, T^2 e^{-\phi/kT} \qquad\qquad\qquad \ldots (2.9)$$

where $\alpha = \dfrac{4\pi q m k^2}{h^3}$ and $\phi = E_B - E_f$. ϕ is known as work function of the metal.

Here, we have used the following formulae :

$$\int\limits_{0}^{\infty} e^{-x} x^{m-1}\, dx = \Gamma(m)$$

$$\int\limits_{-\infty}^{\infty} e^{-bx^2}\, dx = 2\int\limits_{0}^{\infty} e^{-bx^2}\, dx$$

$$= \sqrt{\frac{\pi}{b}}$$

$$\Gamma(1) = 1,\ \Gamma(\tfrac{1}{2}) = \sqrt{\pi},\ \Gamma(m+1) = m\Gamma(m) \qquad \Gamma(m) \text{ is the gamma function.}$$

2.4 VACUUM TUBES

A glass tube is evacuated to a high degree so that there is no air or gas inside it practically. Electron conduction takes place inside this vacuum tube. Electrodes are placed inside this tube to which proper voltages are applied. Depending upon the number of electrodes, vacuum tubes are named as diode, triode, tetrode, pentode, hexode, etc. A diode has two electrodes – one anode and the other cathode. The cathode can be directly or indirectly heated. In direct heating, there exists a potential distribution over the cathode surface.

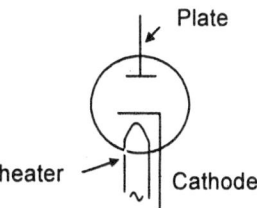

Fig. 2.1. Schematic diagram of a vacuum diode.

In indirect heating, a filament type heater is used and the cathode can be treated as an equipotential surface. The next member of the vacuum tube family is triode which has three electrodes, namely, anode, control grid and the cathode. Control grid is a perforated metallic plate through which a large number of electrons can escape. The number of grids increases for higher members in the vacuum tube family. Tetrode has four electrodes–anode, screen grid, control grid, and cathode. Pentode has three grids–control grid, screen grid and suppressor grid–in addition to the anode and cathode.

The vacuum tubes are also known as valves since they allow the flow of current in one direction only. A valve in the path of a liquid flow controls the flow of the liquid in one direction and not in the opposite direction.

2.5 CHILD–LANGMUIR LAW

The space-charge limited plate current in a vacuum diode varies as the three-half power of the plate voltage. Mathematically,

$$I_P \propto V_P^{3/2} \qquad\qquad\qquad\qquad\qquad ... (2.10)$$

where I_p is the plate current and V_p is the plate voltage of the vacuum diode. This relation is valid for plane parallel geometry of the electrodes and also for cylindrical geometry. The constants of proportionality in two cases are, however, different. This current is space-charge limited, that is, the number of electrons existing in the interelectrode space exceeds the demand of the anode voltage.

DERIVATION

The law will be derived under the following assumptions :

- The electrodes are infinite, plane, parallel equipotential surfaces.
- The number of electrons emitted from the cathode exceeds the demand of the anode potential, i.e., the current is space-charge controlled.
- The electrons start from rest at the cathode surface.
- Electrons only are present between the anode and the cathode.
- The anode voltage is constant.

We treat the problem in one dimension x where the x-axis is directed normally from cathode to anode with the origin on the cathode surface. If $V(x)$ be the potential at any point x in the inter-electrode space, then Poisson's equation can be written as

$$\frac{d^2 V(x)}{dx^2} = -\frac{\rho}{\varepsilon} \qquad \qquad \ldots (2.11a)$$

where ρ = charge density at x, which is negative here and ε = absolute permittivity of vacuum.

The current density J is given by

$$J = -\rho u \qquad \qquad \ldots (2.11b)$$

where u = electron velocity at x. Again,

$$\frac{1}{2} m u^2 = q V(x)$$

$$\therefore \qquad u = \sqrt{\frac{2q}{m} V(x)} \qquad \qquad \ldots (2.12)$$

where m = electron mass and q = magnitude of electronic charge.

From (2.11b) we write

$$\rho = -\frac{J}{u} = -J \sqrt{\frac{m}{2q}} \; V^{-1/2}(x) \qquad \qquad \ldots (2.13)$$

Equation (2.11a) can now be recast as

$$\frac{d^2 V(x)}{dx^2} = \frac{J}{\varepsilon} \sqrt{\frac{m}{2q}} \; V^{-1/2}(x)$$

$$= K J V^{-1/2}(x) \qquad \qquad \ldots (2.14)$$

where $K = \dfrac{1}{\varepsilon} \sqrt{\dfrac{m}{2q}}$.

Multiplying (2.14) by $2\left(\dfrac{dV}{dx}\right)$ and integrating with respect to x we get

$$\int \frac{d}{dx}\left[\left(\frac{dV}{dx}\right)^2\right] dx = 2 K J \int \frac{dV}{V^{1/2}}$$

since J is independent of x. This is because there is no source or sink in the interelectrode space.

$$\therefore \qquad \left(\frac{dV}{dx}\right)^2 = 2 K J . 2 V^{1/2} + C_1 \qquad \qquad \ldots (2.15)$$

where C_1 = integration constant to be evaluated from the boundary condition. On the cathode, the potential is zero. So, at $x = 0$, $V = 0$. Again, since the current is space-charge limited, the electrons on the cathode experience no attraction due to anode voltage. Thus, at $x = 0$, $\dfrac{dV}{dx} = 0$.

$\therefore$ $C_1 = 0$.

From (2.15) we can write

$$\frac{dV}{dx} = 2\sqrt{KJ}\ V^{1/4}$$

or, $$\int_0^{V_p} \frac{dV}{V^{1/4}} = 2\sqrt{KJ} \int_0^d dx$$

since, $V = V_p$ at $x = d$, where 'd' is the anode-cathode separation.

$\therefore$ $$\frac{4}{3} V_p^{3/4} = 2\sqrt{KJ}\ d$$

or, $$K J d^2 = \frac{4}{9} V_p^{3/2}$$ [squaring both sides]

$\therefore$ $$J = \frac{4}{9} \frac{1}{Kd^2} V_p^{3/2}$$

$$= \frac{4}{9} \varepsilon \sqrt{\frac{2q}{m}} \frac{V_p^{3/2}}{d^2} \qquad \qquad \qquad \dots (2.16)$$

putting the expression for K.

$\therefore$ $$J \propto V_p^{3/2}$$

According to Ohm's law, the current–voltage relation of an ohmic device is linear. So, vacuum diode is a non-ohmic device. Here, the current voltage relation is non-linear.

2.6 VACUUM DIODE CHARACTERISTICS

There are two phenomena which govern the current voltage characteristics of a vacuum diode. These are
  (i)   thermionic emission, and
  (ii)   control of current by space charge.

When there are sufficient number of electrons near the cathode, $I_p \propto V_p^{3/2}$. As the anode voltage is increased, a time comes when all the electrons are attracted by the anode voltage and the diode current saturates. Under saturation, the value of saturation current will depend upon the cathode temperature. The higher the cathode temperature (T), the higher is the thermionic emission and the higher is the saturation current. This $I_p - V_p$ characteristics is shown in Fig. 2.2.

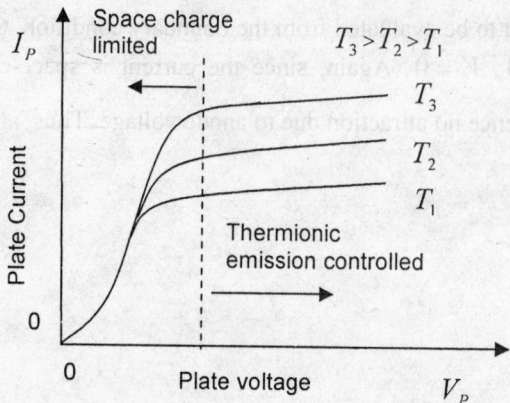

Fig. 2.2. Plate current-plate voltage characteristics of a vacuum diode.

The variation of anode current (I_P) with cathode temperature $T°$ K is shown in Fig. 2.3. When the cathode temperature is low, no emission occurs and there is no emission current. So, $I_P = 0$. When the temperature exceed some critical value, I_P increases according to the law of emission. Now, $I_p \propto T^2 e^{-\phi/kT}$. When the cathode temperature is continuously increased, a stage comes when the emission exceeds the demand of the anode potential and the anode can not attract all the electrons emitted. The anode current saturates. The saturated plate current depends upon the plate voltage. The higher the plate voltage, the more the number of electrons attracted and the higher is the saturated value of plate current.

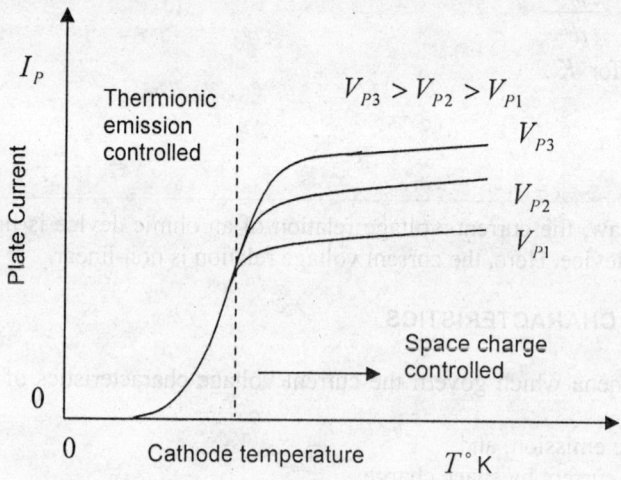

Fig. 2.3. Plate current vs. cathode temperature characteristics of a vacuum diode.

2.7 VACUUM TRIODE

A triode has three electrodes– cathode, control grid and anode. The anode current in a triode can be described by the following equation

$$I_p = K(V_p + \mu V_G)^{\frac{3}{2}} \qquad\qquad \dots (2.17)$$

where I_p = plate current, V_p = plate voltage with respect to cathode, V_G = control grid voltage and μ = amplification factor of the triode. K is a constant of the triode. V_p and V_G are dc voltages and I_p is dc current in the measurement of tube characteristics. A cross-sectional view of the vacuum triode is shown in Fig.2.4. P is the plate, G is the control grid and K is the cathode.

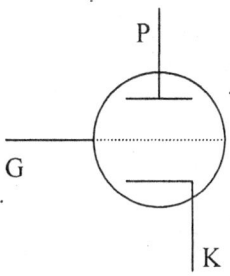

Fig. 2.4. Cross-sectional view of a triode valve.

A schematic diagram of plate current (I_p) versus plate voltage (V_p) with control grid voltage (V_G) as a parameter is shown in Fig.2.5. $V_{G2} > V_{G1} > 0$. The slope of $I_p - V_p$ characteristics gives $\dfrac{1}{r_p}$ where r_p is ac plate resistance of triode. Typical values of V_p lie in the range $(0-250 \text{ V})$ while typical values of the plate current lie in the range $(0-15 \text{ mA})$. The same range for negative grid voltage V_G is $(0,10 \text{ V})$.

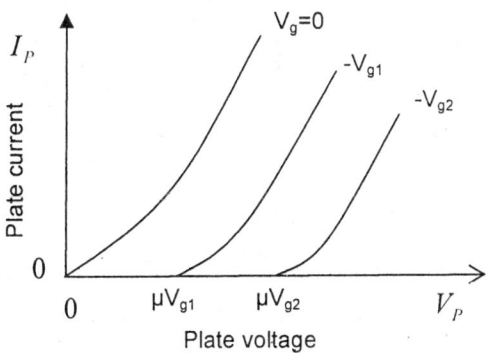

Fig. 2.5. Plate current-plate voltage characteristic of triode.

For a constant V_G, I_p increases with V_p following $\frac{3}{2}$ power law. This is evident from Fig.2.5. The more negative the grid voltage is, the less is the value of plate current. Similarly, a schematic diagram of plate voltage (V_p) versus control grid voltage (V_G) with plate current (I_p)

as a parameter is shown in Fig.2.6. Here, $I_{P2} > I_{P1} > 0$. The slope of $V_P - V_G$ characteristics gives μ, where μ is the amplification factor of the triode. Equation (2.17) gives $V_P + \mu V_G = (\frac{I_P}{K})^{2/3}$. With I_P =constant, this equation gives a straight line. The higher the value of I_P, the higher is the value of intercept on the V_P axis.

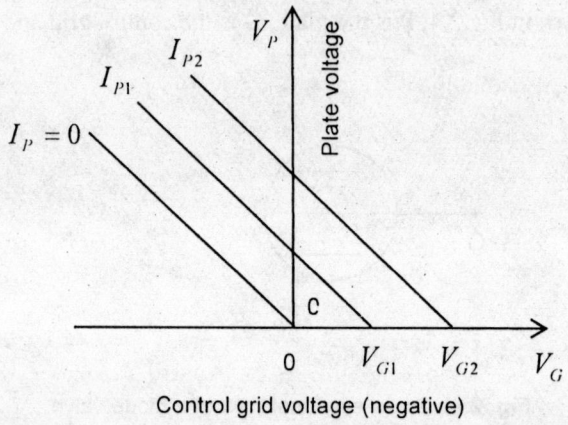

Fig. 2.6. Plate voltage-control grid voltage characteristic of triode. $I_{P2} > I_{P1} > 0$.

The plate current (I_P) as a function of control grid voltage (V_G) characteristics using plate voltage (V_P) as a parameter is shown in Fig.2.7. Here, $V_{P2} > V_{P1} > 0$. The slope of $I_P - V_G$ characteristics gives the transconductance g_m of triode. This characteristics of the triode is also known as transfer characteristics or mutual characteristics of the triode. With V_P = constant, the

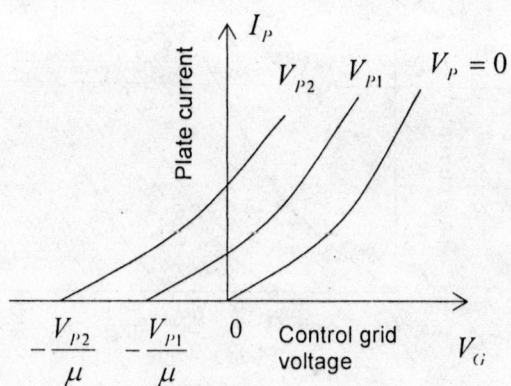

Fig. 2.7. Plate current-control grid voltage characteristic of triode.

plate current I_P varies with the grid voltage V_G following a $\frac{3}{2}$ power law as predicted by eqn. (2.17). The higher the value of V_P the higher is the value of the intercept on the I_P axis.

The triode valve has three coefficients, which describe the relative variations among three pairs of variables v_p, i_p and v_G, which are the total instantaneous values (dc+ac) of the variables, the third variable being held constant. These coefficients are : (i) amplification factor (μ), (ii) plate resistance (r_p) and (iii) transconductance or mutual conductance (g_m).

2.7.1 AMPLIFICATION FACTOR (μ)

It is defined as the negative of the small change in plate voltage to the corresponding change in control grid voltage in order to hold the plate current constant. Thus,

$$\mu = -\frac{\partial v_p}{\partial v_G}\bigg|_{i_p = cons\tan t} \qquad \qquad \ldots (2.18)$$

μ is the ratio of control grid–cathode capacitance and plate–cathode interelectrode capacitance and hence depends upon the geometry of the triode valve.

2.7.2 PLATE RESISTANCE (r_p)

It is defined as the ratio of small change in plate voltage to the corresponding change in plate current when control grid voltage is held constant. Thus,

$$r_p = \frac{\partial v_p}{\partial i_p}\bigg|_{v_G = constant} \qquad \qquad \ldots (2.19)$$

From equation (2.17), using total instantaneous values of current and voltages, we write

$$v_p + \mu v_G = (\frac{i_p}{K})^{\frac{2}{3}}$$

Differentiating with respect to i_p we get

$$\frac{\partial v_p}{\partial i_p}\bigg|_{v_G = constant} = \frac{2}{3}\left(\frac{i_p}{K}\right)^{\frac{-1}{3}}.\frac{1}{K}$$

Thus, $r_p \propto i_p^{\frac{-1}{3}}$. So, r_p varies with plate current.

2.7.3 TRANSCONDUCTANCE (g_m)

It is defined as the ratio of a small change in plate current to the corresponding change in control grid voltage producing this change in plate current when the plate voltage is held constant. Thus,

$$g_m = \frac{\partial i_p}{\partial v_G}\bigg|_{v_p = constant} \qquad \qquad \ldots (2.20)$$

From equation (2.17), using total instantaneous values of the variables and taking partial derivative of i_P with respect to v_G we get

$$\frac{\partial i_P}{\partial v_G}\bigg|_{v_P=constant} = K\frac{3}{2}(v_P + \mu v_G)^{1/2}.\mu$$

$$= \frac{2}{3}K\mu\left(\frac{i_P}{K}\right)^{1/3} \qquad \therefore \quad g_m \propto i_P^{1/3}$$

Thus, transconductance of the triode valve varies with the plate current.

2.7.4 RELATION BETWEEN THREE VALVE COEFFICIENTS

The plate current in a triode valve is a function of plate voltage and control grid voltage. Mathematically, we can write

$$i_P = F(v_P, v_G) \qquad \qquad \ldots (2.21)$$

where F is some function. Taking differential of (2.21)

$$di_P = \frac{\partial F}{\partial v_P}\bigg|_{v_G=constant} dv_P + \frac{\partial F}{\partial v_G}\bigg|_{v_P=constant} dv_G$$

$$= \frac{\partial i_P}{\partial v_P}\bigg|_{v_G=constant} dv_P + \frac{\partial i_P}{\partial v_G}\bigg|_{v_P=constant} dv_G$$

$$= \frac{1}{r_p}dv_P + g_m dv_G$$

Now, if we change v_P and v_G in such a manner that the plate current is held constant (i.e., $di_P = 0$) we can write

$$0 = \frac{1}{r_p}dv_P + g_m dv_G \qquad \text{when } i_P = \text{constant}$$

$$\therefore \quad -\frac{dv_P}{dv_G}\bigg|_{i_P=constant} = r_p g_m$$

$$\therefore \quad \mu = r_p g_m \qquad \qquad \ldots (2.22)$$

2.7.5 SMALL SIGNAL AC EQUIVALENT CIRCUIT OF TRIODE

We consider small signal operation of the triode. We can write

$$i_P = F(v_P, v_G) \qquad \qquad \ldots (2.23)$$

where i_p, v_p and v_G are total instantaneous values (dc+ac) of plate current, and plate and grid voltages respectively. Making a Taylor series expansion of (2.23) we get

$$I_p + \Delta i_p = F\left(V_P + \Delta v_P, V_G + \Delta v_G\right)$$

$$= F\left(V_P, V_G\right) + \frac{\partial F}{\partial v_P}\Delta v_P + \frac{\partial F}{\partial v_G}\Delta v_G + \text{higher order terms.}$$

I_p, V_P and V_G are quiescent values (zero signal) of plate current, plate voltage and grid voltage respectively. The partial derivatives $\dfrac{\partial F}{\partial v_P}$ and $\dfrac{\partial F}{\partial v_G}$ are to be evaluated at the quiescent point (V_P, V_G).

Now, $F\left(V_P, V_G\right) = I_P$.

$$\therefore \qquad \Delta i_p = \frac{\partial i_p}{\partial v_P}\Delta v_P + \frac{\partial i_p}{\partial v_G}\Delta v_G$$

Replacing the small changes Δi_p, Δv_P and Δv_G by the respective ac quantities i_a, v_a and v_c respectively, we get

$$i_a = \frac{1}{r_p}v_a + g_m v_c \qquad\qquad \ldots (2.24)$$

Equation (2.24) can be recast as

$$v_a = r_p i_a - g_m r_p v_c$$

$$= -\mu v_c + r_p i_a \qquad\qquad \ldots (2.25)$$

Equation (2.24) gives rise to the following equivalent circuit of the vacuum triode as shown in Fig.2.8.

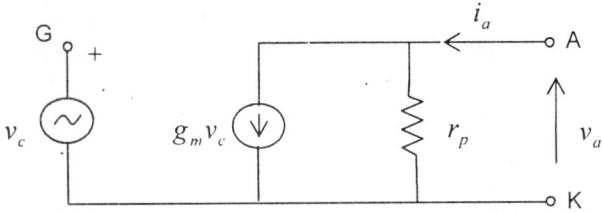

Fig. 2.8. Constant current ac equivalent circuit.

Here, i_a = ac anode current

$\quad\quad\quad v_a$ = ac anode voltage

$\quad\quad\quad v_c$ = ac control grid voltage.

Equation (2.25) gives rise to the following voltage-source equivalent circuit shown in Fig.2.9.

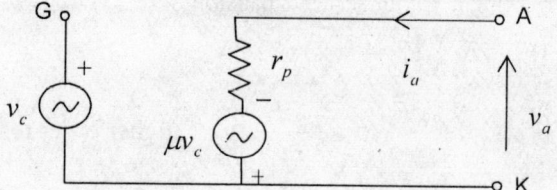

Fig. 2.9. Voltage-source ac equivalent circuit of the triode.

2.7.6 COMMON CATHODE TRIODE AMPLIFIER

In this section, we make an analysis of the common cathode triode amplifier. In common cathode amplifier, the cathode terminal is common to both the input and the output ports.

In Fig.2.10, V_{GG} and V_{PP} are control grid and anode supply voltages respectively. Z_L is the load impedance. R_g is the grid resistance through which V_{GG} is applied to the control grid. R_g is high to avoid power dissipation of the signal at the input. We can write, $Z_L = R_L + jX_L$

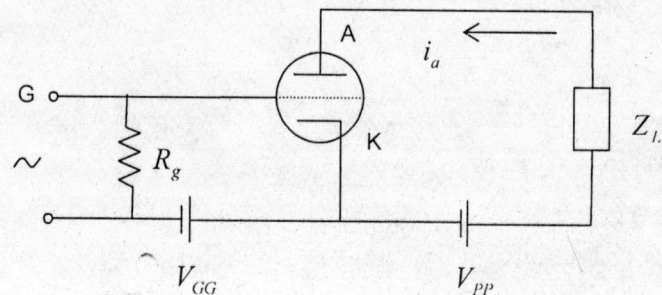

Fig. 2.10. Common cathode triode amplifier circuit

We assume linear class A_1 operation of the triode. In class A_1, 'A' means that plate current flows for the entire cycle of the input ac signal and subscript '1' indicates that grid draws no current. The ac equivalent circuit of the common cathode amplifier is shown in Fig.2.11.

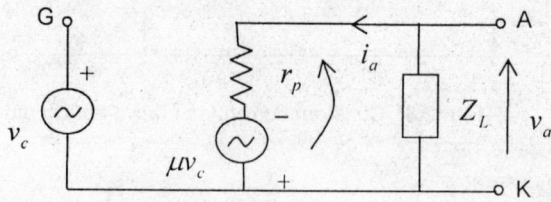

Fig. 2.11. AC equivalent circuit of common cathode amplifier.

The ac plate current is

$$i_a = \frac{\mu v_c}{r_p + Z_L}$$ ... (2.26)

AC voltage developed across the load Z_L is

$$v_a = -i_a Z_L$$

$$= -\frac{\mu v_c}{r_p + Z_L} Z_L$$

$$= -\frac{\mu v_c}{1 + r_p / Z_L}$$ ... (2.27)

The voltage gain of the amplifier is

$$A_v = \frac{v_a}{v_c} = -\frac{\mu}{1 + r_p / Z_L} = -\frac{\mu(R_L + jX_L)}{(r_p + R_L) + jX_L}$$ ... (2.28)

Then,

$$|A_v| = \mu \left[\frac{R_L^2 + X_L^2}{\left(r_p + R_L\right)^2 + X_L^2} \right]^{1/2}$$ ... (2.29)

and $$\varphi = \pi + \tan^{-1}\left(\frac{X_L}{R_L}\right) - \tan^{-1}\left(\frac{X_L}{r_p + R_L}\right)$$ ... (2.30)

φ is the phase angle by which the ac anode voltage lags the ac grid voltage. If the load Z_L is purely resistive then $X_L = 0$ and $Z_L = R_L$. Then,

$$|A_v| = \frac{\mu R_L}{r_p + R_L} \text{ and } \varphi = \pi.$$

If $R_L = 0$ and $Z_L = jX_L$,

$$|A_v| = \frac{\mu X_L}{\left(r_p^2 + X_L^2\right)^{1/2}} \text{ and } \varphi = 3\frac{\pi}{2} - \tan^{-1}\left(\frac{X_L}{r_p}\right).$$

Case I:
Let us consider a purely resistive load.

So, $Z_L = R_L$. Then, $|A_v| = \dfrac{\mu}{1 + r_p / R_L}$.

As R_L increases, $|A_v|$ increases and eventually saturates to μ.

Case II:

Now, we consider purely reactive load.

So, $Z_L = jX_L$. Then, $|A_v| = \dfrac{\mu}{\sqrt{1 + (r_p/X_L)^2}}$.

As X_L increases, $|A_v|$ increases and eventually saturates to μ. The variations of $|A_v|$ and φ with Z_L are shown in Figs. 2.12 and 2.13 respectively.

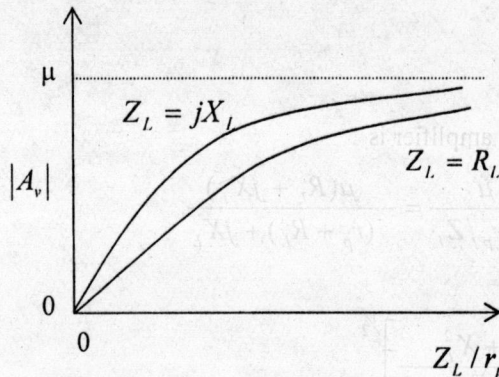

Fig. 2.12. Variation of $|A_v|$ with Z_L/r_p.

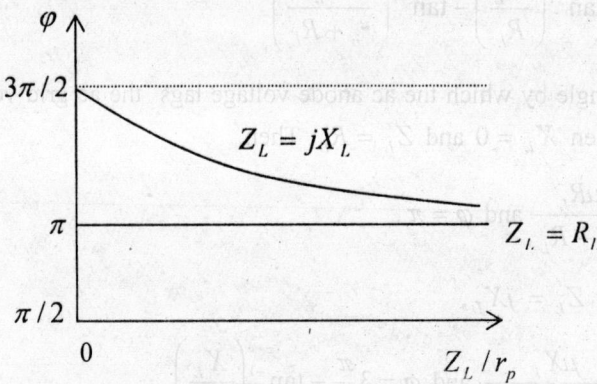

Fig. 2.13. Variation of φ with Z_L/r_p.

2.7.7 HIGH FREQUENCY LIMITATION OF TRIODE AMPLIFIER

The electrodes of a triode valve behave like condenser plates. With three electrodes, there can be three interelectrode capacitances. These are cathode-grid capacitance (C_{gk}), plate-grid capacitance (C_{gp}) and cathode-plate capacitance (C_{pk}). In the low frequency region, the effect of these interelectrode capacitances is negligible since they provide very high reactance in parallel. But in the high frequency region, they provide moderate reactances and their effect should be considered. Taking the interelectrode capacitances into account, the following high frequency ac equivalent circuit of the triode amplifier can be drawn as shown in Fig.2.14.

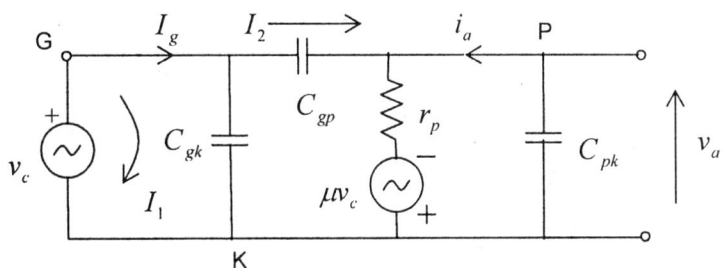

Fig. 2.14. Voltage-source ac equivalent circuit of triode amplifier at high frequencies. G-control grid, P-anode (or plate), K-cathode.

The current passing through C_{gp} is

$$I_2 = \frac{(v_c - v_a)}{\dfrac{1}{j\omega C_{gp}}}$$

$$= (v_c - Av_c)j\omega C_{gp}$$

$$= v_c(1 - A)j\omega C_{gp} \qquad \qquad \ldots (2.31)$$

' A ' is the amplification factor of triode which is complex, in general.

Let $A = A_r + jA_i$, where A_r is the real part and A_i is the imaginary part of A.

The current flowing through C_{gk} is

$$I_1 = j\omega C_{gk} v_c \qquad \qquad \ldots (2.32)$$

Net input grid current is

$$I_g = I_1 + I_2$$

$$= j\omega v_c \left[C_{gk} + (1 - A_r)C_{gp} - jA_i C_{gp} \right] \qquad \qquad \ldots (2.33)$$

Input admittance

$$Y_i = \frac{I_g}{v_c}$$

$$= G_i + jB_i \text{ (say)}$$

where G_i = input conductance and B_i = input susceptance.

Now, $G_i = \omega C_{gp} A_i \qquad \qquad \ldots (2.34)$

and $\quad B_i = \omega[C_{gk} + (1 - \overline{A}_r)C_{gp}]$ $\qquad$ ... (2.35)

Now, we proceed to calculate A. We have assumed $I_2 << i_a$. So, we can neglect the effect of C_{gp} in calculating A. The load impedance Z'_L includes C_{pk}. Thus, $Z'_L = Z_L \| C_{pk}$. Now,

$$A = \frac{v_a}{v_c}$$

$$= -\frac{\mu Z'_L}{r_p + Z'_L}$$

$$= -\mu \frac{R'_L + jX'_L}{(r_p + R'_L) + jX'_L} \cdot \frac{(r_p + R'_L) - jX'_L}{(r_p + R'_L) - jX'_L}$$

$$= -\mu \frac{R'_L(r_p + R'_L) + (X'_L)^2}{(r_p + R'_L)^2 + (X'_L)^2} \cdot -j\mu \frac{r_p X'_L}{(r_p + R'_L)^2 + (X'_L)^2}$$

$$= A_r + jA_i \quad \text{(say)} \qquad \qquad \text{... (2.36)}$$

Then, $\quad G_i = -\mu\omega C_{gp} \dfrac{r_p X'_L}{(r_p + X'_L)^2 + (X'_L)^2}$ $\qquad$ ... (2.37)

and $\quad B_i = \omega[C_{gk} + C_{gp}\{1 + \mu \dfrac{R'_L(r_p + R'_L) + (X'_L)^2}{(r_p + R'_L)^2 + (X'_L)^2}\}]$ $\qquad$ ... (2.38)

Equation (2.37) shows that G_i is negative for an inductive load. This, in turn, implies that the triode valve delivers power to the input circuit. This is a positive feedback and can lead to oscillation in the grid and plate circuit. It results in instability in the device. The negative conductance indicates that the tube generates power instead of consuming power. This tendency towards oscillation poses a limitation on the high frequency operation of the triode amplifier.

2.8 TETRODE

Tetrode is a vacuum tube with four electrodes. These electrodes are – cathode, control grid, screen grid and plate. The screen grid is maintained at a positive potential smaller than the plate voltage. It screens the control grid from the plate and thereby reduces the plate-grid interelectrode capacitance to a low value. A typical anode voltage-anode current characteristics of the tetrode is shown in Fig.2.15. When the positive anode potential is greater than that of the screen grid, the plate collects the primary electrons from the cathode. A few secondary electrons are emitted from the screen grid. But, since the screen grid attracts only a few primary electrons, the number of secondary electrons emitted from it is small and the anode current tends to be nearly constant with increasing plate voltage.

Now, let the plate voltage be gradually reduced. A time comes when the screen grid voltage becomes more positive than the anode voltage. The primary electrons from the cathode produce secondary electrons from the anode surface by collision. These secondary electrons are now attracted by the more positive screen grid. The anode current is now reduced due to the flow of secondary electrons towards the screen grid. Usually, a primary electron can produce more than one secondary electrons. In some cases, even the flow of secondary electrons in the reverse direction can exceed the flow of primary electrons. The anode current, as a result, may be reversed in the extreme case depending upon the secondary electron emissivity of the anode surface and the positivity of the screen grid. In the region 'ab' of the plate current-plate voltage characteristics, the number of secondary electrons emitted from the anode decreases with decreasing plate voltage. So, the reverse flow of electrons towards the screen decreases and hence the plate current increases with decreasing plate voltage.

As the anode potential is reduced monotonically, a stage comes when a cloud of electrons are produced near the anode which is known as virtual cathode. The anode cannot attract all electrons present near it. The anode current reduces with the decrease in anode potential. A typical anode current–anode voltage characteristics of the tetrode is shown in Fig. 2.15.

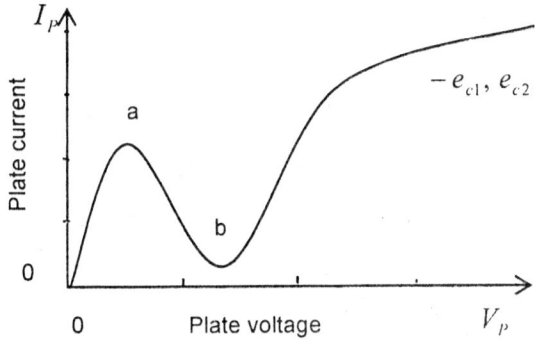

Fig. 2.15. Plate current –plate voltage characteristics of a tetrode.

A typical value of control grid voltage e_{c1} is 3 V and that of the screen grid voltage e_{c2} is 80 V. The tetrode shows a negative resistance (plate current decreasing with increasing plate voltage) in the region 'ab' of the $V_p - I_p$ characteristics. The negative resistance is utilized in designing an oscillator known as Dynatron. A typical tetrode circuit is shown in Fig. 2.16.

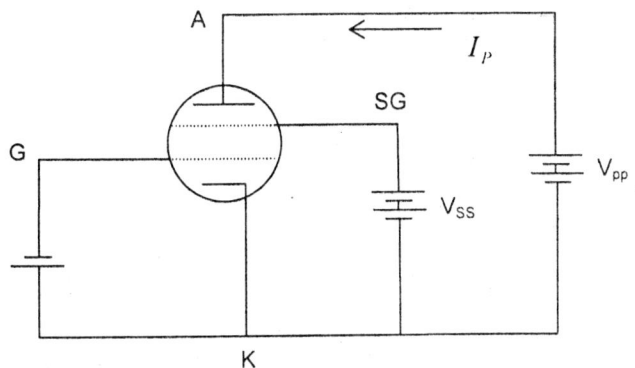

Fig. 2.16. Tetrode characteristics measurement circuit. SG-Screen grid, G-control grid, V_{pp}–plate supply voltage, V_{ss}–Screen supply voltage.

2.9 PENTODE

A pentode is a vacuum tube with five electrodes. These electrodes are cathode, control grid, screen grid, suppressor grid and anode. So, there are three grids in this tube. Usually, the control grid is maintained at a negative potential relative to the cathode. The screen grid is supplied with a positive potential and the suppressor grid is maintained at the cathode potential. The anode is given a positive voltage. The cathode electrons are attracted by the positive potential of the screen grid. The electrons eventually cross the perforated screen grid and get decelerated by the suppressor grid.

In the screen–suppressor space, the electrons feel attraction towards the screen. However, some electrons will pass through the suppressor grid under the attraction of the plate.

When the anode potential is low, electrons in the suppressor–plate region move slowly towards the plate and form a virtual cathode. The anode current increases with plate voltage. When the anode potential is increased, the plate current eventually saturates corresponding to the case when all electrons in the virtual cathode are attracted by the plate. The circuit for pentode characteristics is shown in Fig. 2.17.

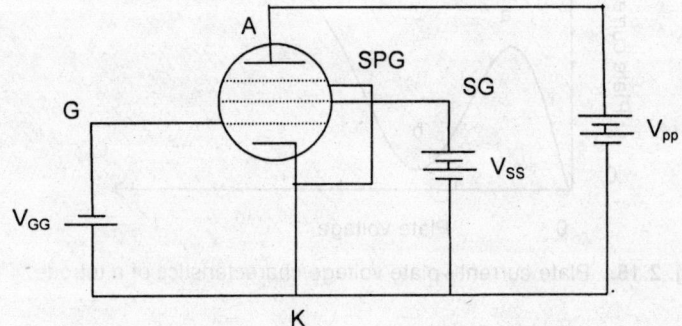

Fig. 2.17. Pentode characteristics circuit. SPG-Suppressor grid, SG- Screen grid.

Plate current–plate voltage characteristics of pentode is shown in Fig. 2.18. e_{c1} is the control grid voltage, e_{c2} is the screen grid voltage which is held constant at a positive value. e_{c3} is the suppressor grid potential.

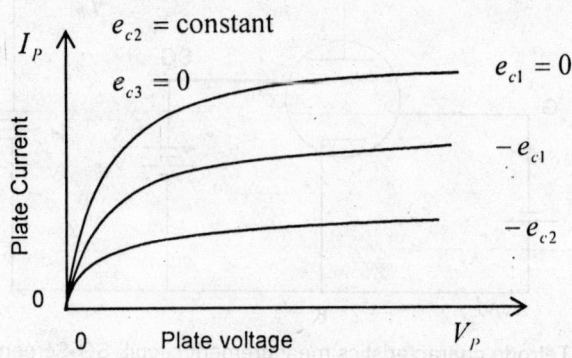

Fig. 2.18. Plate voltage-plate current characteristics of pentode.

SOLVED PROBLEMS

2.1 The amplification factor of a triode is 40 and the plate resistance is 20 kΩ. Find the mutual conductance of the triode.

Solution

Here, $\mu = 40$ and $r_p = 20$ kΩ. We know that

$$\mu = r_p g_m$$

$$\therefore \quad g_m = \frac{\mu}{r_p} = \frac{40}{20 \times 10^3} = 2 \times 10^{-3} \text{ mho.}$$

2.2 Assuming the current to be space charge limited in a vacuum diode, calculate the value of plate voltage required for a plate current of 100 mA. Given that the diode current is 200 mA for a plate voltage of 300 volts.

Solution

From Child-Langmuir's law, we get

$$I_p = K \cdot V_p^{3/2}, \text{ where } K = \text{constant.}$$

Here, $I_p = 200$ mA, $V_p = 300$ volts.

$$\therefore \quad 200 \times 10^{-3} = K(300)^{3/2}$$

$$= K \cdot 3\sqrt{3} \times 10^3$$

$$\therefore \quad K = \frac{200 \times 10^{-6}}{3\sqrt{3}} = \frac{2 \times 10^{-4}}{3\sqrt{3}} \left(A / v^{3/2} \right)$$

When $I_p = 100$ mA, we have

$$V_p = \left(\frac{I_p}{K}\right)^{2/3} = \left[\frac{100 \times 10^{-3} \times 3\sqrt{3}}{2 \times 10^{-4}}\right]^{2/3}$$

$$= \left[\frac{10^3 \times 3\sqrt{3}}{2}\right]^{2/3}$$

$$= \left(\frac{27}{4}\right)^{\frac{1}{3}} \times 100 = \frac{300}{\sqrt[3]{4}} \approx 189 \text{ volts.}$$

2.3 A common cathode triode amplifier with $\mu = 30$, $r_p = 30$ kΩ and a load resistance of 60 kΩ is used to amplify an ac signal of 50 mV. How much is the output voltage?

Solution

Voltage gain $|A_v| = \dfrac{\mu R_L}{r_p + R_L}$

Here, $\mu = 30$, $r_p = 30$ kΩ and $R_L = 60$ kΩ.

$$\therefore \quad |A_v| = \frac{30 \times 60 \times 10^3}{(30 + 60) \times 10^3} = \frac{1800}{90} = 20.$$

Input voltage, $v_i = 50$ mV.
Output voltage,

$$V_0 = |A_v| v_i$$
$$= 20 \times 50 \text{ mV} = 1 \text{ volt.}$$

2.4 The plate current in a vacuum triode is represented by the expression $i_p = 2 \times 10^{-4} \left(v_G + 0.05 v_P \right)^{1.5}$ Amp. where v_G and v_P are the control grid voltage and plate voltage, respectively, in volts. Determine the cut-off grid bias if the plate supply voltage of 200 volts is applied to the tube. Calculate the transconductance of this triode at a plate current of 25 mA.

Solution

$$v_G\Big|_{cut-off} = -\frac{v_P}{\mu}$$

$$\mu = -\frac{\partial v_P}{\partial v_G}\Big|_{i_p=contant}$$

$$i_p = 2 \times 10^{-4} \left(v_G + 0.05 v_P \right)^{1.5}$$

Differentiating partially with respect to v_G,

$$0 = 2 \times 10^{-4} \left(v_G + 0.05 v_P \right)^{0.5} \times (1.5) \left[1 + 0.05(-\mu) \right]$$
or, $\quad 1 - (0.05)\mu = 0$

$$\therefore \quad \mu = \frac{1}{0.05} = 20.$$

When $v_P = 200$ volts, $v_G\Big|_{cut-off} = -\dfrac{200}{20} = -10$ volt.

Now, $\quad g_m = \dfrac{\partial i_p}{\partial v_G}\Big|_{v_P=constant} \quad = 2 \times 10^{-4} \left(v_G + .05 v_P \right)^{0.5} (1.5) \times 1$

$$= 2 \times 10^{-4} (1.5) \left(\frac{i_P}{2 \times 10^{-4}} \right)^{1/3}$$

$$= 3 \times 10^{-4} \left[\frac{25 \times 10^{-3}}{2 \times 10^{-4}} \right]^{1/3}$$

$$= 3 \times 10^{-4} (125)^{1/3} = 15 \times 10^{-4} \text{ mho.}$$

$\therefore \qquad g_m = 1.5$ mmho.

PROBLEMS

2.1 Calculate the emission current density in mA/cm^2 for a tungsten filament at $2500°$ K. Given $\phi = 4.5$ eV and $\alpha = 60$ A/cm$^2(°$K$)^2$. The symbols have their usual meanings.

[**Ans.** 0.156 mA/cm^2]

2.2 In a vacuum diode, the cathode and anode are parallel planes separated by a distance of 5 mm. The electrons are emitted from the cathode with zero initial velocity. Calculate the time of flight of electrons when the applied anode potential is 200 volts.

[**Ans.** 1.19 ns]

2.3 The characteristics of a triode valve are given by $i_P = 0.02(v_P + 10v_G)^{3/2}$ mA. This triode is used as an amplifier with an anode load resistance of 10 kΩ. Calculate the anode current corresponding to a grid bias of -10 volts and anode voltage of 200 volts. Find the values of μ and r_p of the tube at the operating point specified and hence find the gain of the amplifier.

[**Ans.** 20 mA, 10, $\frac{10}{3}$ kΩ, 7.5]

2.4 The plate resistance of a vacuum triode is 20 kΩ at a plate current of 40 mA. Calculate the value of the plate resistance at a plate current 5 mA.

[**Ans.** 40 kΩ]

QUESTIONS

2.1 What are the processes by which electrons are emitted from a metal? What do you mean by 'work function' of a metal?

2.2 Write down Richardson–Dushman equation of thermionic emission. State and explain Child–Langmuir's law for a vacuum diode.

2.3 What are the constants of a triode valve? Define μ, r_p and g_m for a vacuum triode. Establish a relation among them. How do r_p and g_m vary with the plate current?

2.4 Derive the small–signal equivalent circuits of a triode amplifier? Derive an expression for the voltage gain of a common cathode triode amplifier taking a load impedance Z_L. Plot schematically the variation of voltage amplification of the common cathode triode amplifier when the load is (a) purely resistive and (b) purely inductive. Show the nature of output phase variation in both cases.

2.5 Explain mathematically, how high frequency instability appears in a triode amplifier.

2.6 How does the grid-plate interelectrode capacitance in a triode give rise to a negative conductance of the input circuit for an inductive load? Explain mathematically.

2.7 Explain the operation of a tetrode valve. Give a schematic plot of the $I_p - V_p$ characteristics of the tetrode where I_p means anode current and V_p stands for anode voltage. Mention an application of the tetrode.

2.8 Explain the functions of the screen and suppressor grids in a pentode. Draw the plate current–plate voltage characteristics of a pentode and explain the nature of variation physically.

3

Physics of Semiconductor Materials

3.1 CLASSIFICATION OF MATERIALS

Materials can be classified in three heads depending upon their electrical conductivity. These are metal, semiconductor and insulator. Metals are good conductors of electricity. For a good conductor, the resistivity $\rho = 10^{-5}$ Ωcm typically. Insulators are nonconductors of electricity. Typical resistivity of an insulator is $\rho = 10^{14}$ Ωcm. Semiconductors are materials, which have resistivity intermediate between a metal and an insulator. Typical intrinsic resistivity for Germanium (Ge) at $300°$K is $\rho = 50$ Ωcm.

Another method of classification of materials comes from the band theory of solids. In an isolated atom, the energy levels are discrete. When these atoms come close together to form a solid, these energy levels split into large number of quasi-continuous energy levels due to mutual interactions among large number of atoms. These quasi-continuous energy levels form energy bands. Number of atoms in Silicon (Si) is 5×10^{22} /c.c., for Germanium it is 4.4×10^{22} /c.c., and for GaAs it is 2.2×10^{22} /c.c.

In the band structure of solids, there is a conduction band and there is a valence band which are, in general, separated by a forbidden energy gap, which is commonly called as the bandgap. There is no allowed energy state of the electron in the bandgap. This is shown in Fig. 3.1. Depending upon the value of the bandgap (usually expressed in electron volts), the materials are classified as metals, semiconductors and insulators. For an insulator, the bandgap is typically > 3 eV. So, at room temperature or even at much higher temperature, there is no thermal excitation from valence band to conduction band in an insulator. For metals, there are free electrons present in the conduction band at room temperature so that electrical conduction takes place upon application of an electric field across the specimen. In some cases, the valence band and conduction band overlap and there is no bandgap in metals.

In semiconductors, the bandgap is ~ 1 eV. The bandgap for Ge, Si and GaAs at $0°$K are 0.74 eV, 1.16 eV and 1.51 eV respectively. These bandgaps of semiconductors are functions of temperature. The variation can be described by the relation

$$E_g(T) = E_g(0) - \frac{aT^2}{b+T} \qquad \qquad \ldots (3.1)$$

where T is the temperature in $^{\circ}$K, 'a' and 'b' are parameters which fit the plot and $E_g(0)$ is the band gap at $T = 0\,^{\circ}$K. The parameter 'a' has values 0.477 meV/$^{\circ}$K for Ge, 0.473 meV/$^{\circ}$K for Si and 0.541 meV/$^{\circ}$K for GaAs. The values of 'b' in $^{\circ}$K are 235 for Ge, 636 for Si and 204 for GaAs.

There is an energy level called the Fermi level in a solid. In metal, the Fermi level (E_F) lies in the conduction band. In intrinsic (means pure) semiconductor, the Fermi level lies in the middle of the forbidden energy gap. So, in semiconductor the Fermi energy level gives an average location of filled states. A schematic band diagram is shown in Fig.3.1. E_c and E_v represent energies corresponding to the bottom of the conduction band and the top of the valence band. E_g is the band gap energy.

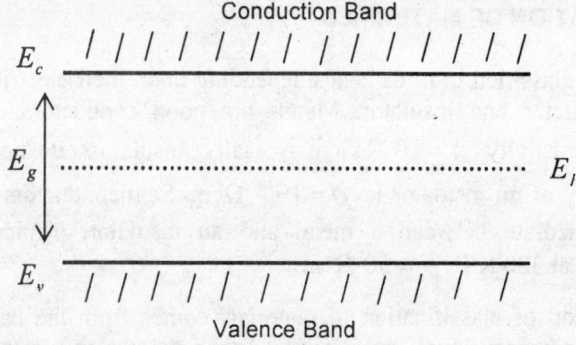

Fig. 3.1. Schematic band diagram.

In metal, all the energy levels up to the Fermi level are filled with electrons at $0\,^{\circ}$K. At finite temperature, Fermi level is the energy level whose probability of occupation is $\frac{1}{2}$. Electrons are spin-half particles. They obey Fermi-Dirac statistics and Pauli exclusion principle. The Fermi–Dirac distribution function is expressed as

$$f(E) = \frac{1}{1 + e^{(E-E_F)/kT}} \qquad \qquad \ldots (3.2)$$

where k = Boltzmann constant and has a value 1.38×10^{-23} J/$^{\circ}$K. The function $f(E)$ gives the probability of occupation of the energy state E at temperature $T\,^{\circ}$K by an electron.

3.2 CONCEPT OF HOLES

Holes are vacancies in an otherwise filled valence band. The group velocity of an electron, treated as a wave, is given by

$$v_g = \frac{\partial \omega}{\partial k} = \frac{1}{h}\frac{\partial}{\partial k}(\hbar \omega)$$

$$= \frac{1}{h}\frac{\partial E}{\partial k} \qquad \qquad \ldots (3.3)$$

k is the magnitude of the wave vector, E is the electron energy and ω is the angular frequency of the wave. $\hbar = h/2\pi$ where h is the Planck's constant. Now, acceleration of the electron inside the band is

$$\frac{dv_g}{dt} = \frac{\partial v_g}{\partial k} \cdot \frac{\partial k}{\partial t}$$

$$= \frac{1}{\hbar}\frac{\partial^2 E}{\partial k^2} \cdot \frac{1}{\hbar}\frac{\partial (k\hbar)}{\partial t} \qquad \qquad \text{... (3.4)}$$

In (3.4), $\dfrac{\partial}{\partial t}(k\hbar) = \dfrac{\partial p}{\partial t}$ where p is the momentum of the electron. $\dfrac{\partial p}{\partial t}$ gives the external force. Then, (3.4) can be written as

$$\alpha = \frac{1}{m^*}\frac{\partial p}{\partial t} \qquad \qquad \text{... (3.5)}$$

where α is the acceleration and m^* is the effective mass of electron. Comparing (3.4) & (3.5) we get

$$m^* = \frac{\hbar^2}{\dfrac{\partial^2 E}{\partial k^2}} \qquad \qquad \text{... (3.6)}$$

m^* thus depends upon the radius of curvature of the $E-k$ curve. At the bottom of the conduction band, $E-k$ curve is concave and m^* is positive. At the top of the valence band, $E-k$ curve is convex and m^* is negative. Negative effective mass means that the particle is accelerated in the opposite direction by the same field. This is equivalent to a positive charge particle. The positive charge is equal in magnitude with that of the electron. This positively charged quasi particle is hole. Thus, holes in the valence band are image charges and not real charges. The vacancies of the electrons (known as holes) in the valence band move under the influence of the applied electric field in a direction opposite to that of the electron.

Semiconductors are important and useful materials in electronics. This is because the conductivity of the semiconductor can be increased as desired by doping foreign atoms into it in small quantities. In absence of doping, the semiconductor is intrinsic (pure). When doping is done, the semiconductor is said to be extrinsic. Depending upon the valency of the of the dopant atoms, the extrinsic semiconductor can be p-type or n-type.

3.3 P-TYPE EXTRINSIC SEMICONDUCTOR

In p-type semiconductor, holes are majority carriers. p-type semiconductor is obtained by doping trivalent impurities like Indium (In), Aluminium (Al), Galium (Ga) etc. The dopant atoms occupy the positions of crystal atoms by displacing them. Each impurity is surrounded by four Ge or Si atoms. The trivalent impurity atom completes three covalent bonds with the Ge or Si atoms. The forth covalent bond is completed with Ge or Si where Ge(or Si) supplies one electron to the dopant atom and in this way the impurity gets ionized. A vacancy is created in the valence band due to the transfer of an electron to the impurity atom. This vacancy can move from one atom to another in the valence band and contributes in the conduction process as a positive charge, which is called hole. Since the trivalent impurity atom accepts one electron from Ge(or Si), it is also called as acceptor impurity. If a small amount of energy is given to this impurity atom, the impurity atom becomes ionized. Typical ionization energy of the acceptor impurity is ~ 0.01 eV. Thus, impurity atoms become ionized at room temperature. This goal will not be reached if we dope arbitrary foreign atoms in Ge or Si. Because, in that case, impurity atoms will not be ionized at room temperature. The covalent bond formation in Si with Al as dopant is shown in Fig. 3.2.

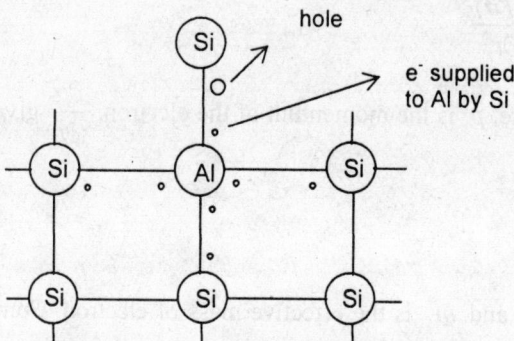

Fig. 3.2. Covalent bond formation with trivalent acceptor Al doped in Si.

Si having an atomic number 14 has an electron distribution $1s^2 2s^2 2p^6 3s^2 3p^2$. There are 4 electrons in the outermost shell (n=3), which is incomplete. These are valence electrons.

In GaAs each Ga atom is surrounded by four As atoms and this forms four covalent bonds. Similarly, each As atom forms four covalent bonds with four Ga atoms. The fifth electron in As is donated to a neighbouring Ga atom. Acceptor impurities like Mg, C, Cd and Li produces acceptor levels at 0.012 eV, 0.019 eV, 0.021 eV and 0.023 eV above the top of the valence band. Donor impurities like Te and Se both produce donor levels in GaAs at 0.0058 eV below the bottom of the conduction band.

3.4 N-TYPE EXTRINSIC SEMICONDUCTOR

In n-type semiconductor, electrons are majority carriers and holes are minority carriers. n-type semiconductor is obtained by doping pentavalent impurity such as Phosphorous (P), Arsenic (As), Antimony (Sb) etc. in Ge or Si. These pentavalent impurity atoms replace one crystal atom and occupy its position in the lattice. A pentavalent impurity atom forms four covalent bonds with four neighbouring crystal atoms (Ge or Si). The fifth electron of the impurity becomes surplus and it has a small ionization energy. This fifth electron be easily detached from its parent atom at room temperature and goes to conduction band. Thus, each impurity atom gives rise to a free electron in the conduction band. That is why such impurity are called donor impurity. Since the charge of the particle donated is negative, the semiconductor is called a n-type semiconductor. Energy levels of donor impurity lie slightly below the bottom of the conduction band in the forbidden energy gap. The ionization energies of P, As and Sb in Ge are 0.012 eV, 0.013 eV and 0.0096 eV respectively.

The Fermi level in n-type semiconductor lie close to the conduction band above the middle of the band gap. At very high temperature, all extrinsic semiconductors tend to be intrinsic and the Fermi level approaches the middle of the bandgap which is the intrinsic level.

The covalent bond formation with pentavalent donor impurity atom Phosphorus (P) doped in Si is shown in Fig. 3.3.

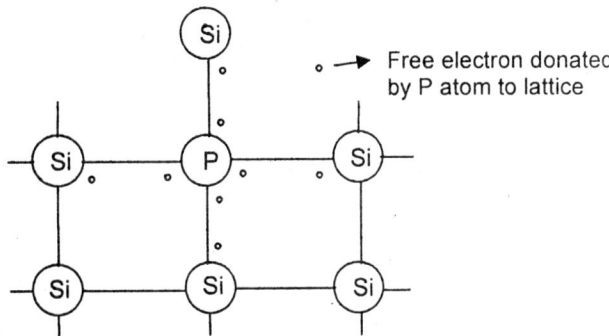

Fig. 3.3. Pentavalent Phosphorous (P) atom doped in Si forming covalent bonds.

3.5 LAW OF MASS ACTION IN SEMICONDUCTOR

The hole concentration in the valence band is given by

$$p = N_v e^{-(E_F - E_v)/kT} \qquad \qquad \text{... (3.7)}$$

where $N_v = 2\left(\dfrac{2\pi m_h kT}{h^2}\right)^{3/2}$ is the effective density of states in the valence band. m_h is the

effective mass of holes. 'h' is Planck's constant and 'k' is Boltzmann constant. E_v is the energy at the top of the valence band.
The electron concentration in the conduction band is given by

$$n = N_c e^{-(E_c - E_F)/kT} \qquad \qquad \text{... (3.8)}$$

where $N_c = 2\left(\dfrac{2\pi m_e kT}{h^2}\right)^{3/2}$ is the effective density of states in the conduction band. m_e is the

effective mass of electrons. E_c is the energy at the bottom of the conduction band. Then,

$$\begin{aligned} np &= N_c N_v e^{-(E_c - E_v)/kT} \\ &= N_c N_v e^{-E_g/kT} \end{aligned} \qquad \text{... (3.9)}$$

where $(E_c - E_v) = E_g$, the band gap.
For intrinsic semiconductor, the hole and electron concentrations are given by

$$p_i = N_v e^{-(E_{Fi} - E_v)/kT} \qquad \qquad \text{... (3.10)}$$

and

$$n_i = N_c e^{-(E_c - E_{Fi})/kT} \qquad \qquad \text{... (3.11)}$$

where the subscript 'i' stands for intrinsic. From (3.10) and (3.11), we get

$$n_i p_i = N_v N_c e^{-E_g/kT}$$ ... (3.12)

But, in intrinsic semiconductor, the electron and hole concentrations are equal. This is because when an electron is excited from valence band to conduction band, a hole is left behind in the valence band. Thus, $n_i = p_i$. Then, in view of (3.9) and (3.12),

$$np = n_i^2$$ ... (3.13)

This is known as law of mass action in semiconductor. This is valid in any semiconductor, intrinsic or extrinsic.

From (3.9) and (3.13), we get

$$n_i = \sqrt{N_c N_v} \, e^{-E_g/2kT}$$ ... (3.14)

When a semiconductor is doped by both acceptor and donor impurity atoms, then for overall charge neutrality of the crystal, we must have concentration of positive charge equal to the concentration of negative charge. So,

$$n + N_a^- = p + N_d^+$$ ... (3.15)

where n is the electron concentration, N_a^- is the ionized acceptor concentration, p is the hole concentration and N_d^+ is the ionized donor concentration. Utilizing the law of mass action as represented by (3.13), we get $n = n_i^2/p$ so that (3.15) becomes

$$\frac{n_i^2}{p} = p + N_d^+ - N_a^-$$

or, $$p^2 - p(N_a^- - N_d^+) - n_i^2 = 0$$ ... (3.16)

$\therefore$ $$p = \frac{1}{2}\left[(N_a^- - N_d^+) + \sqrt{(N_a^- - N_d^+)^2 + 4n_i^2}\right]$$ ... (3.17)

Similarly, $$n = \frac{1}{2}\left[(N_d^+ - N_a^-) + \sqrt{(N_d^+ - N_a^-)^2 + 4n_i^2}\right]$$ ... (3.18)

The negative signs before the square-root in (3.17) & (3.18) are neglected since p and n are number densities and can not be negative.

If $N_d^+ \gg N_a^-$ and $N_d^+ \gg n_i$ then $n = N_d^+$ and $p \approx \dfrac{n_i^2}{N_d^+}$. On the other hand, if $N_a^- \gg N_d^+$ and $N_a^- \gg n_i$ then $p = N_a^-$ and $n \approx \dfrac{n_i^2}{N_a^-}$.

3.6 DIRECT AND INDIRECT BANDGAP SEMICONDUCTOR

In direct bandgap semiconductor, the bottom of the conduction band and the top of the valence band occur at the same value of k, where k is the magnitude of wave vector in the energy-momentum ($E-k$) diagram. Example of this type semiconductor is GaAs, GaSb, InP, InSb, CdS etc. GaAs has a bandgap $E_g = 1.4$ eV at $T = 300\degree$ K.

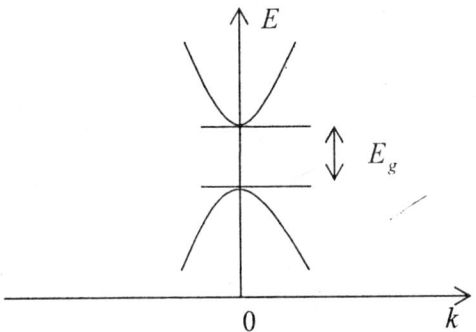

Fig. 3.4. Direct bandgap semiconductor. $E-k$ diagram.

In indirect bandgap semiconductor, the minimum of the conduction band and the maximum of the valence band in the energy-momentum diagram do not occur at the same value of k, the magnitude of wave vector. In order to transfer an electron from the valence band to conduction band, there must be a change in momentum. In order to conserve momentum, the difference in momentum is supplied by scattering agents like phonons. Phonons are quanta of lattice vibration. Examples of indirect bandgap semiconductor are Ge, Si, GaP, AlSb etc.

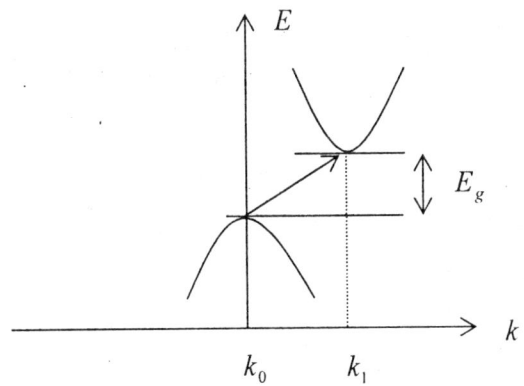

Fig. 3.5. $E-k$ diagram of indirect bandgap semiconductor.

3.7 MOBILITY OF ELECTRONS AND HOLES

Let v_{dn} and v_{dp} be the drift velocities of electrons and holes respectively under the influence of the electric field E. The mobilities of electrons and holes are defined as

$$\mu_n = v_{dn}/E \qquad\qquad\qquad\qquad \text{... (3.19a)}$$
$$\text{and} \quad \mu_p = v_{dp}/E \qquad\qquad\qquad\qquad \text{... (3.19b)}$$

The current densities due to the drift of electrons and holes in a semiconductor are, respectively, written as

$$J_n = nqv_{dn} \qquad \qquad \ldots (3.20a)$$

and $\quad J_p = pqv_{dp} \qquad \qquad \ldots (3.20b)$

where n = electron concentration, p = hole concentration, q = magnitude of electronic charge.

The electrical conductivity (σ) of the semiconductor due to electron and hole motion are

$$\sigma_n = \frac{J_n}{E} = nq\mu_n \qquad \qquad \ldots (3.21a)$$

$$\sigma_p = \frac{J_p}{E} = pq\mu_p \qquad \qquad \ldots (3.21b)$$

respectively. Overall conductivity

$$\begin{aligned} \sigma &= \sigma_n + \sigma_p \\ &= q(n\mu_n + p\mu_p) \end{aligned} \qquad \ldots (3.22)$$

3.8 IMPORTANCE OF SILICON IN TECHNOLOGY

- Silicon (Si) is well abundant in nature.
- Silicon is cheaper than Germanium.
- Leakage current in a reverse biased Si p-n junction is very small ~ nano ampere.
- Si devices are well suited for operation up to 150^0 C.
- Si is suitable for integrated circuits. SiO_2 is a good insulator.

SOLVED PROBLEMS

3.1 The band gap of Silicon at 0° K is 1.16 eV. Calculate the value of band gap at 300° K. Given $a = 0.473 \times 10^{-3}$ eV/$^\circ$K and $b = 636 ^\circ$K. Here 'a' and 'b' are usual parameters.

Solution

Band gap $E_g(T) = E_g(0) - \dfrac{aT^2}{b+T}$

Here, $\quad E_g(0) = 1.16$ eV

$\qquad T = 300^\circ$ K, $a = 0.473 \times 10^{-3}$ eV/$^\circ$K, $b = 636^\circ$K

$\therefore \qquad E_g(T = 300^\circ \text{K}) = 1.16 - \dfrac{0.473 \times 10^{-3}(300)^2}{636 + 300}$

$\qquad \qquad \qquad \qquad = 1.16 - 0.045$

$\qquad \qquad \qquad \qquad = 1.11$ eV

3.2   A silicon sample has an intrinsic carrier concentration of 1.4×10^{10}/ c.c. at room temperature. It is doped by pentavalent impurity with a concentration of 10^{16} / c.c. What will be the value of the hole concentration in the doped material?

Solution

By the law of mass action :

$$n_i^2 = p.n$$

Here, $n_i = 1.4 \times 10^{10}$ /c.c.

n = donor concentration

$= 10^{16}$ / c.c.   assuming all donor atoms are ionized at room temperature.

∴ Hole concentration, $p = \dfrac{n_i^{\,2}}{n} = \dfrac{(1.4 \times 10^{10})^2}{10^{16}}$

$$= 1.96 \times 10^4 \text{ / c.c.}$$

3.3   The intrinsic carrier concentration in Ge at $300\,^{\circ}$K is 2.4×10^{13} /c.c. The band gap of Ge at $300\,^{\circ}$K is 0.66 eV and the same gap at $330\,^{\circ}$K is 0.648 eV. Calculate the intrinsic carrier concentration at $330\,^{\circ}$K.

Solution

Intrinsic carrier concentration

$$
\begin{aligned}
n_i &= \sqrt{N_c\,N_v}\; e^{-E_g/2kT} \\[2mm]
&= 2\left(\frac{2\pi k}{h^2}\right)^{3/2} (m_e\,m_h)^{3/4}\, T^{3/2}\, e^{-E_g/2kT} \\[2mm]
&= F\,T^{3/2}\, e^{-E_g/2kT}
\end{aligned}
$$

where $F = 2\left(\dfrac{2\pi k}{h^2}\right)^{3/2} (m_e\,m_n)^{3/4} = \text{constant.}$

$$
\begin{aligned}
2.4 \times 10^{13} &= F(300)^{3/2}.e^{-0.66/(2 \times 0.025)} \\[2mm]
&= F(300)^{3/2}.e^{-0.66/0.05} \\[2mm]
&= F(300)^{3/2}.e^{-13.2}
\end{aligned}
\qquad \cdots (3.23)
$$

since $kT = 0.025$ eV at $T = 300\,^{\circ}$K.

The intrinsic carrier concentration at $T = 330\,^{\circ}$ K is

$$
\begin{aligned}
n_i &= F(330)^{3/2}.e^{-0.648/0.055} \\[2mm]
&= F(330)^{3/2}.e^{-11.78}
\end{aligned}
\qquad \cdots (3.24)
$$

Dividing equation (3.24) by (3.23) we get

$$n_i \Big/ (2.4 \times 10^{13}) = (330 \Big/ 300)^{3/2} .e^{(13.2 - 11.78)}$$
$$= 1.1536 \times 4.137 = 4.7725$$
$$\therefore \qquad n_i = 1.14 \times 10^{14} / \text{c.c.}$$

3.4 An intrinsic specimen of silicon has a resistivity 230 kΩ cm at room temperature. The mobilities of electrons and holes in silicon at room temperature are 1350 cm^2 / Vs and 480 cm^2 / Vs respectively. Calculate the electron concentration in the specimen.

Solution

Resistivity $\rho = \dfrac{1}{\sigma} = \dfrac{1}{q\left(n\mu_n + p\mu_p\right)}$

Here, $\mu_n = 1350$ cm^2 /Vs $= 0.1350$ m^2 /Vs

$\mu_p = 480$ cm^2 / Vs $= 0.048$ m^2 / Vs

$q = 1.6 \times 10^{-19}$ Coulomb

$\rho = 230$ kΩ cm $= 2.3 \times 10^3$ Ω m

$n = p$

$$\therefore \qquad n = \dfrac{1}{q\left(\mu_n + \mu_p\right)\rho} = \dfrac{1}{1.6 \times 10^{-19} \, (.135 + .048) \, 2.3 \times 10^3}$$
$$= 1.48 \times 10^{16} \, / \text{m}^3$$
$$= 1.48 \times 10^{10} \, / \text{c.c.}$$

PROBLEMS

3.1 The band gap of GaAs at $300°$K is 1.41 eV. Calculate the value of this band gap at $0°$K. given $a = 0.541 \times 10^{-3}$ eV / $°$K and $b = 204°$K where 'a' and 'b' are usual parameter.

, **[Ans. 1.51 eV]**

3.2 The intrinsic carrier concentration in Ge at room temperature $(300°$K) is 2.4×10^{13} /c.c. A Ge sample is doped by acceptor impurity with a concentration of 10^{17} /c.c. . What will be the value of electron concentration in doped sample?

[Ans. 5.76×10^9 /c.c.]

3.3 The resistivity of an intrinsic specimen of Ge at room temperature is 47 Ω cm. The electrons and hole mobilities in this material at room temperature is 3900 cm^2 / Vs and 1900 cm^2 / Vs respectively. What is the hole concentration in this specimen?

[Ans. 2.29×10^{13} /c.c.]

3.4 The intrinsic carrier concentration in Si at $300°K$ is 1.4×10^{10} /c.c. The temperature dependence of the intrinsic carrier concentration (n_i) in Si follows a relation of the form $n_i = AT^{3/2} e^{-E_g/2kT}$, where A = constant, T is the absolute temperature, and E_g is the band gap of the semiconductor. The band gap of Si at $300°K$ has a value of 1.11 eV. Find the value of the constant A.

[**Ans.** 1.18×10^{16} cm^{-3} ^{0}K$^{-3/2}$]

QUESTIONS

3.1 How do you classify solids in accordance with the band theory? Distinguish between metal, semiconductor and insulator. How does the band gap of a semiconductor vary with temperature?

3.2 Define effective mass of an electron in a semiconductor. Give an idea of a hole existing in a semiconductor.

3.3 Define intrinsic and extrinsic semiconductors. Explain, with examples, the formation of p–type and n–type extrinsic semiconductors.

3.4 What is the law of mass action in a semiconductor? Derive an expression for electron concentration in a n–type semiconductor in terms of intrinsic carrier concentration and the doping concentration.

3.5 What do you mean by direct and direct band gap semiconductor? Explain with necessary $E - k$ diagram. Give two examples of indirect band gap semiconductor. Is GaAs a direct band gap semiconductor? Give reasons for your answer.

4

Semiconductor Diodes

4.1 DERIVATION OF SHOCKLEY EQUATION

Consider a p-n junction as shown in Fig. 4.1. All the impurity atoms are ionized at room temperature. Holes are majority carriers in the p-region and electrons are majority carriers in the n-region. So, there exists a concentration gradient across the junction at $x = 0$.

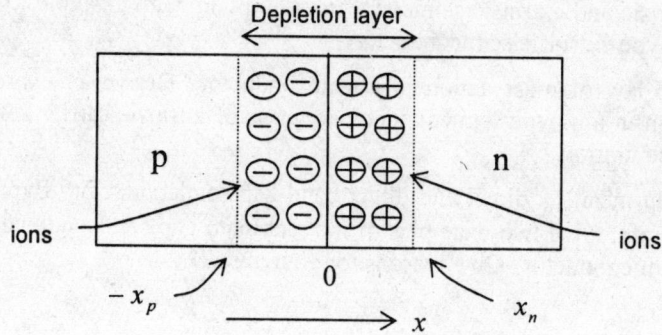

Fig. 4.1. A block diagram of a p-n junction without any bias.

Majority carriers of each side diffuse into the other side of the junction with a view to equalize the concentrations of electrons and holes. The impurity ions shown in Fig. 4.1 are immobile. A layer of negative charges (ions) appears on the p-side and a layer of positive charges (ions) appears on the n-side of the junction as a result of diffusion of mobile carriers. The layers of positive and negative charges on either side of the junction constitute the 'depletion layer'. It is also called as 'space charge layer' or 'transition layer'. The static charges in the depletion layer produces an electric field directed from n-side to p-side through the junction. This electric field opposes further diffusion of majority carriers. So, the diffusion of majority carriers cannot continue indefinitely. The electric field in the depletion layer is, however, in favour of the drift of minority carriers of each side. Electrons are minority carriers in the p-region and holes are minority carriers in the n-region.

Thus, at any instant of time, two types of current flow through the junction:

(i) Diffusion current due to majority carriers migration,

and (ii) drift current due to minority carrier drift motion.

Total hole current density can be expressed as

$$J_p = qp\mu_p E - qD_p \frac{dp}{dx}$$

$$\ldots (4.1)$$

where J_p = hole current density

q = magnitude of electronic charge

p = hole concentration

μ_p = hole mobility

D_p = diffusion coefficient of holes

E = electric field.

Similarly, total electron current density is

$$J_n = qn\mu_n E + qD_n \frac{dn}{dx}$$

$$\ldots (4.2)$$

where, n = electron concentration

μ_n = electron mobility

D_n = diffusion coefficient of electrons.

Other symbols have usual significances as mentioned above.

In thermal equilibrium, no net current can flow through the junction. Again, since there can be no built-up of holes and electrons on either side of the junction, the hole and electron currents must be zero individually. Thus, $J_p = 0$ and $J_n = 0$ in equilibrium.

With $J_p = 0$ we get from (4.1),

$$qD_p \frac{dp}{dx} = qp\mu_p E$$

or, $$\frac{dp}{p} = \frac{\mu_p}{D_p} E dx$$

$\therefore$ $$\int \frac{dp}{p} = \frac{q}{kT} \int E dx$$

$$\ldots (4.3)$$

using Einstein's relation,

$$\frac{D_p}{\mu_p} = \frac{D_n}{\mu_n} = \frac{kT}{q}$$

$$\ldots (4.4)$$

Again, $E = -\dfrac{dV}{dx}$, where V = potential at x. Now, equation (4.3) can be written as,

$$\int_{P_p}^{P_n} \frac{dp}{p} = -\frac{q}{kT} \int_{x=-x_p}^{x_n} dV$$

where p_p = hole concentration at $x = -x_p$ in the p-region,

p_n = hole concentration at $x = x_n$ in the n-region.

Let $V = V_p$ at $x = -x_p$ and $V = V_n$ at $x = x_n$

$$\therefore \quad \ln \frac{p_n}{p_p} = -\frac{q}{kT}\left(V_n - V_p\right)$$

$$= -\frac{q}{kT}\psi$$

where $\psi = V_n - V_p$ is the built-in potential barrier existing across the junction. Then,

$$p_n = p_p e^{-q\psi/kT} \qquad \qquad \text{... (4.5)}$$

Similarly, with $J_n = 0$ in (4.2) we get, as before,

$$\int \frac{dn}{n} = -\frac{q}{kT}\int E dx = \frac{q}{kT}\int dV$$

$$\int_{n_p}^{n_n} \frac{dn}{n} = \frac{q}{kT}\int_{V_p}^{V_n} dV$$

or, $\quad \ln \dfrac{n_n}{n_p} = \dfrac{q}{kT}\psi$

$$\therefore \quad n_p = n_n e^{-q\psi/kT} \qquad \qquad \text{... (4.6)}$$

where n_p = electron concentration in the p-region at $x = -x_p$,

n_n = electron concentration in the n-region at $x = x_n$.

Equations (4.5) and (4.6) are known as Boltzmann equations.

Up to this point, we have considered no application of bias voltage to the p-n junction. Now, let us consider a forward bias V applied to the junction. We assume that

(i) the applied voltage is completely dropped across the depletion layer and no voltage drop exists in the neutral region,

(ii) no generation and recombination occur in the depletion layer,

(iii) minority carrier diffusion lengths are smaller compared with the physical thicknesses of p and n regions, and

(iv) minority carrier injection level is small compared with the majority carrier concentration.

In presence of forward bias V the effective height of the potential barrier is $(\psi - V)$. Due to this lowering of barrier height some excess carriers will diffuse over the barrier.

Let $\Delta p_n(0)$ be the change in hole concentration due to excess hole diffusion at the junction at $x = 0$. By assumption (ii) and (iv) we can write,

$$p_n + \Delta p_n(0) = p_p e^{-q(\psi - V)/kT}$$

$$= p_p e^{-q\psi/kT}.e^{qV/kT} = p_n e^{qV/kT} \quad \text{using (4.5)}$$

$$\therefore \qquad \Delta p_n(0) = p_n\left(e^{qV/kT} - 1\right) \qquad \qquad \text{... (4.7)}$$

The hole current density due to this excess hole diffusion from the p-region to the n-region is given by,

$$J_p = -qD_p\left[\frac{d}{dx}\Delta p_n(x)\right]_{x=0} \qquad \qquad \text{... (4.8)}$$

But, $\Delta p_n(x) = \Delta p_n(0)e^{-x/L_p}$ \qquad \qquad \text{... (4.9)}$

where L_p = diffusion length of holes in the n-region. Equation (4.9) will be derived later on.

Diffusion length of holes is defined as the length of the n-region at which the excess hole concentration injected at $x = 0$ decays by recombination to $\dfrac{1}{e}$ of its original value at $x = 0$.

From (4.9) we write

$$\left[\frac{d}{dx}\Delta p_n(x)\right]_{x=0} = -\frac{1}{L_p}\Delta p_n(0) \qquad \qquad \text{... (4.10)}$$

$$\therefore \qquad J_p = \frac{qD_p}{L_p}\Delta p_n(0) = \frac{qD_p}{L_p}p_n\left(e^{qV/kT} - 1\right) \qquad \qquad \text{... (4.11)}$$

Similarly, the electron current density due to excess electron diffusion from the n-region to the p-region under forward bias is calculated as

$$J_n = \frac{qD_n}{L_n}n_p\left(e^{qV/kT} - 1\right) \qquad \qquad \text{... (4.12)}$$

where L_n = diffusion length of the electrons in the p-region.

Let A = cross sectional area of the p-n junction. Then, the junction current under forward bias is given by

$$I = A(J_p + J_n)$$

$$= Aq\left(\frac{D_p}{L_p}p_n + \frac{D_n}{L_n}n_p\right)\left(e^{qV/kT} - 1\right) \quad \text{using (4.11) and (4.12)}$$

$$= I_s\left(e^{qV/kT} - 1\right) \qquad \qquad \text{... (4.13)}$$

where $I_s = qA\left(\dfrac{D_p}{L_p}p_n + \dfrac{D_n}{L_n}n_p\right)$ is the reverse saturation current of the p-n diode. Equation (4.13) is known as Shockley equation.

Under reverse bias, V should be changed to $-V$ in (4.13). Now, the diode current is given by

$$I = I_s\left(e^{-qV/kT} - 1\right) \qquad \qquad \text{... (4.14)}$$

At room temperature, $\dfrac{kT}{q} = 26\,\text{mV}$. So, when the reverse bias voltage $V > 26$ mV, $e^{-qV/kT} \ll 1$ and $I = I_s$.

The reverse saturation current depends upon temperature since the minority carrier concentration increases with temperature. The reverse saturation current is generated solely by the flow of minority carriers. Silicon has a larger bandgap than Germanium. So, I_s is smaller in Si than in Ge. In any semiconductor, intrinsic or extrinsic,

$$np = \text{constant} = n_i^2, \text{ at a given temperature,}$$

where n_i = intrinsic carrier concentration.

Again, $n_i \sim e^{-E_g/2kT}$

where E_g = bandgap and T = absolute temperature of the semiconductor. Due to larger bandgap in Si, I_s is lower than that of Ge. Hence, the current in Ge diode rises at lower forward bias.

Equation (4.13) describes the current-voltage ($I - V$) characteristics of the Ge p-n junction adequately at low current densities. Departure is found in the case of Si and GaAs p-n junctions. The reasons of departure are as follows:

- surface effect,
- generation and recombination in the depletion region,
- high level injection,
- series resistance of neutral p and n regions.

We consider the recombination occurring in the depletion layer. Since the depletion layer has few mobile carriers, recombination in this region was neglected. The diffusion current is carried by the recombination of excess minority carriers injected into the neutral regions. The thermal generation of electron-hole pairs in the neutral regions and their drift motion through the depletion layer gives rise to reverse saturation current.

Recombination current in the depletion layer,

$$I_{DL} \propto n_i e^{qV/2kT}.$$

Recombination current in the neutral region,

$$I_{NR} \propto p_n e^{qV/kT}$$
$$(or, n_p e^{qV/kT})$$

Now, $p_n n_n = n_i^2$

$$\therefore \quad p_n = \frac{n_i^2}{n_n} = \frac{n_i^2}{n_d}$$

where $n_n = n_d$ = donor concentration in the n-region.

Recombination current in the neutral region,

$$I_{NR} \propto n_i^2 e^{qV/kT}$$

Then,

$$\frac{I_{NR}}{I_{DL}} = \frac{n_i^2 e^{qV/kT}}{n_i e^{qV/2kT}} = n_i e^{qV/2kT}$$

When the injection level is low, i.e., the forward bias (V) is small, and the semiconductor is such that n_i is also small, then $I_{NR} \ll I_{DL}$. In Si, the intrinsic carrier concentration, n_i, is small due to larger bandgap. So, for small values of forward bias (V), I_{DL} dominates over I_{NR} in Si p-n diode. Then, the junction current in Si varies as $e^{qV/2kT}$ for small forward bias V. In general, we can write,

$$I = I_s \left[e^{qV/\eta kT} - 1 \right] \qquad\qquad \dots (4.15)$$

where η is a number which depends upon which recombination current is dominating, I_{DL} or I_{NR}, and $1 \le \eta \le 2$. A typical $I-V$ characteristics of the p-n junction diode is shown in Fig. 4.2.

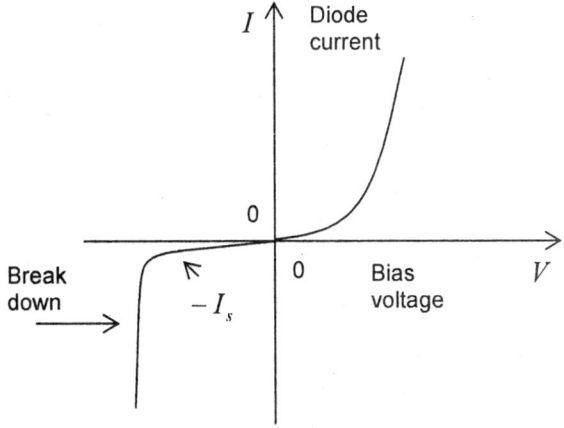

Fig. 4.2. Typical I-V characteristics of the p-n junction diode.

4.1.1 JUNCTION RESISTANCE

The dynamic resistance of the junction is given by,

$$R = \frac{dV}{dI}.$$

From (4.15) we get

$$\frac{dI}{dV} = I_s \cdot \frac{q}{\eta kT} \cdot e^{qV/\eta kT}$$

$$= \frac{q}{\eta kT}\left[I + I_s\right]$$

$$= \frac{1}{\eta V_T}\left(I + I_s\right)$$

where $V_T = \frac{kT}{q} = 26$ mV at $T = 300°$ K.

$$\therefore \qquad R = \frac{\eta V_T}{I + I_s} \qquad\qquad \dots(4.16)$$

Under forward bias, $I \gg I_s$ and

$$R = \frac{\eta V_T}{I}$$

Taking I in mA, we get $R = \frac{26\eta}{I}\,\Omega$.

Under reverse bias, $I = -I_s$. Then,

$$R = \frac{\eta V_T}{I_s - I_s} = \infty.$$

4.2 MINORITY CARRIER INJECTION

Under the application of a forward bias, holes are injected from the p-side into the n-side. Similarly, electrons are injected from n-side into the p-side. These excess holes appear as minority carriers in the n-region. Hence, we call this injection process as the minority carrier injection.

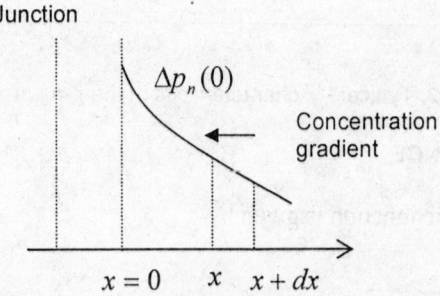

Fig. 4.3.. Minority carrier concentration profile.

Let the origin of the X-axis be located at the depletion layer edge on the n-side. Since these excess holes diffuse with a finite speed and they have a finite life time, there exists a concentration gradient of excess holes in the neutral n-region. This gradient is directed towards the depletion layer boundary.

Let us consider a small width in the neutral n-region lying between the coordinates x and x+dx. Excess holes entering the elementary volume Adx at x per unit time is

$$- D_p \frac{\partial \Delta p(x)}{\partial x}.A.$$

where the negative sign is due to the fact that the flow of holes occur in a direction opposite to the excess concentration gradient. D_p is the diffusion coefficient of holes, A is the cross sectional area of the junction.

Number of holes leaving the volume Adx at $(x + dx)$ per unit time is

$$- D_p A \frac{\partial}{\partial x} \left[\Delta p(x) + \frac{\partial \Delta p(x)}{\partial x}.dx \right]$$

$$= - D_p A \frac{\partial \Delta p(x)}{\partial x} - D_p A \frac{\partial^2 \Delta p(x)}{\partial x^2}.dx$$

Then, the rate of hole accumulation within the volume element Adx is,

= Rate of flow in – Rate of flow out

$$= D_p A \frac{\partial^2 \Delta p(x)}{\partial x^2} dx$$

These excess holes are lost through recombination with the electrons in the n-region. The average life time (τ_p) of holes is defined as

$$R_p = \frac{\Delta p}{\tau_p} \qquad \qquad \dots (4.17)$$

where R_p is the recombination rate per unit volume. Then,

$$D_p A dx \frac{\partial^2 \Delta p(x)}{\partial x^2} = \frac{\Delta p(x)}{\tau_p} A dx$$

or, $$\frac{d^2}{dx^2} \Delta p(x) = \frac{1}{\tau_p D_p} \Delta p(x)$$

or, $$\frac{d^2}{dx^2} \Delta p(x) - \frac{1}{L_p^2} \Delta p(x) = 0$$

where $L_p^2 = \tau_p D_p$ $\qquad \qquad \dots (4.18)$

$$\therefore \qquad \Delta p(x) = C_1 e^{-x/L_p} + C_2 e^{x/L_p}$$

Since $\Delta p(x) \to 0$ as $x \to \infty$, $C_2 = 0$.

$$\Delta p(x) = C_1 e^{-x/L_p} \qquad \qquad \dots (4.19)$$

Again, at $x = 0$, $\Delta p(x) = \Delta p(0)$

$$\therefore \qquad C_1 = \Delta p(0)$$

$$\therefore \qquad \Delta p(x) = \Delta p(0) e^{-x/L_p} \qquad \qquad \text{... (4.20a)}$$

If the width of the depletion layer is not neglected, then equation (4.20a) will be modified as,

$$\Delta p(x) \quad = \quad \Delta p(0) e^{-(x-x_n)/L_p} \qquad \qquad \text{... (4.20b)}$$

where $x > x_n$ and x_n is the depletion layer width in the n-region. A typical minority carrier concentration profile is shown in Fig. 4.3.

4.3 JUNCTION BREAKDOWN

The reverse saturation current (I_s) in a p-n junction diode is independent of reverse bias voltage. With increasing reverse bias, a condition is reached when the reverse current increases abruptly. This increase in reverse current is not predicted by Shockley equation. In the limit, the diode may be damaged. This phenomenon is known as junction breakdown.

Two different mechanisms have been proposed for this breakdown phenomenon. These are

(i) Avalanche breakdown, and
(ii) Zener breakdown.

4.3.1 AVALANCHE BREAKDOWN

This breakdown occurs in p-n junctions with wider depletion layers. Typical depletion layer width is $5 \ \mu$m. The electrons in reverse saturation current are accelerated to high velocities by the electric field existing in the depletion layer. These electrons collide with crystal atoms in the depletion layer and break the covalent bonds. Electron-hole pairs are generated in this process. Since the depletion layer is wider, these electron-hole pairs are accelerated to high energies and produce new electron-hole pairs by collision with the crystal atoms and ionizing them. So, a large number of electron-hole pairs are generated in the depletion layer in a chain reaction. The current builds into an 'avalanche'. The term 'avalanche' means a huge mass of snow and rocks descending down from a mountain. The electrons and holes slide down the potential hill. To ensure many collisions per electron, the depletion layer must be wide enough in comparison with the mean free path of the electrons in the depletion layer. Typical breakdown electric field is $\sim 10^5$ V/cm. Typical breakdown voltage is ~ 40 V $- 50$ V.

The avalanche current can be described by an empirical relation of the form

$$I = I_s \left[1 - \left(\frac{V_R}{V_{BR}} \right)^m \right] \qquad \qquad \text{... (4.21)}$$

where I_s = reverse saturation current

V_R = reverse voltage applied

V_{BR} = Breakdown voltage.

The factor 'm' depends upon the nature of the crystal and $3 \leq m \leq 6$.

4.3.2 ZENER BREAKDOWN

This breakdown happens in p-n junctions with narrow depletion layers. To produce narrow depletion layer widths both the p and n regions are heavily doped. Typical doping concentration for heavy doping is $\sim 10^{17} - 10^{19}$/c.c. Typical depletion layer width is ≤ 10 nm. Due to small depletion layer widths, the distance available for acceleration of charge carriers in the depletion layer is too short to produce ionizing collisions. So, in this p-n junction avalanche ionization can not take place. But, since the width of the depletion layer is small, a small applied voltage can produce a high field. This high field can exert enough force to tear out electrons directly from the covalent bonds. Once this process starts, the current rises to a high value since there are so many atoms available in the depletion layer. The voltage across the diode under this condition becomes a constant so that the diode can act as a voltage regulator. Typical breakdown voltage is $\sim$ 6 V. Typical breakdown field is $\geq$ a few MV/cm. This process of carrier generation by tearing out covalent bonds using high electric fields is analogous to quantum mechanical tunneling effect. This phenomenon is known as Zener breakdown.

4.4 APPLICATIONS OF P-N JUNCTIONS

The p-n junction diode can be used as
- a rectifier–which converts ac power to dc power,
- a clipper and a clamper–a clipper can clip off a part of the input signal without distorting the remaining part. A clamper introduces dc level shifting of a signal,
- a voltage-variable capacitor, known as varactor,
- a voltage-variable resistor, known as varistor,
- a light emitting diode (LED)–conduction band electrons recombine with the valence band holes due to the flow of current in direct gap semiconductors like GaAs. This liberates light energy,
- voltage regulator–such as Zener diodes,
- AND/OR logic gates,
- generation of microwave oscillations – such as in IMPATT and TRAPATT diodes,
- semiconductor diode lasers under special conditions of fabrication.

4.5 JUNCTION CAPACITANCE

The depletion layer contains a layer of negative charge on the p-side and a layer of positive charge on the n-side. If the applied voltage across the depletion layer is varied, the width of the depletion layer also varies. So, the charge contained on the p-side and n-side of the junction also varies. This variation of charge content with the variation in applied voltage gives the concept of capacitance associated with the p-n junction. The junction capacitance can be thought as a parallel plate capacitor having a plate separation equal to the width of the depletion layer.

Consider a p-n junction located at $x = 0$. Let N_a be the acceptor concentration on the p-side and N_d be the donor concentration on the n-side. Let x_p and x_n be the widths of the depletion layers in the p-side and n-side respectively.

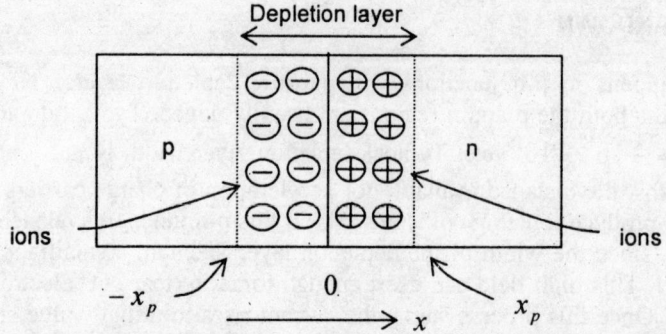

Fig. 4.4. A schematic diagram of the p-n junction.

The space charge density on the p-side is $-qN_a$ and that on the n-side is $+qN_d$, where q is the magnitude of electronic charge. Since the crystal is electrically neutral, we can write

$$qN_d A x_n = qN_a A x_p$$

or, $$\frac{N_d}{N_a} = \frac{x_p}{x_n} \qquad \qquad \text{... (4.22)}$$

Here, A = junction area.. Now, from (4.22) we can write

$$\frac{N_d + N_a}{N_a} = \frac{x_p + x_n}{x_n}$$

$$\therefore \qquad x_n = \frac{N_a}{N_a + N_d} W \qquad \qquad \text{... (4.23)}$$

where $W = x_p + x_n$ is the total width of the depletion layer.

Similarly, from (4.22) we get

$$x_p = \frac{N_d}{N_a + N_d} W \qquad \qquad \text{... (4.24)}$$

Poisson's equation in the depletion layer can be written as

$$-\frac{d^2 V}{dx^2} = \frac{q}{\varepsilon}\left(p - n + N_d - N_a\right) \qquad \qquad \text{... (4.25)}$$

where ε = absolute permittivity of the semiconductor material. Here, $(p - n)$ is small compared with the magnitude of $(N_d - N_a)$ and hence will be neglected. On the p-side of the depletion layer, $N_d = 0$ while on the n-side, $N_a = 0$.

Poisson's equation in either region can be written as

$$\frac{d^2V}{dx^2} = -\frac{q}{\varepsilon}N_d \qquad\qquad \dots (4.26)$$

for $0 \le x \le x_n$, and

$$\frac{d^2V}{dx^2} = \frac{q}{\varepsilon}N_a \qquad\qquad \dots (4.27)$$

for $-x_p \le x \le 0$. Integrating (4.26), we get

$$\frac{dV}{dx} = -\frac{qN_d}{\varepsilon}x + C_1 \qquad\qquad \text{where } C_1 = \text{constant of integration.}$$

At $\quad x = x_n, \dfrac{dV}{dx} = 0.$

$$\therefore \quad C_1 = \frac{qN_d}{\varepsilon}x_n.$$

Then,

$$\frac{dV}{dx} = \frac{qN_d}{\varepsilon}(x_n - x) \qquad\qquad \dots (4.28)$$

Integrating (4.28) with respect to x, we obtain

$$V(x) = \frac{qN_d}{\varepsilon}\left(x_n x - \frac{x^2}{2}\right) + C_2 \qquad\qquad \dots (4.29)$$

where C_2 is integration constant.

Let $V(x = x_n) = V_n$. Then, from (4.29) we get

$$V_n = \frac{qN_d}{\varepsilon}\cdot\frac{x_n^2}{2} + C_2 \qquad\qquad \dots (4.30)$$

Proceeding in a similar manner and noticing that $\dfrac{dV}{dx} = 0$ at $x = -x_p$, we can write from (4.27)

$$\frac{dV}{dx} = \frac{qN_a}{\varepsilon}(x + x_p) \qquad\qquad \dots (4.31)$$

Integrating (4.31) with respect to x, we get

$$V(x) = \frac{qN_a}{\varepsilon}\left(\frac{x^2}{2} + x_p x\right) + C_3 \qquad\qquad \dots (4.32)$$

for $-x_p \le x \le 0$. Here, $C_3 = $ integration constant.

From (4.29), $V(x = 0) = C_2$

From (4.32), $V(x = 0) = C_3$

$$\therefore \qquad C_2 = C_3.$$

Let $V(x = -x_p) = V_p$. Then, from (4.32) one can write

$$V_p \quad = \quad -\frac{qN_a}{\varepsilon}\cdot\frac{x_p^2}{2} + C_3 \qquad\qquad \ldots(4.33)$$

Total effective voltage (i.e., applied voltage plus built-in voltage) across the depletion layer is,

$$V_B \quad = \quad V_n - V_p \quad = \quad \frac{q}{2\varepsilon}\left(N_d x_n^2 + N_a x_p^2\right) \qquad\qquad \ldots(4.34)$$

Substituting the values of x_n and x_p from (4.23) and (4.24) respectively, equation (4.34) can be recast as

$$V_B \quad = \quad \frac{q}{2\varepsilon}\left[N_d\left(\frac{N_a W}{N_a + N_d}\right)^2 + N_a\left(\frac{N_d W}{N_a + N_d}\right)^2\right]$$

$$= \quad \frac{q}{2\varepsilon}W^2\frac{N_a N_d}{N_a + N_d} \qquad\qquad \ldots(4.35)$$

Then, $\quad W = \sqrt{\dfrac{2\varepsilon}{q}}\sqrt{\dfrac{N_a + N_d}{N_a N_d}}\cdot V_B^{\frac{1}{2}} \qquad\qquad \ldots(4.36)$

Amount of charge contained in any side of the depletion layer is

$$Q \quad = \quad qN_d\left(Ax_n\right) \quad = \quad qN_a\left(Ax_p\right)$$

$$= \quad qA\frac{N_a N_d}{N_a + N_d}W \qquad\qquad \ldots(4.37)$$

Junction capacitance,

$$C_J = \frac{dQ}{dV_B}$$

$$= qA\left(\frac{N_a N_d}{N_a + N_d}\right)\frac{dW}{dV_B}$$

$$= qA\left(\frac{N_a N_d}{N_a + N_d}\right)\sqrt{\frac{2\varepsilon}{q}}\sqrt{\frac{N_a + N_d}{N_a N_d}}\frac{1}{2}V_B^{-\frac{1}{2}}$$

$$= A\sqrt{\frac{q\varepsilon}{2}\left(\frac{N_a N_d}{N_a + N_d}\right)}V_B^{-\frac{1}{2}} \qquad\qquad \ldots(4.38)$$

$$\therefore \qquad C_J \propto V_B^{-\frac{1}{2}}.$$

SPECIAL CASE: ONE-SIDED ABRUPT JUNCTION

If either p or n region is heavily doped, then depletion layer will expand primarily into the region of lower doping concentration.

Let $N_a \gg N_d$, i.e., the p-region is heavily doped. Then,

$$x_p = \frac{N_d}{N_a + N_d} W \approx \frac{N_d}{N_a} W \ll W.$$

$$x_n = \frac{N_a}{N_a + N_d} W \approx W \quad \text{since } N_a \gg N_d.$$

So, the depletion layer primarily extends into the n-region. From (4.38), we write

$$C_J = A \sqrt{\frac{q\varepsilon}{2} N_d} \, V_B^{-\frac{1}{2}} \qquad \qquad \text{... (4.39)}$$

Similarly, if the n-region is heavily doped compared with the p-region, $N_d \gg N_a$. Now,

$$C_J = A \sqrt{\frac{q\varepsilon}{2} N_a} \, V_B^{-\frac{1}{2}} \qquad \qquad \text{... (4.40)}$$

In view of (4.38) – (4.40) one can write,

$$\frac{1}{C_J^2} \propto V_B.$$

Now, $V_B = V_a + V_{bi}$ where V_a = applied voltage and V_{bi} = built-in voltage. Plotting $1/C_J^2$ vs. V_a we should get a straight line. When $V_a = -V_{bi}$, $V_B = 0$ and $1/C_J^2 = 0$. From this plot, one can estimate V_{bi}. A typical plot of $1/C_J^2$ as a function of applied voltage (V_a) is shown in Fig. 4.5.

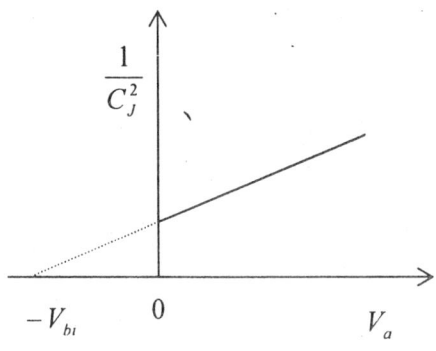

Fig. 4.5. Plot of $1/C_J^2$ as a function of applied voltage (V_a).

V_B is small under forward bias. So, C_J is large. But, the junction capacitance is shunted by small junction resistance under forward bias. Hence, the effect of C_J is not appreciable under forward bias.

V_B is large under reverse bias. So, C_J is small. But, the dynamic resistance of the junction under reverse bias is very high. So, the effect of C_J is pronounced.

For a linearly graded junction, doping concentration $N = gx$, where g is called grade constant. Now,

$$C_J \propto V_B^{-\frac{1}{3}}.$$

A junction is called hyper abrupt junction if

$$C_J \propto V_B^{-n} \quad \text{where} \quad n > \frac{1}{2}. \text{Consider a case where doping concentration} \quad N \propto x^{-\frac{3}{2}}.$$

Then, $N = gx^{-\frac{3}{2}}$ where g = constant. Now, $C_J \propto V_B^{-2}$. This is the case of a varactor diode. When this varactor diode is placed in parallel with L-C resonant circuit, the resonant frequency

$$\omega_r \propto \frac{1}{\sqrt{C_J}} = V_B$$

If V_B is a time varying voltage, the resonant frequency is also time varying. This principle can be applied to achieve frequency modulation.

4.6. DIFFUSION CAPACITANCE

Under forward bias, diffusion of electrons and holes take place. The excess hole concentration decays exponentially in the neutral n-region. The charge stored in the neutral n-region due to this diffusion process is

$$Q_p = \int_{x_n}^{\infty} qA\Delta p_n(x)\, dx$$

$$= qA \int_{x_n}^{\infty} \Delta p_n(0) e^{-(x-x_n)/l_p}\, dx$$

$$= \frac{qA\Delta p_n(0)}{-\dfrac{1}{L_p}} \left[e^{-(x-x_n)/L_p} \right]_{x_n}^{\infty}$$

$$= -qAL_p\Delta p_n(0)[0-1]$$

$$= qAL_p \cdot p_n e^{qV/kT} \qquad \qquad \qquad \text{... (4.41)}$$

using eqn. (4.7) and assuming $e^{qV/kT} \gg 1$.

The diffusion capacitance is given by

$$C_d = \frac{dQ_p}{dV} = qAL_p \cdot P_n \cdot \frac{q}{kT} e^{qV/kT}$$

$$= \frac{q^2 A}{kT} L_p P_n e^{qV/kT} \qquad \qquad \dots (4.42)$$

If we take a $p^+ n$ junction, then excess hole diffusion will dominate over the excess electron diffusion. In this case, the diode forward current

$$I \approx \frac{qAD_p}{L_p} P_n e^{qV/kT} \qquad \qquad \dots (4.43)$$

assuming $V > kT/q$ and $p_n \gg n_p$. Now, equation (4.42) can be written as

$$C_d = \frac{q}{kT} \cdot \frac{qAD_p}{L_p} P_n e^{qV/kT} \cdot \frac{L_p^2}{D_p} = \frac{q}{kT} \cdot I \cdot \tau_p \qquad \qquad \dots (4.44)$$

where $\tau_p = L_p^2 / D_p = $ life time of holes.

4.7 TUNNEL DIODE

In 1958, Leo Esaki invented this diode. A tunnel diode is formed by a degenerate p-n junction. Here, both p and n regions are degenerate. If the doping concentration in a semiconductor is greater than the effective density of states, the semiconductor is said to be degenerate. In this case, the Fermi level lies within the conduction band in the n-type semiconductor and within the valence band in the p-type semiconductor. In the degenerate semiconductor, the doping concentration is greater than 10^{19} /c.c. The symbol of tunnel diode is shown in Fig.4.6(a).

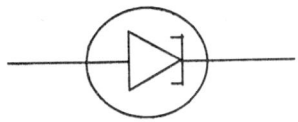

Fig. 4.6. (a) Symbol of tunnel diode.

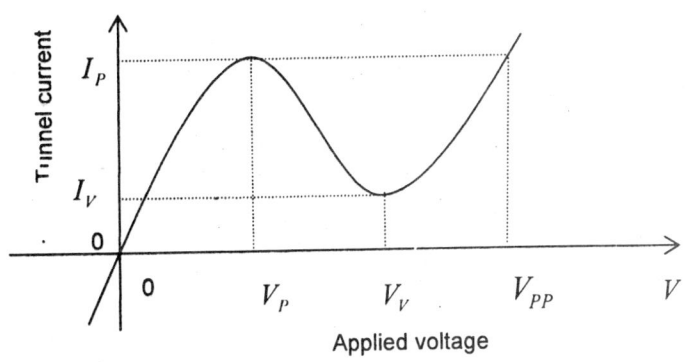

Fig. 4.6. (b) Typical I-V characteristics of a tunnel diode.

In tunnel diode, the current is a multi-valued function of the forward bias voltage. The current increases monotonically under reverse bias. The I-V characteristics is shown in Fig. 4.6(b).

The width of the depletion layer is small due to heavy doping. Here, I_P = peak current, I_V = valley current, V_P = forward voltage corresponding to peak current, V_V = forward voltage at valley current, V_{PP} = injection current voltage. Let us explain the operation of the tunnel diode with the help of energy band diagram.

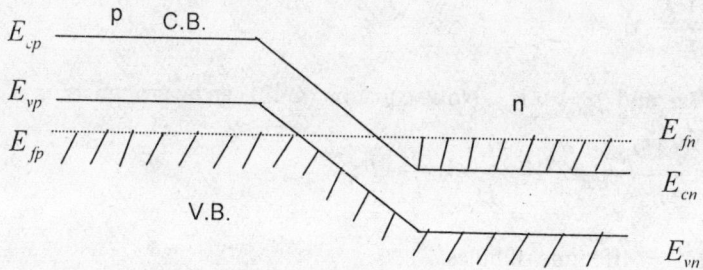

Fig. 4.7. Energy band diagram with no bias voltage applied.

Under no bias voltage applied, the Fermi level is continuous across the junction. The shaded region are filled states. This is shown in Fig. 4.7. E_{cp} and E_{vp} are the energies corresponding to the bottom of the conduction band and the top of the valence band in the p-region respectively. E_{cn} and E_{vn} are similar energies in the n-region. E_{fp} and E_{fn} are Fermi energies in the p- and n-region respectively. C.B. stands for conduction band and V.B. stands for valence band.

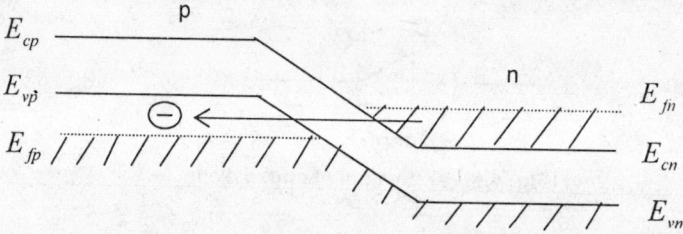

Fig. 4.8. Forward bias applied and tunnelling occurs.

In Fig. 4.8, a forward bias is applied. Electrons in the conduction band of the n-type semiconductor face vacant states in the valence band of the p-type semiconductor. So, tunnelling occurs and current increases. With increasing forward bias, more and more electrons face the empty states in the valence band and tunnel current increases accordingly. With increasing forward bias, a condition is reached when the overlap of the conduction band electrons and the empty states at the top of the valence band becomes maximum. The forward diode current becomes a maximum. This is shown in Fig. 4.9. If forward bias is further increased, this overlap of occupied and unoccupied states decreases and current decreases. This decrease of forward current with increase in forward voltage continues until the diode current reaches a minimum value. Under this condition, the conduction band electrons in the n-region face the band gap. So, no tunnelling occurs. The small current that flows in this condition is due to normal injection current in a p-n diode. The decrease of forward current with increase in forward voltage gives rise to differential negative resistance (DNR).

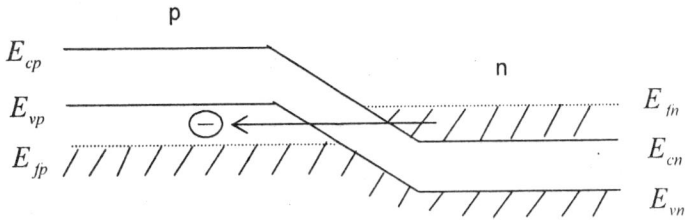

Fig. 4.9. Maximum overlap of filled and empty states. Tunnel current is maximum.

With further increase of forward bias, the potential barrier height is lowered and normal injection current over the barrier in a p-n junction diode flows. This current increases exponentially with voltage. This is shown in Fig. 4.10.

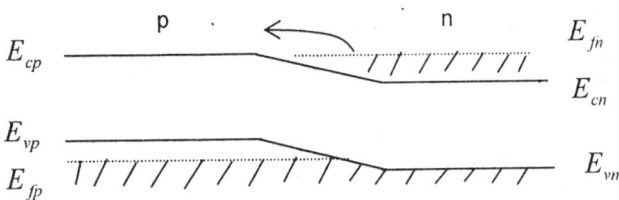

Fig. 4.10. No tunnelling. Normal injection current flowing.

Under reverse bias, as shown in Fig. 4.11, electrons in the valence band in the p-region encounter vacant states in the conduction band of the n-region. So, they tunnel through the barrier. With increasing reverse bias, more and more filled states in the valence band (V.B.) encounter empty states in the conduction band (C.B.). So, reverse current increases monotonically.

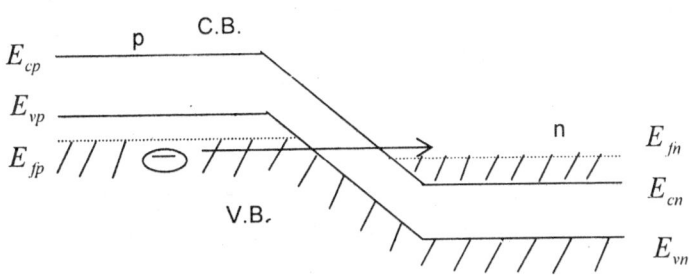

Fig. 4.11. Tunnelling under reverse bias.

The necessary conditions for tunnelling are:

- there must be filled energy states in the side from which the electrons will tunnel,
- there must be vacant energy states in the side to which the electrons will tunnel,

- the potential barrier height should be small and the barrier width should be thin so that finite probability of tunnelling exists,
- the momentum must be conserved in the process.

Typical values of parameters: $I_P \approx 1$ to 100 mA.

Parameter	Ge	Si	GaAs
I_P / I_V	8	4	15
V_P (mv)	55	65	150
V_V (mv)	350	420	500
V_{PP} (mv)	500	-	900

V_{PP} is the injection current voltage defined as the voltage at which the normal injection current equals the peak tunnel current. I_P / I_V is independent of junction area, doping concentration etc. It depends upon the semiconductor material.

Tunnelling process can be of two types:
(i) direct tunnelling and (ii) indirect tunnelling.

- In direct tunnelling, there is no change of momentum of the electron. In the energy momentum diagram ($E - k$ diagram) the conduction band minimum and the valence band maximum occur at the same value of momentum. GaAs, GaSb are examples of direct band gap semiconductors. Probability of direct tunnelling is greater than that of indirect tunnelling.
- In indirect tunnelling, the conduction band minimum and the valence band maximum do not occur at the same value of momentum. For conservation of momentum, the extra momentum must be supplied by phonons. Phonons are quanta of lattice vibration. Examples are Ge and Si.

4.7.1 EQUIVALENT CIRCUIT OF TUNNEL DIODE

The equivalent circuit of the tunnel diode is shown in Fig. 4.12. The equivalent circuit has series

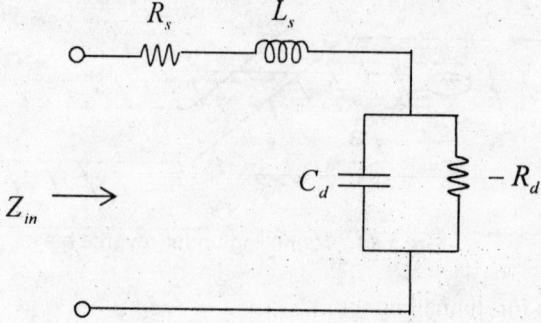

Fig. 4.12. AC equivalent circuit of a tunnel diode.

resistance R_s and inductance L_s. R_s contains the lead resistance, bulk resistance of semiconductors, and ohmic contact resistance. L_s is the lead inductance. C_d is the diode junction capacitance and $-R_d$ is the differential negative resistance (DNR) of the tunnel diode.

The input impedance of the diode is

$$Z_{in} = R_s + j\omega L_s + \frac{-R_d \cdot \dfrac{1}{j\omega C_d}}{-R_d + \dfrac{1}{j\omega C_d}}$$

$$= R_s + j\omega L_s - \frac{jR_d/\omega C_d}{R_d + j/\omega C_d} \cdot \frac{R_d - j/\omega C_d}{R_d - j/\omega C_d}$$

$$= R_s + j\omega L_s - \frac{jR_d^2/\omega C_d + R_d/\omega^2 C_d^2}{R_d^2 + \dfrac{1}{\omega^2 C_d^2}}$$

$$= \left[R_s - \frac{R_d}{1 + (\omega R_d C_d)^2} \right] + j\left[\omega L_s - \frac{\omega C_d R_d^2}{1 + (\omega C_d R_d)^2} \right] \qquad \ldots (4.45)$$

The tunnel diode possesses two characteristic frequencies:

- Resistance (or, Resistive) cut-off frequency, and
- Reactance (or, Reactive) cut-off frequency.

Resistive cut-off frequency (f_r) is the frequency at which the resistive part of diode input impedance is zero. Then,

$$R_s - \frac{R_d}{1 + (\omega_r R_d C_d)^2} = 0$$

or, $$1 + (\omega_r R_d C_d)^2 = \frac{R_d}{R_s}$$

$\therefore$ $$f_r = \frac{\omega_r}{2\pi} = \frac{1}{2\pi R_d C_d} \sqrt{\frac{R_d}{R_s} - 1} \qquad \ldots (4.46)$$

Reactance cut-off frequency (f_x), also known as self-resonant frequency, is the frequency at which the reactive part of diode input impedance is zero. Then,

$$\omega_x L_s - \frac{\omega_x C_d R_d^2}{1 + (\omega_x R_d C_d)^2} = 0$$

or, $\quad 1 + (\omega_x R_d C_d)^2 = \dfrac{C_d R_d^2}{L_s}$

$\therefore \quad f_x = \dfrac{\omega_x}{2\pi} = \dfrac{1}{2\pi C_d R_d} \sqrt{\dfrac{C_d R_d^2}{L_s} - 1}$

$$= \dfrac{1}{2\pi} \sqrt{\dfrac{1}{L_s C_d} - \dfrac{1}{(C_d R_d)^2}} \qquad \qquad \ldots (4.47)$$

DNR of the diode has a minimum value at the point inflection as seen in Fig. 4.6. The maximum value of f_r occurs at $R_d = R_{min}$ and is given by

$$f_r\big|_{max} \;\; = \;\; \dfrac{1}{2\pi R_{min} C_d} \sqrt{\dfrac{R_{min}}{R_s} - 1} \qquad \qquad \ldots (4.48)$$

For $f \geq f_r\big|_{max}$, the diode exhibits no DNR.

At $R_d = R_{min}$, f_x is minimum and

$$f_x\big|_{min} \;\; = \;\; \dfrac{1}{2\pi} \sqrt{\dfrac{1}{L_s C_d} - \dfrac{1}{(R_{min} C_d)^2}} \qquad \qquad \ldots (4.49)$$

The diode can oscillate at a frequency $f_x\big|_{min}$ if $f_r\big|_{max} > f_x\big|_{min}$.

4.7.2 APPLICATIONS OF TUNNEL DIODE

- The tunnel diode finds application as a microwave oscillator. Typical oscillation frequency $\sim 10\,\text{GHz}$. Output power is small.
- It can act as a high speed switching device. Switching time $< 1\,\text{nsec}$.
- It is a low noise device used in microwave amplification.

The disadvantage of the device is that it is a two terminal device and there is no isolation between the input and output terminals. This leads to instability.

Typical values of tunnel diode parameters:

$$R_s \sim 1 \text{ to } 5 \; \Omega, \; L_s \sim 0.1 \text{ to } 4 \text{ nH}, \; C_d \sim 0.35 \text{ to } 100 \text{ pF}, \; R_d \sim 100 \; \Omega.$$

4.7.3 AMPLIFICATION USING TUNNEL DIODE

The tunnel diode amplifier circuit is shown in Fig. 4.13. Here, V_s is the signal voltage. R_L is the load. If i_1 and i_2 be the ac currents in absence and in presence of the tunnel diode, then

$$V_s = i_1 R_L \qquad \qquad \ldots (4.50)$$

and $\quad V_s = i_2 (R_L - R_d) \qquad \qquad \ldots (4.51)$

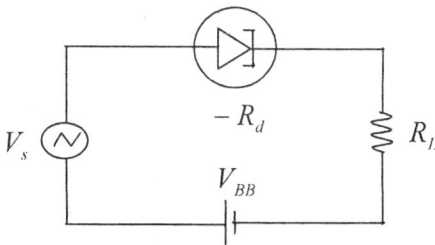

Fig. 4.13. Amplifier circuit with tunnel diode.

Current amplification $= \dfrac{i_2}{i_1} = \dfrac{R_L}{R_L - R_d} > 1$.

4.8 LASER DIODE

The term **Laser** means **L**ight **A**mplification by **S**timulated **E**mission of **R**adiation. The laser diode consists of a p-n junction. Both the p and n regions are heavily doped so that the Fermi levels go into the conduction band in the n-region and into the valence band in the p-region. Thus, electrons are present in the conduction band and holes are present in the valence band. A forward bias is applied across this junction and a current flows through the device. The barrier height is lowered. Electrons flow from the n-region to the p-region. In this process, population inversion takes place at the depletion layer and electrons jump from the conduction band to the valence band in the depletion region. In this transition, a photon is emitted.

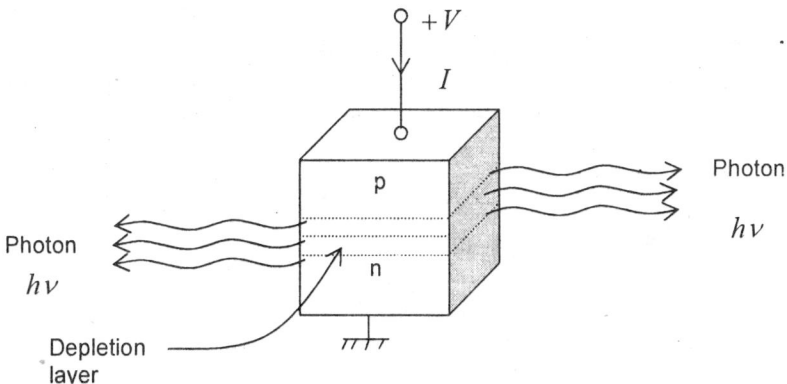

Fig. 4.14. Basic configuration of a diode laser.

A basic configuration of the p-n junction diode laser is shown in Fig. 4.14. The energy band diagram is shown in Figs. 4.15 (a) and 4.15 (b).

In Figs. 4.15 (a) and 4.15 (b), $E_g = $ bandgap, $E_{fn} \rightarrow$ Fermi level on the n-side, $E_{fp} \rightarrow$ Fermi level on the p-side, $E_{fp} - E_{fn} = eV$, where V is the forward bias applied and e is the magnitude of electronic charge.

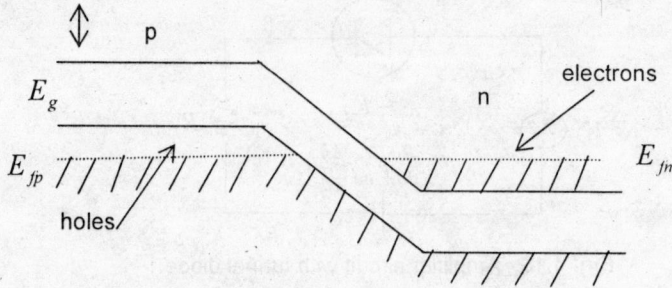

Fig. 4.15. (a) Energy band diagram of the p-n junction laser with no bias applied.

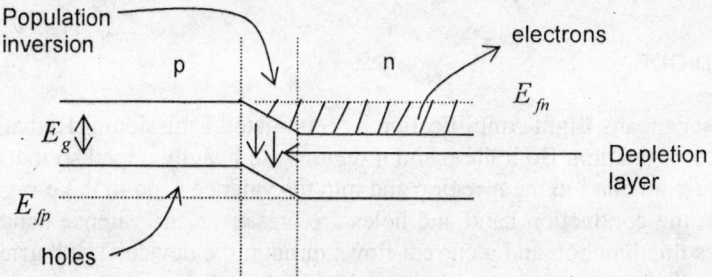

Fig. 4.15. (b) Energy band diagram of the p-n junction laser with a forward bias applied.

A laser is an oscillator operating in the optical frequency region. Then, there must be some mechanism for amplification (gain) and also some mechanism for feedback. The current flowing through the p-n junction provides electrical pumping and produces population inversion. Two end faces of the semiconductor block through which emission occur are polished so that they can act as mirrors. These mirrors provide optical feedback required for lasing. The population inversion gives rise to gain. The lasing action is exhibited by direct band gap semiconductors such as GaAs, $Ga_{1-x}Al_xAs$, $Ga_xIn_{1-x}As_{1-y}P_y$, etc. The wavelength of the radiation emitted from materials like $Ga_{1-x}Al_xAs$ can be varied by varying the atomic fraction x. The wavelength can be varied over the range $0.7\,\mu m$ $-0.9\,\mu m$. For long wavelength operation at $1.3\,\mu m$ or $1.55\,\mu m$, GaInAsP laser diodes are used. Fiber loss is minimum (≈ 0.2 dB/km) at $\lambda = 1.55\,\mu m$ so that $1.55\,\mu m$ lasers are often used in telecommunication.

There are three basic processes in a diode laser– viz. absorption, spontaneous emission and stimulated emission. A photon of proper energy can be absorbed by an atom which is raised from the ground state to an excited state. Spontaneous emission is the source of noise. An atom from an excited state can make a spontaneous downward transition to the ground state by emitting a photon. This radiation is incoherent. This spontaneous emission does not require the presence of any external photon. An atom in an excited state can make a downward transition under the influence of the incident photon which is known as stimulated emission. The stimulated emission occurs at the same frequency and same polarization as the stimulated incident photon. When there are more atoms in the excited state, i.e., when there is population inversion, stimulated emission will dominate.

The absorption and scattering of the radiation inside the medium gives rise to loss. The output transmission is also a kind of loss. Whereas stimulated emission due to population inversion produces gain. Under steady state condition of operation, the loss and gain must balance each other.

4.9 LIGHT-EMITTING DIODE

The light-emitting diode (LED) is essentially a forward-biased p-n junction. The p and n regions are heavily doped compared with the ordinary p-n junctions. The basic mechanism involved in the light emission is the recombination of a conduction band electron with a valence band hole resulting in an emission of light quantum. This is radiative recombination. The light quantum will have an energy approximately equal to the bandgap of the semiconductor. The other way of electron-hole recombination is the nonradiative recombination. In this process, the difference energy appears in the form of heat in the material. This nonradiative recombination takes place at the lattice defects. In order to design an LED the radiative recombination process must dominate over the nonradiative recombination process. The light emission in an LED is the spontaneous emission where the emitted photons have random polarization, have no phase correlation, i.e., incoherent in nature, and have random directions of emission. The current density in the active region of the LED must be high in order to increase the light emission. Ternary compound semiconductors like GaAlAs emit in the short wavelength side while quaternary compounds like InGaAsP emit in the long wavelength side. By varying the composition (atomic fraction) of these compounds, GaAlAs can radiate over 500 nm–1000 nm wavelength range whereas InGaAsP can radiate over the range 1000 nm–1700 nm. The lattice spacing of the semiconductors on either side of the active region must be identical, otherwise defects will be produced at the interfaces resulting in poor light emission.

In the design of a long wavelength LED, a thin InGaAsP layer forms the active region which is placed between a p-type InP layer on one side and a n-type InP layer on the other side. The refractive index of the active layer is greater than those of the p-type InP layer and the n-type InP layer. So, light which is generated in the active layer undergoes total internal reflection at the interfaces of the active layer and InP layers (n & p). This leads to light confinement in the active layer. This optical confinement minimizes loss of the emitted light.

The carriers (electrons and holes) must be confined in the active region in order to achieve high efficiency of light emission. For this, a double heterostructure configuration is designed. The electrons enter the active layer from the n-type InP layer and falls into a potential well in the active layer. This electron faces a barrier at the active layer and the p-type InP interface. The barrier prevents many electrons from surmounting it. Similarly, when holes enter the active layer (InGaAsP) from the p-type InP layer, they face a potential barrier at the active layer and n-type InP interface. Thus, both electrons and holes are confined in the active layer. This is carrier confinement.

LED can have two types of structures –Surface-emitting LED and Edge-emitting LED. A vertical cross sectional view of the long wavelength surface-emitting LED is shown in Fig. 4.16. The direction of current flow and the direction of light emission is the same in this case. Light is emitted through InP substrate which does not absorb light practically. In Edge-emitting LED, light is emitted in a direction perpendicular to the direction of current flow.

The n-type InP substrate and the p^+-InGaAsP layer have lower refractive indices than their adjacent layers n-InP and p-InP respectively. This gradual lowering of refractive index from the active layer forms an optical guide which confines light in the active region. The p^+- InGaAs layer is used to lower the contact resistance of the metal-semiconductor junction. The insulating SiO_2 layer with a hole just below the active layer is used to allow current to pass through the active layer only.

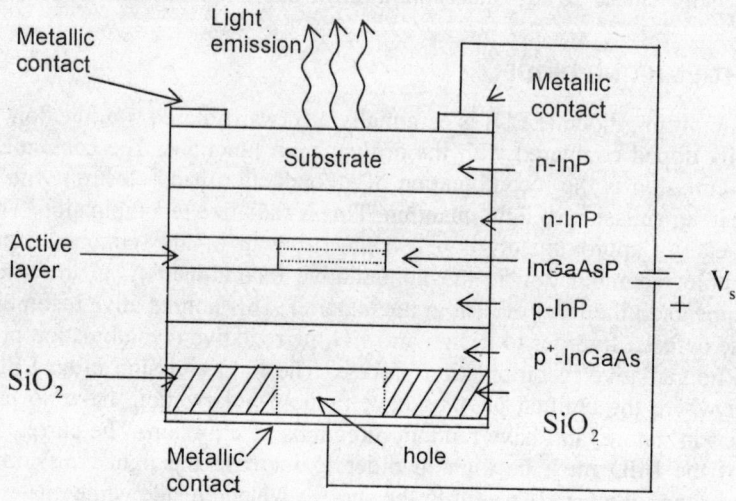

Fig. 4.16. A vertical cross-sectional view of a long wavelength surface-emitting LED. V_s is the bias voltage.

MERITS OF A LASER DIODE OVER LED

The following are the merits of a laser diode (LD) over LED:

- the output power of the LD is higher,
- the light beam of the LD is more directional, i.e., the beam divergence is smaller than that of LED,
- the linewidth of the laser is much smaller than that of an LED. A DFB laser has a typical linewidth of 5 MHz whereas an InGaAsP surface-emitting LED has a typical linewidth of 12.5 Terahertz,
- the modulation bandwidth of the LD is much higher than that of an LED. The direct modulation bandwidth of an LD has reached a value of 30 GHz whereas typical modulation bandwidth of an LED is 50 MHz.

The active region in an LD has a much smaller thickness and a much narrower dimension than that of an LED. There is no reflector in the LED to provide optical feedback. In LD, there is optical feedback which causes stimulated emission to dominate over the spontaneous emission.

4.10 PHOTODIODE

A photodiode is a device which converts the input optical power into corresponding electrical power. All photodiodes are reverse biased. The main use of the photodiode is in the detection of intensity modulated lightwave signals. They are also used as mixers of lightwaves generating difference frequency signals in the microwave and millimeter wave regions. Depending upon the structure, photodiodes (PD) can be divided into three classes:

(i) p-n junction photodiode
(ii) p-i-n photodiode, and

(iii) Avalanche photodiode.

4.10.1 P-N JUNCTION PHOTODIODE

It is the simplest photodiode. The p-n junction is reverse biased as shown in Fig. 4.17. It is well-known that there exists a depletion layer on either side of the junction and this depletion layer is depleted of all mobile carriers. Only ionized impurity atoms exist there. Light is incident into the depletion layer. If ' h ' be the Planck's constant and ν be the frequency of the incident light then the incident photon has an energy $h\nu(=\dfrac{hc}{\lambda})$ where λ is the free-space wavelength of light and c is the vacuum velocity of light. If the photonic energy ($h\nu$) is greater than or equal to the band gap (E_g) of the semiconductor material of the photodiode, then light is absorbed and the photon generates an electron-hole pair. The generated electron moves towards the n-region and the hole moves towards the p-region. The motion occurs under the action of the electric field present in the depletion layer. A current flows in the external circuit. The number of electron-hole pairs generated is proportional to the light intensity. So, the photodiode current is proportional to the incident light intensity. In absence of incident light, a small current flows in the reverse biased p-n junction, which is called dark current. When $h\nu < E_g$, no electron-hole pair is generated. This gives rise to a cut-off wavelength (λ_c) above which no light-induced photocurrent will flow. Thus,

$$\frac{hc}{\lambda_c} = E_g$$

or, $$\lambda_c = \frac{hc}{E_g}$$ ... (4.52)

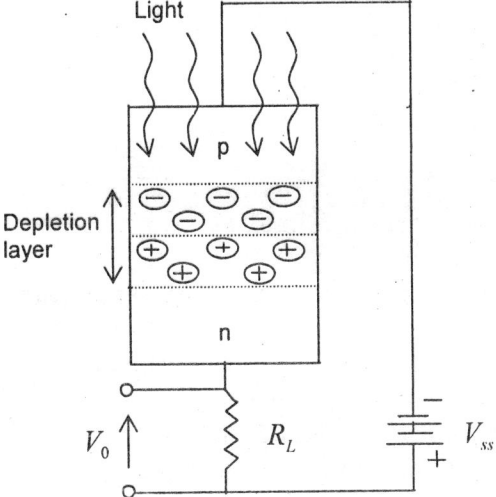

Fig. 4.17. A schematic diagram of the p-n junction photodiode.

In the short wavelength region Si ($E_g = 1.14$ eV at $300°$ K) photodiode is suitable while at long wavelength region InGaAs is preferred as the photodiode material. The bandgap of InGaAs (Indium Gallium Arsenide) depends upon the composition of the semiconductor – particularly, the molecular fraction of In and Ga present in InGaAs will matter. If x = 0.53 in $In_xGa_{1-x}As$ then $E_g = 0.75$ eV.

4.10.2 P-I-N PHOTODIODE

In p-i-n photodiode, an intrinsic semiconductor layer is fabricated in between the p and n regions. This increases the width of the depletion layer. The existence of the electric field in the whole depletion layer makes the photo-generated electrons and holes move fast through the intrinsic layer. This makes the response of the p-i-n photodiode to the incident light considerably faster. The increase in width of the depletion region due to the fabrication of the intrinsic layer leads to an increase in efficiency of the p-i-n photodiode. Since the width of the depletion layer is larger in p-i-n photodiode, the capacitance of the diode is reduced in comparison with the simple p-n junction photodiode. Reduction in device capacitance leads to a higher bandwidth of the p-i-n photodiode.

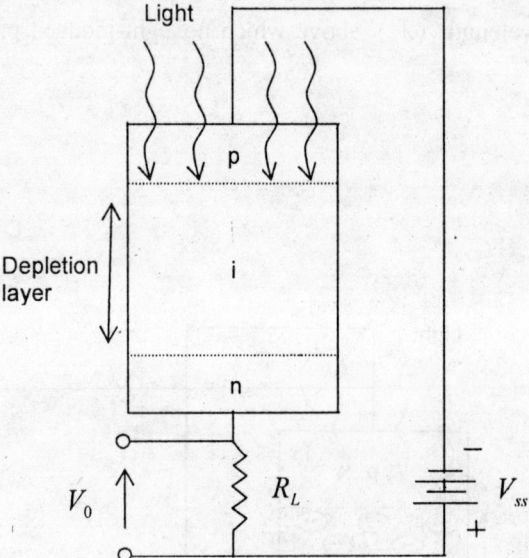

Fig. 4.18. A schematic diagram of the p-i-n photodiode.

In Fig. 4.18, V_0 is the output voltage developed across the load R_L. V_{ss} is the bias voltage applied to the photodiode (PD). 'i' is the intrinsic layer.

The quantum efficiency of the PD is defined as the average number of electron-hole pairs produced per incident photon. If I_L be the load current of the PD generated by the incident light having a power P_l then the number of electron-hole pair generated per unit time is equal to

$\left(\dfrac{I_L}{e}\right)$ and the number of incident photons per unit time is $\left(\dfrac{P_I}{h\nu}\right)$. Here, e is the magnitude of electronic charge. Then, the quantum efficiency (η_q) is given by

$$\eta_q \;=\; \frac{I_L/e}{P_I/h\nu} \;=\; \left(\frac{I_L}{P_I}\right)\cdot\frac{h\nu}{e} \qquad\qquad \ldots (4.53)$$

The responsivity of the photodiode is defined as the output current produced by the PD per unit input optical power. Thus, responsivity (η_r) is given

$$\eta_r = \frac{I_L}{P_I}$$

$$= \frac{\eta_q e}{h\nu} \qquad\qquad \ldots (4.54)$$

using equation (4.53).

4.10.3 AVALANCHE PHOTODIODE

In this photodiode, photocurrent is multiplied and this multiplication occurs by a process known as avalanching or impact ionization. The avalanche photodiode (APD) has a structure like $n^+pi\,p^+$ or its dual $p^+ni\,n^+$. The diode $n^+pi\,p^+$ is known as Read's structure. The symbols p^+ and n^+ mean heavily doped p and n region respectively. The structure of the $n^+pi\,p^+$ APD is shown in Fig. 4.19.

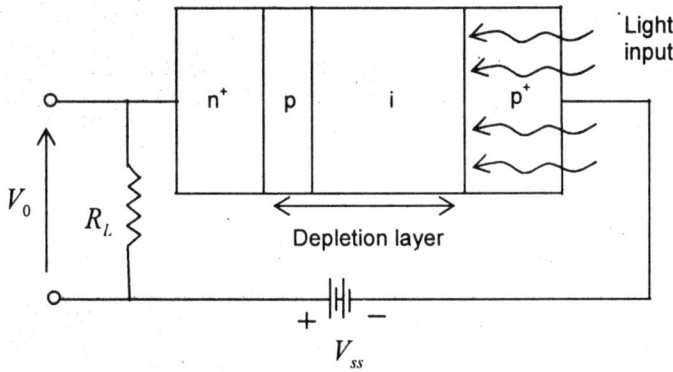

Fig. 4.19. A schematic diagram of the APD. $R_L^{\cdot}$ = load resistance; V_0 = output voltage.

The electric field profile is shown in Fig. 4.20. The electric field near the n^+p junction is sufficiently high and greater than the threshold field (E_{th}) required for impact ionization. The threshold field, E_{th} is $\sim 10^7$ V/m. Light is incident on the intrinsic region (i) through the p^+ region and gets absorbed in the i-region if the photonic energy $(h\nu)$ is equal to or greater than the bandgap energy (E_g) of the semiconductor. Electron-hole pairs are generated by the incident light. The photogenerated electrons drift towards the p-region while the holes drift through the i-region towards the p⁺ region. The electrons which enter the p-region are accelerated to high velocities by the high electric field existing in the p-region. These electrons collide with the crystal atoms thereby generating electron-hole pairs. An electron on the average can produce many electron-hole pairs. A huge number of electron-hole pairs are generated by impact ionization and an avalanche is formed. A large current flows in the external circuit. The current multiplication increases with the increase in reverse bias applied to the APD. The ionization rate of the charge carrier (electron or hole) is defined as the number of electron-hole pairs produced by the ionizing carrier per unit distance. In silicon, the electron ionization rate is much greater than that of the holes. Thus, the avalanching is produced primarily by the electrons. This makes the Si APD less noisy compared with APDs made from Ge, GaAs etc. Current multiplication in an APD depends upon the applied reverse bias.

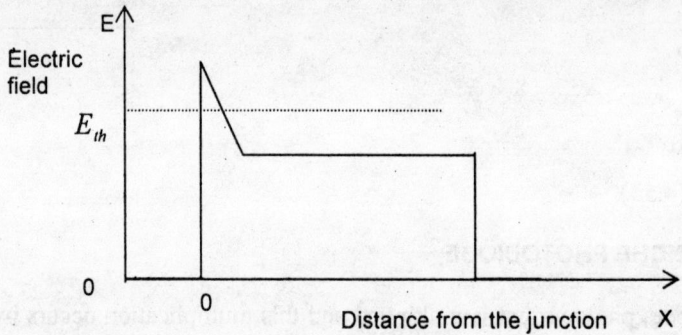

Fig. 4.20. Typical electric field profile of the APD. E_{th} = threshold field.

4.11 MICROWAVE DIODES

Microwaves are electromagnetic waves having frequencies in the range 3 GHz- 30 GHz. The term GHz (gigahertz) means 10^9 Hertz. Electronic valve oscillators like klystron and magnetron can generate microwave power upto about 10 GHz. However, in the high frequency region the efficiency of these devices degrades rapidly. Solid state devices which exploit transit time effect of the carriers can generate microwave power at higher frequencies. Bipolar and field effect transistors can also generate microwaves especially in the lower frequency part up to several GHz. Two terminal microwave devices, known as microwave diodes, such as IMPATT, TRAPATT, BARITT, TUNNETT and Gunn diodes can generate microwave signals. Among these microwave devices, Gunn and IMPATT diodes are frequently used as microwave sources. These two diodes will be discussed in detail in the following subsections. TRAPATT stands for TRApped Plasma Avalanche Triggered Transit. BARITT stands for BARrier-controlled Injection and Transit Time. TUNNETT diode uses TUNNElling and Transit Time properties to generate microwave oscillations.

4.11.1 GUNN DIODE

The Gunn diode was first reported by J. B. Gunn in 1963 whence the name Gunn diode. It is not a junction device but a semiconductor in bulk. Gunn observed microwave oscillations in n-type GaAs and n-type InP bulk devices. The frequency of oscillation was nearly equal to the reciprocal of the carrier transit time. Gunn diode shows differential negative resistance (DNR) across its terminals when a proper bias voltage is applied to it. The drift velocity of the electrons in n-type GaAs as a function of the electric field is shown in Fig. 4.21.

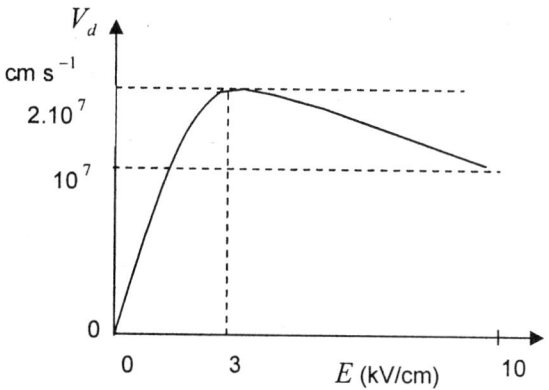

Fig. 4.21. Electron drift velocity (V_d) as a function of electric field (E) in n-type GaAs.

The current density (J) as a function of the electric field inside the device is shown in Fig. 4.22.

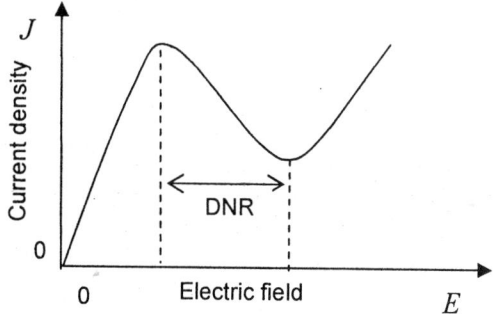

Fig. 4.22. Typical current density (J) variation with electric field (E) in a Gunn diode.

The negative resistance in the Gunn diode is voltage-controlled where the current density is a multivalued function of the electric field. The mechanism responsible for Gunn action was proposed by Ridley, Watkins and Hilsum (RWH) and is known as RWH mechanism. According to RWH mechanism, in the energy-momentum diagram for electrons in the conduction band of some semiconductors, there may exist more than one minimum in energy. These minima of energy are known as conduction band valleys. The electrons in the lower-energy conduction band valley have a smaller effective mass and higher mobility than those in the upper valley where electrons have higher effective mass and lower mobility. When electrons from the lower valley are transferred to the upper valley under the action of the applied electric field, current density decreases with increasing electric field. This is because $J = ne\mu E$ where μ = mobility, n = electron density, e = magnitude of electronic charge and E = electric field. Now, μ decreases in intervalley carrier transfer. So, J decreases with E. The threshold electric field at which this intervalley carrier transfer begins is 3.2 kV/cm in GaAs. The decrease in current density with the increase in electric field gives rise to differential negative resistance (DNR) in a Gunn diode. The energy-momentum ($E - k$) diagram is shown in Fig. 4.23.

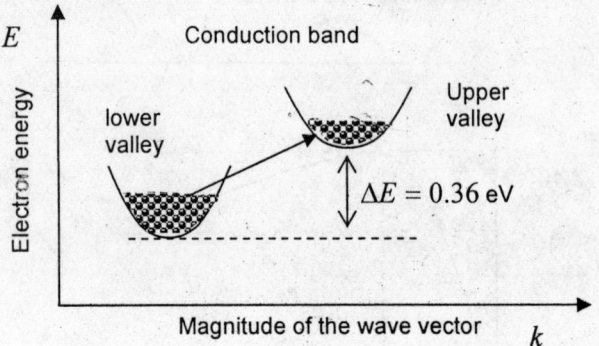

Fig. 4.23. $E - k$ diagram of n-type GaAs.

The following criteria must be satisfied in order that differential negative resistance can be manifested in the Gunn diode:

- The temperature (T) of the device must be low so that the thermal energy kT is much smaller than the intervalley energy separation (ΔE). For n-GaAs, $\Delta E = 0.36$ eV. Under this condition, most of the conduction band electrons will be present in the lower valley.

- The energy difference (ΔE) between the two valleys must be smaller than the band gap energy (E_g) of the semiconductor. This ensures that avalanche multiplication can not occur before or during the RWH process. For GaAs, $E_g = 1.43$ eV at $300°$K.

- The semiconductor in its band structure must possess a low-energy conduction band valley with high electron mobility and a higher-energy conduction band valley with low electron mobility. This is the basic criterion of the Gunn action.

Since, the intervalley electron transport is the fundamental mechanism of Gunn action, the Gunn device is also known as Transferred Electron Device (TED). Gunn oscillator is a low noise device. It is used as a local oscillator in a microwave receiver and a signal source in low-power microwave transmitter.

4.11.2 IMPATT DIODE

The term IMPATT is an acronym for IMPact Avalanche Transit-Time. It is a negative resistance device which can generate power at microwave and millimeter wave frequencies. It is a two terminal device. Read proposed the first IMPATT diode using a $n^+ p\, i\, p^+$ structure in 1958. Here, i stands for intrinsic semiconductor. A schematic diagram of the Read diode is shown in Fig. 4.24(a). A semiconductor device will exhibit negative resistance across its terminals when the terminal ac current lags behind the ac voltage by a phase angle between $\pi/2$ and $3\pi/2$ radians. Maximum negative resistance occurs when this phase lag becomes equal to π radian. The physical mechanisms involved in the operation of the IMPATT diode are,

- the ionization of valence-band atoms of the semiconductor by the impact with energetic electrons, and

- the transit time of the avalanche-generated carriers through the high field depletion region of the device.

When the reverse bias in a *p-n* junction is increased beyond a certain limit, the electrons in the reverse saturation current gain sufficient energy from the dc electric field to ionize the atoms of the crystal by colliding with them. The electron which is liberated through impact-ionization moves to the conduction band leaving behind a hole in the valence band. Holes are the vacancies in an otherwise filled valence band. Once this impact-ionization starts a large number of free electron-hole pairs are generated which form an "Avalanche" moving through the depletion region of the diode. The term "Avalanche" means a large mass of snow with ice and rocks decending down a mountain. The avalanche current density lags behind the ac electric field by $\pi/2$ radian . The terminal current density is further delayed by $\pi/2$ radian since the carriers take some time to pass through the depletion region. This depletion region (*i*) is also called drift region. The electric field ($E(x)$) profile of the Read type IMPATT diode is shown in Fig. 4.24(b).

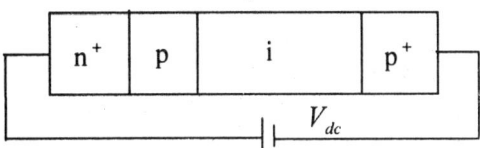

Fig. 4.24. (a) A typical Read diode structure. i → intrinsic region.

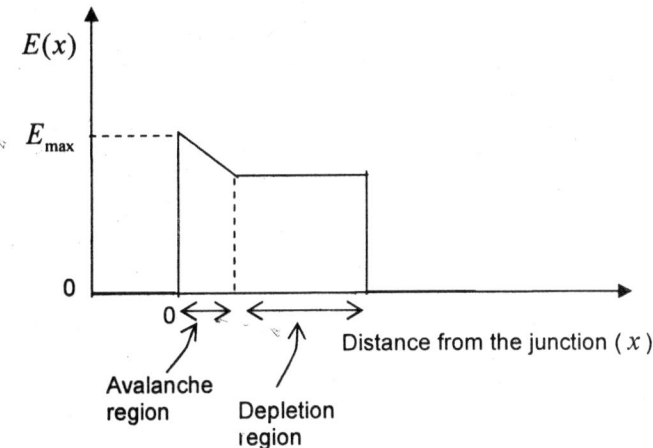

Fig. 4.24. (b) Electric field, $E(x)$, profile inside the IMPATT diode.

Beside the Read's structure ($n^+ p\, i\, p^+$ or $p^+ n\, i\, p^+$), other structures like $p^+ i\, n^+$, $p^+ n\, n^+$, $p^+ p\, n\, n^+$ and even a simple *p-n* junction can exhibit IMPATT action. The IMPATT diode when placed in a microwave cavity can operate as a microwave oscillator. The IMPATT diode finds application as CW (continuous wave) and pulsed power source in microwave and millimeter wave communication systems.

SOLVED PROBLEMS

4.1   The reverse saturation current of a Si p-n junction diode is $1\,\mu$A at $300°$K. Calculate the diode current for a forward bias of 0.5V. Take $\eta = 2$.

Solution

We know that $I = I_s \left(e^{qv/\eta kT} - 1\right)$

Here, $I_s = 1\,\mu A = 10^{-6}$ A

$T = 300°$ K.

Then, $\dfrac{kT}{q} = \dfrac{1.38 \times 10^{-23}\,\text{J}/°\text{K} \times 300°\,\text{K}}{1.6 \times 10^{-19}\,\text{C}}$

$= 25.87$ mV

Here, $v = 0.5V = 500$ mV

$$I = 10^{-6}\left[e^{\frac{500}{2 \times 25.87}} - 1\right] \text{Amp}$$

$= 15.7$ mA.

4.2   The forward current in a Si diode is 25 mA at $20°$C. The reverse saturation current at $20°$C is 1μ A. Calculate the value of the forward bias voltage of the diode. Take $\eta = 2$.

Solution

We have, $I = I_s\left(e^{qv/\eta kT} - 1\right)$

$\therefore \quad e^{qv/\eta kT} = 1 + \dfrac{I}{I_s}$

$\therefore \quad v = \eta\dfrac{kT}{q}\ln\left(1 + \dfrac{I}{I_s}\right)$

Here, $T = 273 + 20 = 293°$ K

$I = 25$ mA

$$I_s = 1 \ \mu A = 10^{-3} \, mA$$

$$\eta = 2$$

Now, $\quad \dfrac{kT}{q} = \dfrac{1.38 \times 10^{-23} \times 293}{1.6 \times 10^{-19}} = 25.27 \text{ mV}$

$\therefore \qquad v = 2 \times 25.27 \ln\left(1 + 25 \times 10^3\right)$

$\qquad\qquad \approx 50.54 \ln 25000$

$\qquad\qquad = 50.54 \times 10.12$

$\qquad\qquad = 511.8 \, mV$

$\qquad\qquad \approx 0.512 \, volt.$

4.3 The forward current in a Si p-n junction diode doubles when the forward bias changes from v_1 to v_2. The operating temperature of the diode is $300°$ K. Assuming $\eta = 2$, calculate the difference in forward bias in the two cases.

Solution

We write, $I_1 = I_s\left(e^{qv_1/\eta kT} - 1\right)$

$\qquad\qquad I_2 = I_s\left(e^{qv_2/\eta kT} - 1\right)$

Now, $\quad I_2 = 2I_1$

$\therefore \qquad \dfrac{I_2}{I_1} = \dfrac{e^{qv_2/\eta kT} - 1}{e^{qv_1/\eta kT} - 1} \approx e^{\frac{q}{\eta kT}(v_2 - v_1)} \qquad \text{since } e^{qv/\eta kT} \gg 1.$

$\therefore \qquad 2 = e^{q(v_2 - v_1)/\eta kT}$

$\qquad\qquad v_2 - v_1 = \eta \dfrac{kT}{q} \ln 2 = 2 \times \dfrac{1.38 \times 10^{-23} \times 300}{1.6 \times 10^{-19}} \ln 2$

$\qquad\qquad\qquad = 2 \times 25.87 \ln 2 = 35.86 \text{ mV}.$

4.4 The reverse saturation current in a Ge p-n diode doubles for every $6°$C rise in temperature. Calculate the rise in temperature required to increase the reverse saturation current by a factor of 20.

Solution

According to the question, $\dfrac{I_{s2}}{I_{s1}} = 2^{\Delta T/6}$

where I_{s1} and I_{s2} are initial and final reverse saturation currents and ΔT is the rise in temperature. Here,

$$\frac{I_{s2}}{I_{s1}} = 20$$

$$\therefore \quad 20 = 2^{\Delta T/6}$$

$$\therefore \quad \Delta T = 6\frac{\ln(20)}{\ln(2)}$$

$$= 25.93^\circ \text{C}$$

4.5 A Ge p-n diode conducts forward currents of 1 mA and 4 mA for forward bias voltages equal to 260 mV and 300 mV respectively. Calculate the operating temperature of the diode. Given $\eta = 1$.

Solution

We have, $I_1 = I_s\left(e^{qv_1/\eta kT} - 1\right)$

$$I_2 = I_s\left(e^{qv_2/\eta kT} - 1\right)$$

$$\therefore \quad \frac{I_2}{I_1} = \frac{e^{qv_2/\eta kT} - 1}{e^{qv_1/\eta kT} - 1} \approx e^{\frac{q}{\eta kT}(v_2 - v_1)}$$

$$\therefore \quad \ln\left(\frac{I_2}{I_1}\right) = \frac{q}{\eta kT}(v_2 - v_1)$$

$$\therefore \quad T = \frac{q}{\eta k}\frac{(v_2 - v_1)}{\ln\left(\dfrac{I_2}{I_1}\right)}$$

Here,

$$v_2 = 300\,\text{mV}$$

$$v_1 = 260\,\text{mV}$$

$$I_2 = 4\,\text{mA}$$

$$I_1 = 1\,\text{mA}$$

$$\eta = 1$$

$$\therefore \quad T = \frac{1.6 \times 10^{-19}}{1.38 \times 10^{-23}} \frac{(300 - 260) \times 10^{-3}}{\ln\left(\dfrac{4}{1}\right)}$$

$$= \frac{1.6 \times 10^{4}}{1.38} \times \frac{40 \times 10^{-3}}{\ln 4}$$

$$= 334.5^\circ \text{ K}$$

4.6 A Si $p^{+}n$ diode has a donor concentration of 10^{15} /c.c in the n-region. The width of the depletion layer is 2μ m. Find the height of the barrier potential if Si has a relative permittivity $\varepsilon_r = 12$.

Solution
From equation (4.35),

$$V_B = \frac{q}{2\varepsilon} W^2 \frac{N_a N_d}{N_a + N_d}.$$

Since it is a $p^{+}n$ diode, $N_a \gg N_d$. So,

$$V_B \approx \frac{q}{2\varepsilon} W^2 N_d.$$

Here, $W = 2\mu m = 2 \times 10^{-6}$ m

$$N_d = 10^{15} / \text{c.c.} = 10^{21} / m^3$$

$$\varepsilon = \varepsilon_0 \varepsilon_r = 12\varepsilon_0$$

$$\varepsilon_0 = \frac{10^{-9}}{36\pi} \text{ F/m.}$$

ε_0 is the absolute permittivity of vacuum.

$$V_B = \frac{1.6 \times 10^{-19} \times 36\pi}{2 \times 12 \times 10^{-9}} \times \left(2 \times 10^{-6}\right)^2 \times 10^{21}$$

$$= \frac{1.6 \times 36\pi \times 4}{24} \times 10^{21+9-19-12}$$

$$= 3 \text{ volt.}$$

4.7 A $n^{+}p$ Ge diode is reverse biased. It has a barrier height of 4 V. The acceptor concentration of the p-region is 10^{15} /c.c. Calculate the width of the depletion region. Given the relative permittivity of Ge is 16.

Solution

Since the given diode is n^+p, we have $N_d \gg N_a$.

From equation (4.36), we write

$$W = \sqrt{\frac{2\varepsilon}{q}} \sqrt{\frac{N_a + N_d}{N_a N_d}} . V_B^{\frac{1}{2}}$$

$$\approx \sqrt{\frac{2\varepsilon}{qN_a}} . V_B^{\frac{1}{2}} \quad \text{since } N_d \gg N_a.$$

Here,

$$\varepsilon_r = 16$$

$$\varepsilon = \varepsilon_r \varepsilon_0 = 16\varepsilon_0$$

$$\varepsilon_0 = \frac{10^{-9}}{36\pi} \, \text{F/m}$$

$$N_a = 10^{15} / \text{c.c.} = 10^{21}/\text{m}^3$$

$$V_B = 4 \, \text{V}$$

$$\therefore \quad W = \sqrt{\frac{2 \times 16 \times 10^{-9}}{36\pi \times 1.6 \times 10^{-19} \times 10^{21}}} (4)^{\frac{1}{2}}$$

$$= \sqrt{\frac{2 \times 10^{-10}}{36\pi}} . 2$$

$$= 0.133 \times 10^{-5} \times 2 = 2.66 \, \mu\text{m}.$$

PROBLEMS

4.1 Calculate the reverse saturation current of a Si p-n diode operated at $300° \text{K}$ if the diode conducts a forward current of 2 mA for a forward bias of 350 mV. Given $\eta = 2$.

[**Ans.** $2.3\mu\text{A}$]

4.2 Find the percentage change in forward current of a Ge p-n diode when the forward diode voltage increases by 15 mV. The diode is operated at $300° \text{K}$ and $\eta = 1$.

[**Ans.** 78.5%]

4.3   A Ge p-n diode conducts a forward current of 50 mA for a forward diode voltage of 0.25 V. The reverse saturation current of the diode is 10μ A. Calculate the operating temperature of the diode. Given $\eta = 1$.

[**Ans.** 340.3° K]

4.4   A Si n^+p diode has an acceptor concentration of 5×10^{15} / c.c. in the p-region. The width of the depletion layer is 1μ m. Si has a relative permittivity $\varepsilon_r = 12$. Find the height of the potential barrier at the junction.

[**Ans.** 3.77 V]

4.5   The reverse saturation current in a Si p-n junction doubles for every $10°$ C rise in temperature. Calculate the rise in temperature required to increase the reverse saturation current by a factor of 50.

[**Ans.** 56.4° C]

QUESTIONS

4.1   Describe physically the current transport dynamics in an unbiased p-n junction. What are Boltzmann equations for a p-n junction.

4.2   Derive Shockley equation describing the current flow in a biased p-n junction. What are the reasons of departure of the real I-V characteristics from that predicted by Shockley equation for a silicon p-n diode? Derive the current-voltage equation for a silicon p-n diode under small forward bias.

4.3   What are the corrections to be applied to a silicon p-n junction when attempting to apply the equation

$$I = I_s \left(e^{qv/kT} - 1 \right)$$

Modify this equation considering the recombination current in the depletion region. Derive an expression for junction resistance for a forward biased p-n junction.

4.4   What do you mean by minority carrier injection? Define the term 'diffusion length' of electron. Derive an expression for the decay of excess minority carriers, injected at the junction, in the neutral region of a p-n junction.

4.5   What do you mean by junction breakdown? What are the different mechanisms which can explain breakdown of a p-n junction? Explain 'Avalanche breakdown' and 'Zener breakdown' in the context of a p-n junction.

4.6   Derive an expression for junction capacitance for a p^+n abrupt junction in which the p-region is heavily doped compared with the n-region. How do you measure the value of built-in potential from a measurement of junction capacitance? Explain.

4.7 What do you mean by a graded junction? How does the junction capacitance of a p^+n abrupt junction vary with the barrier potential? What is a hyper-abrupt junction? What is a varactor diode? Mention an application of the varactor diode.

4.8 What is the diffusion capacitance of a p-n junction? Show that the diffusion capacitance of a p^+n junction is directly proportional to the forward current of the diode.

4.9 What is a Tunnel diode? Draw and explain its I-V characteristics with the help of energy band diagrams.

4.10 What are resistance and reactance cut-off frequencies of a Tunnel diode? Derive expressions for them. What are the applications of a tunnel diode? How do you obtain amplification in a Tunnel diode?

4.11 What do you mean by the term 'LASER'? What is a diode laser? Give a diagram of a diode laser and explain how lasing occurs in the diode with the help of energy-band diagram.

4.12 What is an LED? Give a cross-sectional view of a surface-emitting LED and explain its operation. What are the merits of a laser diode over an LED?

4.13 What is a photodiode? Mention different classes of photodiodes. Explain the principle of operation of a p-n junction photodiode. What is cut-off wavelength of the photodiode?

4.14 Describe the structure of a p-i-n photodiode. What is the advantage of a p-i-n photodiode over a p-n junction photodiode?

4.15 What is an Avalanche photodiode? Explain the working principle of an APD with the help of a block diagram. Why do you prefer silicon APD over a Ge APD?

4.16 What do you mean by 'Microwaves'? Name some solid state microwave sources. What is a Gunn diode? Explain how differential negative resistance occurs in a Gunn diode.

4.17 What is RWH mechanism? Explain with the help of E-k diagram. What are the basic conditions which must be fulfilled by the device in order to exhibit Gunn action? Why is the Gunn diode also called as a TED?

4.18 What do you mean by the term 'IMPATT' diode? Explain the physical mechanisms involved in the operation of a Read diode. How does the Read diode manifest a differential negative resistance? Mention an application of the IMPATT diode.

5

Solid-State
Three Electrode Devices

5.1 BIPOLAR JUNCTION TRANSISTOR

Transistor is an acronym for 'Transfer resistor'. In normal transistor operation, the emitter-base junction is forward biased while the collector-base junction is reverse biased. So, the resistance of the input emitter-base diode is small and the resistance of the output collector-base diode is high. Thus, there occurs a resistance transformation in the transistor. This is just opposite of the vacuum triode in which the input grid circuit resistance is much higher than that of the output anode circuit.

Bipolar junction transistor (BJT) was invented in 1951. A transistor has three terminals – emitter, base and collector. Since there are two p-n junctions in a BJT, it can have two structures – viz., p-n-p and n-p-n. The symbols of a BJT are shown in Figs. 5.1(a) and 5.1(b).

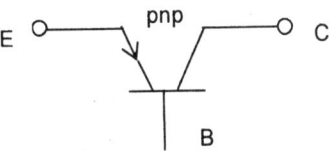

Fig. 5.1. (a) Symbol of pnp BJT

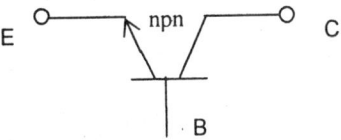

Fig. 5.1. (b) Symbol of npn BJT

The arrow on the emitter head gives the direction of current flow when the emitter-base junction is forward biased. The direction of arrow is different in pnp and npn BJT.

Depending upon which terminal is common to both the input and output circuits of a BJT, there can be three modes of operation of the BJT. These are:

- Common Base (CB)
- Common Emitter (CE)
- Common Collector (CC)

5.2 TRANSISTOR PARAMETERS

We now define some quantities associated with the transistor operation.

(i) DC ALPHA

It is the ratio of the dc collector current (I_c) to the dc emitter current (I_E).

$$\alpha_{dc} = -\frac{I_c}{I_E} \qquad \qquad \ldots (5.1)$$

Here, I_c and I_E are of opposite signs. So, α_{dc} is positive. The current entering a transistor terminal is considered to be positive while the current leaving a transistor terminal is taken as negative. This is a convention. In general, $\alpha_{dc} > 0.95$. α_{dc} is the dc current gain in the CB mode. $\alpha_{dc} < 1$.

(ii) SMALL-SIGNAL ALPHA(α')

It is defined as the ratio of small change in collector current to the small change in emitter current which produces the change in collector current.

$$\alpha' = \frac{\partial I_c}{\partial I_E}\bigg|_{V_{CB}=\text{constant}} \qquad \qquad \ldots (5.2)$$

α' is the negative of the short circuit CB current gain.

(iii) LARGE SIGNAL ALPHA(α)

It is the ratio of the negative of the collector current increment from cut-off to the emitter current change from cut-off.

$$\therefore \qquad \alpha = -\frac{I_c - I_{CBO}}{I_E - 0} = -\frac{I_c - I_{CBO}}{I_E} \qquad \qquad \ldots (5.3)$$

I_c and I_E have opposite signs. So, α is positive. Again, since the cut-off current $I_{CBO} \ll I_c$ under normal operating condition,

$$\alpha \approx -\frac{I_c}{I_E} = \alpha_{dc}.$$

At cut-off, $I_E = 0$, $I_c = I_{CBO}$ and $I_B = -I_{CBO}$.

α depends upon I_E, V_{CB} and temperature.

(iv) DC BETA

It is the ratio of the dc collector current to the dc base current of a transistor.

$$\beta_{dc} = \frac{I_c}{I_B} \qquad \qquad \ldots (5.4)$$

Since I_c and I_B have the same sign. So, β_{dc} is positive. β_{dc} is the dc current gain in the CE mode.

(v) SMALL-SIGNAL BETA(β')

It is defined as the ratio of a small change in collector current to the corresponding change in base current for a given collector-emitter voltage.

$$\beta' = \frac{\partial I_c}{\partial I_B}\bigg|_{V_{CE}=\text{constant}} \qquad \ldots (5.5)$$

β' is the negative of the small-signal CE current gain.

(vi) LARGE-SIGNAL BETA

It is the ratio of collector-current increment from cut-off to the corresponding base current increment from cut-off.

$$\beta = \frac{I_c - I_{CBO}}{I_B - (-I_{CBO})} \qquad \ldots (5.6)$$

5.2.1 RELATION BETWEEN α_{dc} AND β_{dc}

Taking all the 3 terminal currents as flowing into the terminals in a BJT, we can write

$$I_E + I_B + I_C = 0 \qquad \ldots (5.7)$$

or, $-I_E = I_B + I_C$

or, $-\dfrac{I_E}{I_C} = 1 + \dfrac{I_B}{I_C}$

or, $\dfrac{1}{\alpha_{dc}} = 1 + \dfrac{1}{\beta_{dc}}$

$\therefore \qquad \alpha_{dc} = \dfrac{\beta_{dc}}{1 + \beta_{dc}} \qquad \ldots (5.8)$

Again,

$$\frac{1}{\beta_{dc}} = \frac{1}{\alpha_{dc}} - 1 = \frac{1 - \alpha_{dc}}{\alpha_{dc}}$$

$\therefore \qquad \beta_{dc} = \dfrac{\alpha_{dc}}{1 - \alpha_{dc}} \qquad \ldots (5.9)$

If $\beta_{dc} = 100$, $\alpha_{dc} = 0.99$.

From (5.6) we can write

$$I_C - I_{CBO} = \beta I_B + \beta I_{CBO}$$

$$\therefore \quad I_C = \beta I_B + (1 + \beta)I_{CBO} \qquad \qquad \dots (5.10)$$

5.3 I-V CHARACTERISTICS IN CB MODE

The circuit configuration in CB mode of the BJT is shown in Fig. 5.2. Here, I_E is the emitter current, I_C is the collector current and I_B is the base current. V_{EB} is the emitter voltage and V_{CB} is the collector voltage. V_{EE} and V_{CC} are emitter supply and collector supply voltages respectively. R_E is the resistance in the input circuit and R_C is the resistance in the collector circuit.

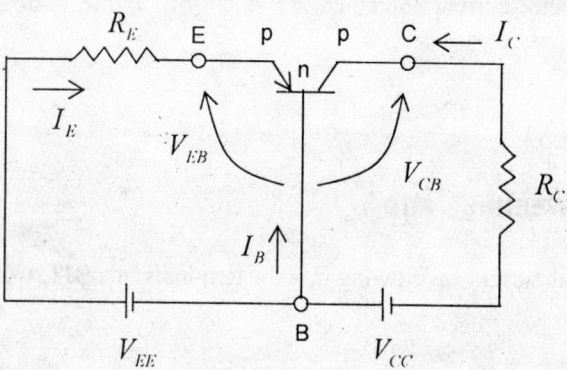

Fig. 5.2. Circuit diagram for measuring I-V characteristics in CB mode.

Here, I_E is positive while I_B and I_C are negative. The current which enters a terminal is taken as positive. The current which leaves a terminal is taken as negative. V_{CC} is the collector supply voltage and V_{EE} is the emitter supply voltage.

5.3.1 INPUT CHARACTERISTICS

Suppose that the collector circuit is open. Under this condition, the $I_E - V_{EB}$ characteristics is that of a p-n junction diode. Now, as the magnitude of V_{CB} increases the collector junction depletion layer widens into the base. This reduces the effective base width. Reduction of base width leads to an increase in minority carrier concentration gradient $\left(\dfrac{dp}{dx}\right)$ across the emitter-base junction of a pnp transistor. So, the emitter current increases with increasing $|V_{CB}|$. The $V_{EB} - I_B$ characteristics is shown in Fig. 5.3 with V_{CB} as a parameter. Typical values of I_E lies in the range $(0 - 5 \text{ mA})$ and typical values of V_{EB} lies in the range $(0 - 0.75 \text{ V})$ for a Si BJT.

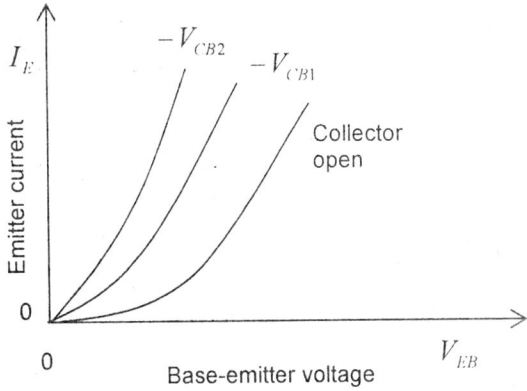

Fig. 5.3. Input characteristics of the pnp BJT in CB mode. V_{CB} is a parameter here with $V_{CB2} > V_{CB1}$.

5.3.2 OUTPUT CHARACTERISTICS IN CB MODE

There are three distinct regions of the output characteristics of the BJT in CB mode. These are

(i) active region
(ii) saturation region, and
(iii) cut-off region.

(i) In the active region, the emitter-base junction of the BJT is forward biased and the collector base junction is reverse biased. When $I_E = 0$, I_C is equal to the reverse saturation current (I_{CBO}) in the reverse-biased collector-base junction. So, $I_C = I_{CBO}$. When $I_E \neq 0$, $I_C = -\alpha I_E$ and I_C is independent of V_{CB}. When $|V_{CB}|$ is large, the base width decreases due to the widening of the collector junction depletion layer. So, recombination current in the base decreases. Now, since $I_E + I_B + I_C = 0$ and I_E is held constant, a reduction in $|I_B|$ leads to an increase in $|I_C|$. Thus, $|I_C|$ increases very slowly with increasing $|V_{CB}|$. V_{CB} and I_C for a Si pnp BJT can have typical values in the range $(-5\,\text{V}, 0.2\,\text{V})$ and $(0, -5\,\text{mA})$ in the active region respectively. I_E has typical values in the range $(0, 5\,\text{mA})$ in the active region.

(ii) In the saturation region, the emitter-base junction and the collector-base junction both are forward biased. The current through the collector-base junction increases with increasing forward bias of this junction. This collector current flows in opposite direction to the emitter current. When V_{CB} becomes large, I_C will flow in the reverse direction and become positive.

(iii) In the cut-off region, both the emitter-base junction and the collector-base junction are reverse biased. So, only reverse saturation currents flow through these junctions.

The output characteristics of a pnp BJT in CB mode is shown in Fig. 5.4.

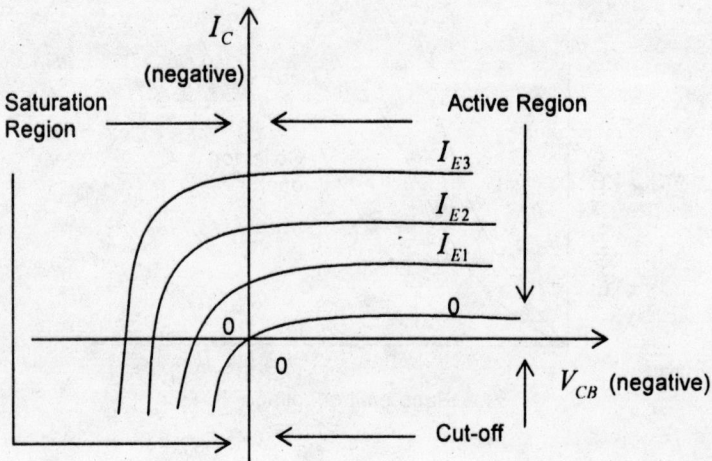

Fig. 5.4. Output characteristics of a p-n-p Si BJT in CB mode. Emitter current, I_E, is a parameter. $I_{E3} > I_{E2} > I_{E1} > 0$.

5.4 I-V CHARACTERISTICS IN CE MODE

The circuit configuration in CE mode of a pnp BJT is shown in Fig. 5.5. I_B is the base current and I_C is the collector current. V_{BE} is the base voltage with respect to emitter and V_{CE} is the collector voltage with respect to the emitter. V_{BE} and I_B are quantities of the input circuit and V_{CE} and I_C are quantities of the output circuit. In a pnp transistor, the currents I_B and I_C are negative, under normal operating conditions, since these currents flows out from the corresponding terminals. V_{BE} and V_{CE} are negative under normal operating conditions.

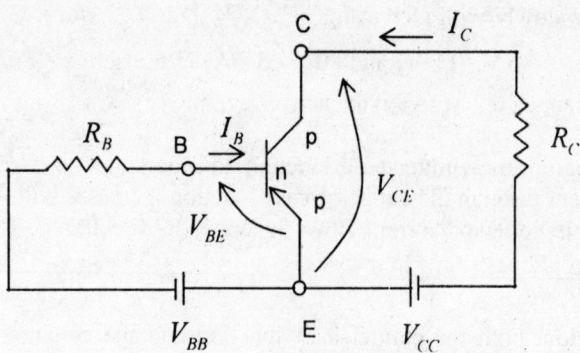

Fig. 5.5. Circuit for I-V characteristics measurement of a pnp transistor in CE configuration.

Here, R_B is the input resistance which controls the base current. R_C is the load resistance which controls the collector voltage (V_{CE}). Here, B stands for base, E stands for emitter and C stands for collector. V_{CC} is the collector supply voltage and V_{BB} is the base supply voltage.

5.4.1 INPUT CHARACTERISTICS

When the collector is open, the input $V_{BE} - I_B$ characteristics is that of a forward biased diode. Now, if the collector circuit is closed and a collector voltage (V_{CE}) is applied, a depletion layer appears at the reverse biased collector-base junction. Since the base is lightly doped, the depletion layer penetrates into the base with increasing $|V_{CE}|$. The effective base width, thus, decreases leading to a decrease in recombination base current.

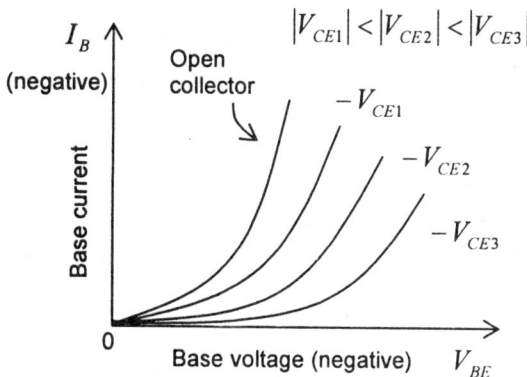

Fig. 5.6. Input I-V characteristics of a pnp transistor in CE mode of operation.

The $V_{BE} - I_B$ characteristics of a Si pnp transistor in CE mode is shown in Fig. 5.6. with V_{CE} as a parameter. $|V_{CE3}| > |V_{CE2}| > |V_{CE1}|$. Typical ranges of values of V_{BE} and I_B lie in the range $(0, -0.75\,\text{V})$ and $(0, -0.5\,\text{mA})$ respectively for operation in the active region. V_{CE} has typical values in the range $(0, -5\,\text{V})$.

5.4.2 OUTPUT CHARACTERISTICS

Here, I_C is the collector current and V_{CE} is the collector voltage measured with respect to the emitter. The output $V_{CE} - I_C$ characteristics of a pnp Si transistor is shown in Fig. 5.7 using input current I_B as a parameter. V_{CE} and I_C have typical ranges of values $(0, -6\,\text{V})$ and $(0, -10\,\text{mA})$ respectively for operation in the active region. I_B has typical values in the range $(0, -0.5\,\text{mA})$.

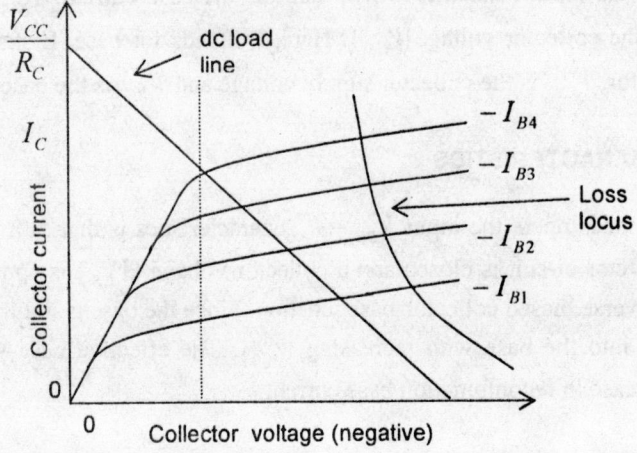

Fig. 5.7. Output characteristics of a pnp transistor operated in CE mode.
$$I_{B4} > I_{B3} > I_{B2} > I_{B1}.$$

Now, from (5.10) we have

$$I_C = \beta I_B + (1 + \beta) I_{CBO}$$

where $\beta = \dfrac{\alpha}{1-\alpha}$ is the large signal β. I_{CBO} is the reverse saturation current of the base-emitter diode when reverse biased. When $I_B = 0, I_C = (1 + \beta) I_{CBO} = I_{CEO}$ (say). This is the collector current in CE mode when the base is open (i.e., $I_B = 0$). When $|V_{CE}|$ increases, the collector junction depletion layer expands into the base and the recombination base current decreases. Since, $I_C + I_B + I_E = 0$, a decrease in I_B leads to an increase in I_C (with $I_E =$ constant). So,

$$\alpha = -\frac{I_C - I_{CBO}}{I_E}$$ increases. Now, a small increase in α produces a greater change in

$\beta \left(= \dfrac{\alpha}{1-\alpha}\right)$. So, I_C increases with the increase in collector voltage $|V_{CE}|$ when I_B is held

constant. This is the active region of the transistor in CE mode. When the emitter-base junction and the collector-base junction both are forward biased, the emitter current of the pnp transistor causes a collector current which flows out from the collector. On the other hand, when the collector-base junction is forward biased, a collector current flows into the collector terminal, which opposes the former current. So, the resultant collector current is the difference of two currents. Now, with increasing $|V_{CE}|$, the forward-biased collector junction current increases and hence the overall collector current decreases. The $V_{CE} - I_C$ characteristics tend to merge together and approaches the origin. This is the saturation region.

When both the emitter-base junction and the collector-base junction are reverse biased, $I_C = I_{CO}$, the reverse saturation current. I_{CO} flows out from the collector terminal and so I_{CO} is itself negative. Also, $I_C + I_B + I_E = 0$. But, $I_E = 0$ under cut-off condition. So, $I_B = -I_C = -I_{CO}$. I_B flows into the base terminal, so I_B is positive. The transistor is said to be cut-off under this condition.

5.5 JUNCTION FIELD EFFECT TRANSISTOR (JFET)

The junction field effect transistor is a special type of transistor where the current is controlled by the application of an electric field. JFET has three terminals, which are source, drain and the gate. A control voltage is applied to the gate terminal. A schematic diagram of the JFET in the form of a rectangular bar is shown in Fig. 5.8. Here, S is the source, D is the drain and G is the gate terminal. The region of the semiconductor bar through which the majority carriers flow from source to drain is called the channel of the JFET. If the semiconductor bar is a n-type material, the JFET is a n-channel JFET. On the other hand, if the semiconductor bar is a p-type material, the JFET is said to be a p-channel JFET. On either side of the channel, a pair of p-n junction is formed which are known as gates. If the JFET is a n-channel one, p^+ regions are formed on either side as gates. Heavy doping of the p-region (p^+) ensures that the depletion layer expands primarily into the n-channel when a reverse bias is applied to the gate terminal. Similarly, n^+p junctions are formed as gates in the case of p-channel JFET. Here, the depletion layer will expand into the p-channel when a reverse bias is applied to gate terminals. The depletion layer expands into the region where the doping concentration is lower. The source (S) terminal is analogous with the emitter of a BJT. The terminal drain (D) is similar with the collector terminal of the BJT and the terminal gate (G) resembles with the base terminal of the BJT.

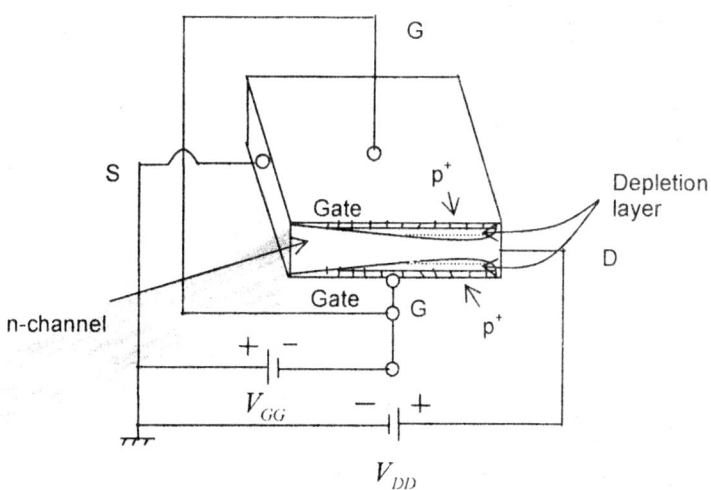

Fig. 5.8. Schematic diagram of an n-channel JFET.

In Fig. 5.8, V_{DD} is the drain supply voltage and V_{GG} is the gate supply voltage.

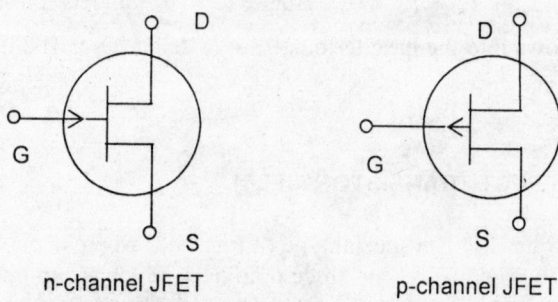

n-channel JFET   p-channel JFET

Fig. 5.9. Symbols of n-channel and p-channel JFETs.

The symbols of n-channel and p-channel JFET are shown in Fig. 5.9. The arrow on the gate terminal shows the direction of current flow when the gate junction (p^+n or n^+p) is forward biased. When p^+n junction is forward biased the control current enters the gate (G) terminal.

5.5.1 JFET OPERATION

The reverse bias applied to the gate terminals determines the width of the depletion layers of the p^+n (or n^+p) junctions. The higher the value of the reverse bias, the wider is the depletion layer. Since this depletion layer intrudes into the channel of the JFET, the effective width of the channel available for conduction gets reduced. The effective cross section of the channel reduces and hence the channel resistance increases as the reverse gate voltage is increased. The drain current is thus controlled by the reverse bias of the gate junctions. In effect, the electric field associated with the gate junctions controls the JFET current and hence the name field effect transistor.

The drain current (I_D) vs. drain voltage (V_{DS}) characteristics of the JFET with gate voltage (V_{GS}) as a parameter is shown in Fig. 5.10. Let us first consider the case with $V_{GS} = 0$. Now, if we set $V_{DS} = 0$, the drain current $I_D = 0$. Now, let us apply a small drain voltage (V_{DS}). A small current flows from the source to the drain. This current produces a continuously increasing potential drop along the channel with respect to the source which becomes maximum at the drain end. For a n-channel JFET, V_{DS} is positive. The channel drop provides a reverse bias to the p^+-n junction for the n-channel JFET and this reverse bias increases continuously along the channel. The reverse bias becomes a maximum at the drain end. The width of the depletion layer associated with the reverse-biased gate junction will increase accordingly towards the drain end and will be a maximum at the drain end. So, the effective channel width will decrease as we move from the source towards the drain. The effective channel width will become minimum at the drain end. The I_D vs. V_{DS} characteristics is nearly linear at the beginning when the drain voltage is small. This is because the semiconductor now behaves like a Ohmic resistance. With increasing V_{DS} the channel cross-section decreases and hence effective channel resistance increases. So, I_D increases less rapidly with V_{DS} and the $I_D - V_{DS}$ characteristics bends gradually towards the V_{DS} axis. Eventually, a value of V_{DS} is reached when the current I_D levels off (i.e., almost parallel to the V_{DS} axis). The channel is said to attain the "Pinch–off" condition. If V_{DS} is increased further, the drain current I_D does not change appreciably. If V_{DS} is increased monotonically a stage is reached when the reverse-biased p^+n junction breaks down and I_D increases sharply.

Now, consider the case when $V_{GS} \neq 0$. If V_{GS} is negative it will provide an additional reverse bias to the p^+-n gate junction of the n-channel JFET. So, pinch-off will be attained at smaller values of V_{DS} compared with the case when $V_{GS} = 0$. The break down will also occur at a smaller value of V_{DS}. On the other hand, if we apply a small positive V_{GS}, the effective reverse bias of the gate junction will be smaller. So, the pinch-off will be obtained at a higher value of V_{DS} and the breakdown will occur at a higher value of V_{DS} compared with the case when $V_{GS} = 0$. This is shown in Fig. 5.10.

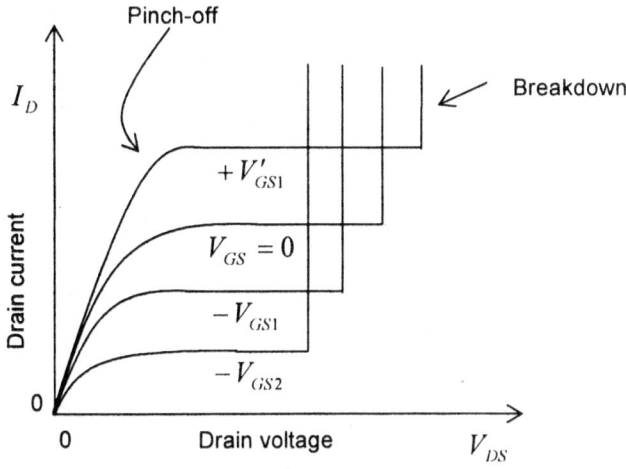

Fig. 5.10. Drain Current (I_D) vs. drain voltage (V_{DS}) characteristics of the

n- channel JFET. V_{GS} = applied gate voltage. $V_{GS2} > V_{GS1} > 0$.

Typical values of drain voltage and drain current lie in the range (0,10 V) and (0, 10 mA) respectively. Gate voltage has typical values in the range (-3 V, 0.5 V).

5.5.2 PINCH–OFF VOLTAGE CALCULATION

We consider the effect of the gate junctions on either side of the channel to be the same. This is because the channel structure is assumed to be symmetric with respect to the source–drain central line ($y = 0$, $z = 0$ in Fig. 5.11). The simple channel structure is shown in Fig. 5.11.

Let a be the metallurgical half-width of the channel. Also, let $W(x)$ be width of the depletion layer and $h(x)$ be the width of the clear channel at x. Then,

$$W(x) = a - h(x) \qquad \qquad \text{... (5.11)}$$

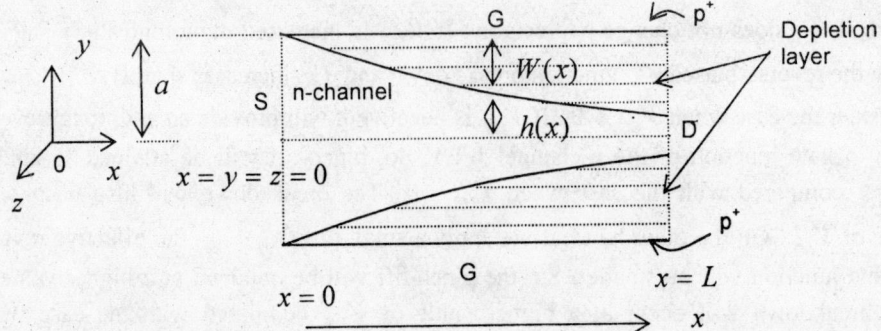

Fig. 5.11. Two dimensional channel structure of an n-channel JFET.

Again, from the theory of p⁺-n junction we have

$$W(x) = \sqrt{\frac{2\varepsilon}{qN_d}}\left[\psi - V_{GX}\right]^{\frac{1}{2}} \qquad \dots (5.12)$$

where ε is the absolute permittivity of the semiconductor material, q is the magnitude of electronic charge and N_d is the donor concentration of the n-channel. ψ is the built-in potential of the p⁺n junction. Since the p⁺ region is heavily doped, the depletion layer primarily expands into the n-channel. V_{GX} is the gate bias voltage measured at the point x. For reverse bias, $V_{GX} < 0$.

At the drain end, $x = L$ and $V_{GX}(x = L) = V_{GD}$ (say). V_{GD} is the gate voltage measured with respect to the drain. For the n-channel JFET, we take $V_{GD} < 0$ where the gate junctions are reversed biased. Moreover, $|\psi| << |V_{GD}|$.

$$\therefore \qquad W(x = L) \approx \sqrt{\frac{2\varepsilon}{qN_d}}\left[-V_{GD}\right]^{\frac{1}{2}} \qquad \dots (5.13)$$

neglecting ψ compared with $|V_{GD}|$. Here, L is the length of the channel measured from source to drain. When the channel is pinched-off, the width of the clear channel at the drain end becomes very small so that we put $h(x = L) \approx 0$. Then, $W(x = L) \approx a$.

Let $V_P = -V_{GD}$ which is the pinch-off voltage.

From (5.13) we get

$$a = \sqrt{\frac{2\varepsilon}{qN_d}}.V_P^{1/2} \qquad \dots (5.14)$$

$$\therefore \qquad V_P = \frac{qN_d}{2\varepsilon}a^2 \qquad \dots (5.15)$$

This is the expression for the Pinch-off voltage of the n-channel JFET.

CALCULATION OF PINCH-OFF CURRENT

The magnitude of the drain current density

$$J_D = \sigma \frac{dV_x}{dx} \qquad \ldots (5.16)$$

where σ is the channel conductivity and V_x is the voltage at x measured with respect to the source. If L_z be the channel dimension in the Z-direction then the drain current (I_D) is given by

$$I_D = J_D.2h(x)L_z \qquad \ldots (5.17)$$

Now, $\quad V_{GS} = V_{Gx} + V_{xS} \qquad \ldots (5.18)$

where the second subscript indicates that the voltage is measured with respect to the point referred by this second subscript. Here, 'S' means source, 'G' means gate and 'x' is a point along the x-axis. Also, $V_{xS} \equiv V_x$. Then,

$$V_{Gx} = V_{GS} - V_{xS} = V_{GS} - V_x \qquad \ldots (5.19)$$

Substituting the value of $\sqrt{\dfrac{2\varepsilon}{qN_d}}$ from (5.14) into (5.12), we get

$$W(x) = a\left[\frac{(\psi - V_{Gx})}{V_P}\right]^{\frac{1}{2}} \qquad \ldots (5.20)$$

Again, from (5.11) we write

$$h(x) = a - W(x)$$

$$= a\left[1 - \left(\frac{\psi - V_{Gx}}{V_P}\right)^{\frac{1}{2}}\right] \qquad \ldots (5.21a)$$

Assuming $|\psi| << |V_{Gx}|$ and using (5.19) we have

$$h(x) = a\left[1 - \left(\frac{V_x - V_{GS}}{V_P}\right)^{\frac{1}{2}}\right] \qquad \ldots (5.21b)$$

From (5.17) we write

$$I_D = 2\sigma L_z a \frac{dV_x}{dx}\left[1 - \left(\frac{V_x - V_{GS}}{V_P}\right)^{\frac{1}{2}}\right]$$

$$\therefore \quad I_D dx = 2\sigma L_z a \left[1 - \left(\frac{V_x - V_{GS}}{V_P} \right)^{\frac{1}{2}} \right] dV_x \qquad \ldots (5.22)$$

Integrating (5.22) with respect to x, we get

$$I_D x = 2a\sigma L_z \left[\int_0^{V_{DS}} dV_x - \frac{1}{\sqrt{V_P}} \int_0^{V_{DS}} \sqrt{V_x - V_{GS}} \, dV_x \right]$$

$$\therefore \quad I_D L = 2a\sigma L_z \left[V_{DS} - \frac{1}{\sqrt{V_P}} \left\{ \frac{(V_x - V_{GS})^{\frac{3}{2}}}{3/2} \right\}_0^{V_{DS}} \right]$$

$$= 2a\sigma L_z \left[V_{DS} - \frac{2}{3\sqrt{V_P}} (V_{DS} - V_{GS})^{\frac{3}{2}} + \frac{2}{3\sqrt{V_P}} (-V_{GS})^{\frac{3}{2}} \right]$$

$$\therefore \quad I_D = G_0 V_P \left[\frac{V_{DS}}{V_P} + \frac{2}{3} \left(-\frac{V_{GS}}{V_P} \right)^{\frac{3}{2}} - \frac{2}{3} \left(\frac{V_{DS} - V_{GS}}{V_P} \right)^{\frac{3}{2}} \right] \qquad \ldots (5.23)$$

where $G_0 = \dfrac{2a\sigma L_z}{L}$. When $V_{GS} = 0$ we get $I_D \approx G_0 V_{DS}$ for very small values of V_{DS}. Thus, G_0 is the channel conductance with no gate voltage applied and with small drain voltage.

At pinch-off,

$$-V_{GD} = V_P$$

i.e., $\quad -\left(V_{GS} + V_{SD} \right) = V_P$

or, $\quad -\left(V_{GS} - V_{DS} \right) = V_P \qquad$ [since $V_{SD} = -V_{DS}$]

$$\therefore \quad V_P = V_{DS} - V_{GS} \qquad \ldots (5.24)$$

Let $I_D = I_P$, the pinch-off current at pinch-off. From (5.23), we write

$$I_P = G_0 V_P \left[\frac{V_{DS}}{V_P} + \frac{2}{3} \left(\frac{-V_{GS}}{V_P} \right)^{\frac{3}{2}} - \frac{2}{3} \right] \quad \text{using (5.24)}$$

$$= G_0 V_P \left[\frac{V_{GS} + V_P}{V_P} + \frac{2}{3} \left(\frac{-V_{GS}}{V_P} \right)^{\frac{3}{2}} - \frac{2}{3} \right] \quad \text{using (5.24)}$$

$$= G_0 V_P \left[\frac{V_{GS}}{V_P} + \frac{2}{3} \left(-\frac{V_{GS}}{V_P} \right)^{3/2} + \frac{1}{3} \right] \qquad \text{... (5.25)}$$

This is the expression for Pinch-off current of the JFET.

When $V_{GS} = 0$, $I_P = \dfrac{G_0 V_P}{3} = I_{PO}$ (say).

The transconductance of the JFET operated in the saturation region beyond pinch-off is

$$g_m = \frac{\partial I_P}{\partial V_{GS}}$$

$$= G_0 \left[1 - \left(\frac{-V_{GS}}{V_P} \right)^{\frac{1}{2}} \right] \qquad \text{... (5.26)}$$

Thus, g_m is a function of gate voltage, V_{GS}.

Is Ohm's law valid at and beyond Pinch-off?

The channel resistance increases with the increase of drain current. Thus, channel resistance (R) becomes a function of drain current. The current voltage characteristics in this region is hence nonlinear and therefore Ohm's law does not hold good in this region.

When the drain voltage is very small and the depletion layer width is small, the channel is fully open. Now, Ohm's law is valid.

What happens to the channel when the drain voltage is increased beyond Pinch-off?

$I_D = \dfrac{V_{DS}}{R} =$ constant, beyond pinch-off. If V_{DS} is increased, channel resistance R also increases

to keep I_D constant. Increase of channel resistance means reduction of channel cross-section. The minimum-cross-section region of the channel, which appears near the drain end, extends towards the source when drain voltage is increased beyond pinch-off.

5.5.3 SMALL-SIGNAL CIRCUIT MODEL OF THE JFET

The total instantaneous drain current i_D can be expressed as

$$i_D = F(v_{GS}, v_{DS}) \qquad \text{... (5.27)}$$

where v_{GS} = total (dc+ac) instantaneous gate voltage,

v_{DS} = total (dc+ac) instantaneous drain voltage,

and $F(v_{GS}, v_{DS})$ is some function of the variables in parenthesis. Now, if v_{GS} and v_{DS} both are varied, the corresponding variation in i_D can be written as

$$I_D + \Delta i_D = F(V_{GS} + \Delta v_{GS}, V_{DS} + \Delta v_{DS})$$

$$= F\left(V_{GS}, V_{DS}\right) + \frac{\partial F}{\partial v_{GS}}\bigg|_{V_{DS}} \Delta v_{GS} + \frac{\partial F}{\partial v_{DS}}\bigg|_{V_{GS}} \Delta v_{DS} \qquad \dots (5.28)$$

Here, we have made a Taylor series expansion of (5.27) and retained the first order terms. I_D is the operating point drain current and $I_D = F\left(V_{GS}, V_{DS}\right)$. V_{GS} and V_{DS} are operating point values of gate and drain voltage respectively. Δv_{GS} and Δv_{DS} are the small variations in V_{GS} and V_{DS} respectively, from their operating point values. Then, from (5.28) we get

$$\Delta i_d = \frac{\partial i_D}{\partial v_{GS}}\bigg|_{V_{DS}} \Delta v_{GS} + \frac{\partial i_D}{\partial v_{DS}}\bigg|_{V_{GS}} \Delta v_{DS}$$

$$= g_m \Delta v_{GS} + \frac{1}{r_d} \Delta v_{DS} \qquad \dots (5.29)$$

where $g_m = \dfrac{\partial i_D}{\partial v_{GS}}\bigg|_{V_{DS}}$ = transconductance (or mutual conductance) of the JFET and

$\dfrac{1}{r_d} = \dfrac{\partial i_D}{\partial v_{DS}}\bigg|_{V_{GS}}$ where r_d = drain resistance. Replacing Δv_{GS} by the corresponding ac gate voltage v_{gs}, Δv_{DS} by the corresponding ac drain voltage v_{ds}, and Δi_D by the corresponding ac drain current i_d, we get

$$i_d = g_m v_{gs} + \frac{1}{r_d} v_{ds} \qquad \dots (5.30)$$

If v_{gs} and v_{ds} are varied in such a way that i_d remains constant then $\Delta i_D = 0$. Now, from (5.29) we can write

$$g_m \Delta v_{GS} + \frac{1}{r_d} \Delta v_{DS} = 0$$

$$\therefore \qquad -\frac{\Delta v_{DS}}{\Delta v_{GS}}\bigg|_{I_D} = g_m r_d \qquad \dots (5.31)$$

The amplification factor (μ) of the JFET is defined as, $\mu = -\dfrac{\partial v_{DS}}{\partial v_{GS}}\bigg|_{I_D}$.

$$\therefore \qquad \mu = g_m r_d \qquad \dots (5.32)$$

This relation is analogous with that of a vacuum triode. Equation (5.30) can be rewritten as

$$v_{ds} = -\mu v_{gs} + r_d i_d \qquad \dots (5.33)$$

Equation (5.30) indicates the following circuit:

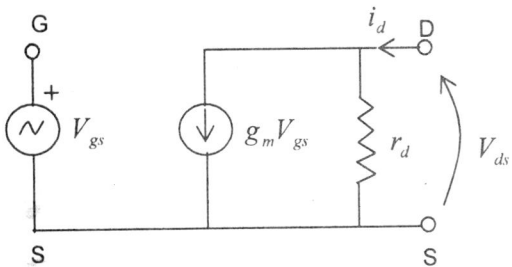

Fig. 5.12. Current–source ac equivalent circuit of JFET.

Equation (5.33) indicates the following equivalent circuit as shown in Fig. 5.13.

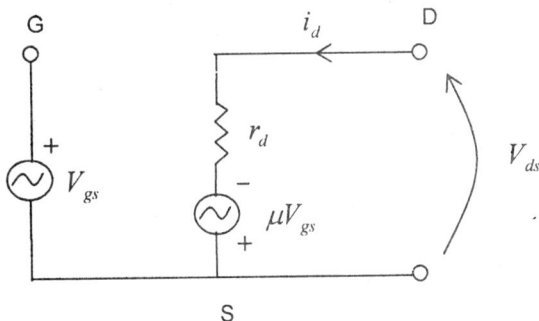

Fig. 5.13. Voltage-source ac equivalent circuit of JFET.

5.5.4 COMPARISON OF BIPOLAR JUNCTION TRANSISTOR (BJT) AND JFET

- BJT is a bipolar device where both electrons and holes take part in conduction. But, JFET is a unipolar device where the majority carriers (either electrons or holes) take part in conduction.

- Input resistance of a JFET is fairly high ~ Megaohms. This is because the control voltage is applied to a reverse–biased p-n junction. Input resistance of a BJT is much lower.

- JFET is less noisy than a BJT since only one type of carrier take part in conduction in JFET and there is no recombination noise in JFET.
- BJT is controlled by current fed to the base terminal whereas JFET is controlled by voltage applied to the gate terminal.

- Fabrication of JFET in integrated circuit is simple.

5.6 METAL-OXIDE-SEMICONDUCTOR FIELD EFFECT TRANSISTOR (MOSFET)

This transistor has also three terminals- Source, Gate and the Drain. An oxide layer (silicon dioxide) is fabricated between the gate and the channel. SiO_2 is a good insulator. So, the gate terminal is insulated from the semiconductor channel and hence MOSFET is also known as Insulated Gate Field effect Transistor (IGFET). In short form, MOSFET is also called as Metal-oxide-semiconductor Transistor (MOST). The metallic gate electrode and the semiconductor channel form two plates of a parallel plate capacitor separated by an insulator (SiO_2) and hence form a capacitor. Due to gate insulation, the input resistance of MOSFET is very high ~ several giga-ohm, giga means 10^9. Depending upon what type of carrier is involved in the conduction of MOSFET, the devices are classified as p-channel or n-channel MOSFET. Again, depending upon whether the channel conductivity increases or decreases by applying a proper voltage to the gate terminal, MOSFETs can be classified into two categories, viz., Enhancement-type MOSFET and Depletion-type MOSFET.

5.6.1 ENHANCEMENT MOSFET

A schematic device structure of the p-channel Enhancement MOSFET is shown in Fig. 5.14. Here, the silicon substrate is n-type. Two heavily-doped p-regions (p^+) form the source and the drain. Metallic contacts are made to these p^+ regions. An insulating thin layer of SiO_2 is deposited over the top surface of the semiconductor while the bottom surface of the substrate is grounded.

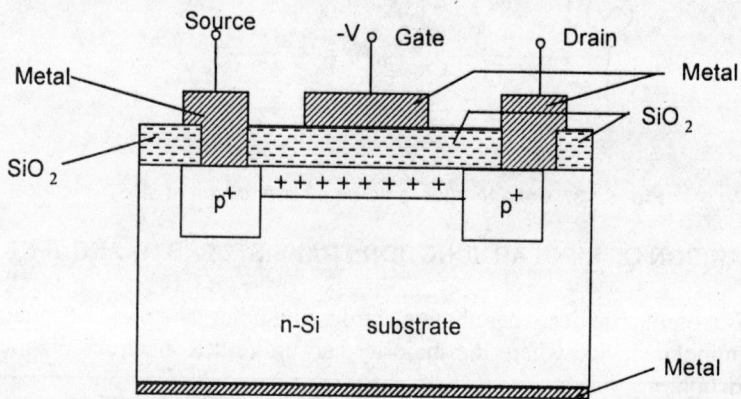

Fig. 5.14. A schematic structure of Enhancement MOSFET.

A metallic electrode is fabricated above the SiO_2 layer over the gate region. The SiO_2 layer also provides isolation between the source and the gate on one hand and the drain and the gate on the other hand.

When a negative voltage is applied to the gate terminal, positive charges are induced in the semiconductor just below the SiO_2 layer. These positive charges form a p-channel between the source and the drain. This is called the inversion layer. Now, if a negative voltage is applied to the drain with respect to the source, current will flow from source to drain due to the flow of holes through the p-channel. The drain current tends towards a saturation with increasing V_{DS}. The higher the magnitude of the negative voltage applied to the gate, the more is the amount of positive charge induced in the channel and the drain current increases with the increase of the magnitude of the negative gate voltage. The I_D vs. V_{DS} characteristics of the enhancement MOSFET is shown in Fig 5.15. A typical transfer characteristics (I_D vs. V_{GS}) is shown in Fig. 5.16. V_{DS} and I_D have typical values in the range $(0, -10\,\text{V})$ and $(0, -10\,\text{mA})$ respectively. V_{GS} has typical values in the range $(0, -5\,\text{V})$.

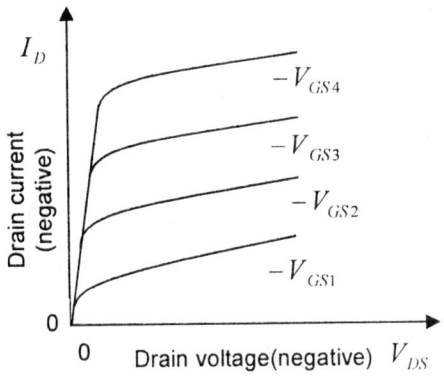

Fig.5.15. Typical drain characteristics of Enhancement MOSFET.

$$(V_{GS4} > V_{GS3} > V_{GS2} > V_{GS1}).$$

Since the drain current leaves the drain terminal in the external circuit, I_D is negative in Fig. 5.15 and 5.16.

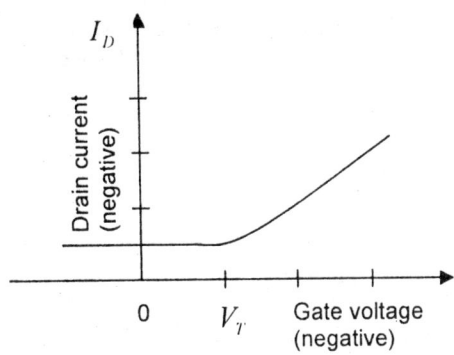

Fig. 5.16. Typical transfer characteristics of Enhancement MOSFET.

V_{DS} = Constant.

There is a threshold value of gate voltage (V_T) below which the channel conductivity is very small and the drain current has a very small value typically $\sim$ nA. When $|V_{GS}| > |V_T|$, I_D increases rapidly. V_T is called the threshold voltage of the enhancement MOSFET.

5.6.2 DEPLETION MOSFET

Let us consider an n-channel MOSFET where heavily-doped n regions on a p-type Si substrate forms the source and the drain. A schematic structure of an n-channel MOSFET is shown in Fig. 5.17.

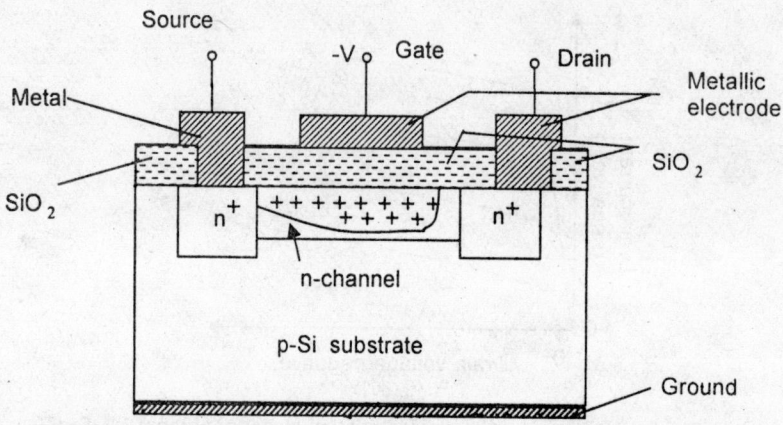

Fig. 5.17. Schematic diagram of an n-channel MOSFET structure.

The semiconductor between the source and the drain is lightly doped by donor impurity which forms the n-channel. When no gate voltage is applied and a positive voltage is applied to the drain with respect to the source, electrons will flow from the source to the drain through the n-channel and considerable drain current flows. This is because the n-channel has some conductivity due to dopping. Now, if a negative gate voltage is applied, positive charges are produced in the channel by induction. This, in turn, reduces the conductivity of the channel by charge neutralization. This results in a reduction of drain current. Thus, the drain current decreases with increasing magnitude of negative gate voltage. Since carrier density in the n-channel gets reduced or depleted in this mode of operation, the device is called a depletion MOSFET. Again, since the ohmic voltage drop near the drain end is maximum, the effective gate voltage near the drain end is also maximum. Hence, the induced positive charge in the channel near the drain end is maximum and the channel depletion is also greater near the drain end. On the other hand, if a positive gate voltage is applied, negative charges are induced in the n-channel. This increases the conductivity of the n-channel and so the drain current increases with increasing positive gate voltage. This mode of operation is similar to the enhancement mode of the MOSFET is shown in Fig. 5.18, while the transfer characteristics is shown in Fig. 5.19. The symbols of MOSFET are shown in Figs. 5.20 (a), (b) and (c).

In the n-channel depletion MOSFET, the drain current enters the drain terminal. So, I_D is positive. Again, since the current carriers are electrons, the drain voltage (V_{DS}) is positive. Typical values of V_{DS} and I_D lie in the range (0, 10 V) and (0, 10 mA). V_{GS} have typical values in the range (−5 V, 5 V).

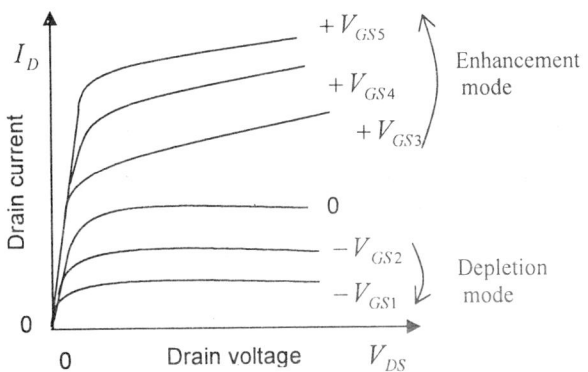

Fig. 5.18. Drain characteristics of an n-channel MOSFET having both depletion and enhancement region.

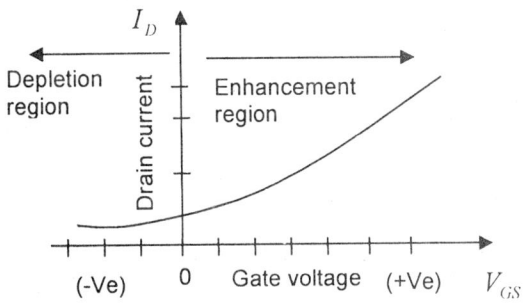

Fig. 5.19. Transfer characteristics of n-channel MOSFET operated in depletion or enhancement mode.

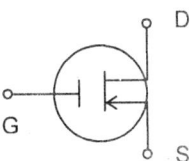

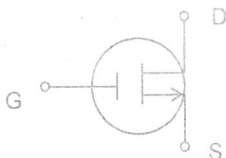

Fig. 5.20 (a) Symbol of p-channel MOSFET (enhancement or depletion type) S→Source, D→Drain, G→Gate.

Fig. 5.20 (b) Symbol of n-channel MOSFET (enhancement or depletion type) S→Source, D→Drain, G→Gate.

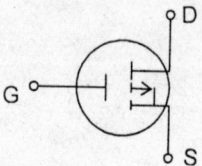

Fig. 5.20 (c) Symbol of enhancement MOSFET.
S→Source, D→Drain, G→Gate.

The source is usally connected with the substrate.

5.6.3 P-CHANNEL AND N-CHANNEL MOSFET: A COMPARISON

The fabrication of p-channel enhancement MOSFET is easier than n-channel MOSFET. But, p-channel devices often get positively charged contaminants in SiO_2 layer during the fabrication process. These positively charged contaminants gather in the gate-SiO_2 interface because the channel contains positive charges and the gate voltage is negative. So, the channel is not affected by the contaminant. On the other hand, in the n-channel enhancement device the positively charged contaminants gather in the SiO_2 and n-channel interface. This is because the gate voltage is positive and the charge induced in the channel is negative. These contaminant positive charges attract electrons from the n-channel and affects the conductivity of the channel.

The hole mobility is much smaller than the electron mobility in Si. So, p-channel MOSFETs will have higher ON resistance than similar n-channel MOSFETs. In order to get same resistance, the p-channel MOSFET must have wider cross-sectional area. Hence, the n-channel MOSFETs are more compact leading to higher packing density in an IC. Again, since the device capacitance is directly proportional to the junction area, the RC time constant of the p-channel device is higher than the n-channel MOSFET. So, the n-channel MOSFET will be faster than the p-channel MOSFET so far as the speed of operation is concerned. Due to the same reason, n-channel MOSFETs can be operated at higher frequencies than the p-channel MOSFET.

MOSFETs are used in the design of logic gates, registers and memory arrays. But, due to internal capacitances the speed of MOSFET device is smaller than the BJT device. The MOSFETs, however, consume much less power than BJTs. The MOSFETs have also higher packing density in ICs. Digital gates such as NAND or NOR gates, can be designed with the help of MOSFETs only without requiring the fabrication of other components like diodes, resistors, capacitors etc..The drain characteristics below pinch-off is fairly nonlinear and the drain resistance varies with the gate voltage. The MOSFET is used as a voltage-variable resistor (VVR) or voltage-dependent resistor (VDR). This VVR is used for automatic gain control (AGC) in multistage amplifier circuit.

5.7 SEMICONDUCTOR-CONTROLLED RECTIFIER (SCR)

SCR is a four layer pnpn device. It has 3 pn junctions and 3 terminals known as Anode (A), Gate (G), and Cathode (K). Since, the device is fabricated on a silicon substrate, it is also known as silicon-controlled rectifier (SCR). The device conducts in one direction when the anode is positive with respect to the cathode and a proper gate bias is applied. The device does not conduct in the reverse direction unless it breaks down by the application of a large reverse voltage exceeding the breakdown voltage. The family of semiconductor devices which are used for switching and control of power is termed as 'Thyristor'. SCR is member of this Thyristor group. The block diagram and symbol of an SCR is shown in Figs. 5.21 (a) and 5.21 (b) respectively.

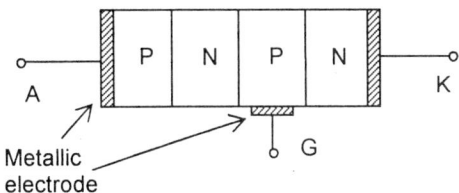
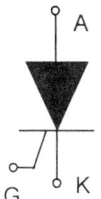

Fig. 5.21(a) Schematic block diagram of an SCR   **Fig. 5.21(b)** Symbol of an SCR.
                 A→Anode, K→Cathode, G→Gate.

. The device has two states – one conducting or ON state and the other blocking or OFF state. It can handle large current in the ON state and can withstand high voltage in the OFF state. So, it finds aplication as a power switch in power circuits. The device has a low ON resistance. So, the voltage drop across the device is small in the conducting state. Typically current-voltage characteristic of an SCR is shown in Fig. 5.22.

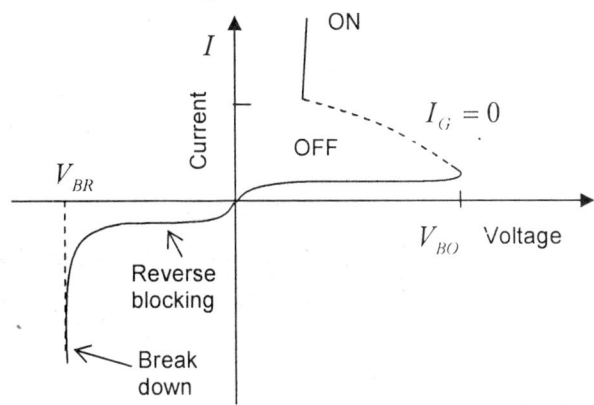

Fig. 5.22. Typical current-voltage characteristics of an SCR.
V_{BO} = Break-over voltage; V_{BR} = Break-down voltage.

To explain the operation of the SCR, we can think the SCR consisting of two interconnected transistors as shown in Fig. 5.23. The p-layer connected with the anode and the n-layer connected with the cathode act as the emitters of two transistors. The inner two layers, n and p, from the left side of Fig. 5.21 (a) are common to both transistors.

Q_2 and Q_1 are two transistors in Fig. 5.23. The base of Q_2 is connected with the collector of Q_1. Similarly, the collector of Q_2 is connected with the base of Q_1. The three junctions J_1, J_2 and J_3 are shown in Fig. 5.23. If a positive voltage is applied to the anode (A) with respect to the cathode (K), J_1 and J_3 junctions are forward biased while J_2 junction is reverse biased. Since,

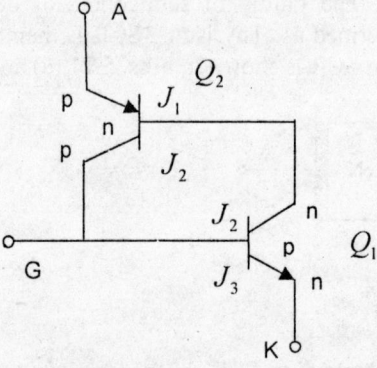

Fig. 5.23. Interconnected transistor model of SCR.

J_1 junction is forward biased and J_2 junction is reverse biased, the transistor Q_2 is in the active region. Similarly, the transistor Q_1 is also biased in the active region. Now, unless a bias current is supplied to Q_1, the device remains nonconducting. Let us apply a current pulse to the base (G) of transistor Q_1. Q_1 conducts and its collector current flows. This collector current supplies the base current in Q_2. So, Q_2 begins to conduct and the resulting collector current in Q_2 supplies the large base current in Q_1. Eventually, both Q_1 and Q_2 becomes saturated. Operation in the saturation region makes all the three junctions J_1, J_2 and J_3 forward biased. Typical voltage drop in forward-biased J_1 and J_3 junctions are 0.7 volt each. The voltage drop in the J_2 junction is 0.2 V (approximately) which is of opposite sign to those developed across J_1 and J_3. The net voltage drop across the SCR in the ON state is typically $(0.7 + 0.7 - 0.2) = 1.2$ volt. In general, this voltage drop lies between 1 and 2 volt depending upon the device current. There is a value of forward voltage, V_{BO}, known as break-over voltage such that if a voltage greater than V_{BO} is applied, the SCR is switched ON by the increased leakage current without the application of any gate current. When a negative voltage is applied to the anode, the pn junctions J_1 and J_3 are reverse biased and the tranistors Q_1 and Q_2 are cut-off. When this reverse voltage attains a critical value, V_{BR}, the device breaks down due to avalanching and the SCR will be damaged. When the SCR current falls below the holding current, Q_1 and Q_2 come out of saturation and enter into active region. If the SCR current is further reduced, the transistors Q_2 and Q_1 are cut-off by a cumulative collector and base current reduction. The SCR is then cut-off.

The SCR is used in dc to ac inverter circuits and also in high power rectifiers. It is used as a high power switch. It is used in controlling the speed of electrical motors and heat of an electrical furnace. Light emission from bulbs can be controlled by using SCR. The SCR can be used in over-voltage protection circuit in which useful equipments are procted from the effect of over-voltages.

5.8 DIAC

It is a three layer pnp device. It has two pn junctions. It has two terminals viz., anode and cathode. In this regard, this device is anlogous with a diode. Although it has two pn junctions, it is quite different from a bipolar transistor. This is because it has no base terminal. It can conduct in both directions and act as a bidirectional switch. The two pn junctions form a pair of avalanche diodes. When a positive voltage is applied to the anode, the pn junction connected with the anode is forward biased but the other np junction is reverse biased. So, the device is nonconducting. But, when the applied voltage is high enough to provide sufficient reverse bias to the np diode, the np diode breaks down due to avalanching. A large current flows through the diac and the diac is ON. In the ON state, the diac shows a negative resistance because the current increases with decreasing voltage. When a negative voltage is applied to the anode with respect to the cathode, the same break down takes place. The block diagram, symbol and I-V characteristics of the diac is shown in Figs. 5.24 (a), 5.24 (b) and 5.25 respectively.

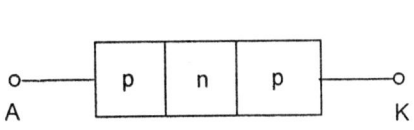

Fig. 5.24(a) block diagram of a diac.

Fig. 5.24(b) Symbol of diac.

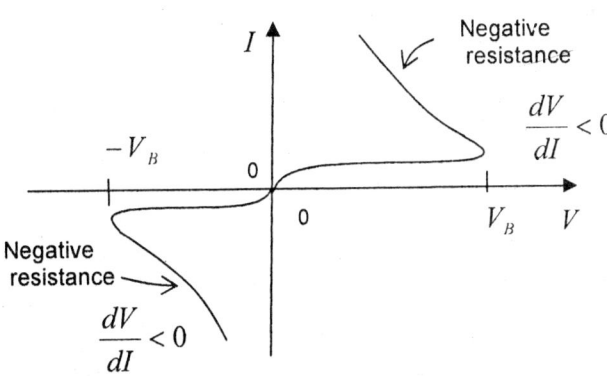

Fig. 5.25. Typical $I - V$ characteristics of a diac.

V_B is the break down voltage of the diac.

The diac can be used to trigger a triac by supplying a gate pulse current. The firing angle of the input ac voltage can be controlled by a diac in a phase-controlled circuit.

5.9 TRIAC

It is a bilateral switching device having 3 terminals. It can conduct in both directions. It is equivalent to two semiconductor controlled rectifiers (SCR) connected in antiparallel configuration. A cross-sectional view of the triac is shown in Fig. 5.26 (a) while the symbol of triac is shown in fig. 5.26 (b). The current-voltage characteristics of the triac is shown in Fig. 5.27.

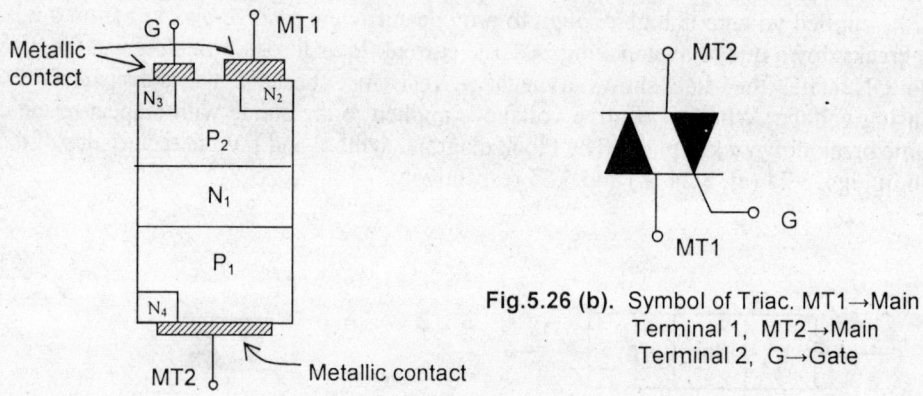

Fig.5.26 (b). Symbol of Triac. MT1→Main Terminal 1, MT2→Main Terminal 2, G→Gate

Fig.5.26 (a). Cross-sectional view of Triac structure.

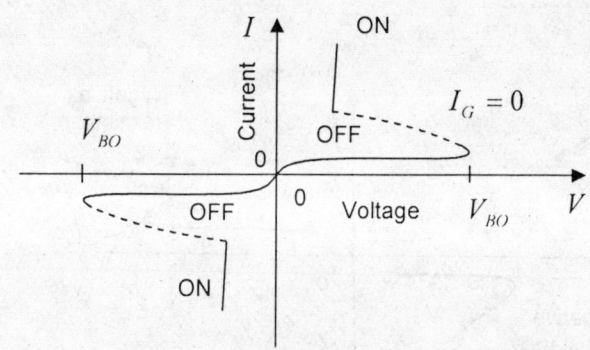

Fig.5.27. Typical current-voltage characteristics of triac.
V_{BO} →Breakover voltage, I_G →gate current.

Since the triac conducts in both direction, the two terminals are designated as main terminal MT1 and main terminal MT2. The third terminal called as gate (G) is located near the main terminal MT1.

Let us suppose that no current is applied to the gate terminal. Now, if a voltage is applied to the terminal MT1 (or MT2) with respect to MT2 (or MT1) whose magnitude is less than the breakover voltage (V_{BO}), the triac does not conduct. This is the OFF state of the triac. If the applied voltage exceeds the breakover voltage (V_{BO}), the triac conducts. This is ON state of triac.

Applying a positive or a negative voltage to the gate terminal, the triac can be triggered and the firing of the triac (i.e., transition from OFF state to ON state) can be controlled. An SCR conducts only in forward direction but the triac conducts in both directions.

Let a positive voltage be applied to MT2 terminal with respect to MT1. Also, let a positive voltage be applied to the gate G with respect to MT1. Now, $P_1N_1P_2N_2$ with a positive gate voltage at P_2 behaves like a SCR which is triggered by the gate voltage to the ON state. On the other hand, if a negative gate voltage is applied relative to MT1 then the SCR $P_1N_1P_2N_3$ conducts since P_2N_3 junction becomes forward biased. The conduction of P_2N_3 junction produces a current flowing in the P_2 region. Now, the $P_1N_1P_2N_2$ SCR becomes ON with a gate current flowing into the P_2 layer.

Now, let a negative voltage be applied to MT2 with respect to MT1. The SCR $P_2N_1P_1N_4$ is nonconducting. If a positive gate voltage is applied with respect to MT1, a current will flow from the gate into the P_2 region. Now, consider the operation of the transistor $N_2P_2N_1$. The collector junction N_2P_2 is reverse biased whereas the emitter base junction N_1P_2 is forward biased and there is a base current in P_2. So, an emitter current will flow through the layer N_1. Then, the SCR $P_2N_1P_1N_4$ with a current flowing through the N_1 layer becomes ON.

Let us consider the case when the gate voltage is negative with respect to MT1 terminal and MT2 is negative with respect to MT1. Consider the operation of the transistor $N_3P_2N_1$. Here, the emitter-base junction N_1P_2 is reverse-biased. Now, with MT1 positive with respect to gate (G) a current flows into P_2 layer which acts as the base current. So, the transistor $N_3P_2N_1$ conducts and an emitter current flows through N_1 layer. The SCR $P_2N_1P_1N_4$ with a current flowing through N_1 turns ON. This is mentioned in section 5.7.

SOLVED PROBLEMS

5.1 A transistor in an amplifier circuit has a collector current which is 99% of the emitter current. If the emitter current is 10 mA, how much is the base current? For operation in the CB mode, find the value of α_{dc}. Calculate the value of β_{dc} of this transistor when operated in CE mode.

Solution

Let I_C = collector current and I_E = emitter current.

Now, $I_C = 0.99 I_E$, where $I_E = 10$ mA

$\therefore$ $I_C = 9.9$ mA

Now, $I_E = I_B + I_C$ (for both pnp and npn BJT)

$\therefore$ $I_B = I_E - I_C$

$\qquad = 10 - 9.9 = 0.1$ mA.

$$\alpha_{dc} = \frac{I_C}{I_E} = \frac{9.9}{10} = 0.99 .$$

$$\beta_{dc} = \frac{\alpha_{dc}}{1 - \alpha_{dc}} \qquad \text{[from equation (5.9)]}$$

$$= \frac{0.99}{1 - 0.99} = 99.$$

5.2 In a Ge transistor used in a CB amplifier circuit, 99.6% of the emitter-injected carriers reach the collector. The current flowing out from the collector is 10 mA and the leakage current is 2μA. Find the value of α_{dc} and large signal α. What type of transistor it is? Find the leakage current in CE mode of operation.

Solution

$$I_C = 99.6\% I_E$$

$$= 0.996 I_E$$

$$\alpha_{dc} = \frac{I_C}{I_E} = 0.996$$

Large signal $\alpha = -\dfrac{I_C - I_{CBO}}{I_E}$ (I_C and I_E are in opposite directions)

$$= 0.996 - \frac{2 \times 10^{-6}}{10^{-2}}$$

$$= 0.996 - 0.0002 = 0.9958.$$

Since I_C flows out of the collector terminal, it is a pnp transistor.

$$I_B = I_E - I_C = I_C \left(\frac{1 - \alpha_{dc}}{\alpha_{dc}} \right) = 10 \times \frac{0.004}{0.996} = 0.0401 \, \text{mA}$$

$$\beta = \frac{I_C - I_{CBO}}{I_B + I_{CBO}} = \frac{10 - 0.002}{0.040 + 0.002} \approx 238$$

Leak current in CE mode, $I_{CEO} = (1 + \beta) I_{CBO}$

$$= 478 \, \mu\text{A}.$$

5.3 An n-channel Si JFET has metallurgical full width of the channel equal to 2μm and a doping concentration of 10^{16}/c.c. of the n-channel. Calculate the pinch-off voltage of the JFET. Given relative permittivity of Si is 12.

Solution

Full width of the channel, $2a = 2 \, \mu$m

$$\therefore \qquad a = 1 \mu\text{m} = 10^{-6} \, \text{m}$$

$$N_d = 10^{16}/c.c. = 10^{22}/m^3$$

$$\varepsilon_r = 12$$

Absolute permittivity $\varepsilon = \varepsilon_r \varepsilon_0 = 12 \times \dfrac{10^{-9}}{36\pi}$ F/m

$$q = 1.6 \times 10^{-19} \text{ Coulomb}$$

From equation (5.15), we get

$$V_P = \frac{qN_d}{2\varepsilon} a^2$$

$$= \frac{1.6 \times 10^{-19} \times 10^{22} \times \left(10^{-6}\right)^2}{2 \times 12 \times \dfrac{10^{-9}}{36\pi}}$$

$$= 2.4\pi = 7.54 \text{ V}.$$

5.4   An n-channel Si JFET has a pinch-off current of 10 mA with the gate shorted to source. Find the value of pinch-off current when a gate voltage of -3 V is applied. Given the pinch-off voltage of the JFET is 5 volt.

Solution
From equation (5.25), we have

$$I_P = G_0 V_P \left[\frac{V_{GS}}{V_P} + \frac{2}{3} \left(\frac{-V_{GS}}{V_P} \right)^{3/2} + \frac{1}{3} \right]$$

$$I_{Po} = \frac{G_0 V_P}{3} \quad \text{where } I_{PO} = \text{Pinch-off current with zero gate voltage.}$$

$$I_P = 3 I_{Po} \left[\frac{V_{GS}}{V_P} + \frac{2}{3} \left(\frac{-V_{GS}}{V_P} \right)^{3/2} + \frac{1}{3} \right]$$

Here, $I_{Po} = 10\,mA, V_{GS} = -3V, V_P = 5V$.

$$I_P = 30 \left[\frac{-3}{5} + \frac{2}{3} \left(\frac{3}{5} \right)^{3/2} + \frac{1}{3} \right]$$

$$= 30 \times 0.0431 = 1.295 \text{ mA}.$$

5.5 An n-channel Si JFET has a pinch-off current of 12 mA when no gate voltage is applied. The pinch-off voltage of the JFET is 6 volt. Calculate the transconductance of the JFET at a gate voltage at -3 volt when the JFET is operated in the saturation region.

Solution

From equation (5.26), we write

$$g_m = G_0 \left[1 - \left(\frac{-V_{GS}}{V_P} \right)^{1/2} \right].$$

Again, $I_{Po} = 12$ mA, $V_P = 6$ V, $V_{GS} = -3$ V

$$\therefore \quad G_0 = \frac{3 \times 12 \times 10^{-3}}{6} = 6 \times 10^{-3}$$

$$\therefore \quad g_m = 6 \times 10^{-3} \left[1 - (3/6)^{\frac{1}{2}} \right]$$

$$= 6 \times 10^{-3} \left[1 - \frac{1}{\sqrt{2}} \right] = 1.757 \text{ mmho.}$$

PROBLEMS

5.1 A Si transistor has $\beta = 100$ and $I_{CBO} = 0.4\,\mu$A. A dc current of $50\,\mu$A enters the base of this transistor. How much is the dc collector current? What type of BJT is this?

[**Ans.** 5.04 mA, npn]

Hints: Use equation (5.10).

5.2 What will be the percentage of error in the calculation of collector current in question 5.1 if the leakage current is neglected?

[**Ans.** 0.8%]

Hints: Neglecting I_{CBO}, $I_C = \beta I_B$. Find error in I_C. Then, find percentage error.

5.3 An n-channel JFET made from Si has a pinch-off voltage of 6 volt. The doping concentration of the n-channel is 10^{15}/c.c. Calculate the metallurgical half-width of the channel. Given the relative permittivity of Si is 12.

[**Ans.** 2.82 μm]

Hints: From equation (5.14), $a = \left[\dfrac{2\varepsilon V_P}{qN_d} \right]^{\frac{1}{2}}$

5.4 An n-channel Si JFET has values of pinch-off currents of 10 mA with no gate voltage and 4 mA with a gate voltage of − 2 volts respectively. Find the pinch-off voltage of the JFET.

[**Ans.** 6.2 V]

Hints: Use equation (5.25) and get V_p.

5.5 An n-channel Si JFET has a pinch-off current 8 mA with gate shorted to source. The JFET has a pinch-off voltage of 5 volt. Calculate the gate voltage required for the JFET to exhibit a transconductance of 2 mmho when operated in the saturation region.

[**Ans.** − 1.7 V]

Hints: Use equation (5.26) and get

$$V_{GS} = -V_p \left[1 - \frac{g_m}{G_0} \right]^2 \text{ where } G_0 = \frac{3I_{P0}}{V_p}.$$

QUESTIONS

5.1 What do you mean by the term 'Transistor'? What are different types of transistors? What are different modes of operation of a transistor? Define the following terms: (i) DC alpha (ii) small signal α and (iii) large signal α.

5.2 What do you mean by (i) DC beta (ii) small signal β and (iii) large signal β of a transistor. Derive a relation between α_{dc} and β_{dc} of a transistor. Find an expression for the leakage current of a transistor operating in CE mode.

5.3 Draw a circuit for measuring I-V characteristics of a pnp transistor in CB mode. Show schematic variations of current with voltage in input and output characteristics in this mode. Explain the nature of variations with reasons.

5.4 Explain, using necessary measurement circuit diagram, the variations of current with voltage in the input and output characteristics of a pnp transistor in CE mode. What is a dc load line? How do you define the operating point of a CE amplifier?

5.5 What is a Junction field effect transistor? Why is the name 'field effect'? Describe with the help of a neat diagram the structure of an n-channel JFET.

5.6 Explain the operation of a JFET. Draw schematically the I-V characteristics of a JFET. What do you mean by Pinch-off?

5.7 Derive an expression for the Pinch-off voltage of an n-channel JFET having a metallurgical half-width of the channel ' a ' and a doping concentration N_d.

5.8 Derive an expression for pinch-off current of an n-channel JFET in terms of channel conductance, pinch-off voltage and gate voltage. Find an expression for transconductance of the JFET in the saturation region. Is Ohm's law valid beyond pinch-off? What happens to the channel when drain voltage is increased beyond pinch-off?

5.9 Establish a current source and a voltage source small-signal equivalent circuit of a JFET. Distinguish between a BJT and a JFET.

5.10 What is a MOSFET? Why is it called a IGFET? What are the different classes of MOSFET? Explain with a schematic structural diagram the operation of an enhancement MOSFET.

5.11 What is a depletion MOSFET? Give the principle of operation of the depletion MOSFET. Make a comparison of a p-channel and an n-channel MOSFET. Mention some applications of the MOSFET.

5.12 What is meant by the term 'SCR'? Describe with the help of a block diagram the structure of an 'SCR'. Explain the switching action of an 'SCR'. Write down some applications of an 'SCR'.

5.13 What is a Diac? Explain the bi-directional switching action of a Diac.

5.14 What is a Triac? Give the structural details of a Triac using a block diagram. Write down the I-V characteristics of a Triac and explain its operation.

Small-Signal Single Stage Bipolar Junction Transistor Amplifier

6.1 TRANSISTOR BIASING

Application of a proper base current in a transistor is called biasing. More specifically, transistor biasing means to establish a suitable operating point in a transistor. The operating point is defined by collector voltage, collector current and base current in the CE mode of operation of the transistor.

The operating point can shift due to the change of temperature. Again, replacement of a bipolar junction transistor (BJT) by one of the same type can lead to a change in the operating point. This is known as production spread. It is desired that the operating point should be stable. For amplifier operation, the operating point should be confined within the linear region of the transistor characteristics. Any improper operation can lead to an overheating of the transistor resulting in a damage of the transistor.

Change in temperature can affect the transistor cut-off current I_{CBO}, base emitter voltage V_{BE} and dc current gain h_{FE} in CE mode. Under proper biasing, these adverse effects can be minimized.

Transistor biasing can be classified into two heads:

(i) Fixed bias, and (ii) Self bias.

6.1.1 FIXED BIAS

Firstly, we consider a fixed bias circuit. This is shown in Fig.6.1.

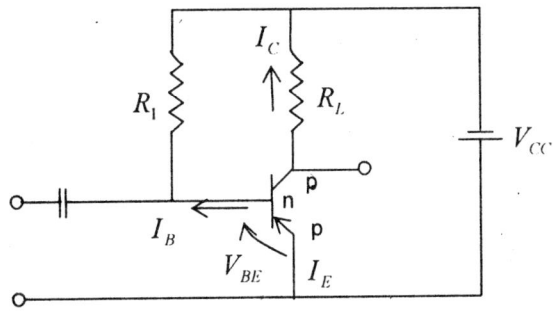

Fig. 6.1. Fixed bias circuit

143

From the circuit in Fig. 6.1, we can write

$$I_E = I_B + I_C \qquad \qquad \text{... (6.1)}$$

$$I_C = \alpha I_E + I_{CBO} \qquad \qquad \text{... (6.2)}$$

Substituting (6.1) in (6.2) we get

$$I_C = \alpha(I_B + I_C) + I_{CBO}$$

or, $\qquad (1 - \alpha)I_C = \alpha I_B + I_{CBO}$

$$\therefore \qquad I_C = \frac{\alpha}{1 - \alpha} I_B + \frac{I_{CBO}}{1 - \alpha}$$

$$= h_{FE} I_B + (1 + h_{FE}) I_{CBO} \qquad \qquad \text{... (6.3)}$$

We define stability factors S, S_h and S_v as

$$S = \frac{\partial I_C}{\partial I_{CBO}}, \; S_h = \frac{\partial I_C}{\partial h_{FE}}, \; S_v = \frac{\partial I_C}{\partial V_{BE}}$$

Large value of stability factor means more unstable circuit. Thus, actually S, S_h and S_v are instability factors. From (6.3) we can get

$$\frac{\partial I_C}{\partial I_{CBO}} = 1 + h_{FE}$$

$$\therefore \qquad S = 1 + h_{FE} \qquad \qquad \text{... (6.4)}$$

Since h_{FE}, the dc current gain in CE mode, is large, S is also large. So, this circuit has poor temperature stability.

Again, taking $\dfrac{\partial}{\partial h_{FE}}$ of (6.3) we get

$$\frac{\partial I_C}{\partial h_{FE}} = I_B + I_{CBO} \qquad \qquad \text{... (6.5)}$$

$$\therefore \qquad S_h = I_B + I_{CBO} \qquad \qquad \text{... (6.6)}$$

From (6.3), we write

$$I_B = \frac{I_C - (1 + h_{FE}) I_{CBO}}{h_{FE}}$$

$$\therefore \qquad I_B + I_{CBO} = \frac{I_C - I_{CBO}}{h_{FE}}$$

Replacing the partial derivatives of (6.5) by small changes of the corresponding variables, we get

$$\frac{\Delta I_C}{\Delta h_{FE}} = \frac{I_C - I_{CBO}}{h_{FE}} \approx \frac{I_C}{h_{FE}} \qquad \text{since } I_{CBO} << I_C$$

$$\therefore \qquad \frac{\Delta I_C}{I_C} = \frac{\Delta h_{FE}}{h_{FE}} \qquad\qquad\qquad \dots (6.7)$$

In practice, Δh_{FE} can range from +100% to –50%. So, $\Delta h_{FE}/h_{FE}$ is large. The circuit has, therefore, poor stability.

The input circuit equation can be written as

$$V_{BE} = V_{CC} + I_B R_1 \qquad\qquad\qquad \dots (6.8)$$

$$\therefore \qquad I_B = \frac{V_{BE} - V_{CC}}{R_1}$$

Here, V_{cc} is inherently negative, that is, $V_{cc} = -|V_{cc}|$.

Since $|V_{BE}| << |V_{cc}|$, $I_B \approx \dfrac{-V_{CC}}{R_1}$. Thus, I_B is fixed by V_{CC} and R_1. Hence, the name fixed bias circuit.

From (6.8) and (6.3) we have

$$V_{BE} - V_{CC} = R_1 \frac{I_C - (1 + h_{FE}) I_{CBO}}{h_{FE}} \qquad\qquad \dots (6.9)$$

Taking $\dfrac{\partial}{\partial V_{BE}}$ of (6.9) we get

$$1 = \frac{R_1}{h_{FE}} S_V \qquad \therefore S_V = \frac{h_{FE}}{R_1} \qquad\qquad \dots (6.10)$$

S_V has a large value compared with self-bias circuit. From this view point, this circuit has poor stability.

6.1.2 SELF BIAS

We now consider a self-bias circuit as shown in Fig.6.2(a). This self bias circuit employs a potential divider for providing a base current and a $R_E - C_E$ emitter bias for automatic control of the bias point. R_1, R_2 forms a voltage dividing network. The supply voltage V_{CC} appears across R_1, R_2 series combination. Applying Thevenin's theorem across R_1 we replace the R_1, R_2 biasing network by a voltage source V_B in series with a resistance $R_p = R_1 \| R_2$. This is shown in Fig.6.2(b). Here,

$$V_B = \frac{V_{cc}}{R_1 + R_2} R_1.$$

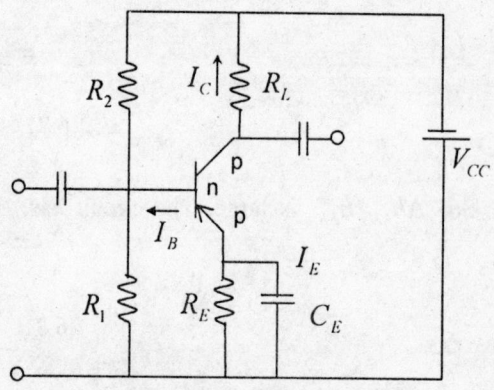

Fig. 6.2. (a) Self-bias circuit.

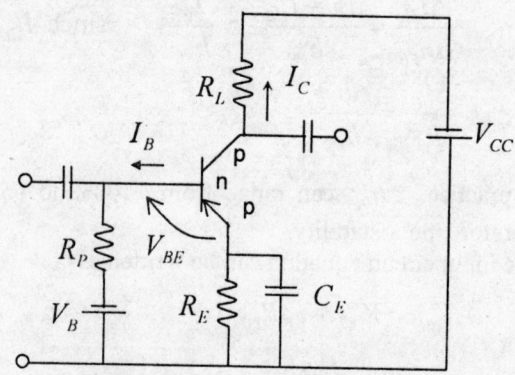

Fig. 6.2.(b) Thevenin's equivalent circuit of Fig. 6.2(a).

R_E provides negative current feedback. If I_C increases in anyway, the potential drop across R_E increases. So, the effective input voltage V_{BE} decreases. I_C is brought back to its initial value. The current equations can be written as

$$I_E = I_B + I_C \qquad \qquad \dots (6.11)$$

and $\quad I_C = h_{FE} I_B + (1 + h_{FE}) I_{CBO} \qquad \qquad \dots (6.12)$

The base-emitter equation from Fig. (6.3) can be written as

$$V_B = -I_E R_E + V_{BE} - I_B R_P$$

or, $\quad V_B - V_{BE} = -(I_C + I_B) R_E - I_B R_P$

$$= -I_C R_E - (R_E + R_P) I_B$$

$$= -I_C R_E - \frac{(R_E + R_P)}{h_{FE}} \left[I_C - (1 + h_{FE}) I_{CBO} \right] \quad \text{substituting for } I_B \text{ from (6.12)}$$

$$= -I_C \left[R_E + \frac{R_E + R_P}{h_{FE}} \right] + \frac{(R_E + R_P)}{h_{FE}} (1 + h_{FE}) I_{CBO} \qquad \dots (6.13)$$

Taking $\dfrac{\partial}{\partial I_{CBO}}$ of equation (6.13) we get

$$0 = -S \left[R_E + \frac{R_E + R_P}{h_{FE}} \right] + \frac{(R_E + R_P)}{h_{FE}} (1 + h_{FE})$$

$\therefore \quad S = \dfrac{(R_E + R_P)}{h_{FE}} (1 + h_{FE}) \cdot \dfrac{h_{FE}}{R_E (1 + h_{FE}) + R_P}$

$$= \frac{(R_E + R_P)(1 + h_{FE})}{R_E (1 + h_{FE}) + R_P} \qquad \qquad \dots (6.14)$$

If $R_E \to 0$, $S \to (1 + h_{FE})$ which is the same as in fixed bias case. If $R_P \to 0$, $S \to 1$. This is the deal case. But, if R_P is very low, it will shunt the input signal. High values of R_E leads to dissipation of much dc power from V_{CC}.

Taking $\dfrac{\partial}{\partial h_{FE}}$ of equation (6.13), we get

$$0 = -S_h[R_E + \frac{R_E + R_P}{h_{FE}}] + (I_C - I_{CBO})\frac{(R_E + R_P)}{h_{FE}^2}$$

$$\therefore \quad S_h = \frac{I_C - I_{CBO}}{R_E(1 + h_{FE}) + R_P}\frac{R_E + R_P}{h_{FE}} \cong \frac{R_E + R_P}{R_E(1 + h_{FE}) + R_P}\frac{I_C}{h_{FE}}, \text{ since } I_C \gg I_{CBO.}$$

$$\therefore \quad \frac{\Delta I_C}{\Delta h_{FE}} = \frac{I_C}{h_{FE}}\frac{R_E + R_P}{R_E(1 + h_{FE}) + R_P}, \text{ replacing } S_h \text{ by small increments.}$$

$$\therefore \quad \frac{\Delta I_C}{I_C} = \frac{\Delta h_{FE}}{h_{FE}}\frac{R_E + R_P}{(R_E + R_P) + h_{FE}R_E} < \frac{\Delta h_{FE}}{h_{FE}}.$$

Hence the fractional change in collector current due to production spread is less than that in the fixed bias case. The stability is thus improved in this self bias case.

Taking $\dfrac{\partial}{\partial V_{BE}}$ of equation (6.13), we get

$$-1 = -S_V[R_E + \frac{R_E + R_P}{h_{FE}}]$$

$$\therefore \quad S_V = \frac{h_{FE}}{R_E(1 + h_{FE}) + R_P}.$$

6.1.3 COLLECTOR BIAS CIRCUIT

Here, a resistance R_F connects the collector to the base and provides the base current I_B. The

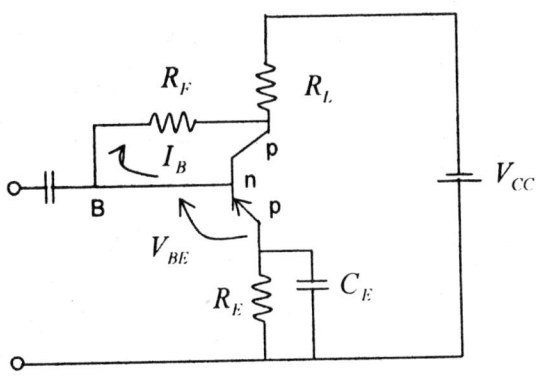

Fig. 6.3. Collector bias circuit.

circuit is shown in Fig. 6.3. Let I_C = collector current, I_E = emitter current and I_B = base current. All these currents are dc.

The circuit equation can be written as

$$V_{CC} = -I_E R_E + V_{BE} - I_B R_F - (I_B + I_C) R_L \qquad \ldots (6.15)$$

Here, V_{CC} and V_{BE} are themselves negative. Now, $I_E = I_B + I_C$ and $I_B = \dfrac{I_C - (1 + h_{FE}) I_{CBO}}{h_{FE}}$.

Putting the expressions for I_E and I_B in terms of I_C in (6.15) we get

$$I_C = \frac{(V_{BE} - V_{CC}) h_{FE}}{R_F + (R_E + R_L)(1 + h_{FE})} + \frac{I_{CBO}(1 + h_{FE})(R_E + R_L + R_F)}{R_F + (R_E + R_L)(1 + h_{FE})} \qquad \ldots (6.16)$$

The stability factor

$$S = \frac{\partial I_C}{\partial I_{CBO}} = \frac{(1 + h_{FE})(R_E + R_L + R_F)}{R_F + (R_E + R_L)(1 + h_{FE})}$$

Depending upon the value of R_F, S can have limiting values. When $R_F \gg (R_E + R_L)(1 + h_{FE})$, $S = (1 + h_{FE})$ which is the same as in the fixed bias circuit. On the other hand, if $R_F \ll (R_E + R_L)(1 + h_{FE})$ then $S = 1 + \dfrac{R_F}{R_E + R_L}$. However, R_F cannot be very small since smaller R_F means higher base current. Too small R_F will push the BJT to operate in the saturation region.

To find $S_h = \dfrac{\partial I_C}{\partial h_{FE}}$, we take the derivative of equation (6.15) with respect to h_{FE} and substitute for $(V_{BE} - V_{CC})$ in terms of I_C and then simplifying we get

$$S_h = \frac{(R_F + R_E + R_L)}{R_F + (R_E + R_L)(1 + h_{FE})} \cdot \frac{I_C - I_{CBO}}{h_{FE}}$$

$$\approx \frac{(R_F + R_E + R_L)}{R_F + (R_E + R_L)(1 + h_{FE})} \cdot \frac{I_C}{h_{FE}} \qquad [\text{since } I_{CBO} \ll I_C] \qquad \ldots (6.17)$$

Replacing the partial derivatives in S_h by finite small changes, we write

$$\frac{\Delta I_C}{\Delta h_{FE}} = \frac{(R_F + R_E + R_L)}{R_F + (R_E + R_L)(1 + h_{FE})} \cdot \frac{I_C}{h_{FE}}$$

or, $$\frac{\Delta I_C}{I_C} = \frac{(R_F + R_E + R_L)}{R_F + (R_E + R_L)(1 + h_{FE})} \cdot \frac{\Delta h_{FE}}{h_{FE}} \qquad \ldots (6.18)$$

Now, if $R_F \gg (R_E + R_L)(1 + h_{FE})$, $\dfrac{\Delta I_C}{I_C} = \dfrac{\Delta h_{FE}}{h_{FE}}$. Again, since $\dfrac{\Delta h_{FE}}{h_{FE}}$ lies in the range

$+100\%$ to -50% typically, $\dfrac{\Delta I_C}{I_C}$ can have high values. So, in this case the circuit will have poor stability.

On the other hand, if $R_F \ll (R_E + R_L)(1 + h_{FE})$ then $\dfrac{\Delta I_C}{I_C} = \dfrac{\Delta h_{FE}}{h_{FE}} \cdot \dfrac{1}{1 + h_{FE}}$. In this case,

variation in I_C due to a variation in h_{FE} is reduced by the factor $\dfrac{1}{1 + h_{FE}}$. This indicates a high

stability of the collector bias circuit in this case.

To find the stability factor $S_V = \dfrac{\partial I_C}{\partial V_{BE}}$, we differentiate I_C in equation (6.15) with respect to

V_{BE} and get

$$S_V = \frac{h_{FE}}{R_F + (R_E + R_L)(1 + h_{FE})} \qquad \ldots (6.19)$$

If $R_F \gg (R_E + R_L)(1 + h_{FE})$, $S_V = \dfrac{h_{FE}}{R_F}$.

If $R_F \ll (R_E + R_L)(1 + h_{FE})$, $S_V = \dfrac{h_{FE}}{(R_E + R_L)(1 + h_{FE})}$.

In the practical case, the value of S_V will be smaller in the second case when $R_F \ll (R_E + R_L)(1 + h_{FE})$. This leads to a more stable biasing circuit.

What is the reason of this improved stability?

The reason is that R_F provides a negative voltage feedback which improves the stability. If I_C increases due to any reason, the collector voltage V_{CE} decreases. Decrease in V_{CE} produces a reduction in base current which, in turn, tends to restore I_C to its original value.

6.2 h-PARAMETER EQUIVALENT CIRCUIT OF BJT

Assume linear operation of the BJT. There are four variables:

i_I = total (dc+ac) instantaneous input current

v_{IG} = total (dc+ac) instantaneous input voltage

i_O = total (dc+ac) instantaneous output current

v_{OG} = total (dc+ac) instantaneous output voltage

G stands for the common electrode. Now, we can write

$$v_{IG} = f(v_{OG}, i_I, i_O) \qquad \ldots (6.20)$$

and $\quad i_O = f'(v_{IG}, i_I, v_{OG}) \qquad \ldots (6.21)$

Here, one variable is expressed as a function of 3 other variables. Eliminating i_O from (6.20) and (6.21) we can write

$$v_{IG} = f_1(v_{OG}, i_I) \qquad \ldots (6.22)$$

Similarly, eliminating v_{IG} from (6.20) and (6.21) we get

$$i_O = f_2(v_{OG}, i_I) \qquad \ldots (6.23)$$

In (6.22) and (6.23), f_1 and f_2 are some functions. Now, let us express the total instantaneous values of voltage and current as the sum of dc and ac values explicitly. Thus,

$$v_{IG} = V_{IG} + v_{ig}$$

$$i_I = I_I + i_i$$

$$i_O \doteq I_O + i_o \qquad\qquad v_{OG} = V_{OG} + v_{og}$$

where V_{IG}, I_I, I_O and V_{OG} are the dc values of input voltage, input current, output current and output voltage respectively. While $v_{ig}, i_i, i_o,$ and v_{og} are the ac components of input voltage, input current, output current and output voltage respectively. 'g' indicates common terminal in ac operation.

Equation (6.22) can be written as

$$V_{IG} + v_{ig} = f_1[(V_{OG} + v_{og}), (I_I + i_i)] \qquad \ldots (6.24)$$

$$\approx f_1(V_{OG}, I_I) + \left(\frac{\partial f_1}{\partial v_{OG}}\right)_{I_I, V_{OG}} \cdot v_{og} + \left(\frac{\partial f_1}{\partial i_I}\right)_{I_I, V_{OG}} \cdot i_i \qquad \ldots (6.25)$$

neglecting squares and higher powers of i_i and v_{og} under small-signal approximation from the Taylor series expansion about the point (I_I, V_{OG}).

Equating the dc values from both sides of (6.25) we get

$$V_{IG} = f_1(V_{OG}, I_I).$$

Let $\quad \left.\dfrac{\partial f_1}{\partial v_{OG}}\right|_{I_I, V_{OG}} = \left.\dfrac{\partial v_{IG}}{\partial v_{OG}}\right|_{I_I, V_{OG}} = h_{rg}$ (say)

and $\quad \left.\dfrac{\partial f_1}{\partial i_I}\right|_{I_I, V_{OG}} = \left.\dfrac{\partial v_{IG}}{\partial i_I}\right|_{I_I, V_{OG}} = h_{ig}$ (say)

Now, equation (6.25) can be written as

$$v_{ig} = h_{ig} i_i + h_{rg} v_{og}$$ ... (6.26)

Then, $h_{ig} = \dfrac{v_{ig}}{i_i}\bigg|_{v_{og}=0}$ and $h_{rg} = \dfrac{v_{ig}}{v_{og}}\bigg|_{i_i=0}$.

h_{ig} is short-circuit input impedance and h_{rg} is the open-circuit reverse voltage transfer ratio. Expanding in a Taylor's series and neglecting squares and higher powers of ac quantities, we can recast equation (6.23) as

$$I_O + i_o = f_2\left(V_{OG} + v_{og}, I_I + i_i\right)$$

$$= f_2(V_{OG}, I_I) + \frac{\partial f_2}{\partial v_{OG}}\bigg|_{I_I, V_{OG}} \cdot v_{og} + \frac{\partial f_2}{\partial i_I}\bigg|_{I_I, V_{OG}} \cdot i_i$$ ... (6.27)

Now, $I_O = f_2(V_{OG}, I_I)$.

Also, let $\dfrac{\partial f_2}{\partial v_{OG}}\bigg|_{I_I, V_{OG}} = \dfrac{\partial i_O}{\partial v_{OG}}\bigg|_{I_I, V_{OG}} = h_{og}$ (say)

and $\dfrac{\partial f_2}{\partial i_I}\bigg|_{I_I, V_{OG}} = \dfrac{\partial i_O}{\partial i_I}\bigg|_{I_I, V_{OG}} = h_{fg}$ (say).

Then, from (6.27) we write

$$i_o = h_{fg} i_i + h_{og} v_{og}$$ ... (6.28)

From (6.28) we write $h_{fg} = \dfrac{i_o}{i_i}\bigg|_{v_{og}=0}$ and $h_{og} = \dfrac{i_o}{v_{og}}\bigg|_{i_i=0}$.

h_{fg} is the short-circuit forward current transfer ratio and h_{og} is the open-circuit output admittance. Thus, h_{ig} is dimensionally an impedance, h_{og} is admittance, while h_{fg} and h_{rg} are dimensionless. So, h-parameters have different dimensions. Hence, the name hybrid parameter model. The equivalent h-parameter model of BJT is obtained as follows:

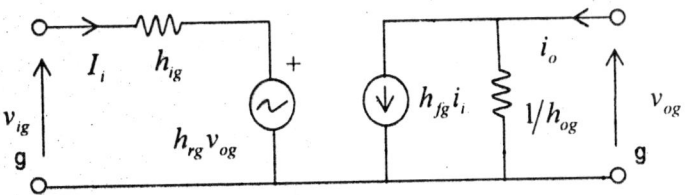

Fig. 6.5. Small-signal h-parameter ac equivalent circuit of a BJT in common G (G ≡ E, B, C) mode.

In common emitter (CE) mode, the h-parameter equivalent circuit of the BJT is shown in Fig. 6.6.

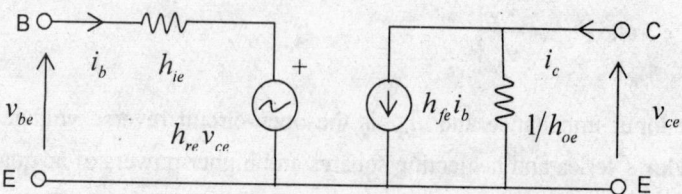

Fig. 6.6. h-parameter ac equivalent circuit of the BJT in CE mode.

To derive the circuit in Fig. 6.6 from Fig. 6.5., we put $v_{ig} = v_{be}$, $i_i = i_b$, $v_{og} = v_{ce}$, $i_o = i_c$, $h_{rg} = h_{re}$, $h_{fg} = h_{fe}$, $h_{ig} = h_{ie}$ and $h_{og} = h_{oe}$. This is because the input terminal in CE configuration is base (b), common terminal is emitter (e) and output terminal is collector (c).

Similarly, the h-parameter ac equivalent circuit of the BJT in the CB mode can be derived from Fig. 6.5 by substituting $v_{ig} = v_{eb}$, $i_i = i_e$, $v_{og} = v_{cb}$, $i_o = i_c$, $h_{rg} = h_{rb}$, $h_{fg} = h_{fb}$, $h_{ig} = h_{ib}$ and $h_{og} = h_{ob}$. This circuit is shown in Fig. 6.7.

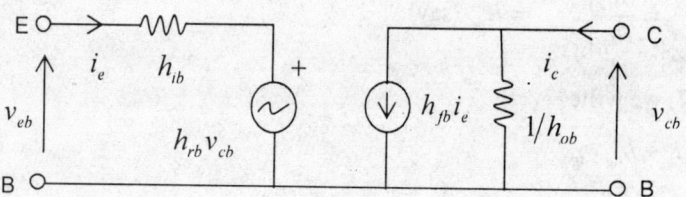

Fig. 6.7. h-parameter ac equivalent circuit of the BJT in CB mode.

For the common collector (CC) mode, we have from Fig. 6.5 $v_{ig} = v_{bc}$, $i_i = i_b$, $v_{og} = v_{ec}$, $i_o = i_e$, $h_{rg} = h_{rc}$, $h_{fg} = h_{fc}$, $h_{ig} = h_{ic}$ and $h_{og} = h_{oc}$. The corresponding ac equivalent circuit is shown in Fig. 6.8.

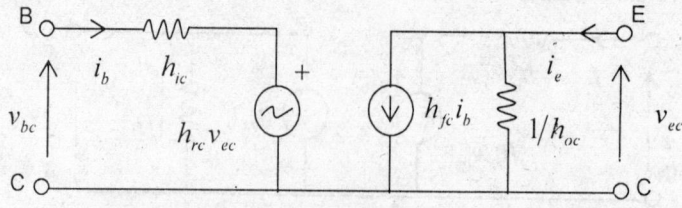

Fig. 6.8. h-parameter ac equivalent circuit of a BJT in CC mode.

In CC mode, the load R_L is connected with the emitter and the collector is ac grounded. The load current is the emitter current.

6.3 RELATIONSHIP BETWEEN h-PARAMETERS OF DIFFERENT MODES

In this section, we derive the relations between different types of h-parameters in CE, CB and CC modes.

6.3.1 RELATIONS BETWEEN CB AND CE h-PARAMETERS

Let us find out the relations between h-parameters of CB and CE modes. For this, we rearrange the CB circuit so that it look like a CE circuit. These are shown in Fig. 6.9(a) and 6.9(b). A standard CE circuit is shown in Fig. 9(c).

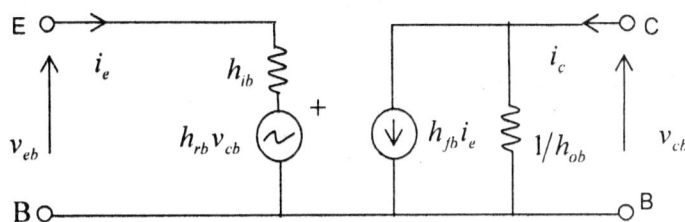

Fig. 6.9. (a) Small signal CB circuit.

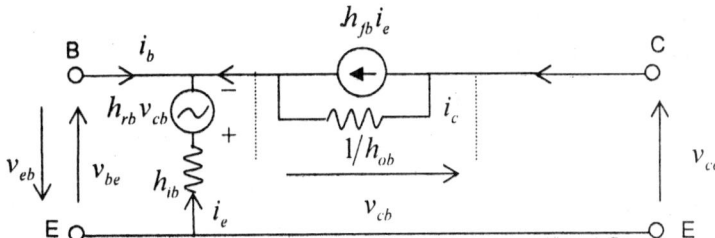

Fig. 6.9. (b) Rearrangement of CB circuit shown in Fig. 6.9(a) in the form of CE circuit.

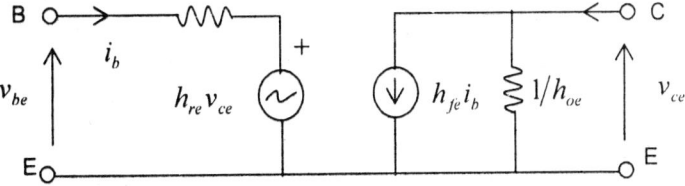

Fig.6.9.(c) Standard CE circuit.

From the equivalence of two circuits as shown in Figs. 6.9(a) and 6.9(b), we write

$$v_{be} = -v_{eb} \qquad \qquad \ldots (6.29)$$

Applying Kirchhoff's voltage law (KVL) and Kirchhoff's current law (KCL) to the two circuits, we get

$$v_{eb} = h_{ib}i_e + h_{rb}v_{cb} \qquad \qquad \ldots (6.30)$$

$$i_c = h_{fb}i_e + h_{ob}v_{cb} \qquad \qquad \ldots (6.31)$$

$$i_b + i_e + i_c = 0 \qquad \qquad \ldots (6.32)$$

$$v_{ce} = v_{be} + v_{cb} \qquad \qquad \ldots (6.33)$$

Putting (6.29) in (6.30) and rearranging we get

$$v_{be} + h_{ib}i_e + h_{rb}v_{cb} + 0 = 0 \qquad \qquad \ldots (6.34)$$

Again equations (6.31)–(6.33) can be rearranged as

$$0 + h_{fb}i_e + h_{ob}v_{cb} - i_c = 0 \qquad \qquad \ldots (6.35)$$

$$0 - i_e + 0 - i_c = i_b \qquad \qquad \ldots (6.36)$$

$$v_{be} + 0 + v_{cb} + 0 = v_{ce} \qquad \qquad \ldots (6.37)$$

Solving equations (6.34)–(6.37) by the method of determinants we get

$$v_{be} = \frac{1}{D}\begin{vmatrix} 0 & h_{ib} & h_{rb} & 0 \\ 0 & h_{fb} & h_{ob} & -1 \\ i_b & -1 & 0 & -1 \\ v_{ce} & 0 & 1 & 0 \end{vmatrix} \qquad \qquad \ldots (6.38)$$

where

$$D = \begin{vmatrix} 1 & h_{ib} & h_{rb} & 0 \\ 0 & h_{fb} & h_{ob} & -1 \\ 0 & -1 & 0 & -1 \\ 1 & 0 & 1 & 0 \end{vmatrix}$$

$$= \begin{vmatrix} h_{fb} & h_{ob} & -1 \\ -1 & 0 & -1 \\ 0 & 1 & 0 \end{vmatrix} - \begin{vmatrix} h_{ib} & h_{rb} & 0 \\ h_{fb} & h_{ob} & -1 \\ -1 & 0 & -1 \end{vmatrix}$$

$$= +h_{fb} + 1 + \left(h_{ib}h_{ob} - h_{rb}\{h_{fb} + 1\}\right)$$

$$= 1 + h_{fb} + \Delta h^b - h_{rb} \qquad \qquad \ldots (6.39)$$

where $\Delta h^b = h_{ib} h_{ob} - h_{rb} h_{fb}$.

Equation (6.38) yields

$$v_{be} = \frac{1}{D} \left[i_b \begin{vmatrix} h_{ib} & h_{rb} & 0 \\ h_{fb} & h_{ob} & -1 \\ 0 & 1 & 0 \end{vmatrix} - v_{ce} \begin{vmatrix} h_{ib} & h_{rb} & 0 \\ h_{fb} & h_{ob} & -1 \\ -1 & 0 & -1 \end{vmatrix} \right]$$

$$= \frac{1}{D} \left[i_b h_{ib} - v_{ce} \left(h_{rb} - \Delta^h \right) \right]$$

$$= \frac{1}{D} \left[h_{ib} i_b + \left(\Delta^h - h_{rb} \right) v_{ce} \right] \qquad \ldots (6.40)$$

The usual CE circuit as shown in Fig.9(c) has an equation

$$v_{be} = h_{ie} i_b + h_{re} v_{ce} \qquad \ldots (6.41)$$

Comparing (6.40) and (6.41) we get

$$h_{ie} = \frac{h_{ib}}{D} \approx \frac{h_{ib}}{1 + h_{fb}} \qquad \ldots (6.42)$$

neglecting $(\Delta h^b - h_{rb})$ compared with $(1 + h_{fb})$ since $\left| \left(\Delta^{hb} - h_{rb} \right) \right| << (1 + h_{fb})$. Also,

$$h_{re} = \frac{\Delta^h - h_{rb}}{D} \approx \frac{\Delta^h - h_{rb}}{1 + h_{fb}} \qquad \ldots (6.43)$$

Again solving equations (6.34)–(6.37) we get

$$i_c = \frac{1}{D} \begin{vmatrix} 1 & h_{ib} & h_{rb} & 0 \\ 0 & h_{fb} & h_{ob} & 0 \\ 0 & -1 & 0 & i_b \\ 1 & 0 & 1 & v_{ce} \end{vmatrix}$$

$$= \frac{1}{D} \left[-i_b \begin{vmatrix} 1 & h_{ib} & h_{rb} \\ 0 & h_{fb} & h_{ob} \\ 1 & 0 & 1 \end{vmatrix} + v_{ce} \begin{vmatrix} 1 & h_{ib} & h_{rb} \\ 0 & h_{fb} & h_{ob} \\ 0 & -1 & 0 \end{vmatrix} \right]$$

$$= \frac{1}{D} \left[-h_{fb} i_b - \Delta^{hb} i_b + h_{ob} v_{ce} \right]$$

$$= \frac{1}{D} \left[-\left(h_{fb} + \Delta^{hb} \right) i_b + h_{ob} v_{ce} \right] \qquad \ldots (6.44)$$

The usual output CE circuit (as shown in Fig.9(c)), equation can be written as

$$i_c = h_{fe} i_b + h_{oe} v_{ce} \qquad \qquad \dots (6.45)$$

Comparing equation (6.44) with (6.45) we get

$$h_{fe} = -\frac{h_{fb} + \Delta^{hb}}{D} \approx -\frac{h_{fb}}{1 + h_{fb}} \qquad \qquad \dots (6.46a)$$

and

$$h_{oe} = \frac{h_{ob}}{D} \approx \frac{h_{ob}}{1 + h_{fb}} \qquad \qquad \dots (6.46b)$$

neglecting the term $\left(\Delta^{hb} - h_{rb}\right)$ from the expression for D.

6.3.2 CE h-PARAMETERS IN TERMS OF CB h-PARAMETERS

The CB h-parameters in terms of CE h-parameters can be obtained by simply interchanging 'b' and 'e' in the expressions (6.42), (6.43), (6.46a) and (6.46b). The starting circuit should be the CE circuit and the final circuit should be converted CB circuit. The resultant expressions are

$$h_{ib} = \frac{h_{ie}}{1 + h_{fe}} \qquad \qquad \dots (6.47a)$$

$$h_{rb} = \frac{\Delta^{he} - h_{re}}{1 + h_{fe}} \qquad \qquad \dots (6.47b)$$

$$h_{fb} = -\frac{h_{fe}}{1 + h_{fe}} \qquad \qquad \dots (6.47c)$$

$$h_{ob} = \frac{h_{oe}}{1 + h_{fe}} \qquad \qquad \dots (6.47d)$$

6.3.3 h-PARAMETERS OF CB AND CC MODES

As mentioned in section 6.3.1, we first rearrange the CB circuit in CC form. This is shown in Figs. 6.10(a) and 6.10(b). A standard CC circuit is shown in Fig.6.10(c).
From the equation of two circuits in Figs.6.10(a) and 6.10(b), we can write

$$v_{eb} = h_{rb} v_{cb} + h_{ib} i_e \qquad \qquad \dots (6.48)$$

$$i_c = h_{fb} i_e + h_{ob} v_{cb} \qquad \qquad \dots (6.49)$$

$$i_b + i_e + i_c = 0 \qquad \qquad \dots (6.50)$$

$$v_{ec} = v_{bc} + v_{eb} \qquad \qquad \dots (6.51)$$

Again, $v_{cb} = -v_{bc}$ \qquad \qquad \dots (6.52)

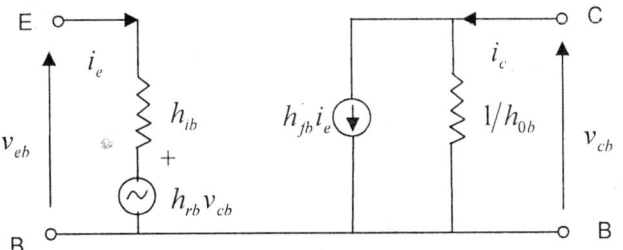

Fig.6.10. (a) Small-signal low-frequency CB circuit.

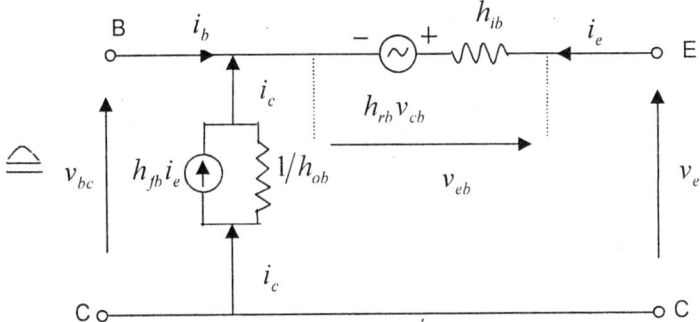

Fig. 6.10. (b) Rearrangement of CB circuit in CC form.

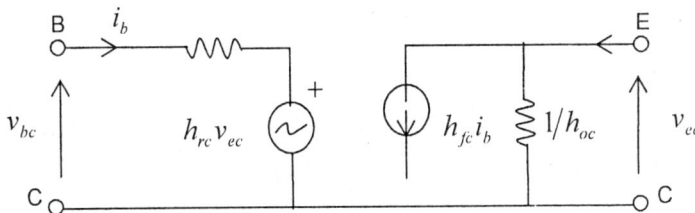

Fig.6.10. (c) Standard CC circuit.

Rearranging equations (6.48)–(6.51) we write

$$- h_{rb}\, v_{bc} + h_{ib}\, i_e - v_{eb} + 0 = 0 \qquad \qquad \ldots (6.53)$$

$$- h_{ob}\, v_{bc} + h_{fb}\, i_e + 0 - i_c = 0 \qquad \qquad \ldots (6.54)$$

$$0 - i_e + 0 - i_c = i_b \qquad \qquad \ldots (6.55)$$

$$v_{bc} + 0 + v_{eb} + 0 = v_{ec} \qquad \qquad \ldots (6.56)$$

Solving equations (6.53) – (6.56) we get

$$v_{bc} = \frac{1}{D} \begin{vmatrix} 0 & h_{ib} & -1 & 0 \\ 0 & h_{fb} & 0 & -1 \\ i_b & -1 & 0 & -1 \\ v_{ec} & 0 & 1 & 0 \end{vmatrix} \qquad \qquad \dots (6.57)$$

where

$$D = \begin{vmatrix} -h_{rb} & h_{ib} & -1 & 0 \\ -h_{ob} & h_{fb} & 0 & -1 \\ 0 & -1 & 0 & -1 \\ 1 & 0 & 1 & 0 \end{vmatrix}$$

Now,

$$D = -\begin{vmatrix} -h_{rb} & h_{ib} & -1 \\ 0 & -1 & 0 \\ 1 & 0 & 1 \end{vmatrix} + \begin{vmatrix} -h_{rb} & h_{ib} & -1 \\ -h_{ob} & h_{fb} & 0 \\ 1 & 0 & 1 \end{vmatrix}$$

$$= -h_{rb} + 1 + (h_{fb} + h_{ib} h_{ob} - h_{fb} h_{rb})$$

$$= 1 + h_{fb} + \Delta^{hb} - h_{rb}$$

$$\approx 1 + h_{fb} \qquad \qquad \dots (6.58)$$

since $(1 + h_{fb}) >> (\Delta^{hb} - h_{rb})$. From (6.57) we get

$$v_{bc} = \frac{1}{D} \left[i_b \begin{vmatrix} h_{ib} & -1 & 0 \\ h_{fb} & 0 & -1 \\ 0 & 1 & 0 \end{vmatrix} - v_{ec} \begin{vmatrix} h_{ib} & -1 & 0 \\ h_{fb} & 0 & -1 \\ -1 & 0 & -1 \end{vmatrix} \right]$$

$$= \frac{1}{D} \left[i_b h_{ib} + v_{ec} (1 + h_{fb}) \right] \qquad \qquad \dots (6.59)$$

Consider the input circuit equation of the CC configuration as shown in Fig.10(c). The corresponding equation is

$$v_{bc} = h_{ic} i_b + h_{rc} v_{ec} \qquad \qquad \dots (6.60)$$

Comparing equations (6.59) and (6.60) we get

$$h_{ic} = \frac{h_{ib}}{D} \approx \frac{h_{ib}}{1 + h_{fb}} \qquad \qquad \dots (6.61)$$

and $\quad h_{rc} = \dfrac{(1+h_{fb})}{D} \approx 1$ $\hspace{6cm}$ ... (6.62)

Again, solving equations (6.53)–(6.56) we get

$$i_e = \frac{1}{D} \begin{vmatrix} -h_{rb} & 0 & -1 & 0 \\ -h_{ob} & 0 & 0 & -1 \\ 0 & i_b & 0 & -1 \\ 1 & v_{ec} & 1 & 0 \end{vmatrix}$$

$$= \frac{1}{D}\left[-i_b \begin{vmatrix} -h_{rb} & -1 & 0 \\ -h_{ob} & 0 & -1 \\ 1 & 1 & 0 \end{vmatrix} + v_{ec} \begin{vmatrix} -h_{rb} & -1 & 0 \\ -h_{ob} & 0 & -1 \\ 0 & 0 & -1 \end{vmatrix} \right]$$

$$= \frac{1}{D}\left[-i_b(1-h_{rb}) + h_{ob} v_{ec} \right] \hspace{3cm} ... (6.63)$$

Now, consider the output circuit equation of the CC configuration as shown in Fig.10(c). The corresponding equation is

$$i_e = h_{fc} i_b + h_{oc} v_{ec} \hspace{4cm} ... (6.64)$$

comparing (6.64) with (6.63) we get

$$h_{fc} = -\frac{(1-h_{rb})}{D} \approx -\frac{(1-h_{rb})}{1+h_{fb}} \approx -\frac{1}{(1+h_{fb})} \hspace{2cm} ... (6.65)$$

since $h_{rb} \ll 1$, and

$$h_{oc} = \frac{h_{ob}}{D} = \frac{h_{ob}}{1+h_{fb}} \hspace{4cm} ... (6.66)$$

6.3.4 CC PARAMETERS IN TERMS OF CE PARAMETERS

The CC parameters h_{ic}, h_{rc}, h_{fc}, and h_{0c} in terms of CB parameters are given in section 6.3.2. Now, substitution of the CB parameters in terms of CE parameters will give the desired result. Thus,

$$1 + h_{fb} = 1 - \frac{h_{fe}}{1+h_{fe}} = \frac{1}{1+h_{fe}} \hspace{3cm} ... (6.67)$$

$$h_{ic} = \frac{h_{ib}}{1+h_{fb}} = \frac{h_{ie}}{1+h_{fe}}.(1+h_{fe}) = h_{ie} \hspace{2cm} ... (6.68)$$

using equations (6.47a) and (6.67). Also, $\hspace{1cm} h_{rc} = 1$

$$h_{fc} = -\frac{1}{(1+h_{fb})} = -(1+h_{fe}) \qquad \qquad \dots (6.69)$$

using equations (6.65) and (6.67).

$$h_{oc} = \frac{h_{ob}}{1+h_{fb}} = \frac{h_{oe}}{1+h_{fe}}(1+h_{fe}) = h_{oe} \qquad \qquad \dots (6.70)$$

using equations (6.47d) and (6.67).

6.4 GENERALIZED ANALYSIS OF CE, CB AND CC AMPLIFIERS

The circuit diagrams of CE, CB and CC amplifiers are shown in Figs. 6.11, 6.12, and 6.13 respectively.

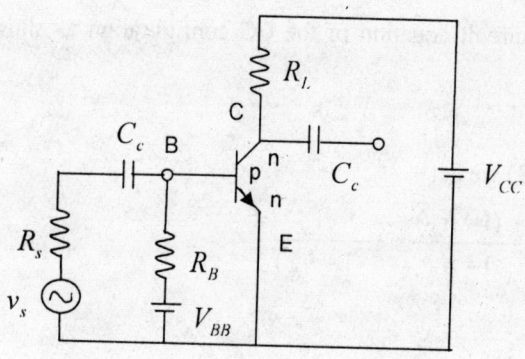

Fig. 6.11. Circuit diagram of a CE amplifier.

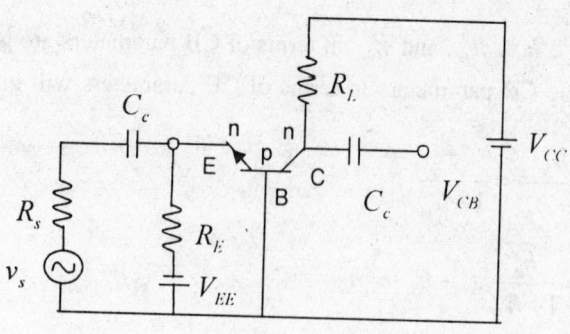

Fig. 6.12. Circuit diagram of a CB amplifier.

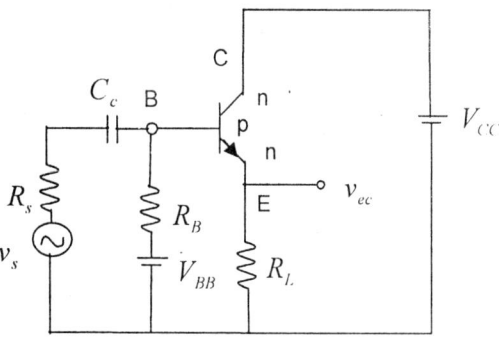

Fig. 6.13. Circuit diagram of CC amplifier.

The small-signal ac equivalent circuit of these amplifiers is shown in Fig. 6.14.

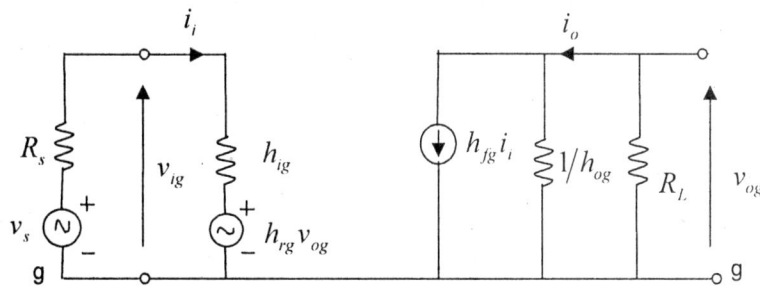

Fig. 6.14. Small-signal ac equivalent circuit of the amplifier in common 'g' mode.
g → e for CE mode, g → b for CB mode, g → c for CC mode.

i_i = input current, i_o = output current, v_{ig} = input voltage, v_{og} = output voltage, v_s = signal voltage, R_s = internal resistance of the signal source, R_L = load resistance.

The following circuit equations can be written from the above equivalent circuit.

$$v_s = \left(h_{ig} + R_s\right)i_i + h_{rg}v_{og} \qquad \ldots (6.71)$$

$$i_o = h_{fg}i_i + h_{og}v_{og} \qquad \ldots (6.72)$$

$$v_{og} = -i_o R_L \qquad \ldots (6.73)$$

Multiplying (6.72) by R_L and adding with (6.73) we get

$$0 = h_{fg}R_L i_i + \left(1 + h_{og}R_L\right)v_{og} \qquad \ldots (6.74)$$

$$\therefore \quad \frac{v_{og}}{i_i} = -\frac{h_{fg}R_L}{1 + h_{og}R_L} \qquad \ldots (6.75)$$

The current gain of the amplifier is

$$A_i = \frac{i_o}{i_i} = -\frac{v_{og}}{R_L i_i} \quad \text{[using (6.73)]}$$

$$= \frac{h_{fg}}{1 + h_{og} R_L} \qquad \qquad \dots (6.76)$$

Input resistance of the amplifier is

$$R_i = \frac{v_{ig}}{i_i} = \frac{v_s - R_s i_i}{i_i} \quad \text{[from the input circuit]}$$

$$= \frac{v_s}{i_i} - R_s \qquad \qquad \dots (6.77)$$

substituting (6.75) for v_{og} in terms of i_i in (6.71) we get

$$v_s = (h_{ig} + R_s) i_i - h_{rg} \cdot \frac{h_{fg} R_L}{1 + h_{og} R_L} i_i$$

$$= i_i \left[R_s + \frac{h_{ig} + R_L \; \Delta^{hg}}{1 + h_{og} R_L} \right]$$

where $\quad \Delta^{hg} = h_{ig} h_{og} - h_{rg} h_{fg}$.

$$\therefore \quad \frac{v_s}{i_i} = R_s + \frac{h_{ig} + R_L \; \Delta^{hg}}{1 + h_{og} R_L} \qquad \qquad \dots (6.78)$$

From (6.77) we get, using (6.78),

$$R_i = \frac{h_{ig} + R_L \Delta^{hg}}{1 + h_{og} R_L} \qquad \qquad \dots (6.79)$$

The voltage gain of the amplifier is

$$A_v = \frac{v_{og}}{v_{ig}} = -\frac{R_L i_o}{R_i i_i}$$

$$= -\frac{R_L}{R_i} A_i = -R_L \cdot \frac{h_{fg}}{(1 + h_{og} R_L)} \frac{(1 + h_{og} R_L)}{h_{ig} + R_L \Delta^{hg}}$$

$$= -\frac{h_{fg} R_L}{h_{ig} + R_L \; \Delta^{hg}} \qquad \qquad \dots (6.80)$$

The output resistance of the amplifier is the resistance measured at the output terminals with the source (v_s) eliminated and removing the load R_L.

Putting $v_s = 0$ in (6.71) we get

$$i_i = -\frac{h_{rg}\, v_{og}}{h_{ig} + R_s} \qquad \qquad \dots (6.81)$$

Putting (6.81) in (6.72) we get

$$i_o = h_{fg}\left(-\frac{h_{rg} v_{og}}{h_{ig} + R_s}\right) + h_{og} v_{og}$$

$$= v_{og} \cdot \frac{\Delta^{hg} + h_{og}\, R_s}{h_{ig} + R_s}$$

The output resistance

$$R_o = \frac{v_{og}}{i_o}\Big|_{v_s = 0}$$

$$= \frac{h_{ig} + R_s}{\Delta^{hg} + h_{og}\, R_s} \qquad \qquad \dots (6.82)$$

The power gain of the amplifier is given by

$$A_p = \frac{\text{output power}}{\text{input power}} = \frac{i_o^2 R_L}{i_i^2 R_i}$$

$$= A_i^2 \frac{R_L}{R_i} = \frac{h_{fg}^2 R_L}{(1 + h_{og} R_L)^2}\frac{(1 + h_{og} R_L)}{(h_{ig} + R_L\, \Delta^{hg})}$$

$$= \frac{h_{fg}^2 R_L}{(1 + h_{og} R_L)(h_{ig} + R_L\, \Delta^{hg})} \qquad \qquad \dots (6.83)$$

This power gain becomes maximum when the source resistance matches the input resistance of the amplifier, and the load resistance matches the output resistance of the amplifier. Here, the term 'matches' is equivalent to the term 'equal'.

Therefore, for input matching we get

$$R_s = R_i = \frac{h_{ig} + R_L\, \Delta^{hg}}{1 + h_{og} R_L} \qquad \qquad \dots (6.84)$$

For output matching, we get

$$R_L = R_o = \frac{h_{ig} + R_s}{\Delta^{hg} + h_{og} R_s} \qquad \dots (6.85)$$

From (6.84), we get

$$R_s + R_s R_L h_{og} = h_{ig} + R_L \Delta^{hg} \qquad \dots (6.86)$$

Similarly, from (6.85), we get

$$R_s + h_{ig} = R_s R_L h_{og} + R_L \Delta^{hg} \qquad \dots (6.87)$$

Solving (6.86) and (6.87) we get

$$R_s = \sqrt{\frac{h_{ig} \Delta^{hg}}{h_{og}}} \qquad \dots (6.88)$$

and

$$R_L = \sqrt{\frac{h_{ig}}{h_{og} \Delta^{hg}}} \qquad \dots (6.89)$$

Substituting the values of R_s and R_L in (6.83) from (6.88) and (6.89) respectively, we get the maximum available power gain (MAPG) as

$$\text{MAPG} = \frac{h_{fg}^2 \sqrt{h_{ig} \Big/ (h_{og} \Delta^{hg})}}{\left(1 + \sqrt{\dfrac{h_{ig} h_{og}}{\Delta^{hg}}}\right)\left(h_{ig} + \sqrt{\dfrac{h_{ig} \Delta^{hg}}{h_{og}}}\right)}$$

$$= \frac{h_{fg}^2}{\left(\sqrt{h_{ig} h_{og}} + \sqrt{\Delta^{hg}}\right)^2} \qquad \dots (6.90)$$

SPECIAL CASES:

(i) CE AMPLIFIER

Here, $i_i = i_b$, $i_o = i_c$, $v_{ig} = v_{be}$ and $v_{og} = v_{ce}$ where i_b = base current, i_c = collector current, v_{be} = base-emitter voltage and v_{ce} = collector voltage. Emitter (e) is the common terminal. These voltages and currents are ac quantities.

Equations (6.76), (6.79), (6.80), (6.82), (6.83) and (6.90) can be rewritten as

$$A_i = \frac{h_{fe}}{1 + h_{oe} R_L} \qquad \dots (6.91)$$

$$R_i = \frac{h_{ie} + R_L \, \Delta^{he}}{1 + h_{oe} R_L} \qquad \qquad \text{... (6.92)}$$

$$A_v = -\frac{h_{fe} R_L}{h_{ie} + R_L \, \Delta^{he}} \qquad \qquad \text{... (6.93)}$$

$$R_o = \frac{h_{ie} + R_s}{\Delta^{he} + h_{oe} \cdot R_s} \qquad \qquad \text{... (6.94)}$$

$$A_p = \frac{h_{fe}^2 R_L}{\left(1 + h_{oe} R_L\right)\left(h_{ie} + R_L \, \Delta^{he}\right)} \qquad \qquad \text{... (6.95)}$$

$$\text{MAPG} = \frac{h_{fe}^2}{\left(\sqrt{h_{ie} h_{oe}} + \sqrt{\Delta^{he}}\right)^2} \qquad \qquad \text{... (6.96)}$$

where $\Delta^{he} = h_{ie} h_{oe} - h_{fe} h_{re}$.

CE amplifier has high current and voltage gains. It has a high power gain. Its input resistance is much lower than the output resistance. There is a phase difference of $180°$ between the input and output voltages. Since h_{re} is small, this amplifier has a very small feedback from output to input.

(ii) CB AMPLIFIER

Here, $i_i = i_e$, $i_0 = i_c$, $v_{ig} = v_{eb}$, $v_{og} = v_{cb}$, where i_e = emitter current, i_c = collector current, v_{eb} = emitter voltage, v_{cb} = collector voltage. Voltages are measured with respect to the common terminal base (b). These currents and voltages are all ac quantities. Equations (6.76), (6.79), (6.80), (6.82), (6.83) and (6.90) for CB mode will take the following forms:

$$A_i = \frac{h_{fb}}{1 + h_{ob} R_L} \qquad \qquad \text{... (6.97)}$$

$$R_i = \frac{h_{ib} + R_L \, \Delta^{hb}}{1 + h_{ob} R_L} \qquad \qquad \text{...(6.98)}$$

$$A_v = -\frac{h_{fb} R_L}{h_{ib} + R_L \, \Delta^{hb}} \qquad \qquad \text{... (6.99)}$$

$$R_o = \frac{h_{ib} + R_s}{\Delta^{hb} + h_{ob} R_s} \qquad \qquad \text{... (6.100)}$$

$$A_p = \frac{h_{fb}^2 R_L}{\left(1 + h_{ob} R_L\right)\left(h_{ib} + R_L \, \Delta^{hb}\right)} \qquad \qquad \text{... (6.101)}$$

$$MAPG = \frac{h_{fb}^2}{\left(\sqrt{h_{ib}h_{ob}} + \sqrt{\Delta^{hb}}\right)^2} \qquad \text{... (6.102)}$$

where $\Delta^{hb} = h_{ib}\,h_{ob} - h_{rb}\,h_{fb}$.

CB amplifier has a high voltage gain but a current gain less than unity. It has a very low input resistance and a high output resistance. The power gain is much smaller than the CE amplifier. It has a small feedback from output to input due to the small value of h_{rb}. It can be used to match a low impedance source to a high impedance load.

(iii) CC AMPLIFIER

In CC mode of operation, equations (6.76), (6.79), (6.80), (6.82), (6.83) and (6.90) will be converted to following forms :

$$A_i = \frac{h_{fc}}{1 + h_{oc}\,R_L} \qquad \text{... (6.103)}$$

$$R_i = \frac{h_{ic} + R_L\,\Delta^{hc}}{1 + h_{oc}R_L} \qquad \text{... (6.104)}$$

$$A_v = -\frac{h_{fc}\,R_L}{h_{ic} + R_L\,\Delta^{hc}} \qquad \text{... (6.105)}$$

$$R_o = \frac{h_{ic} + R_s}{\Delta^{hc} + h_{oc}\,R_s} \qquad \text{... (6.106)}$$

$$A_p = \frac{h_{fc}^2\,R_L}{(1 + h_{oc}\,R_L)(h_{ic} + R_L\,\Delta^{hc})} \qquad \text{... (6.107)}$$

$$MAPG = \frac{h_{fc}^2}{(\sqrt{h_{ic}h_{oc}} + \sqrt{\Delta^{hc}})^2} \qquad \text{... (6.108)}$$

where $\Delta^{hc} = h_{ic}h_{oc} - h_{rc}h_{fc}$.

The CC amplifier is also known as emitter follower. The output voltage in CC amplifier is the emitter voltage. Moreover, since the voltage gain is nearly unity, the emitter voltage is nearly equal to the input voltage. Thus, the emitter follows the input signal and hence the name emitter follower. It has a high current gain but its voltage gain is a little less than unity. Obviously, the power gain is much smaller than the CE amplifier. It has a high input resistance and low output resistance. It can act as an impedance transforming device. It can match a high impedance source to a low impedance load. The impedance transformation property is opposite to that of the CB amplifier.

6.5 HIGH FREQUENCY EQUIVALENT CIRCUIT OF THE BJT

To find out the high frequency model of the bipolar junction transistor, we must include the emitter-base and collector-base capacitances. Let C_{bc} be the depletion layer capacitance of the reverse-biased collector-base junction and C_{be} be the diffusion capacitance of the forward biased emitter-base junction. The collector-base junction resistance appearing in parallel with C'_{bc} is very high so that it is neglected. Similarly, the collector-emitter resistance is also very high which appears in parallel with the output load and so it is neglected. The base-emitter junction resistance r_{bc} will be present in this model. Here, base means internal base terminal. There is a resistance between this internal base point and the external base terminal. The equivalent circuit is shown in Fig. 6.15.

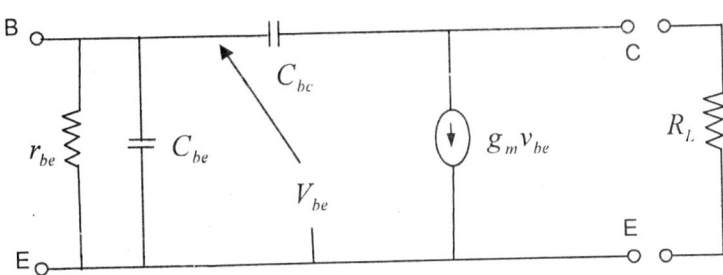

Fig. 6.15. High frequency equivalent circuit of BJT.

g_m is the transconductance of the BJT. R_L is the load resistance. The capacitance C_{bc} provide feedback at high frequencies. Let A_v be the voltage gain between the points B and C. Applying Miller's theorem the effect of the capacitor C_{bc} can be obtained by connecting a capacitance $C_{bc}(1 - A_v)$ at the input. The effect of the capacitance C_{bc} on the output can be obtained by connecting a capacitance $C_{bc}(1 - \dfrac{1}{A_v})$ at the output. The equivalent circuit obtained by applying Miller's theorem is shown in Fig. 6.16.

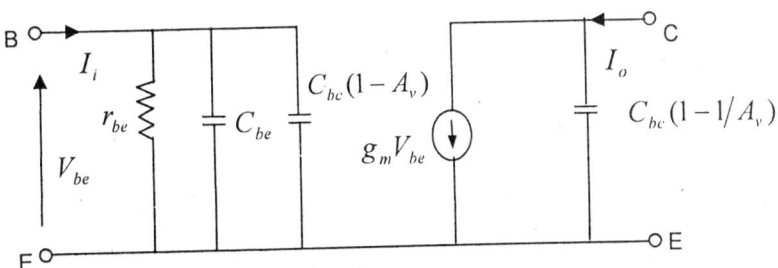

Fig. 6.16. High frequency equivalent circuit of a BJT by using Miller theorem.

In the low frequency region, all capacitances are neglected. Under this condition, we have

$$V_{be} = r_{be} I_i$$

... (6.109)

Short-circuit ($R_L = 0$) output current I_o in the low frequency region is given by

$$I_o = g_m V_{be} = g_m r_{be} I_i$$

... (6.110)

The short circuit current gain at low frequencies is

$$h_{fe} = \frac{I_o}{I_i} = g_m r_{be}$$

... (6.111)

When the output is short circuited ($R_L = 0$), we have $A_v = 0$. Then, Miller capacitance at the output, $C_{bc}(1 - \frac{1}{A_v})$, provide a virtual short circuit.

Now, at high frequencies,

$$V_{be} = \frac{I_i}{g_{be} + j\omega(C_{be} + C_{bc})}$$

... (6.112)

where $g_{be} = 1/r_{be}$. The output load current at high frequencies is $I_L = -I_0$ which flows from emitter to collector. The short circuit current gain at high frequencies is

$$A_i = \frac{I_L}{I_i} = -\frac{I_o}{I_i} = \frac{-g_m V_{be}}{I_i}$$

$$= -\frac{g_m}{g_{be} + j\omega(C_{be} + C_{bc})}$$

$$= -\frac{g_m r_{be}}{1 + j\omega r_{be}(C_{be} + C_{bc})}$$

$$= -\frac{h_{fe}}{1 + j(\omega / \omega_\beta)}$$

... (6.113)

where $\omega_\beta = \dfrac{1}{r_{be}(C_{be} + C_{bc})}$

... (6.114a)

and $A_i\big|_{\omega \to 0} = g_m r_{be} = h_{fe}$

... (6.114b)

h_{fe} is the short circuit CE current gain at low frequencies. The negative sign in equation (6.113) appears since the direction of load current is opposite to the direction of collector current. The opposite convention may also be taken.

At $\omega = \omega_\beta$, $|A_i| = h_{fe}/\sqrt{2}$. Thus, ω_β is the radian frequency at which the short-circuit current gain falls to $1/\sqrt{2}$ of its low frequency value. ω_β is called beta cut-off frequency of the BJT in radian.

6.6 R-C COUPLED AMPLIFIER

The circuit diagram of a single stage R-C Coupled amplifier is shown in Fig. 6.17.

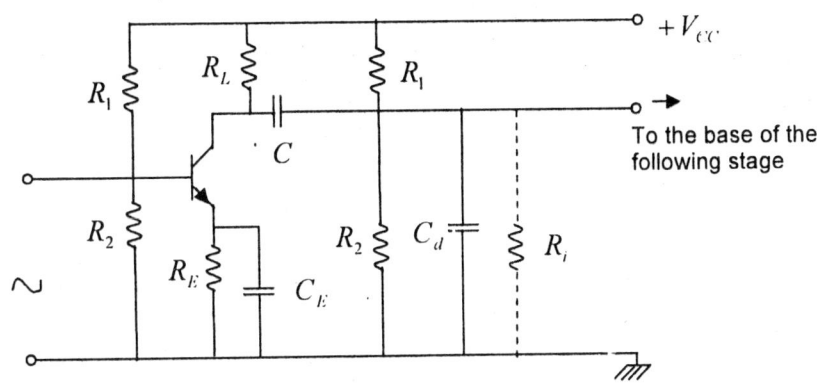

To the base of the
following stage

Fig. 6.17. Circuit diagram of an R-C Coupled amplifier.

Here, R_1, R_2 are biasing resistances. C is the coupling capacitor. R_i is the input resistance of the following stage. C_d is the effective input capacitance of the following stage.
The operation of this amplifier is divided into three regions:

    (i) Low frequency range – here C and C_E produce some voltage drop while C_d provides an open circuit.

  (ii) Mid frequency range – here all capacitances, C, C_E and C_d have no effect on amplifier operation. C and C_E provide short circuit while C_d provide an open circuit.

 (iii) High frequency range – here, C and C_E provide a short circuit but C_d will exhibit finite reactance.

MID FREQUENCY RANGE

The ac equivalent circuit of the amplifier in this range is shown in Fig. 6.18. h_{re} and h_{oe} being small are neglected from the equivalent circuit. Here, $R_i \cong h_{ie}$. Let $R_{eq} = R_1 \| R_2 \| h_{ie}$. Typically, $h_{ie} \sim 1$ kΩ. Assuming R_1, R_2 much higher than h_{ie}, $R_{eq} \cong h_{ie}$.

Now, $V_o = -h_{fe} I_i \dfrac{R_L h_{ie}}{R_L + h_{ie}}$ ... (6.115)

$$V_i = h_{ie} \, I_i \qquad \qquad \ldots (6.116)$$

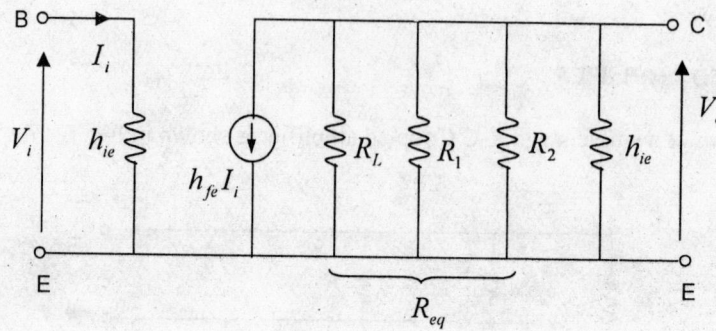

Fig. 6.18. Mid frequency equivalent circuit of the R-C Coupled amplifier.

Mid frequency voltage gain (A_M) is

$$A_M = \frac{V_o}{V_i} = -\frac{h_{fe} R_L}{R_L + h_{ie}} \qquad \qquad \ldots (6.117a)$$

$$= |A_M| \angle \theta_M \qquad \qquad \ldots (6.117b)$$

There is a phase difference of $180°$ between the input and the output. The phase angle $< \theta_M = 180°$.

LOW FREQUENCY RANGE

The ac equivalent circuit of this amplifier over this range is shown in Fig. 6.19.

Let $\quad Z_e = R_E \parallel \dfrac{1}{j\omega C_E} = \dfrac{R_E}{1 + j\omega C_E \, R_E}$

Now, the input impedance of the amplifier is $R_i = h_{ie} + (1 + h_{fe}) Z_e$.

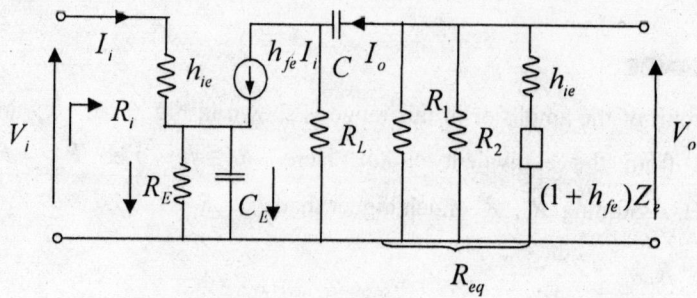

Fig. 6.19. Low frequency ac equivalent circuit.

Let I_o = current through R_{eq}. Here, $R_{eq} = R_1 \parallel R_2 \parallel [h_{ie} + (1 + h_{fe})Z_e] \cong h_{ie} + (1 + h_{fe})Z_e$

assuming R_1, R_2 much larger than $[(1 + h_{fe})Z_e + h_{ie}]'$. If R_1, R_2 be comparable with $[(1 + h_{fe})Z_e + h_{ie}]$ then R_{eq} will be the equivalent resistance of the parallel combination of R_1, R_2 and $[h_{ie} + (1 + h_{fe})Z_e]$.

Now, $\quad I_o = h_{fe} I_i \dfrac{R_L}{R_L + R_{eq} + \dfrac{1}{j\omega C}}$
$\qquad\qquad\qquad\qquad\qquad\qquad$ ... (6.118)

Output voltage(V_o) is

$\qquad V_o = -I_o R_{eq}$
$\qquad\qquad\qquad\qquad\qquad\qquad$ ... (6.119)

Input voltage (V_i) is

$\qquad V_i = I_i [h_{ie} + (1 + h_{fe})Z_e]$
$\qquad\qquad\qquad\qquad\qquad\qquad$ ... (6.120)

Low frequency voltage gain (A_L) is

$$A_L = \frac{V_o}{V_i} = -\frac{I_o R_{eq}}{V_i}$$

$$= -\frac{h_{fe} I_i R_L [h_{ie} + (1 + h_{fe})Z_e]}{I_i [h_{ie} + (1 + h_{fe})Z_e] [R_L + h_{ie} + (1 + h_{fe})Z_e + \dfrac{1}{j\omega C}]}$$

$$= -\frac{h_{fe} R_L}{(R_L + h_{ie})[1 + \dfrac{(1 + h_{fe})Z_e}{R_L + h_{ie}} + \dfrac{1}{j\omega C(R_L + h_{ie})}]}$$

$$= \frac{A_M}{1 + \dfrac{(1 + h_{fe})}{R_L + h_{ie}} \dfrac{R_E}{1 + j\omega C_E R_E} + \dfrac{1}{j\omega C(R_L + h_{ie})}}$$

$\therefore \quad \dfrac{A_L}{A_M} = \dfrac{1}{1 + \dfrac{\eta}{1 + j\dfrac{\omega}{\omega_E}} + \dfrac{1}{j\dfrac{\omega}{\omega_L}}}$
$\qquad\qquad\qquad\qquad\qquad\qquad$ ... (6.121)

where $\quad \omega_E = \dfrac{1}{R_E C_E}, \omega_L = \dfrac{1}{C(R_L + h_{ie})}$ and $\eta = \dfrac{(1 + h_{fe})R_E}{R_L + h_{ie}}$.

Equation (6.121) can be written as

$$\frac{A_L}{A_M} = \frac{1}{\left[1 + \dfrac{\eta}{1 + \dfrac{\omega^2}{\omega_E^2}}\right] - j\left[\dfrac{\omega_L}{\omega} + \dfrac{\eta \dfrac{\omega}{\omega_E}}{1 + \dfrac{\omega^2}{\omega_E^2}}\right]} \qquad \ldots (6.122)$$

Suppose there is no emitter bias applied. Then, $R_E = 0$ and $C_E = 0$. So, $\omega_E = \infty$ and $\eta = 0$. Under this condition,

$$\frac{A_L}{A_M} = \frac{1}{1 - j\dfrac{\omega_L}{\omega}} \qquad \ldots (6.123)$$

$$\therefore \qquad \frac{|A_L|}{|A_M|} = \frac{1}{\sqrt{1 + \left(\dfrac{\omega_L}{\omega}\right)^2}} \qquad \ldots (6.124)$$

At $\omega = \omega_L$, $\dfrac{|A_L|}{A_M} = \dfrac{1}{\sqrt{2}}$.

ω_L is called the lower half-power frequency in radian, when the gain falls to $1/\sqrt{2}$ of its mid-frequency value. We write, $A_L = |A_L| \angle \theta_L$. The phase angle of the low-frequency gain function is $\angle \theta_L = \pi + \tan^{-1}\left(\dfrac{\omega_L}{\omega}\right)$. At $\omega = \omega_L$, $\angle \theta_L = \pi + \dfrac{\pi}{4} = \dfrac{5\pi}{4}$. When $\omega \to 0$, $\angle \theta_L \to \dfrac{3\pi}{2}$.

To find the effect of $R_E - C_E$ bias circuit on amplifier operation, we put $\dfrac{\omega}{\omega_E} = x$. Then, we can get from eqn. (6.122)

$$\frac{|A_L|}{|A_M|} = \frac{1}{\left[(1 + \dfrac{\eta}{1 + x^2})^2 + \dfrac{\eta^2 x^2}{(1 + x^2)^2}\right]^{1/2}} \qquad \ldots (6.125)$$

Here, we are not considering the effect of capacitance C. At the half power point, we have

$$\frac{|A_L|}{|A_M|} = \frac{1}{\sqrt{2}}.$$

Then, $1 + \dfrac{\eta^2}{(1 + x^2)^2} + \dfrac{2\eta}{1 + x^2} + \dfrac{\eta^2 x^2}{(1 + x^2)^2} = 2$

or, $\quad \dfrac{\eta^2(1+x^2)}{(1+x^2)^2} + \dfrac{2\eta}{1+x^2} = 1$

or, $\quad \eta^2 + 2\eta - 1 = x^2$

$\therefore \quad x = \sqrt{\eta^2 + 2\eta - 1}$

If $\omega = \omega'$ be the half-power frequency due to the effect of $R_E - C_E$ alone, then

$$\omega' = x\omega_E = \omega_E \sqrt{\eta^2 + 2\eta - 1} \qquad \qquad \dots (6.127)$$

HIGH FREQUENCY RANGE

The ac equivalent circuit in the high frequency range is shown in Fig. 6.20.

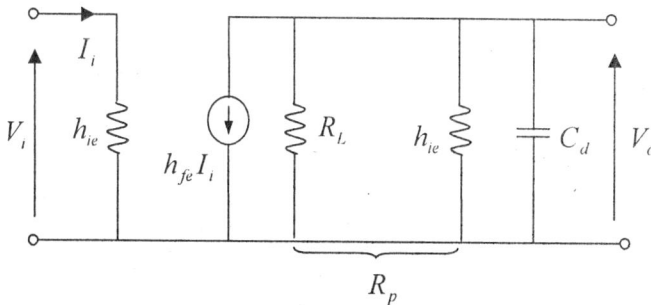

Fig. 6.20. High frequency ac equivalent circuit of the R-C coupled amplifier.

C_d is the effective input capacitance of the following stage.

Here, $\quad V_i = I_i h_{ie}$ $\qquad \qquad \dots (6.128)$

Let $\quad R_p = R_L \| h_{ie} = \dfrac{R_L h_{ie}}{R_L + h_{ie}}$

Output voltage $V_o = -h_{fe} I_i \dfrac{R_p}{1 + j\omega C_d R_p}$ $\qquad \qquad \dots (6.129)$

High frequency voltage gain of the R-C coupled amplifier is

$$A_H = \dfrac{V_o}{V_i}$$

$$= -\dfrac{h_{fe} R_p}{h_{ie}(1 + j\omega C_d R_p)}$$

$$= -\frac{h_{fe}\,R_L}{(R_L + h_{ie})(1 + \dfrac{j\omega}{\omega_H})}$$

$$= \frac{A_M}{1 + j\dfrac{\omega}{\omega_H}} \qquad\qquad \text{... (6.130a)}$$

$$= |A_H| < \theta_H \qquad\qquad \text{... (6.130b)}$$

where $\quad \omega_H = \dfrac{1}{C_d\,R_p}.$

Now, $\quad \dfrac{|A_H|}{|A_M|} = \dfrac{1}{\sqrt{1 + \left(\dfrac{\omega}{\omega_H}\right)^2}} \qquad\qquad \text{... (6.131)}$

The phase angle of the high frequency gain function is

$$\angle\theta_H = \pi - \tan^{-1}(\omega/\omega_H) \qquad\qquad \text{... (6.132)}$$

At $\quad \omega = \omega_H, \quad \dfrac{|A_H|}{|A_M|} = \dfrac{1}{\sqrt{2}}$

Thus, ω_H is the upper half-power frequency of the R-C coupled amplifier at which the gain falls

to $\quad 1\!\!\Big/\!\!\sqrt{2}\quad$ of its mid-frequency value. At $\omega = \omega_H$, $\angle\theta_H = \pi - \dfrac{\pi}{4} = \dfrac{3\pi}{4}$. As $\omega \to \infty$,

$\angle\theta_H \to \pi\!\!\Big/\!\!2$. The frequency response of the R-C coupled amplifier is shown in fig. 6.21 (a) while the phase characteristics is shown in Fig. 6.21 (b).

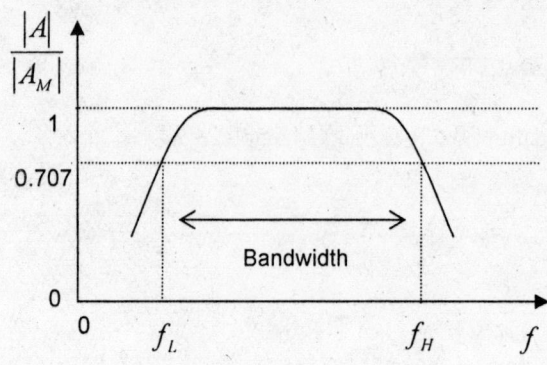

Fig. 6.21. (a) frequency response of the R-C coupled amplifier.

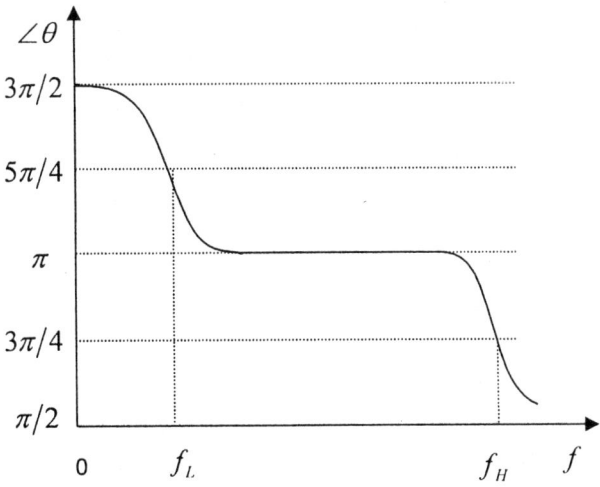

Fig. 6.21. (b) Phase characteristics of the R-C Coupled amplifier.

The gain–bandwidth product of the R-C coupled amplifier is given by

$$| A_M | \times (\omega_H - \omega_L) \frac{1}{2\pi}$$

$$\cong | A_M | \frac{\omega_H}{2\pi} \quad \text{(since } \omega_L << \omega_H)$$

$$= \frac{h_{fe} R_L}{(h_{ie} + R_L)} \frac{1}{2\pi C_d R_p}$$

$$= \frac{h_{fe} R_L}{(h_{ie} + R_L)} \frac{1}{2\pi C_d} \frac{(h_{ie} + R_L)}{R_L h_{ie}}$$

$$= \frac{h_{fe}}{2\pi h_{ie} C_d} = \text{constant.} \qquad \qquad \dots (6.133)$$

Thus, the gain-bandwidth product of the R-C coupled amplifier is a constant independent of external parameters.

SOLVED PROBLEM

6.1 Consider the following amplifier circuit using a Si npn transistor which has a dc current gain equal to 100.
(a) **Find** out the dc base current, dc collector current, and **dc collector voltage.**

(b) If the input signal is a $10\,\text{mV}$, $10\,\text{kHz}$ ac find the output voltage (V_o) using simplified h–parameter model of the transistor. Given $h_{fe} = 100$ and $h_{ie} = 2\,\text{k}\Omega$.

(c) What will be the effect on gain of this amplifier if the $100\,\mu\text{F}$ emitter bypass capacitor is removed?

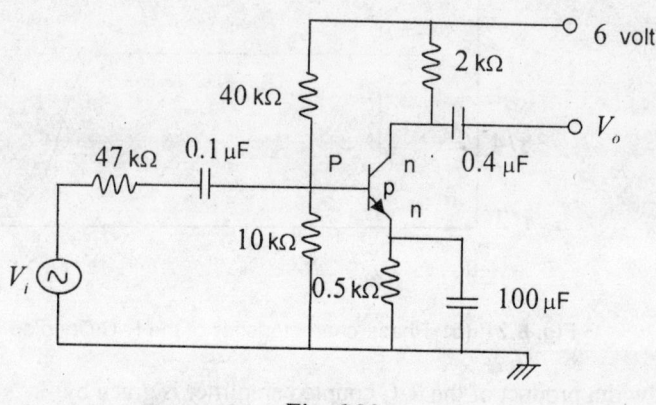

Fig. 6.22.

Solution

Here, $h_{FE} = 100$. Then, $I_C = h_{FE} I_B = 100 I_B$

Applying Thevenin's theorem at the point P in Figure 6.22, the equivalent voltage source (V_{BB}) is

$$V_{BB} = \frac{6.10.10^3}{(40+10).10^3} = \frac{6}{5} = 1.2 \text{ volt}$$

The internal resistance of this equivalent voltage source is

$$R_B = \frac{(40.10).10^3.10^3}{(40+10).10^3} = 8 \text{ k}\Omega$$

The reduced base emitter circuit under dc condition looks the following:

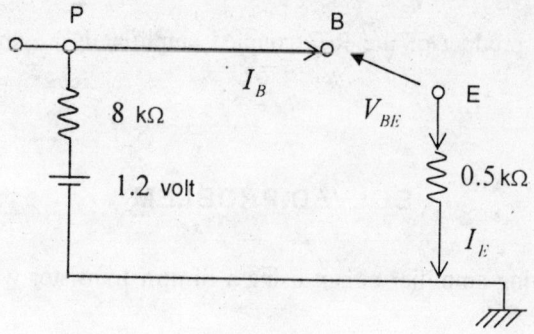

Fig. 6.23

This circuit gives a dc equation of the form ,

$$1.2 = 8 \times I_B + V_{BE} + I_E (0.5), \text{ where } I_B, I_C, I_E \text{ are in mA,}$$

$$I_E = I_C + I_B = 101 I_B \text{ since } I_C = 100 I_B$$

$$V_{BE} = 0.6 \text{ V}$$

$$\therefore \quad 1.2 = 58.5 I_B + 0.6$$

$$I_B = \frac{0.6}{58.5} \text{mA} = 10.25 \,\mu\text{A}.$$

$$I_C = 100 I_B = 1.025 \text{ mA}$$

The dc equation for the collector circuit is written as

$$0.5 I_E + V_{CE} + I_C \cdot 2 = 6 \quad \text{where } I_E \text{ and } I_C \text{ are in mA.}$$

$$\therefore \quad V_{CE} = 6 - 2 I_C - 0.5(I_C + I_B)$$

$$= 6 - 2.5 I_C - 0.5 I_B$$

$$= 6 - 2.57$$

$$= 3.43 \text{ volt.}$$

(b) The simplified ac equivalent circuit is shown below.

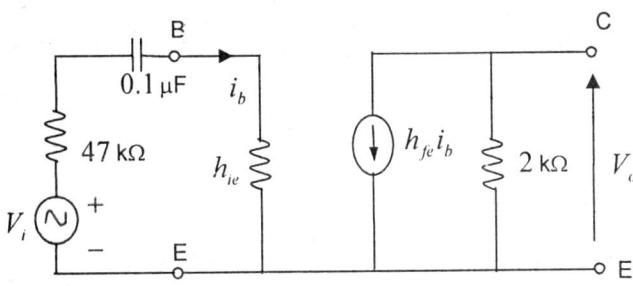

Fig. 6.24

$$i_b = \frac{V_i}{(47+2) \times 10^3 - j \dfrac{10^3}{2\pi}} \approx \frac{10 \times 10^{-3}}{49 \times 10^3} \approx 0.204 \,\mu\text{A}$$

Since , $V_i = 10 \,\text{mV}$, $\omega = 2\pi \times 10^4 \text{radian}$, $\dfrac{1}{\omega C} = \dfrac{1}{2\pi.10^4.10^{-7}} = \dfrac{10^3}{2\pi} \Omega$ and $\dfrac{1}{2\pi} \ll 49$.

$$V_o = -h_{fe}i_b(2 \times 10^3)$$
$$= -100 \cdot (0.204 \times 10^{-6})2 \times 10^3$$
$$= -40.8 \times 10^{-3} = -40.8\,\text{mV}$$

(c) If the emitter resistor $0.5\,\text{k}\Omega$ is unbypassed, then h_{ie} will be replaced by $\left[h_{ie} + (1 + h_{fe})R_E\right]$ in the equation for base current determination. The bypass capacitor of $100\,\mu\text{F}$ is removed in this case.

Thus, $i_b = \dfrac{V_i}{47 \times 10^3 + [2 + (1+100) \times 0.5]10^3 - j\dfrac{10^3}{2\pi}}$

$$\approx \frac{10.10^{-3}}{[49 + 50.5].10^3}$$

$$= \frac{10}{99.5}10^{-6} \approx 0.1\ \mu\text{A}$$

$\therefore \qquad V_0 = -h_{fe}i_b(2 \times 10^3)$
$$= -100(0.1 \times 10^{-6}) \times 2 \times 10^3$$
$$= -20 \times 10^{-3} \qquad = -20\,\text{mV}.$$

6.2　Calculate the voltage gain, input impedance and output impedance of a single stage CE transistor amplifier having a load resistance of $1\,\text{k}\Omega$. Given: $h_{ie} = 1\,\text{k}\Omega$, $h_{fe} = 50$, $h_{re} = 2 \times 10^{-4}$, $h_{oe} = 2 \times 10^{-5}$ mho. Neglect the internal impedance of the source.

Solution

$$R_L = 1\,\text{k}\Omega = 10^3\,\Omega,\ R_s = 0$$

$$\Delta^{he} = h_{ie}h_{oe} - h_{fe}h_{re}$$
$$= 10^3(2 \times 20^{-5}) - 50(2 \times 10^{-4})$$
$$= 0.02 - 0.01 = 0.01$$

Voltage gain $A_v = -\dfrac{h_{fe}R_L}{h_{ie} + R_L\Delta^{he}}$

$$\cong \frac{50 \times 10^3}{10^3 + 10^3 \times (.01)}$$

$$= -\frac{50}{1 + .01} = -\frac{50}{1.01} = -49.5$$

Input impedance $R_i = \dfrac{h_{ie} + R_L \, \Delta^{he}}{1 + h_{oe} \, R_L}$

$= \dfrac{10^3 + 10^3 \, (.01)}{1 + 2 \, . \, 10^{-5} \times 10^3} = \dfrac{1.01}{1.02} \times 10^3 = 0.99 \text{ k}\Omega$

Output impedance $R_o = \dfrac{h_{ie} + R_s}{\Delta^{he} + h_{oe} \, R_s}$

$= \dfrac{h_{ie}}{\Delta^{he}}$ (since $R_s = 0$)

$= \dfrac{10^3}{0.01} = 100 \text{ k}\Omega$

6.3   Calculate the maximum available power gain of a CE amplifier having a load resistance of 1 kΩ. The transistor parameters are as follows:

$h_{ie} = 1 \text{ k}\Omega$, $h_{fe} = 50$, $h_{re} = 2 \times 10^{-4}$ and $h_{oe} = 2 \times 10^{-5}$ mho.

Solution

$\text{MAPG} = \dfrac{h_{fe}^2}{\left(\sqrt{h_{ie} \, h_{oe}} + \sqrt{\Delta^{he}} \right)^2}$

$h_{ie} h_{oe} = 10^3 . \left(2 \times 10^{-5} \right) = 2 \times 10^{-2} = 0.02$

$\Delta^{he} = h_{ie} h_{oe} - h_{fe} h_{re}$

$= 2 \times 10^{-2} - 50 \times \left(2 \times 10^{-4} \right)$

$= 0.02 - 0.01 = 0.01$

$\therefore \text{ MAPG } = \dfrac{50 \times 50}{(\sqrt{2} + 1)^2 . 10^{-2}}$

$= \dfrac{25 \times 10^4}{5.82} = 4.29 \times 10^4.$

6.4   An R-C coupled CE amplifier has a load resistance of 1 kΩ and an input capacitance of 1 nF at the emitter-base junction of the following stage. Given, $h_{ie} = 1 \text{ k}\Omega$.

(a) Calculate the upper half-power frequency of this amplifier.

(b) Assuming that the lower half-power frequency is much smaller than the upper half-power frequency, calculate the mid frequency gain of this amplifier when the gain–bandwidth product is 8 MHz.

Solution

Given, $h_{ie} = 1\,k\Omega = 10^3\,\Omega$.

$C_d = 1\,nF = 10^{-9}\,F$

$R_L = 10^3\,\Omega$.

Now, $R_P = \dfrac{R_L h_{ie}}{R_L + h_{ie}} = \dfrac{10^3 \times 10^3}{(1+1) \times 10^3} = 500\,\Omega$.

Upper half-power frequency

$$f_H = \frac{1}{2\pi R_P C_d} = \frac{1}{2\pi . 500 . 10^{-9}}$$

$$= \frac{10^6}{\pi} = \frac{1}{\pi}\,MHz = 0.318\,MHz.$$

(b) $|A_M| \times f_H = 8\,MHz$

$\therefore$ $|A_M| = 8 \times \pi = 25.1$

6.5   (a) Calculate the lower 3 dB frequency of the following amplifier circuit.

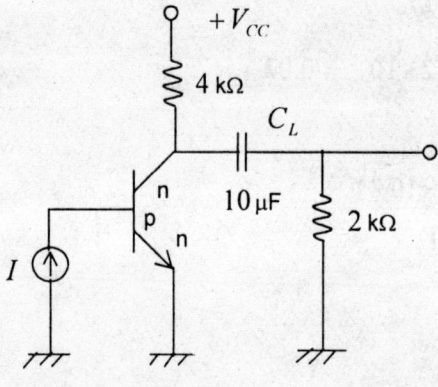

Fig. 6.25.

(b) Calculate the percentage tilt in the output if the input current I is a square   wave of frequency 200 Hz.

(c) What is the lowest frequency square wave which will suffer less than 2% tilt?

Solution

5.(a) Neglecting $h_{re}v_{oe}$ and h_{oe} from the equivalent circuit and then applying Thevenin's theorem to the points (a, b) we get the following circuits,

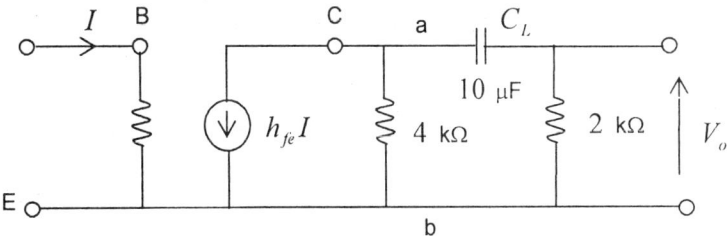

Fig. 6.26. (a) AC equivalent circuit.

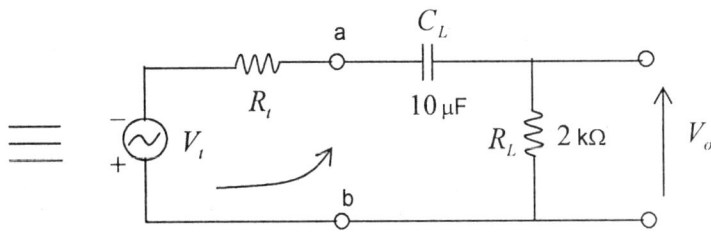

Fig. 6.26. (b) Thevenin equivalent circuit.

Total impedance of the Thevenin equivalent circuit is

$$Z = R_t + R_L - jX_C$$

where $X_C = \dfrac{1}{\omega C_L}$, R_t = Thevenin resistance and $R_L = 2$ kΩ = load impedance, ω = signal frequency. Now,

$$V_o = \frac{V_t R_L}{R_t + R_L - jX_C} = \frac{V_t \cdot \dfrac{R_L}{R_t + R_L}}{1 - j\dfrac{X_C}{R_t + R_L}} = \frac{V'}{1 - j\dfrac{\omega_L}{\omega}}.$$

where $V' = V_t \dfrac{R_L}{R_t + R_L}$ and $\omega_L = \dfrac{1}{(R_t + R_L)C_L}$.

$$V_t = h_{fe}I \times 4 = 4h_{fe}I \quad \text{where } I \text{ is in mA.}$$
$$R_t = 4\,\text{k}\Omega$$

This is similar with a high-pass filter response.

Lower 3-dB frequency $f_L = \dfrac{\omega_L}{2\pi} = \dfrac{1}{2\pi(R_L + R_t)C_L}$

$$= \frac{1}{2\pi.10^{-5}.(2+4).10^3} = \frac{100}{12\pi} = 2.65 \text{ Hz.}$$

5.(b) For calculating the tilt of the square wave, consider the high pass characteristics of the R-C circuit in response to a step input.

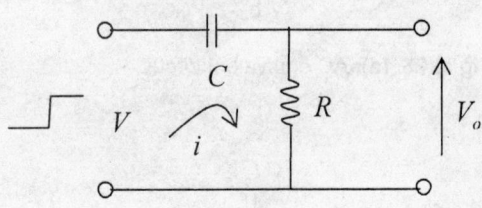

Fig. 6.27.

The circuit equation is

$$V_o + \frac{1}{C} \int i \, dt = V$$

or, $Ri + \dfrac{1}{C} \int i \, dt = V$ [since $V_o = Ri$]

or, $R\dfrac{di}{dt} + \dfrac{i}{C} = 0$ [differentiating]

$\therefore$ $i(t) = i_o e^{-1/RC}$ where i_o = constant of integration.

Then, $V_o = Ri = R i_o e^{-1/CR}$. For $t \ll RC$, $V_o = R i_o \left(1 - \dfrac{t}{CR}\right)$

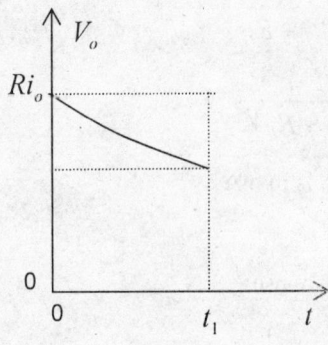

Fig. 6.28.

$$\therefore \qquad V_o(t = t_1) = Ri_o\left(1 - \frac{t_1}{RC}\right)$$

Tilt of the square wave (i.e., sag) in time t_1 is

$$S = \frac{Ri_o - V_o(t = t_1)}{Ri_o} = \frac{t_1}{RC}$$

For a square wave of period T, the ON time is $T/2$. So, the percentage of tilt of the square wave

is $S = \dfrac{T/2}{RC} \times 100\%$ [putting $t_1 = T/2$]

$$= \frac{1}{2fRC} \times 100\%$$

Comparing the standard RC circuit with the given circuit, we write $\dfrac{1}{RC} = \omega_L$.

Here, $f = 200\,\text{Hz}$.

$$\therefore \qquad S = \frac{\omega_L}{2f} \times 100\% = \frac{\pi f_L}{f} \times 100\%$$

$$= \pi \cdot \frac{100}{12\pi} \cdot \frac{1}{200} \times 100\% = 4.166\%$$

5.(c) Here, $S = 2\% = 0.02$

$$\therefore \qquad \frac{\pi f_L}{f} = 0.02 . \text{ Hence, } f = \frac{\pi f_L}{0.02} = \pi \times \frac{100}{12\pi} \times 50 = 416.6 \text{ Hz.}$$

PROBLEMS

6.1   A single stage CE amplifier using a Si npn transistor employs potential divider bias and an emitter bias as shown in Fig. 6.2(a). Resistances R_1, R_2 form the potential divider and R_E is the emitter resistor. Determine the quiescent values of I_B, I_C and V_{CE}.

Given : $h_{FE} = 50$, $V_{CC} = 10\,\text{V}$, $R_L = 1.5\,\text{k}\Omega$, $R_E = 100\,\Omega$, $R_2 = 50\,\text{k}\Omega$ and $R_1 = 5\,\text{k}\Omega$. R_L is the load resistance. Other symbols have their usual meanings.

[**Ans.** 32 µA, 1.6 mA, 7.43 V]

6.2   A CE transistor has a beta cut-off frequency equal to 1 MHz. The h_{fe} parameter at low frequency has a value of 150. Find the value of short-circuit current gain at 3 MHz.

[**Ans.** 47.4]

6.3 A transistor amplifier operating in the CE mode has the following h-parameters: $h_{ie} = 800\ \Omega$, $h_{re} = 5 \times 10^{-4}$, $h_{fe} = 50$, and $h_{oe} = 8 \times 10^{-5}$ mho. If the load resistance is 5 kΩ, calculate the current gain, input resistance, voltage gain and output resistance of this amplifier. Neglect internal resistance of the source.

[**Ans.** 35.7, 710.7 Ω, 251.2, 20.5 kΩ]

6.4 A silicon npn transistor with $h_{FE} = 50$ is to be operated as a CE amplifier with a base resistor (R_B) connected from the collector power supply to the base point in order to provide a fixed bias. The operating point collector current, $I_C = 1$ mA, and collector voltage $V_{CE} = 4$ V. The load resistance is 2 kΩ. Assuming a base–emitter voltage $V_{BE} = 0.7$ volt, find the value of the base resistor R_B. The circuit is shown in Fig. 6.1.

[**Ans.** 265 kΩ]

QUESTIONS

6.1 What do you mean by the term 'biasing' of a transistor? What is the necessity of proper biasing of a transistor? Give an example of a fixed bias circuit? Calculate the stability factors of this circuit.

6.2 Define stability factors of a transistor bias circuit. Calculate the stability factors of a fixed-bias circuit and a self-bias circuit. Compare the stability factors of these circuits.

6.3 Analyze a self-bias circuit using a potential divider (R_1, R_2) bias and a $R_E - C_E$ biasing circuit connected with the emitter of the transistor. Derive expressions for the stability factors for this circuit. How does the $R_E - C_E$ parallel combination improves the stability of the bias point of the transistor? Explain physically.

6.4 Derive from fundamental considerations the h-parameter equivalent circuit of a transistor for small-signal low frequency operation. How does this model get modified in the high frequency region?

6.5 Draw the circuit diagram of a CE amplifier with a resistive load R_L. Derive expressions for voltage gain, current gain, input resistance and output resistance of this amplifier. Find out an expression for maximum available power gain of a CE amplifier.

6.6 Draw the circuit diagram of a CC amplifier. Why is it called an emitter follower? Derive expressions for voltage gain, current gain, input resistance and output resistance of the CC amplifier. Mention a use of CC amplifier.

6.7 Draw a neat circuit diagram of a R–C coupled transistor amplifier. Derive an expression for high frequency gain of this amplifier. Assuming that the emitter is grounded, find out an expression for low frequency gain of this amplifier.

6.8 Prove that the gain–bandwidth product of the R–C coupled amplifier is a constant. Plot schematically the frequency response and the phase characteristics of the R–C coupled amplifier.

7

Special Purpose
Transistor Amplifiers

7.1 CASCADED AMPLIFIER

In this section, we will discuss cascaded amplifiers. These are cascode amplifier and Darlington pair. We will also describe cascaded RC couple amplifiers.

7.1.1 CASCODE AMPLIFIER

It is a cascade of CE–CB amplifiers. At the input of the cascade there is a CE amplifier and a CB amplifier is connected with the output of the CE amplifier. The input impedance of the CB amplifier acts as the load resistance of the CE amplifier. Since this load resistance is small, the voltage gain of the CE amplifier is small (typically less than unity). There is a capacitance (C_{bc}) between the collector and base of the CE (Q_1) amplifier which provides feedback at high frequencies. Since the voltage gain (A) of the CE amplifier is small ($|A| < 1$), the Miller capacitance at the input, $(1 - A)C_{bc}$, becomes equal to C_{bc} effectively. Thus, the effect of feedback capacitance C_{bc} is greatly reduced. This makes the cascode amplifier more suitable at high frequencies. The bandwidth of the cascode amplifier is large. The circuit of the cascode amplifier is shown in Fig. 7.1.

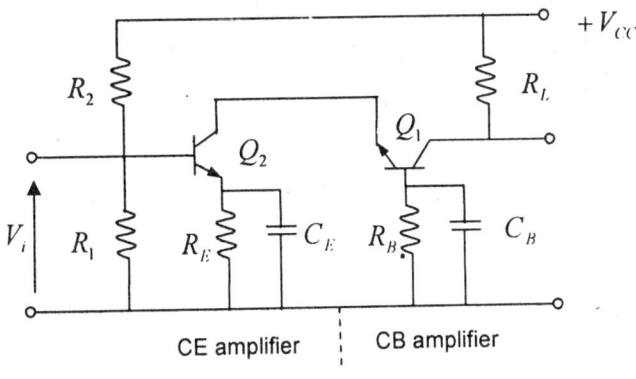

Fig. 7.1. Circuit diagram of cascode amplifier.

The current gain of the CB amplifier (Q_1) is

$$A_2 = \frac{h_{fb}}{1 + h_{ob} R_L}$$

$$= \frac{-h_{fe}/(1 + h_{fe})}{1 + \dfrac{h_{oe}}{(1 + h_{fe})} R_L} = -\frac{h_{fe}}{1 + h_{fe} + h_{oe} R_L}$$

$$= -\frac{h_{fe}}{1 + h_{fe}} \qquad \qquad \ldots (7.1)$$

The input impedance of the CB stage (Q_2) is

$$R_{i2} = \frac{h_{ib} + R_L \Delta^{hb}}{1 + h_{ob} R_L} = \frac{h_{ie} + R_L \Delta^{he}/(1 + h_{fe})}{1 + \dfrac{h_{oe}}{1 + h_{fe}} R_L}$$

$$= \frac{h_{ie} + R_L \Delta^{he}}{1 + h_{fe} + h_{oe} R_L} \cong \frac{h_{ie}}{1 + h_{fe}} \qquad \qquad \ldots (7.2)$$

Since, $R_L \Delta^{he} \ll h_{ie}$ and $h_{oe} R_L \ll (1 + h_{fe})$. Also, $\Delta^{hb} = \Delta^{he}/(1 + h_{fe})$.

Current gain of the CE amplifier (Q_1) is

$$A_1 = \frac{h_{fe}}{1 + h_{oe} R_{i2}}$$

$$\cong h_{fe} \qquad \qquad \ldots (7.3)$$

since $h_{oe} R_{i2} \ll 1$. Here, R_{i2} acts as the load resistance of the CE amplifier.

Overall current gain of he cascode amplifier is

$$A_c = A_1 A_2 = -\frac{h_{fe}^2}{(1 + h_{fe})} \cong -h_{fe} \qquad \qquad \ldots (7.4)$$

The input impedance of the cascode amplifier is

$$R_{i1} = \frac{h_{ie} + R_{i2} \Delta^{he}}{1 + h_{oe} R_{i2}}$$

$$= \frac{h_{ie} + \dfrac{h_{ie} \Delta^{he}}{(1 + h_{fe})}}{1 + \dfrac{h_{oe} h_{ie}}{(1 + h_{fe})}}$$

$$= \frac{h_{ie}(1 + h_{fe}) + h_{ie}\Delta^{he}}{(1 + h_{fe}) + h_{oe}h_{ie}}$$

$$\cong h_{ie} \qquad\qquad\qquad \dots (7.5)$$

since $\Delta^{he} \ll (1 + h_{fe})$ and $h_{oe}h_{ie} \ll (1 + h_{fe})$.

Voltage gain of the CB amplifier of the cascode is

$$A_{v2} = -\frac{h_{fb}R_L}{h_{ib} + R_L\Delta^{hb}}$$

$$= \frac{h_{fe}R_L}{h_{ie} + R_L\Delta^{he}} \qquad\qquad \dots (7.6)$$

Voltage gain of the CE amplifier (Q_1) of the cascode is

$$A_{v1} = -\frac{h_{fe}R_{i2}}{h_{ie} + R_{i2}\Delta^{he}}$$

$$= -\frac{h_{fe}h_{ie}}{h_{ie}(1 + h_{fe}) + h_{ie}\Delta^{he}}$$

$$\cong -\frac{h_{fe}}{1 + h_{fe}} \qquad\qquad \dots (7.7)$$

since $\Delta^{he} \ll (1 + h_{fe})$. R_{i2} acts as the load of the CE stage.

The overall voltage gain of the cascode is

$$A_{vc} = A_{v1}A_{v2} = -\frac{h_{fe}^2 R_L}{(1 + h_{fe})(h_{ie} + R_L\Delta^{he})}$$

$$\cong -\frac{h_{fe}R_L}{h_{ie}} \qquad\qquad \dots (7.8)$$

since $h_{fe} \gg 1$ and $R_L\Delta^{he} \ll h_{ie}$.

The output impedance of the CE amplifier (Q_1) of the cascode is

$$R_{o1} = \frac{h_{ie} + R_s}{h_{oe}R_s + \Delta^{he}} \qquad\qquad \dots (7.9)$$

where R_s is the internal resistance of the input signal source. R_{o1} acts as the internal resistance of the source for the CB amplifier of the cascode. Then, the output resistance of the cascode is

$$R_{o2} = \frac{h_{ib} + R_{o1}}{h_{ob} R_{o1} + \Delta^{hb}}$$

$$= \frac{h_{ie} + R_{o1}(1 + h_{fe})}{h_{oe} R_{o1} + \Delta^{he}}$$

$$= \frac{h_{ie}(h_{oe} R_s + \Delta^{he}) + (1 + h_{fe})(h_{ie} + R_s)}{h_{oe}(h_{ie} + R_s) + \Delta^{he}(h_{oe} R_s + \Delta^{he})}$$

$$\cong \frac{(1 + h_{fe})}{h_{oe}} \qquad \qquad \dots (7.10)$$

neglecting the second term of the denominator which is small. Also, $(h_{oe} R_s + \Delta^{he}) \ll (1 + h_{fe})$, so the first term is neglected from the numerator. Cascode amplifier is useful for the amplification of video signals having bandwidth in the MHz region.

7.1.2 DARLINGTON PAIR

The circuit diagram of a Darlington pair is shown in Fig. 7.2. This circuit is used to realize high input resistance in BJT circuits–much higher than the input resistance of an emitter follower. Darlington pair is essentially a cascade of two emitter followers where the second emitter follower's input resistance act as the emitter load of the first emitter follower.

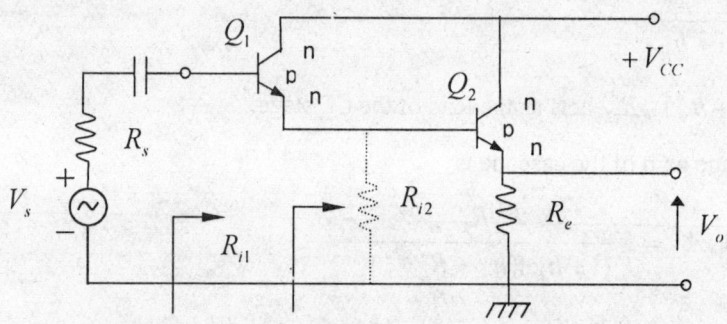

Fig. 7.2. Circuit diagram of a Darlington pair (dc biasing not shown)

The current gain of the output transistor Q_2 is

$$A_2 = \frac{h_{fc}}{1 + h_{oc} R_e} \cong -(1 + h_{fe}) \qquad \qquad \dots (7.11)$$

since $h_{oc} R_e = h_{oe} R_e \ll 1$.

Input resistance of the second emitter follower Q_2 is

$$R_{i2} = \frac{h_{ic} + R_e \Delta^{hc}}{1 + h_{oc} R_e}$$

$$= \frac{h_{ie} + R_e(1 + h_{fe})}{1 + h_{oe} R_e}$$

$$\cong (1 + h_{fe}) R_e \qquad \qquad \ldots (7.12)$$

since $\Delta^{hc} \cong (1 + h_{fe})$, $h_{oe} R_e \ll 1$ and $h_{ie} \ll R_e(1 + h_{fe})$. Now, R_{i2} is the emitter load of the first emitter follower Q_1. The current gain of this emitter follower Q_1 is

$$A_1 = \frac{h_{fc}}{1 + h_{oc} R_{i2}}$$

$$= -\frac{(1 + h_{fe})}{1 + h_{oe}(1 + h_{fe}) R_e} \qquad \qquad \ldots (7.13)$$

The overall current gain of the Darlington pair is

$$A_D = A_2 A_1$$

$$= \frac{(1 + h_{fe})^2}{1 + h_{fe} h_{oe} R_e} \qquad \qquad \ldots (7.14)$$

since $h_{oe} R_e \ll 1$.

The input resistance of the Darlington pair is

$$R_{i1} = \frac{h_{ic} + R_{i2} \Delta^{hc}}{1 + h_{oc} R_{i2}}$$

$$= \frac{h_{ie} + (1 + h_{fe})^2 R_e}{1 + h_{oe}(1 + h_{fe}) R_e}$$

$$\cong \frac{h_{ie} + (1 + h_{fe})^2 R_e}{1 + h_{fe} h_{oe} R_e} \qquad \qquad \ldots (7.15)$$

In practice, $h_{ie} \ll (1 + h_{fe})^2 R_e$.

Then,

$$R_{i1} \cong \frac{(1 + h_{fe})^2 R_e}{1 + h_{fe} h_{oe} R_e} \qquad \qquad \ldots (7.16)$$

h_{ie} of the transistors Q_1 and Q_2 are inversely proportional to their collector currents. The collector current of Q_2 is $(1 + h_{fe})$ times greater than that of Q_1. Then, $h_{ie1} = (1 + h_{fe})h_{ie2}$. h_{ie1} and h_{ie2} are h_{ie} parameters for Q_1 and Q_2 respectively.

The voltage gain of the output emitter follower Q_2 is

$$A_{v2} = -\frac{h_{fc}R_e}{h_{ic2} + R_e\Delta^{hc2}}$$

$$= \frac{(1 + h_{fe})R_e}{h_{ie2} + (1 + h_{fe})R_e} \qquad \ldots (7.17)$$

Since h_{fe} of Q_1 and Q_2 do not vary much with collector current, h_{fe} has been taken to be the same for Q_1 and Q_2.

The voltage gain of the first emitter follower Q_1 is

$$A_{v1} = -\frac{h_{fc}R_{i2}}{h_{ie1} + R_{i2}\Delta^{hc1}}$$

$$= \frac{(1 + h_{fe})^2 R_e}{h_{ie1} + (1 + h_{fe})^2 R_e} \qquad \ldots (7.18)$$

The voltage gain of the Darlington pair is

$$A_{vD} = A_{v1}A_{v2}$$

$$= \frac{(1 + h_{fe})^3 R_e^2}{\left[h_{ie2} + (1 + h_{fe})R_e\right]\left[h_{ie1} + (1 + h_{fe})^2 R_e\right]}$$

$$= \frac{(1 + h_{fe})^2 R_e^2}{\left[h_{ie2} + (1 + h_{fe})R_e\right]^2} \qquad \ldots (7.19)$$

since $h_{ie1} = (1 + h_{fe})h_{ie2}$. Obviously, $A_{vD} < 1$.

We will now calculate the output impedance of the Darlington pair. The output resistance of first emitter follower Q_1 is

$$R_{o1} = \frac{h_{ic} + R_s}{h_{oc}R_s + \Delta^{hc}}$$

$$= \frac{h_{ie1} + R_s}{h_{oe}R_s + (1 + h_{fe})}$$

$$\cong \frac{h_{ie1} + R_s}{(1 + h_{fe})} \qquad\qquad \ldots (7.20)$$

since $h_{oe}R_s << (1 + h_{fe})$. R_{o1} is much smaller than the input resistance R_{i1}.

The output resistance of the second emitter follower Q_2 is the output resistance (R_o) of the Darlington pair. Here, R_{o1} acts as the source internal resistance for the emitter follower Q_2.

Then,

$$R_o = \frac{h_{ic2} + R_{o1}}{h_{oc2}R_{o1} + \Delta^{hc2}}$$

$$= \frac{h_{ie2} + \dfrac{h_{ie1} + R_s}{1 + h_{fe}}}{h_{oe}R_{o1} + (1 + h_{fe})}$$

$$\cong \frac{h_{ie2}(1 + h_{fe}) + h_{ie1} + R_s}{(1 + h_{fe})^2}$$

since $h_{oe}R_{o1} << (1 + h_{fe})$.

Then, $R_o = \dfrac{2.h_{ie2}(1 + h_{fe}) + R_s}{(1 + h_{fe})^2}$ \qquad\qquad $\ldots (7.21)$

Thus, R_o is much smaller than R_{o1} and hence $R_o << R_{i1}$ since $R_{o1} << R_{i1}$.

7.1.3 RC COUPLED AMPLIFIERS IN CASCADE

When the gain from a single stage amplifier is not sufficient, we require a multistage amplifier in the form of a cascade. The output of one amplifier is connected to the input of the next amplifier. A cascade of n-RC coupled amplifiers is shown in Fig.7.3 using a block diagram.

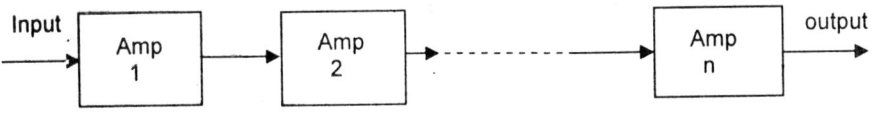

Fig. 7.3. Typical cascade of n amplifiers.

Since the lower half power frequency (ω_L) is much smaller than the upper half power frequency (ω_H), the bandwidth (BW) of the single stage RC coupled amplifier is

$$BW = \omega_H - \omega_L \cong \omega_H \qquad\qquad \ldots (7.22)$$

The high frequency gain of a single stage RC coupled amplifier is written as

$$A_H = \frac{A_M}{1 + j\dfrac{\omega}{\omega_H}} \qquad \qquad \dots (7.23)$$

where A_M is the mid frequency gain of the amplifier. The overall gain of the cascade of n RC coupled amplifiers is

$$A_{HC} = A_H^n = \frac{A_M^n}{\left(1 + j\dfrac{\omega}{\omega_H}\right)^n} \qquad \qquad \dots (7.24)$$

Now, $\quad |A_{HC}| = \dfrac{|A_M^n|}{\left[1 + \dfrac{\omega^2}{\omega_H^2}\right]^{n/2}} \qquad \qquad \dots (7.25)$

The upper half power frequency of the cascade is ω_{HC} defined by the condition

$$|A_{HC}|_{\omega=\omega_{HC}} = \frac{1}{\sqrt{2}}|A_M^n| \qquad \qquad \dots (7.26)$$

using eqn. (7.25), we write

$$\frac{|A_M^n|}{\left[1 + \dfrac{\omega_{HC}^2}{\omega_H^2}\right]^{n/2}} = \frac{1}{\sqrt{2}}|A_M^n|$$

$$\therefore \quad \left[1 + \left(\frac{\omega_{HC}}{\omega_H}\right)^2\right]^n = 2$$

or, $\quad \left(\dfrac{\omega_{HC}}{\omega_H}\right)^2 - 2^{\frac{1}{n}} - 1$

$$\therefore \quad \omega_{HC} = \omega_H \sqrt{2^{1/n} - 1} \qquad \qquad \dots (7.27)$$

For a single amplifier $n = 1$ and $\omega_{HC} = \omega_H$. Thus, the overall bandwidth of the cascaded RC coupled amplifier is reduced and reduction occurs by the multiplying factor $\sqrt{2^{1/n} - 1}$ from that of a single-stage isolated amplifier. This penalty in bandwidth is paid for achieving a higher gain $\left(A_M^n\right)$.

7.2 AMPLIFIER CLASSIFICATION

There are various ways of classifying amplifiers. Depending upon the nature of the load impedance, amplifiers are classified as RC coupled amplifier, transformer coupled amplifier, and tuned amplifier. A tuned amplifier uses a tuned circuit as the load. In the other two amplifiers, the type of output coupling is important. Depending upon the range of signal frequency to be amplified, amplifiers are classified as audio, radio and video amplifers. In audio amplifier, signal frequencies lie in the audible range typically up to frequencies $\sim 20\,\text{kHz}$. Radio frequency amplifiers amplify signals having frequency in the rf region ($300\,\text{kHz}$-$30\,\text{kHz}$). Video frequency amplifliers amplify video signals (e.g., in TV) having typical frequencies in the MHz region. The amplifiers amplifying signals in the frequency range $30\,\text{kHz}$-$300\,\text{kHz}$ are low frequency amplifiers.

Depending upon whether we desire to have high output voltage or high output power, amplifiers can be classified as voltage and power amplifiers. When the received signal voltage is low, it requires amplification in one or more stages of voltage amplifiers. Voltage amplifiers are basically small signal amplifiers. The amplified voltage in a radio receiver is finally applied to a power amplifier which gives a power-amplified signal at the output. Power amplifiers are large signal amplifies. In a radio receiver, the input signal is very weak. It is first voltage-amplified in more than one stages of voltage amplifiers and finally the voltage amplifier output is power-amplified in a power amplifier and fed to a loud speaker.

Depending upon the operation of the amplifier, (particularly the period of conduction) it can be classified into class-A, class-AB, class-B and class-C categories. In class-A operation, the active device (i.e., the transistor) conducts for the entire cycle of the input signal. In class-B operation, the active device conducts for half cycle ($180°$) of the input signal. In class-AB operation, the device conducts for more than half cycle but less than full cycle of the input signal. In class-C operation, the device conducts for less than half cycle of the input signal. In addition to these four classes, a class-D operation is sometimes defined. In class-D operation, the amplifier is switched on and off. It generates a series of pulses at the output.

7.3 CLASS-A AF POWER AMPLIFIER

In class-A operation, the transistor conducts for the full cycle (i.e., $360°°$) of the input signal. The operating point of the transistor remains in the active region over the entire cycle of operation. The circuit diagram of a class-A power amplifier is shown in Fig. 7.4.

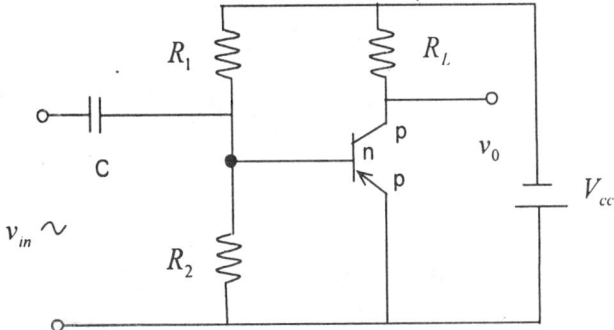

Fig. 7.4. Schematic circuit diagram of a class-A power amplifier.

The efficiency (η) of the amplifier is defined as the ratio of ac power delivered to the load to the dc power input to the amplifier.

Let V_L = peak value of load voltage.

I_L = peak value of load current.

V_L and I_L are sinusoidal since the input signal v_{in} is assumed to be sinusoidal. The ac power delivered to the load is

$$P_{ac} = \frac{1}{2} V_L I_L \qquad \qquad \ldots (7.28)$$

Let V_{CC} = supply voltage

I_0 = operating point current through the transistor.

DC power input to the amplifier is

$$P_{dc} = V_{CC} I_0 \qquad \qquad \ldots (7.29)$$

When there is no input signal, the ac load current

$$I_L = 0. \text{ So, } P_{ac} = 0. \text{ Then, } \eta = \frac{P_{ac}}{P_{dc}} = 0.$$

In order to get maximum swing of the quiescent operating point, we place the operating point in the middle of the dc load line. Secondly, in order to maximize the useful ac power (P_{ac}), the input signal amplitude must be large enough to drive the transistor to the extremeties of the active region. The extremeties of the active region are in the boundary of saturation on one side and in the boundary of cut-off on the other side.

When the transistor is just cut-off, $I_L = 0$ and when the transistor is just saturated $I_L = V_{CC}/R_L$. So, for maximum efficiency, the load current amplitude should be

$$I_L = \frac{1}{2}\left(\frac{V_{CC}}{R_L} - 0\right) = \frac{V_{CC}}{2R_L}.$$

Again, the operating point dc current is

$$I_0 = \frac{1}{2}\frac{V_{CC}}{R_L} \qquad \qquad \ldots (7.30)$$

since the dc operating point is located at the middle of the load line. Now, the maximum efficiency (η) is given by

$$\eta_{max} = \frac{P_{ac}}{P_{dc}} = \frac{\frac{1}{2}\left(\frac{V_{CC}}{2R_L}\right)^2 R_L}{V_{CC}\left(\frac{V_{CC}}{2R_L}\right)} = \frac{1}{4} = 25\%$$

So, class-A power amplifier has a poor efficiency having a maximum value of 25%. The advantage of class-A power amplifier is that since its operation is confined in the active region, the distortion of this amplifier is small. In this case, since the load R_l is directly connected with the collector the dc current flows through R_l producing dc power loss. If the load is coupled by means of a transformer we can minimize this dc power loss and increase the efficiency.

7.4 TRANSFORMER-COUPLED CLASS-A POWER AMPLIFIER

Here, the load is couple to the amplifier by means of an audio transformer. The circuit diagram is shown in Fig. 7.5.

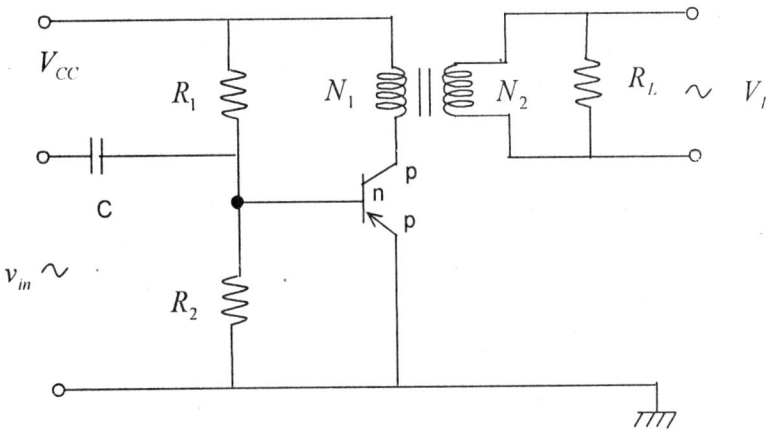

Fig. 7.5. Schematic circuit diagram of a transformer-coupled class-A AF power amplifier.

Here, N_1 and N_2 are the number of turns of the primary and secondary of the transformer. The power loss in the biasing resistances is neglected since it is small. When there is no input signal, the quiescent operating point collector voltage is V_{CC}. V_{CC} is the collector supply voltage which is itself negative for the pnp transistor. The reflected resistance in the primary circuit produced by the secondary circuit of the output transformer is $\left(\dfrac{N_1}{N_2}\right)^2 R_L$, where R_L is the load resistance in the secondary circuit. Then, the quiescent operating point collector current is

$$I_0 = \frac{V_{CC}}{n^2 R_L} \qquad\qquad \text{... (7.31)}$$

where $n = \dfrac{N_1}{N_2}$ is the turns ratio. The dc power input to the amplifier is

$$P_{dc} = V_{CC}.I_0 = \frac{V_{CC}^2}{n^2 R_L} \qquad\qquad \text{... (7.32)}$$

This power loss occurs in the collector of the transistor because there is no external resistance in the collector circuit. When the transistor touches the saturation point, the collector voltage becomes nearly zero. Thus, the collector voltage reduces from V_{CC} to 0 during the increase of collector current due to the rise in the input signal. On the other side, the collector voltage can rise up to $2V_{CC}$ when the transistor touches the cut-off point. Thus, the maximum collector voltage

swing will have an amplitude equal to $\frac{1}{2}(2V_{CC} - 0) = V_{CC}$. The operating point is located at the

middle of the ac load line. Since there is no collector resistance, the collector current is not limited by the resistance in the collector circuit. Thus, the collector current reaches a maximum level of $2I_0$ when the collector voltage becomes zero. The only limitation in collector current comes from the transformer saturation effect. The excursion of the operating point occurs from $(V_{CE} = 0, I_C = 2I_0)$ to $(V_{CE} = V_{CC}, I_C = 0)$. The ac load voltage (V_L) is given by

$\quad V_L$ = Voltage amplitude in the primary per turns ratio (n).

$$= \frac{V_{CC}}{n} \qquad \qquad \dots (7.33)$$

$\therefore$ The maximum ac power delivered to the load is

$$P_{ac} = \frac{1}{2}\frac{V_L^2}{R_L} = \frac{V_{CC}^2}{2n^2 R_L} \qquad \qquad \dots (7.34)$$

The maximum efficiency of transformer-coupled class-A power amplifier is

$$\eta_{\max} = \frac{P_{ac}}{P_{dc}} = \frac{V_{CC}^2}{2R_L n^2} \cdot \frac{R_L n^2}{V_{CC}^2}$$

$$= \frac{1}{2} = 50\%.$$

This improvement in efficiency originates from the fact that there is no collector resistance in this transformer-coupled class-A power amplifier and so there is no dc power loss in it. This is identical with the collector circuit efficiency of the amplifier since the total power loss occurs in the collector of the transistor.

7.5 CLASS-B PUSH-PULL AF AMPLIFIER

Operation of the amplifier is said to be class-B when the amplifier is biased so that the transistor conducts for $180°$ of the input signal cycle. In class-B operation, the transistors are biased at cut-off which is taken to be 0 V. To be more precise, the cut-off bias may be taken as -0.7 V for Si pnp transistor. The push-pull configuration uses two transistors – one of which conducts current during the positive half cycle of the input signal while the other conducts current during the negative half cycle of the input signal. Since one part of the circuit conducts during the high level (i.e., positive half cycle) of the input signal it is called pushing. Again, since the other part of the circuit conducts during the low level (i.e., negative half cycle) of the input signal it is called pulling. Hence the name push-pull amplifier. The load current flows during the full cycle of the input signal. The class-B push-pull amplifier circuit is shown in Fig.7.6.

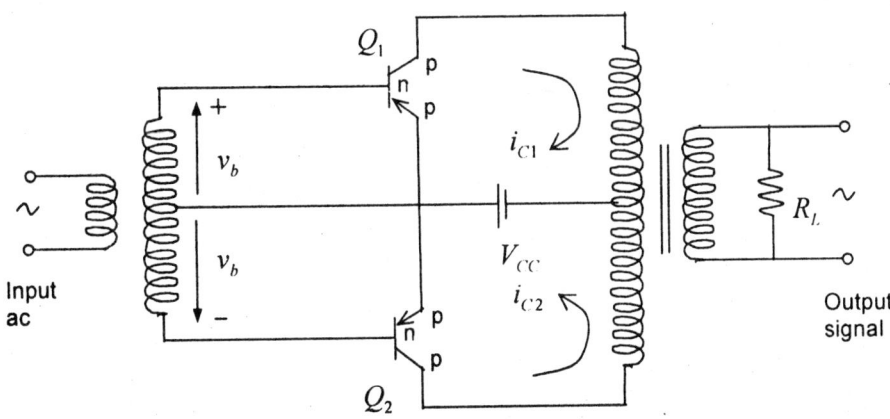

Fig. 7.6. Class-B push-pull amplifier circuit.

The two transistors Q_1 and Q_2 are assumed to be identical. V_{CC} is the collector supply voltage and R_L is load resistance. The ac currents i_{C1} and i_{C2} due to Q_1 and Q_2 respectively flow for every alternate half cycle as shown in Fig. 7.7.

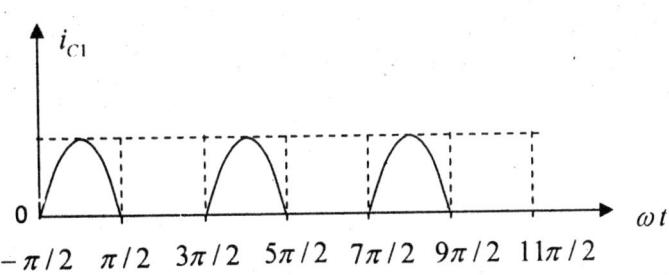

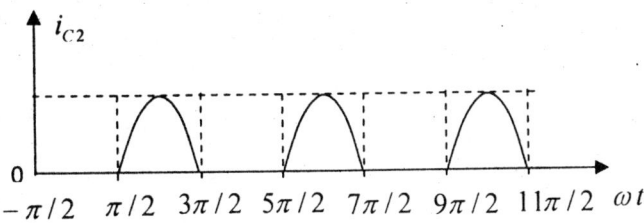

Fig. 7.7. Waveforms of collector currents i_{C1} and i_{C2} through Q_1 and Q_2 respectively.

Power series representation of i_{C1} and i_{C2} can be expressed as

$$i_{C1} = a_1 v_b + a_2 v_b^2 + a_3 v_b^3 + \ldots\ldots\ldots \qquad \ldots (7.35)$$

$$i_{C2} = -a_1 v_b + a_2 v_b^2 - a_3 v_b^3 + \ldots\ldots\ldots \qquad \ldots (7.36)$$

since the input voltage for Q_1 is v_b while that for Q_2 is $-v_b$. $a_1, a_2, a_3, \ldots..$etc. are power series coefficients. Taking $v_b = V_i \cos \omega t$ we can write

$$i_{C1} = A_0 + A_1 \cos \omega t + A_2 \cos 2\omega t + A_3 \cos 3\omega t + \ldots\ldots \qquad \ldots (7.37)$$

$$i_{C2} = A_0 - A_1 \cos \omega t + A_2 \cos 2\omega t - A_3 \cos 3\omega t + \ldots\ldots \qquad \ldots (7.38)$$

where $A_0, A_1, A_2, \ldots\ldots$ etc. are new constants.

The currents i_{C1} and i_{C2} through the primary of the output transformer flow in opposite directions. The primary current is then given by

$$i_{C1} - i_{C2} = 2(A_1 \cos \omega t + A_3 \cos 3\omega t + \ldots\ldots) \qquad \ldots (7.39)$$

So, there is no net dc current flowing through the output transformer. This eliminates the possibility of core saturation of the transformer. This allows the use of smaller and cheaper transformer for the output circuit.

We assume the transformer to be ideal. The primary and secondary magnetomotive forces are then equal. Thus,

$$i_L N_2 = (i_{C1} - i_{C2}).2N_1 \qquad \ldots (7.40)$$

where i_L = load current, N_2 = number of turns in the secondary, and $2N_1$ is the total number of turns in the primary of the output transformer.

$$\therefore \quad i_L = \left(\frac{2N_1}{N_2}\right) \cdot 2(A_1 \cos \omega t + A_3 \cos 3\omega t + \ldots\ldots) \qquad \ldots (7.41)$$

The load current is thus free from all even harmonics and the dc current. This is a great advantage of the push-pull connection. Even harmonic distortion is thus eliminated from the output of the push-pull amplifier. The only distortion present is only due to the odd harmonics. The current through the collector supply source V_{CC} is given by

$$i_{C1} + i_{C2} = 2(A_0 + A_2 \cos 2\omega t + \ldots\ldots) \qquad \ldots (7.42)$$

So, there is no signal frequency (ω) current through the power supply V_{CC}. This implies that there can be no regeneration due to positive feedback through an impedance common to the power source and other stages.

The dc collector current for each transistor can be found out from the waveforms of i_{C1} and i_{C2} shown in Fig. 7.7, and can be written as

$$I_{dc} = \frac{1}{2\pi} \int\limits_{-\pi/2}^{\pi/2} I_p \cos \omega t \cdot d\omega t$$

$$= \frac{I_p}{\pi}$$

where I_p = peak collector current of each transistor. Then, the total dc power input to the two transistors is

$$P_{dc} = 2 \cdot V_{CC} \cdot I_{dc} = \frac{2}{\pi} V_{CC} I_p \qquad \qquad \dots (7.43)$$

We assume linear operation of the amplifier. Let the transistors be driven into saturation $(V_{CE} = 0)$ on one swing and into cut-off $(V_{CE} = V_{CC})$ on the other swing of the input ac. Then, the output voltage amplitude of each transistor is $\frac{V_{CC}}{2}$. Now, the total ac power output is

$$P_0 = \frac{1}{2} V_p I_p \times 2 = V_p I_p \qquad \qquad \dots (7.44)$$

where V_p = output voltage amplitude of the amplifier for each transistor.

To get maximum output power, we put $V_p = \frac{V_{CC}}{2}$. Then,

$$P_0\big|_{max} = \frac{V_{CC}}{2} I_p \qquad \qquad \dots (7.45)$$

The collector circuit efficiency of the class-B push-pull amplifier is

$$\eta = \text{ac power delivered / total dc power input}$$

$$= \frac{V_p I_p}{2 V_{CC} I_p} \cdot \pi$$

$$= \frac{\pi V_p}{2 V_{CC}} \qquad \qquad \dots (7.46)$$

η has a maximum value given by

$$\eta_{max} = \frac{V_{CC} \pi}{2 V_{CC} \cdot 2} = \frac{\pi}{4} = 78.5\%$$

Now, we summarize the advantages of the push-pull class-B amplifier: These are

- Push-pull connection eliminates the even harmonic distortion of the output power.
- The possibility of core saturation is eliminated since no net dc current flows through the output transformer.
- No flow of signal frequency current through the collector supply source ensures that no regeneration is possible through the internal impedance of the common power source of more than one stage.

Advantages of use of a transformer are the following:

- The use of the output transformer avoids the flow of dc collector current through the load impedance (R_L) and hence minimizes dc power loss.
- The use of the output transformer makes it possible to present proper collector impedance to the transistors for any arbitrary load. So, maximum power transfer to the load is possible through impedance matching. This is done by selecting proper turns ratio of the transformer.

7.6 TUNED RF CLASS-B POWER AMPLIFIER

Tuned class-B power amplifiers use a tuned circuit as the load. This amplifier is used to amply radio frequency (RF) signals. The tuned circuit is resonant at the signal frequency (fundamental component). The tuned circuit response falls off rapidly as we move away from the resonance frequency. So, the harmonics of the signal frequency are eliminated. The circuit diagram of the tuned class-B power amplifier is shown in Fig. 7.8. Here, RFC is radio frequency choke which blocks the input signal from passing through the base bias source V_{B0} to be applied to the base of the transistor. The coupling capacitor C_C in the input circuit prevents the dc bias voltage V_{B0} from being short circuited and allows the input RF signal to pass through.

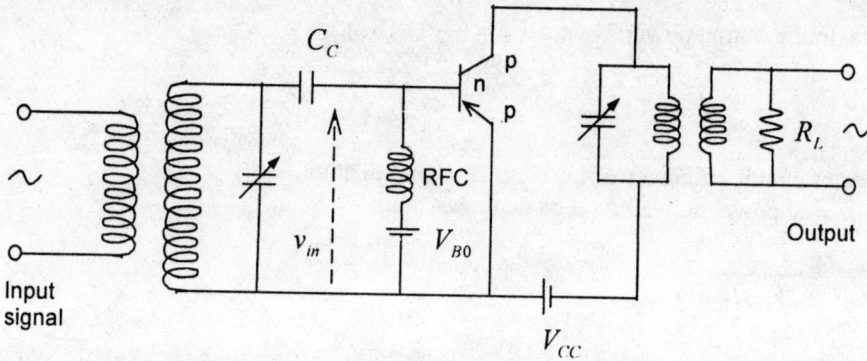

Fig. 7.8. Tuned Class-B power amplifier.

A transfer characteristics of the transistor is shown in Fig. 7.9. The linear portion of this characteristics is extraploted to meet the V_{BE} axis at V_{B0} which is the projected cut-off of the transistor. The transistor is biased at projected cut-off (V_{B0}).

ANALYSIS

Let g_m be the transconductance of the pnp transistor. With sinusoidal input, the collector current (i_C) can be written as

$$i_C = 0 \qquad \text{for } 0 \leq \omega t \leq \pi$$

$$i_C = g_m(v_{in} + V_{B0}) \qquad \text{for } \pi \leq \omega t \leq 2\pi$$

... (7.47)

where v_{in} is the input ac voltage appearing at the base of the transistor. Assuming V_{B0} to be small, we have taken the conduction angle of the transistor equal to $180°$ consistent with class B operation.

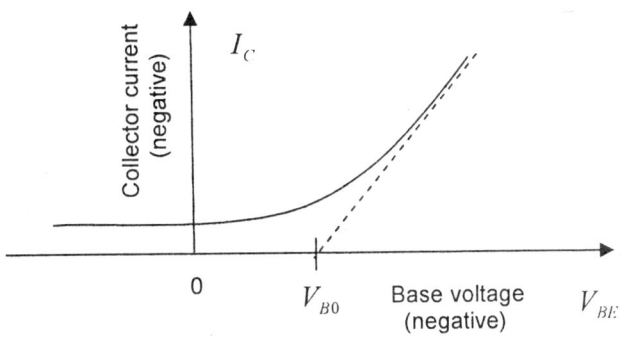

Fig. 7.9. Transfer characteristics of the pnp transistor.

Let

$$v_{in} = V_i \sin \omega t \qquad \qquad \text{... (7.48)}$$

Making a Fourier analysis of the collector current i_C as given by (7.43), the fundamental component of collector current is obtained as

$$I_1 = \frac{1}{\pi} \int_{\pi}^{2\pi} g_m (V_i \sin \omega t + V_{B0}) \sin \omega t \, d\omega t$$

$$= g_m \left[\frac{V_i}{2} - \frac{2}{\pi} V_{B0} \right] \qquad \qquad \text{... (7.49)}$$

The dc collector current is

$$I_{dc} = \frac{1}{2\pi} \int_{\pi}^{2\pi} g_m (V_i \sin \omega t + V_{B0}) d\omega t$$

$$= -g_m \left[\frac{V_i}{\pi} - \frac{V_{B0}}{2} \right] \qquad \qquad \text{... (7.50)}$$

The negative sign in (7.50) indicates that the current leaves the collector terminal in the pnp transistor. Let R_d = dynamic resistance of the tuned circuit. The ac power output at the fundamental frequency is

$$P_0 = \frac{1}{2} I_1^2 R_d \qquad \qquad \text{... (7.51)}$$

The dc power input to the transistor is

$$P_{dc} = V_{CC} I_{dc} \qquad \qquad \qquad \dots (7.52)$$

where V_{CC} is the collector supply voltage which is negative. The collector circuit efficiency (η) is given by

$$\eta = \frac{\text{ac power output}}{\text{dc power input}}$$

$$= \frac{\dfrac{1}{2} I_1^2 R_d}{V_{CC} I_{dc}}$$

$$= \frac{g_m}{2} \cdot \frac{\left(\dfrac{V_i}{2} - \dfrac{2}{\pi} V_{B0} \right)^2 R_d}{|V_{CC}| \left[\dfrac{V_i}{\pi} - \dfrac{V_{B0}}{2} \right]} \qquad \qquad \dots (7.53)$$

Alternatively, $\eta = \dfrac{1}{2} \cdot \dfrac{I_1}{I_{dc}} \cdot \dfrac{I_1 R_d}{V_{CC}}$ where I_{dc} and V_{CC} are both negative and $I_1 R_d$ is positive. If $V_C|_{min}$ be the minimum collector voltage reached during the swing then $I_1 R_d = \left| \left(V_{CC} - V_C|_{min} \right) \right|$.

Then, $\quad \eta = \dfrac{1}{2} \cdot \dfrac{I_1}{I_{dc}} \cdot \dfrac{\left| \left(V_{CC} - V_C|_{min} \right) \right|}{V_{CC}} \qquad \qquad \dots (7.54)$

Now, $\quad \left| \dfrac{I_1}{I_{dc}} \right| = \dfrac{\dfrac{V_i}{2} - \dfrac{2}{\pi} V_{B0}}{\dfrac{V_i}{\pi} - \dfrac{V_{B0}}{2}} \qquad \qquad \dots (7.55)$

To find the maximum value of η, we neglect V_{B0} since it is small. Further, we take $V_C|_{min} = 0$ as the theoretical limit. Then,

$$\eta_{max} = \frac{1}{2} \cdot \frac{\pi}{2} = \frac{\pi}{4} = 78.5\%$$

This amplifier is suitable for the power amplification of amplitude modulated signals.

7.7 CLASS-C POWER AMPLIFIER

In class-C operation, the transistor conducts for less than $180°$ of the input signal. The base of the transistor is biased below cut-off. This is done by applying a positive voltage to the base of a p-n-p transistor. The transistor conducts only when the net base voltage (dc + ac) goes above the built in potential barrier which is 0.7 V (approximately) for silicon transistors. The collector current waveform is nonsinusoidal. But, since a tank circuit is used as the collector load the effect of harmonics is eliminated from the output. This is because a tank circuit response has a peak at the resonant frequency and it falls sharply to very small levels beyond a narrow band of frequencies. The harmonic frequencies thus produce no voltage drop across the tuned load. The term 'tank circuit' means a parallel resonant LC circuit. The word 'tank' is used to mean the reservoir of energy. A schematic circuit diagram of a class-C radio frequency power amplifier is shown in Fig. 7.10. The collector current waveform is shown in Fig. 7.11. This amplifier has a very high dc to rf conversion efficiency. This high efficiency owes its origin to the fact that the device conducts for short time only when the input signal voltage is near its peak and the device does not conduct during the rest of the input cycle.

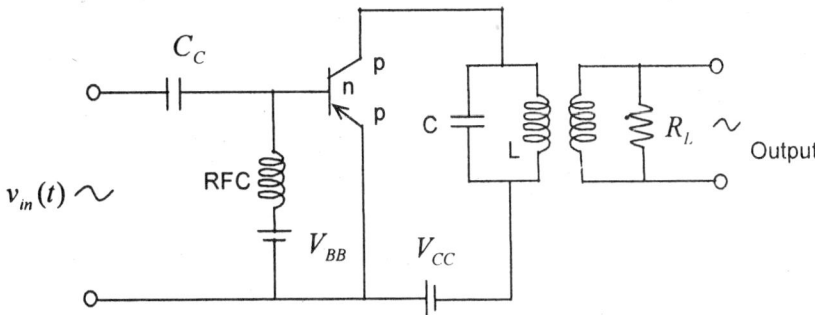

Fig. 7.10. Schematic circuit diagram of a class-C power amplifier.

In Fig. 7.10, RFC means radio frequency choke. It blocks the passage of rf signal through it but allows dc to pass through. The input rf signal should not pass through the dc source V_{BB}. The base-emitter voltage is $v_{be} = V_{BB} + v_{in}(t)$. Taking cosinusoidal variation, $v_{in}(t)$ is negative for $\dfrac{T}{4} < t < \dfrac{3T}{4}$. The pnp Si transistor in Fig.7.10 conducts when $v_{be} \le -0.7$ V .

With proper choice of the origin, the collector current can be represented as

$$i_C = I_m \cos \omega t \qquad \text{for} -\theta_C < \omega t < \theta_C \qquad \qquad ... (7.56)$$

$$= 0 \qquad \text{for } \theta_C < \omega t < 2\pi - \theta_C \qquad \qquad ... (7.57)$$

where $2\theta_C$ is the conduction angle of the transistor. The dc collector current (I_{dc}) is given by

$$I_{dc} = \frac{1}{2\pi} \int_{-\theta_C}^{\theta_C} I_m \cos \omega t \cdot d\omega t$$

$$= \frac{I_m}{\pi} \sin \theta_C \qquad \qquad ... (7.58)$$

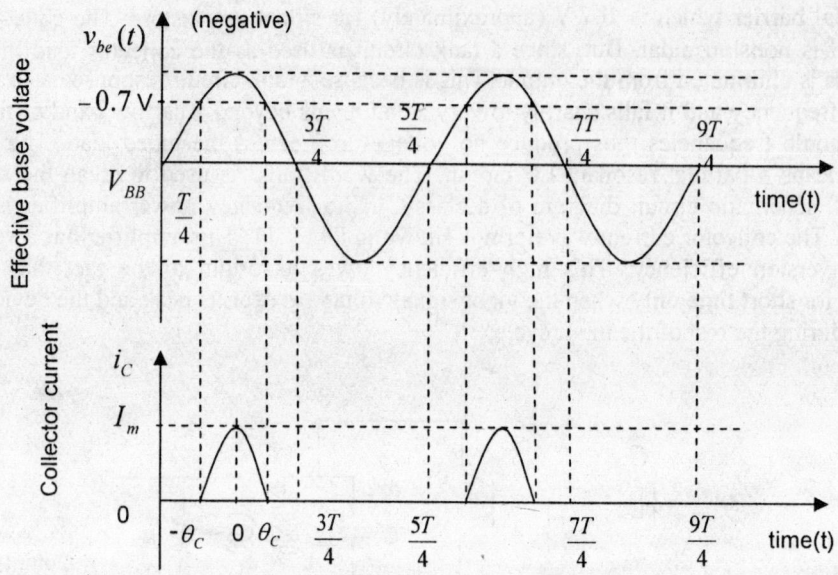

Fig. 7.11. Effective base voltage (V_{be}) and collector current (i_c) waveform.
T = time period of the cosinusoidal input ac signal.

The fundamental component of collector current is

$$I_1 = \frac{1}{\pi} \int_0^{2\pi} I_C \cos\omega t \cdot d\omega t$$

$$= \frac{1}{\pi} \int_{-\theta_c}^{\theta_c} I_m \cos^2\omega t \cdot d\omega t$$

$$= \frac{I_m}{\pi}\left[\theta_C + \frac{1}{2}\sin 2\theta_C\right] \qquad \ldots (7.59)$$

AC power developed across the tuned load is

$$P_0 = \frac{1}{2}I_1^2 R_d \qquad \ldots (7.60)$$

where R_d = dynamic resistance of the tank circuit. DC power supplied by the collector supply source is

$$P_{dc} = V_{CC}I_{dc} \qquad \ldots (7.61)$$

The collector circuit efficiency (η) of the class-C power amplifier is

$$\eta = \frac{P_0}{P_{dc}} = \frac{1}{2}\frac{I_1^2 R_d}{V_{CC}I_{dc}} = \frac{1}{2}\frac{I_1}{I_{dc}} \cdot \frac{I_1 R_d}{V_{CC}} \qquad \ldots (7.62)$$

If $V_C|_{min}$ be the minimum collector voltage reached during a cycle, then $I_1 R_d = V_{CC} - V_C|_{min}$, neglecting the effect of harmonics. Now,

$$\frac{I_1}{I_{dc}} = \frac{\theta_C + \frac{1}{2}\sin 2\theta_C}{\sin \theta_C} = \frac{\theta_C}{\sin \theta_C} + \cos \theta_C \qquad \qquad \ldots (7.63)$$

Now, $\quad \eta = \frac{1}{2}\left[\frac{\theta_C}{\sin \theta_C} + \cos \theta_C\right]\left[1 - \frac{V_C|_{min}}{V_{CC}}\right] \qquad \qquad \ldots (7.64)$

To find an upper limit of the value of η, we set $V_C|_{min} = 0$. Now, when $\theta_C \to 0$, $\eta \to 1$. As an example, if $\theta_C = 30°$ then $\eta = 95.6\%$ with $V_C|_{min} = 0$. If we take $\theta_C = 60°$ then $\eta = 85.4\%$.

The conduction angle of the class-B tuned power amplifier is $180°$. The upper limit of the efficiency for a class-B tuned power amplifier can be found out from that for a class-C power amplifier by setting the conduction angle $2\theta_C = 180°$ in eqn. (7.64). In this case, $\theta_C = 90°$ and

$$\eta = \frac{1}{2}\left[\frac{\pi/2}{\sin \pi/2} + \cos \pi/2\right]\left[1 - \frac{V_C|_{min}}{V_{CC}}\right]$$

$$= \frac{\pi}{4} = 78.5\% \text{ with } V_C|_{min} = 0.$$

7.8 SINGLE TUNED AMPLIFIER

This is a narrow-band radio frequency amplifier. The load is in the form of a tuned circuit. A tuned circuit is a parallel resonant LC circuit. This amplifier is used for the amplification of AM signals. Single-tuned amplifiers are used in radio transmitters and receivers. A schematic circuit diagram of a single-tuned rf amplifier is shown in Fig. 7.12.

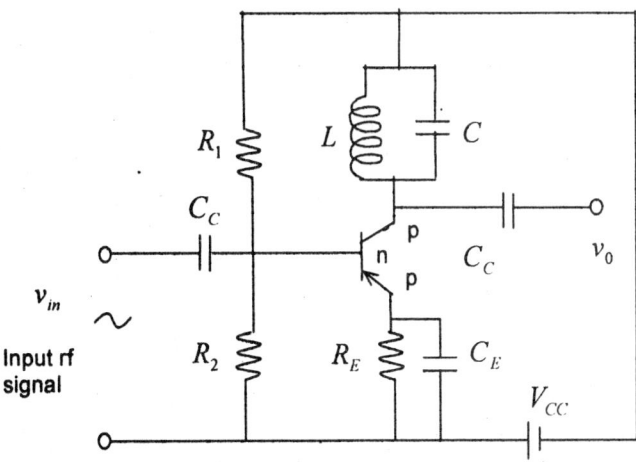

Fig. 7.12. Schematic circuit diagram of a single-tuned amplifier.

Here, R_1, R_2 and R_E-C_E combination provide necessary bias for class-A operation. C_C is the coupling capacitance which blocks the dc and passes the ac to the output. The operation of this amplifier is under class-A. The high frequency ac equivalent circuit is shown in Fig 7.13. Since it is an rf amplifier, the junction capacitances of the transistor must be taken into account in the analysis. The tuned circuit has a resonant frequency $\omega_0 = 1/\sqrt{LC}$, where L is the inductance and C is the capacitance of the tuned circuit.

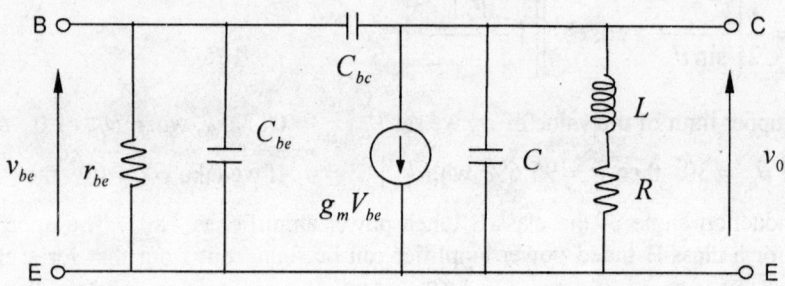

Fig. 7.13. High frequency ac equivalent circuit of the single-tuned amplifier.

Here, v_{be} is the base-emitter voltage, r_{be} is the input resistance between the base and the emitter, C_{be} and C_{bc} are capacitances of the base-emitter and collector-base junctions. g_m is the transconductance of the transistor. R is the small resistance of the inductor L. v_0 is the output voltage. C_{bc} provide a feedback between the collector and base at high frequencies. Applying Miller theorem, we can take the feedback effect of C_{bc} into the input and output circuits as shown in Fig. 7.14.

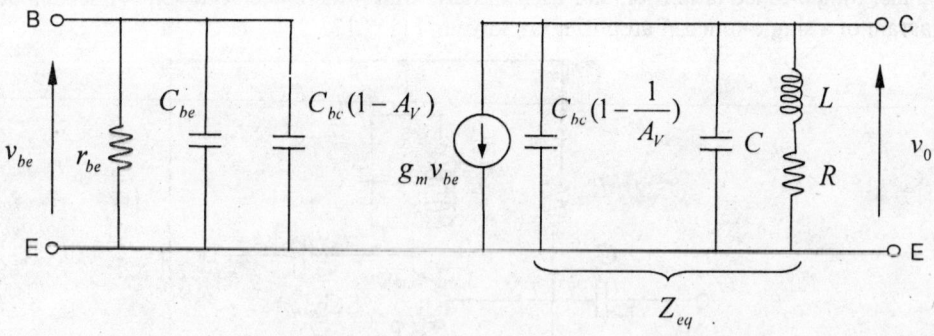

Fig. 7.14. High frequency ac equivalent circuit reduced by applying Miller theorem.

A_V =voltage gain between the collector (C) and the base (B). We take $|A_V| \gg 1$. So, $C_{bc}\left(1 - \dfrac{1}{A_V}\right) \simeq C_{bc}$. Let $C' = C_{bc} + C$. The equivalent impedance of the output load is

$$Z_{eq} = (R + j\omega L) \| \frac{1}{j\omega C'}$$

$$= \frac{(R + j\omega L)\dfrac{1}{j\omega C'}}{(R + j\omega L) + \dfrac{1}{j\omega C'}}$$

$$\cong \frac{\dfrac{L}{RC'}}{1 + j\dfrac{\omega L}{R}\left(1 - \dfrac{1}{\omega^2 LC'}\right)} \quad \text{assuming } \frac{\omega_0 L}{R} = Q \gg 1 .$$

$$\therefore \quad Z_{eq} = \frac{R_t}{1 + j\dfrac{\omega}{\omega_0}Q\left(1 - \dfrac{\omega_0^2}{\omega^2}\right)} \qquad \qquad \dots (7.65)$$

where $R_t = \dfrac{L}{RC'} = $ dynamic resistance of the tuned circuit. It is the resistance of the tuned circuit

at the resonance frequency ω_0. $Q = \dfrac{\omega_0 L}{R}$ is the Q-factor of the tuned circuit.

Now, $\quad Z_{eq} = \dfrac{R_t}{1 + jQ\left(\dfrac{\omega}{\omega_0} - \dfrac{\omega_0}{\omega}\right)}$ $\qquad \qquad \dots (7.66)$

The output voltage of the tuned amplifier is

$$V_0 = -g_m V_{be} Z_{eq} \qquad \qquad \dots (7.67)$$

The voltage gain of the single-tuned amplifier is

$$A = \frac{V_0}{V_{be}} - g_m Z_{eq}$$

$$= \frac{A_0}{1 + jQ\left(\dfrac{\omega}{\omega_0} - \dfrac{\omega_0}{\omega}\right)} \qquad \qquad \dots (7.68)$$

where $A_0 = -g_m R_t$ is the voltage gain at $\omega = \omega_0$. Let $A = |A| < \phi$. Then,

$$|A| = \frac{|A_0|}{\left[1 + Q^2\left(\frac{\omega}{\omega_0} - \frac{\omega_0}{\omega}\right)^2\right]^{1/2}} \qquad \qquad \ldots (7.69)$$

and

$$\phi = \pi - \tan^{-1}\left[Q\left(\frac{\omega}{\omega_0} - \frac{\omega_0}{\omega}\right)\right] \qquad \qquad \ldots (7.70)$$

At half-power points, the voltage gain falls to $\frac{1}{\sqrt{2}}$ of its peak value. Let $|A| = \frac{|A_0|}{\sqrt{2}}$ at $\omega = \omega'$.

Now, since $|\omega - \omega_0| \ll \omega_0$, we put $\omega + \omega_0 \simeq 2\omega_0$ and $\omega\omega_0 \simeq \omega_0^2$. Then,

$$\frac{\omega}{\omega_0} - \frac{\omega_0}{\omega} = 2\left(\frac{\omega}{\omega_0} - 1\right)$$

Under this condition,

$$A = \frac{A_0}{1 + j2Q\left(\frac{\omega}{\omega_0} - 1\right)} \cdot \qquad \qquad \ldots (7.71)$$

Then, $2Q\left(\frac{\omega'}{\omega_0} - 1\right) = \pm 1$

$$\therefore \qquad \frac{\omega'}{\omega_0} = 1 \pm \frac{1}{2Q}$$

Let these two roots of ω' corresponding to half-power points be denoted as ω_1' and ω_2' where

$$\omega_1' = \omega_0\left(1 - \frac{1}{2Q}\right) \qquad \qquad \ldots (7.72)$$

and

$$\omega_2' = \omega_0\left(1 + \frac{1}{2Q}\right) \qquad \qquad \ldots (7.73)$$

$$\therefore \qquad \omega_2' - \omega_1' = \frac{\omega_0}{Q} \qquad \qquad \ldots (7.74)$$

This is the bandwidth of the single-tuned amplifier. A schematic plot of the normalized voltage gain as a function of frequency is shown in Fig. 7.15.

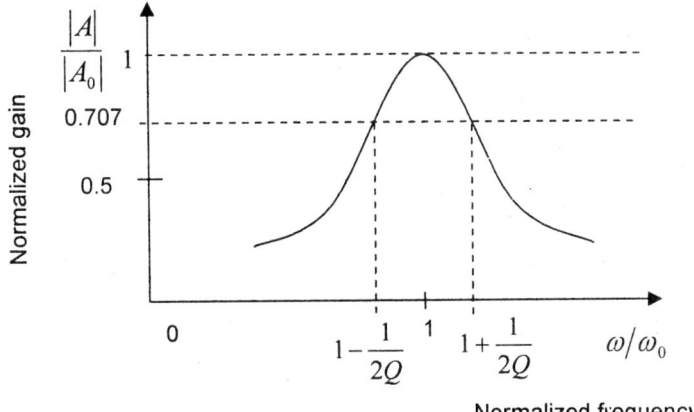

Fig. 7.15. Schematic plot of normalized voltage gain as a function of the normalized frequency.

At $\omega = \omega_0$, $\phi = \pi$.

At $\omega = \omega_1'$, $\phi = \pi + \tan^{-1}(1) = \dfrac{5\pi}{4}$.

At $\omega = \omega_2'$, $\phi = \pi - \tan^{-1}(1) = \dfrac{3\pi}{4}$.

At $\omega = \infty$, $\phi = \pi - \tan^{-1}(\infty) = \pi - \dfrac{\pi}{2} = \dfrac{\pi}{2}$.

At $\omega \to 0$, $\phi \to \dfrac{3\pi}{2}$.

The phase shift (ϕ) produced by this amplifier is plotted in Fig. 7.16 schematically.

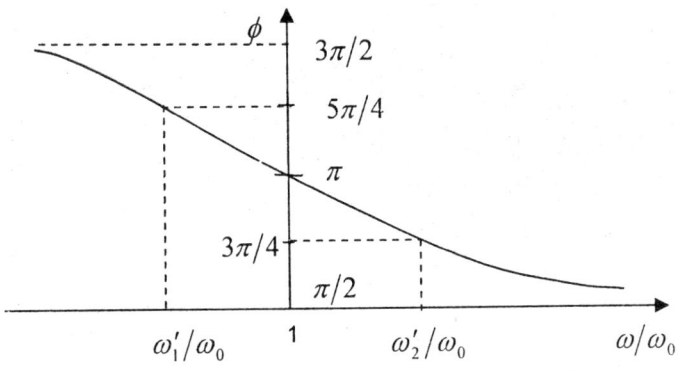

Fig. 7.16. Schematic plot of the phase angle (ϕ) of the single-tuned amplifier as a function of ω / ω_0.

The gain-bandwidth product is

$$|A_0| \times BW = g_m R_t \cdot \frac{\omega_0}{Q} = g_m \frac{L}{RC'} \cdot \frac{\omega_0}{Q}$$

$$= g_m \frac{L}{RC'} \cdot \frac{R}{L} = \frac{g_m}{C'} \qquad \dots (7.75)$$

Thus, the gain-bandwidth product is independent of ω_0 and Q. It only depends upon the capacitance of the output circuit. When C' is a minimum, the gain-bandwidth product is maximum. C'_{min} is determined by the collector-base capacitance C_{bc} and stray capacitances in the circuit.

7.9 DOUBLE TUNED RF AMPLIFIER

The double tuned rf amplifier shown in Fig. 7.17 is a CE amplifier with a $L_1 C_1$ tuned circuit at the collector and a mutually coupled tuned circuit $L_2 C_2$ which delivers the output. R_1, R_2 potential divider together with the R_E, C_E emitter-bias provide necessary bias for class-A operation of the amplifier.

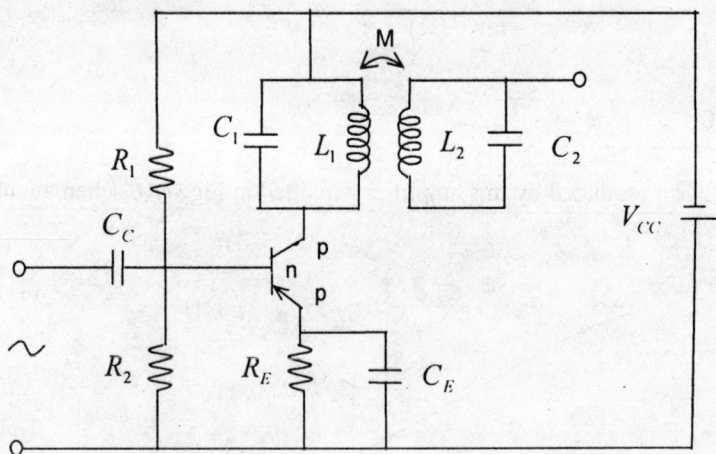

Fig. 7.17. Schematic circuit diagram of a double-tuned rf amplifier.

The ac equivalent circuit of the double-tuned amplifier is shown in Fig. 7.18. i_b is the base current. h_{re} and h_{oe} being small are neglected.

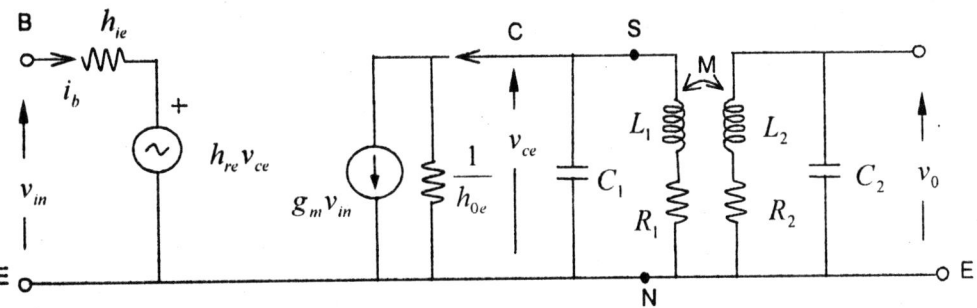

Fig. 7.18. AC equivalent circuit of the double-tuned rf amplifier.

R_1 and R_2 are the resistances of the primary coil and secondary coil respectively. The primary

capacitance C_1 includes the Miller capacitance $C_{bc}\left(1 - \dfrac{1}{A_v}\right) \simeq C_{bc}$ assuming $|A_v| \gg 1 . A_v$ is

the voltage gain of the amplifier. C_{bc} is the collector-base capacitance.

Appling Thevenin's theorem at the terminals SN, we put the ac equivalent circuit in the following form as shown in Fig. 7.19

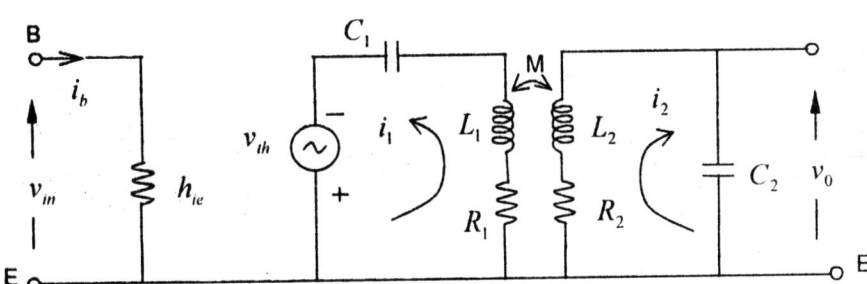

Fig. 7.19. Thevenin's equivalent of the circuit shown in Fig.7.18.

The Thevenin equivalent voltage source

$$v_{th} = \frac{g_m v_{in}}{j\omega C_1} \qquad \qquad \dots (7.76)$$

Let i_1 and i_2 be the primary and secondary currents and M be the mutual inductance of the coupled circuits.

Then, $\left(R_1 + j\omega L_1 - \dfrac{j}{\omega C_1} \right) i_1 + j\omega M i_2 = v_{th}$ $\qquad \dots (7.77)$

For the secondary circuit, we write

$$\left(R_2 + j\omega L_2 - \frac{j}{\omega C_2} \right) i_2 + j\omega M i_1 = 0 \qquad \ldots (7.78)$$

Let $\omega_0 = \dfrac{1}{\sqrt{L_1 C_1}} = \dfrac{1}{\sqrt{L_2 C_2}}$ be the resonant angular frequency of the tuned circuits, and

$Q_1 = \dfrac{\omega_0 L_1}{R_1}$ and $Q_1 = \dfrac{\omega_0 L_1}{R_1}$ be the Q-factors of the primary and secondary circuits.

Now, $\quad \left(R_1 + j\omega L_1 - \dfrac{j}{\omega C_1} \right) = R_1 \left(1 + jQ_1\Omega \right) \qquad \ldots (7.79)$

and $\quad \left(R_2 + j\omega L_2 - \dfrac{j}{\omega C_2} \right) = R_2 \left(1 + jQ_2\Omega \right) \qquad \ldots (7.80)$

where $\quad \Omega = \dfrac{\omega}{\omega_0} - \dfrac{\omega_0}{\omega}$. Solving (7.77) and (7.78) we get

$$i_1 = \frac{v_{th}}{R_1 \left(1 + jQ_1\Omega \right) + \dfrac{\omega^2 M^2}{R_2 \left(1 + jQ_2\Omega \right)}} \qquad \ldots (7.81)$$

and $\quad i_2 = \dfrac{- j\omega M v_{th}}{R_1 R_2 \left(1 + jQ_1\Omega \right)\left(1 + jQ_2\Omega \right) + \omega^2 M^2} \qquad \ldots (7.82)$

The output voltage of the double tuned amplifier is

$$v_0 = \frac{1}{j\omega C_2} i_2$$

$$= \frac{-\dfrac{M}{C_2} v_{th}}{R_1 R_2 \left(1 + jQ_1\Omega \right)\left(1 + jQ_2\Omega \right) + \omega^2 M^2}$$

$$= \frac{- g_m v_{in} M}{j\omega C_1 C_2 R_1 R_2} \cdot \frac{1}{\left(1 + jQ_1\Omega \right)\left(1 + jQ_2\Omega \right) + \dfrac{\omega^2 M^2}{R_1 R_2}} \qquad \ldots (7.83)$$

Now, $\quad M^2 = k^2 L_1 L_2$

then, $\quad \dfrac{\omega^2 M^2}{R_1 R_2} = \left(\dfrac{\omega}{\omega_0} \right)^2 k^2 Q_1 Q_2$

$$\frac{M}{C_1 C_2 R_1 R_2} = \frac{kQ_1Q_2}{\sqrt{C_1 C_2}} = \omega_0 k \sqrt{R_1 R_2} \left(Q_1 Q_2\right)^{3/2}$$

since, $Q_1 = \dfrac{1}{\omega_0 C_1 R_1}$ and $Q_2 = \dfrac{1}{\omega_0 C_2 R_2}$.

Now, eqn. (7.83) can be written as

$$v_0 = \frac{j g_m v_{in} \left(\dfrac{\omega_0}{\omega}\right) k (Q_1 Q_2)^{3/2} \sqrt{R_1 R_2}}{1 + j(Q_1 + Q_2)\Omega - Q_1 Q_2 \Omega^2 + \left(\dfrac{\omega}{\omega_0}\right)^2 k^2 Q_1 Q_2} \qquad \ldots (7.84)$$

If $A_v = \dfrac{v_0}{v_{in}}$ be the voltage gain of the double-tuned amplifier then

$$|A_v|^2 = \frac{g_m^2 k^2 R_1 R_2 (Q_1 Q_2)^3}{\left(1 + k^2 Q_1 Q_2 - Q_1 Q_2 \Omega^2\right)^2 + (Q_1 + Q_2)^2 \Omega^2} \qquad \ldots (7.85)$$

putting $\dfrac{\omega_0}{\omega} \simeq 1$ in (7.84). Rearranging eqn. (7.85), we get

$$|A_v|^2 = \frac{g_m^2 k^2 R_1 R_2 (Q_1 Q_2)^3}{\left(1 + k^2 Q_1 Q_2\right)^2 \left[1 + \dfrac{\Omega^4 Q_1^2 Q_2^2 + \Omega^2 \left(Q_1^2 + Q_2^2 - 2 Q_1^2 Q_2^2 k^2\right)}{\left(1 + k^2 Q_1 Q_2\right)^2}\right]}$$

$$= g_m^2 R_1 R_2 (Q_1 Q_2)^3 \; \frac{\dfrac{k^2}{\left(k^2 + \dfrac{1}{Q_1 Q_2}\right)^2 Q_1^2 Q_2^2}}{1 + \dfrac{\Omega^4 + \Omega^2 \left(\dfrac{1}{Q_1^2} + \dfrac{1}{Q_2^2} - 2k^2\right)}{\left(k^2 + \dfrac{1}{Q_1 Q_2}\right)^2}} \qquad \ldots (7.86)$$

Assuming $\dfrac{1}{Q_1^2} + \dfrac{1}{Q_2^2} \simeq \dfrac{2}{Q_1 Q_2}$ in (7.86) we get

$$\left|A_v^2\right| = \cfrac{g_m^2 R_1 R_2 Q_1 Q_2 \cfrac{k^2}{\left(k^2 + \cfrac{1}{Q_1 Q_2}\right)^2}}{1 + \cfrac{\Omega^4 + \Omega^2 \cdot 2\left(\cfrac{1}{Q_1 Q_2} - k^2\right)}{\left(k^2 + \cfrac{1}{Q_1 Q_2}\right)^2}} \qquad \dots (7.87)$$

Now, depending upon the value of the coefficient of coupling (k) three different cases will arise.

Case-I. $k^2 = \dfrac{1}{Q_1 Q_2}$. The coupling in this case is said to be critical coupling

Now, $\left|A_v\right|^2 = \dfrac{g_m^2 R_1 R_2 (Q_1 Q_2)}{4k^2\left(1 + \dfrac{\Omega^4}{4k^4}\right)}$

$$= \dfrac{g_m^2 R_1 R_2 (Q_1 Q_2)^2}{4\left(1 + \dfrac{\Omega^4}{4}(Q_1 Q_2)^2\right)} \qquad (7.88)$$

putting $k^2 = \dfrac{1}{Q_1 Q_2}$. When $\omega = \omega_0$, we have $\Omega = 0$ and $\left|A_v\right|^2_{\omega=\omega_0} = \dfrac{g_m^2 R_1 R_2 (Q_1 Q_2)^2}{4} = A_{v_0}$

(say).

Let $\omega = \omega'$ when $\left|A_v\right|^2_{\omega=\omega'} = \dfrac{1}{2}\left|A_v\right|^2_{\omega=\omega_0}$. Then, ω' will be the half-power frequency.

Now, $1 + \dfrac{\Omega^4}{4}(Q_1 Q_2)^2 = 2$ at $\omega = \omega'$.

$\therefore \quad \left(\dfrac{\omega'}{\omega_0} - \dfrac{\omega_0}{\omega'}\right)^2 Q_1 Q_2 = 2$

Again, $\left(\dfrac{\omega'}{\omega_0} - \dfrac{\omega_0}{\omega'}\right) \simeq \dfrac{2}{\omega_0}(\omega' - \omega_0)$

$\therefore \quad \dfrac{2}{\omega_0}(\omega' - \omega_0) = \pm\sqrt{\dfrac{2}{Q_1 Q_2}}$

Then, $(\omega' - \omega_0) = \pm\dfrac{\omega_0}{2}\sqrt{\dfrac{.2}{Q_1Q_2}}$ ...(7.89)

The $\pm$ sign in (7.89) corresponds to the upper and lower half-power points ω_2' and ω_1' respectively.

$\therefore$ Bandwidth of the double-tuned amplifier under critical coupling is

$$\omega_2' - \omega_1' = \omega_0\sqrt{\dfrac{2}{Q_1Q_2}}$$ ... (7.90)

where ω_2' and ω_1' are upper and lower half-power frequencies corresponding to the two roots of (7.89). If $Q_1 = Q_2 = Q$ (say), then

$$\omega_2' - \omega_1' = \sqrt{2}\dfrac{\omega_0}{Q}$$ ... (7.91)

This bandwidth is $\sqrt{2}$ times greater than that of a single-tuned amplifier. The amplifier has flat bandwidth under critical coupling. It is used as intermediate frequency (IF) amplifier in radio receivers. The gain vs. frequency plot is schematically in Fig. 7.20.

Case-II. $k^2 < \dfrac{1}{Q_1Q_2}$. The amplifier is said to be undercoupled.

The top of the frequency response curve rounds off.

Case-III. $k^2 > \dfrac{1}{Q_1Q_2}$. The amplifier is overcoupled.

A double hump appears in the response. To find that $|A_v|$ have maximum values at the humps, the denominator in (7.87) should be a minimum.

Then,

$$\dfrac{d}{d\Omega}\left[\Omega^4 + 2\left(\dfrac{1}{Q_1Q_2} - k^2\right)\Omega^2\right] = 0$$

$\therefore$ $\quad 4\Omega^3 - 4\Omega\left(k^2 - \dfrac{1}{Q_1Q_2}\right) = 0$

$\therefore$ $\quad \Omega = 0$ or, $\Omega = \pm\sqrt{k^2 - \dfrac{1}{Q_1Q_2}}$

Now, $\Omega \simeq 2\left(\dfrac{\omega - \omega_0}{\omega_0}\right)$. Then,

$$\omega = \omega_0 \left[1 \pm \frac{1}{2} \sqrt{k^2 - \frac{1}{Q_1 Q_2}} \right] \qquad \qquad \dots (7.92)$$

The peaks of the response occur at two roots of ω. At $\omega = \omega_0$, there occurs a dip. The response of the double-tuned amplifier is shown in Fig. 7.20

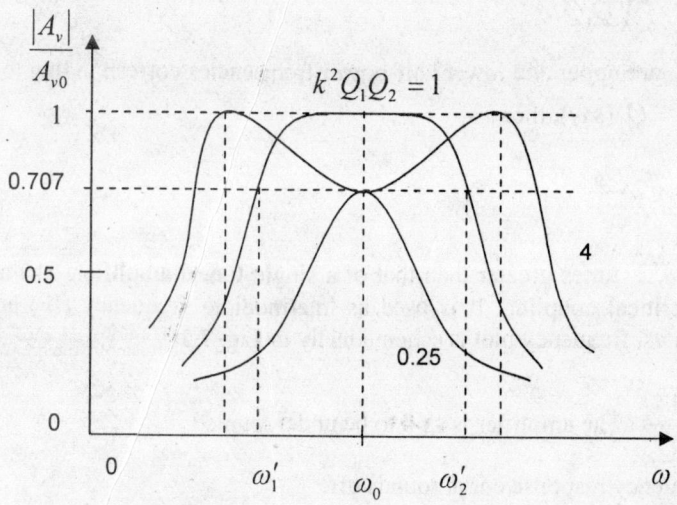

Fig. 7.20. Typical frequency response of the double-tuned amplifier.

The gain-bandwidth product of the double-tuned amplifier is

$$|A_v|_{\omega=\omega_0} \times BW = \frac{g_m}{2} \sqrt{R_1 R_2} Q_1 Q_2 \cdot \frac{\omega_0}{2\pi} \sqrt{\frac{2}{Q_1 Q_2}}$$

$$= \frac{g_m}{\sqrt{2}} \sqrt{R_1 R_2 Q_1 Q_2} \cdot \frac{\omega_0}{2\pi}$$

$$= \frac{g_m}{2\sqrt{2}\,\pi} \cdot \frac{1}{\sqrt{C_1 C_2}} \qquad \qquad \dots (7.93)$$

putting $Q_1 = \dfrac{1}{\omega_0 C_1 R_1}$ and $Q_2 = \dfrac{1}{\omega_0 C_2 R_2}$.

The minimum values of C_1 and C_2 are set by the output capacitance of the amplifier and the input capacitance of the next stage.

SOLVED PROBLEMS

7.1   A class-A power amplifier has resistance of $20\ \Omega$ connected with the collector and it has a power supply of $12\ V$. The quiescent operating point collector voltage is $6\ V$. Calculate the maximum efficiency of this amplifier. Also, find the collector circuit efficiency.

Solution

Here, the load resistance $R_L = 20\Omega$

Supply voltage $V_{CC} = 12\ V$.

Operating point collector voltage, $V = 6\ V$.

$\therefore$ Operating point collector current, $I_0 = \dfrac{V_{CC} - V}{R_L} = \dfrac{12 - 6}{20} = \dfrac{6}{20} = 0.3\ \text{Amp}.$

Total dc power input, $P_{dc} = V_{CC}I_0 = 12 \times 0.3 = 3.6\ \text{Watt}.$

AC power developed across the load

$$P_{ac} = \frac{1}{2}\frac{V_L^2}{R_L} \quad \text{where } V_L = \text{peak load voltage.}$$

Maximum possible amplitude of load voltage is

$$V_L\big|_{\text{max}} = \frac{12 - 0}{2} = 6\ V$$

$\therefore \qquad P_{ac}\big|_{\text{max}} = \dfrac{1}{2}\cdot\dfrac{6 \times 6}{20} = 0.9\ W.$

$\therefore$ Maximum efficiency of the amplifier , $\eta_{\text{max}} = \dfrac{P_{ac}\big|_{\text{max}}}{P_{dc}} = \dfrac{0.9}{3.6} = \dfrac{1}{4} = 25\%\,.$

Power loss at the collector,

$$P_D = P_{dc} - I_0^2 R_L = 3.6 - (0.3)^2 \times 20 = 3.6 - 1.8 = 1.8\ W$$

$\therefore \qquad$ Maximum collector circuit efficiency

$$\eta_C\big|_{\text{max}} = \frac{P_{ac}}{P_D} = \frac{0.9}{1.8} = \frac{1}{2} = 50\%\,.$$

7.2   In class-B push-pull AF power amplifier, the peak collector current is $500\ mA$ and the collector supply voltage is $10\ V$. The efficiency of this amplifier is 60%. Calculate the ac power delivered to the load and the power dissipation of the amplifier.

Solution

Here, peak collector current, $I_P = 0.5\,\text{A}$.

Supply voltage, $V_{CC} = 10\,\text{V}$.

Efficiency, $\eta = 0.6$.

DC power input, $P_{dc} = V_{CC} \cdot \dfrac{2}{\pi} I_P = 10 \cdot \dfrac{2}{\pi} \cdot 0.5 = \dfrac{10}{\pi}$ watt.

$\therefore$ AC power delivered, $P_{ac} = \eta P_{dc} = 0.6 \times \dfrac{10}{\pi} = \dfrac{6}{\pi}$ watt $= 1.91\,\text{W}$

Power dissipation, $P_D = P_{dc} - P_{ac} = \dfrac{10}{\pi} - \dfrac{6}{\pi} = \dfrac{4}{\pi} = 1.27$ watt.

7.3 A class-B tuned RF power amplifier is driven by an input signal $v_{in}(t) = 5\sin(10^6 \pi t)$ volts. The projected cut-off of the transistor has a magnitude of 0.7 V. The dynamic resistance of the tuned load is 20 Ω and the peak collector current is 300 mA. Calculate the collector circuit efficiency of the amplifier.

Solution

From eqn. (7.55), we write

$$\frac{I_1}{I_{dc}} = \frac{\dfrac{V_i}{2} - \dfrac{2}{\pi} V_{B0}}{\dfrac{V_i}{\pi} - \dfrac{V_{B0}}{2}}$$

Here, $V_i = 5\,\text{V}$

$\qquad V_{B0} = 0.7\,\text{V}$

Then, $\dfrac{I_1}{I_{dc}} = \dfrac{\dfrac{5}{2} - \dfrac{2}{\pi} \times 0.7}{\dfrac{5}{\pi} - \dfrac{0.7}{2}} = 1.65$

Peak collector current, $I_1 = 300\,\text{mA}$.

Dynamic resistance, $R_d = 20\Omega$.

Collector circuit efficiency,

$$\eta = \frac{1}{2} \frac{I_1}{I_{dc}} \cdot \frac{I_1 R_d}{V_{CC}} = \frac{1}{2} \times 1.65 \times \frac{0.3 \times 20}{10} \qquad = 0.3 \times 1.65 = 49.5\%.$$

7.4   A class-C rf power amplifier is driven by an input signal which saturates the pnp transistor used. The collector supply voltage is 6 volt and the saturation collector voltage of the transistor is 0.2 volt. If the conduction angle of the amplifier is $45°$ then calculate its efficiency.

Solution

Here, $V_{CC} = 6$ volt.

$$V_C\big|_{min.} = 0.2 \text{ volt.}$$

$$2\theta_C = 45° = \frac{\pi}{4} \text{ radian}$$

$\therefore$ $\theta_C = \dfrac{\pi}{8}$ radian

Then, $\eta = \dfrac{1}{2}\left[\dfrac{\theta_C}{\sin\theta_C} + \cos\theta_C\right]\left[1 - \dfrac{V_C\big|_{min.}}{V_{CC}}\right]$

$$= \frac{1}{2}\left[\frac{(\pi/8)}{\sin(\pi/8)} + \cos(\pi/8)\right]\left[1 - \frac{0.2}{6}\right]$$

$$= 0.975 \times \frac{29}{30} = 94.2\%$$

7.5   A AM signal having a carrier frequency of 550 kHz and a bandwidth of 10 kHz is to be amplified by a single tuned amplifier. The tuned circuit has a dynamic resistance of 10 kΩ. Find the component values of the tuned circuit of the amplifier. Neglect collector-base junction capacitance in comparison with external capacitance.

Solution

Here, carrier frequency, $f_0 = \dfrac{\omega_0}{2\pi} = 550$ kHz.

This is the resonance frequency of the tuned circuit.

Bandwidth of the signal = Bandwidth of the tuned circuit = 10 kHz

$\therefore$ Quality factor, $Q = \dfrac{f_0}{\text{bandwidth}} = \dfrac{550}{10} = 55$

Dynamic resistance, $R_t = \dfrac{L}{RC} = 10^4 \Omega$

Now, $Q = \dfrac{\omega_0 L}{R}$

$\therefore$ $\dfrac{L}{R} = \dfrac{Q}{\omega_0} = \dfrac{Q}{2\pi f_0} = \dfrac{55}{2\pi \cdot 550 \times 10^3} = \dfrac{10^{-4}}{2\pi}$ sec.

Then, $C = \dfrac{L}{R} \cdot 10^{-4} = \dfrac{10^{-8}}{2\pi} = 1.59 \, \text{nF}.$

Again, $\omega_0^2 LC = 1$

$\therefore \qquad L = \dfrac{1}{\omega_0^2 C} = \dfrac{1}{4\pi^2 f_0^2 C} = \dfrac{1}{4\pi^2} \dfrac{1}{(550)^2 \cdot 10^6} \cdot 2\pi \times 10^8 = \dfrac{1}{2\pi} \cdot \dfrac{1}{(55)^2} \, \text{Henry} = 52.6 \, \mu\text{H}$

Now, $\dfrac{L}{R} = \dfrac{10^{-4}}{2\pi} \qquad \therefore \quad R = 2\pi L \cdot 10^4 = \dfrac{10^4}{(55)^2} = 3.3\,\Omega.$

So, the tuned circuit of the amplifier must have an inductance of $52.6\,\mu\text{H}$ with a coil resistance of $3.3\,\Omega$. The capacitor will have a value of $1.59\,\text{nF}$.

7.6 A double tuned rf amplifier has identical resonant circuits having a resonance frequency of $600\,\text{kHz}$. Calculate the bandwidth of this amplifier when critically coupled. The Q-value of the resonant circuits is 20. Calculate the separation of double hump peaks when it is overcoupled with $kQ = 2$ where k is the coupling coefficient of the two circuits.

Solution

Bandwidth under critical coupling $= \dfrac{f_0}{Q}\sqrt{2}$.

Here, $f_0 = 600\,\text{kHz}$ and $Q = 20$.

Then, $BW = \dfrac{\sqrt{2} \cdot 600}{20} = 30 \cdot \sqrt{2} = 42.3$ kHz.

From eqn. (7.92), the double hump peaks occur at

$$\omega = \omega_0 \left[1 \pm \frac{1}{2}\sqrt{k^2 - \frac{1}{Q^2}} \right]$$

since $Q_1 = Q_2 = Q$. Then, the double hump peak separation

$$\omega_2 - \omega_1 = \omega_0 \sqrt{k^2 - \frac{1}{Q^2}} = \frac{\omega_0}{Q}\sqrt{k^2 Q^2 - 1}$$

Here, $kQ = 2$.

$\therefore \qquad \omega_2 - \omega_1 = \dfrac{\omega_0}{Q}\sqrt{3}$

In terms of frequency in Hz, we have

$$f_2 - f_1 = \frac{f_0}{Q}\sqrt{3} = \sqrt{3} \cdot \frac{600}{20} = 30\sqrt{3} = 30 \times 1.732 = 51.96 \text{ kHz.}$$

PROBLEMS

7.1   The conduction angle of a class-C rf power amplifier has a value of $75°$. The peak voltage across the tank circuit attained during a cycle is 5 volt. Calculate the efficiency of the amplifier if the collector supply voltage is 6 volts.

[Ans. 77.85%]

7.2   The collector voltage of each transistor in class-B push-pull AF power amplifier varies in the range $0.5 - 9.5$ V. The peak collector supply voltage is 10 V. calculate the efficiency of this amplifier.

[Ans. 70.68%]

7.3   The number of primary and secondary turns of a transformer with a center-tapped primary push-pull AF power amplifier is 200 and 100 respectively. The peak collector current of each transistor is 600 mA and the peak collector voltage swing of each transistor is half of the supply voltage. Calculate the output ac power delivered across the load if the magnitude of the collector supply voltage is 10 V. Also calculate the efficiency and the power dissipation of the amplifier.

[Ans. 1.5 W, 39.3%, 2.32 W]

7.4   A transformer- coupled class-A power amplifier has a load resistance of 50 Ω connected with the secondary of the transformer. The collector supply voltage is 12 V. The number of turns in the primary and the secondary is the same. To minimize distortion the collector voltage is restricted to be in the range (2 V, 22 V). Calculate the efficiency of the amplifier. Find the collector circuit efficiency.

[Ans. 8.68%, 8.68%]

7.5   A single-tuned amplifier has a resonance frequency of 1 MHz. The tuned circuit has a Q-value of 100 and the inductive arm has a reactance of $1 k\Omega$ at resonance. Calculate the bandwidth of the amplifier. Find the component values of the tuned circuit. Calculate the dynamic resistance of the tuned circuit. Assume the coil resistance to be very small compared with the inductive resistance at resonance and neglect collector-base capacitance.

[Ans. 10 kHz, $L = 159.1$ µH, $C = 159.1$ pF, $R = 10\,\Omega$, $R_t = 100\,k\Omega$]

Hints: $\Delta f = \dfrac{f_0}{Q}$, $\omega_0^2 LC = 1,$

$$\omega_0 L = 10^3\,\Omega, \quad Q = \frac{\omega_0 L}{R},$$

$$R_t = \frac{L}{RC}$$

QUESTIONS

7.1 Draw a schematic circuit diagram of a cascode amplifier. Derive expressions for its voltage gain, current gain, and input and output impedance. What is its main use?

7.2 What do you mean by a 'Darlington pair'? Explain using a circuit diagram. Derive expressions for its current gain, voltage gain, input resistance and output resistance. Write down its application.

7.3 Suppose 'n' identical RC coupled amplifiers are cascaded to produce a high level amplification. Calculate the overall bandwidth of the cascaded amplifier if a single stage has a bandwidth of B Hz.

7.4 What do you mean by voltage and power amplifiers? How do you classify power amplifiers on the basis of its operation? Give a circuit diagram of the direct coupled class-A AF power amplifier. Calculate the maximum possible efficiency of this amplifier.

7.5 Describe the operation of a transformer-coupled class-A power amplifier. Derive an expression for the efficiency of this amplifier. Why is the maximum efficiency of the transformer-coupled class-A amplifier greater than that of a direct coupled amplifier?

7.6 Write down the principle of operation of a class-B push-pull AF amplifier. Derive an expression for collector circuit efficiency of this amplifier. What is the maximum value of this efficiency? What are the advantages of the push-pull amplifier?

7.7 Draw a schematic circuit diagram of a tuned RF class-B power amplifier and explain its operation. Find out an expression for the collector circuit efficiency of this amplifier. What is the upper limit of this efficiency? Where do you use this amplifier?

7.8 What is the class-C operation of a power amplifier? Derive an expression for the collector circuit efficiency of a class-C power amplifier. What is the reason of its high efficiency? How does this efficiency vary with the magnitude of the conduction angle of the transistor?

7.9 What do you mean by a RF amplifier? Analyze a single-tuned amplifier to find its voltage gain and bandwidth. Calculate its gain-bandwidth product.

7.10 Draw the circuit diagram of a double-tuned RF amplifier. Derive an expression for the voltage gain of this circuit. Find out an expression for the bandwidth of this amplifier when it is critically coupled. How does the frequency response get modified when the amplifier is over coupled? Calculate the frequency separation between the double hump peaks of a overcoupled double-tuned amplifier.

8

Feedback in Amplifiers

8.1 FEEDBACK DEFINITION AND CLASSIFICATION

Feedback is the process in which a portion of the output signal of an amplifier is coupled back to the input through some network. This process modifies the characteristics of the amplifier. A block diagram of the feedback process is shown below: .

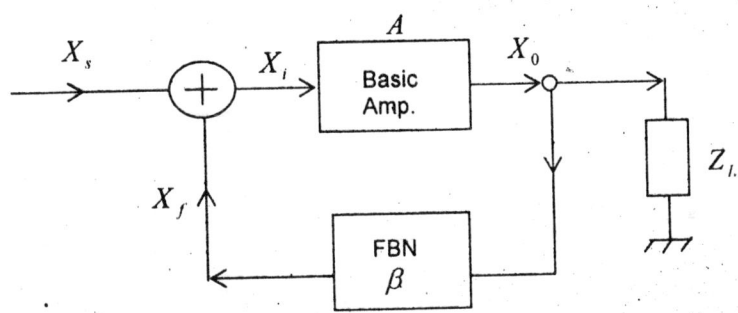

Fig. 8.1. A schematic block diagram of the feedback amplifier.

Here, X_s = input signal

X_0 = output signal

X_f = feedback signal

X_i = net input signal.

FBN is the feedback network. From the block diagram,

$$X_i = X_s + X_f \qquad \ldots (8.1)$$

$$X_0 = AX_i \qquad \ldots (8.2)$$

$$X_f = \beta X_0 \qquad \ldots (8.3)$$

where A is the open-loop gain of the amplifier and β is the feedback factor. The feedback factor is defined as the fraction of the output signal that is fed back to the input. β can, in general, be a complex quantity and can depend on signal frequency. The signal means voltage or current.

Now, the open-loop gain, $A = \dfrac{X_0}{X_i}$. The gain of the amplifier with feedback is

$$A_f = \frac{X_0}{X_s} = \frac{AX_i}{X_s} = \frac{A(X_s + X_f)}{X_s}$$

$$= A\left[1 + \frac{X_f}{X_s}\right] = A\left[1 + \frac{\beta X_0}{X_s}\right]$$

$$= A\left[1 + \beta A_f\right]$$

$$\therefore \quad A_f(1 - A\beta) = A$$

or, $\qquad A_f = \dfrac{A}{1 - A\beta}$ ... (8.4)

The equation (8.4) can be derived in another way.

Suppose the loop is open. The output signal is then AX_s. Now, connect the feedback loop. The input changes by the feedback voltage $A\beta X_s$. This change in input produces a change of $(A.A\beta X_s =)A^2\beta X_s$ in the output. Again, this change in output makes a change in feedback signal equal to $(\beta.A^2\beta X_s =)A^2\beta^2 X_s$. Now, this is the change in input signal. This change in input produces a change $(A.A^2\beta^2 X_s =)A^3\beta^2 X_s$ in the output. This process is cumulative and the net output signal in presence of feedback can be written as

$$X_0 = AX_s + A^2\beta X_s + A^3\beta^2 X_s + \ldots\ldots$$

$$= AX_s\left[1 + A\beta + A^2\beta^2 + \ldots\ldots\right]$$

$$= AX_s \cdot \frac{1}{1 - A\beta}$$ ... (8.5)

since, $|A\beta| < 1$.

Then, the gain of the amplifier with feedback is

$$A_f = \frac{X_0}{X_s} = \frac{A}{1 - A\beta}$$ ... (8.6)

A_f is also known as closed-loop gain. The gain suffered by the signal in passing through the amplifier and the feedback network is $A\beta$. That is why $A\beta$ is called the loop gain.

The feedback can be divided into two heads depending upon its effect on the output of the amplifier. These are

(i) Positive or, regenerative feedback, and
(ii) Negative or, degenerative feedback.

If $\left|A_f\right| > \left|A\right|$, that is $\left|1 - A\beta\right| < 1$, the feedback is positive. The output signal in this case is greater than the output signal in absence of feedback . On the other hand, if $\left|A_f\right| < \left|A\right|$, that is $\left|1 - A\beta\right| > 1$, the feedback is negative. The magnitude of the output signal in this case is less than that of the output signal in absence of feedback.

The feedback can be divided into two heads depending upon whether the feedback signal depends upon the output voltage or current. These are

(i) voltage feedback, and
(ii) current feedback.

If the signal (voltage or current) fed back to the input is proportional to the output voltage of the amplifier, the feedback is said to be a voltage feedback. On the other hand, if the signal fed back to the input is proportional to the output current of the amplifier, it is called the current feedback. The feedback is said to be series feedback or shunt feedback depending upon whether the feedback signal is introduced in series or in shunt with the input signal respectively.

8.2 EFFECT OF NEGATIVE FEEDBACK

Negative feedback reduces closed-loop gain of the feedback amplifier. But, at the expense of gain reduction some useful properties are obtained.

We define the sensitivity of the gain of the amplifier as the ratio of the fractional change in closed-loop gain to the fractional change in open-loop gain.

Now,

$$A_f = \frac{A}{1 - A\beta}$$

or, $\ln A_f = \ln A - \ln(1 - A\beta)$

Taking differentials,

we get

$$\frac{dA_f}{A_f} = \frac{dA}{A} - \frac{-\beta dA}{1 - A\beta}$$

$$= \frac{dA}{A}\left[1 + \frac{A\beta}{1 - A\beta}\right]$$

$$= \frac{dA}{A} \cdot \frac{1}{1 - A\beta}$$

$$\therefore \quad S = \frac{\left|\dfrac{dA_f}{A_f}\right|}{\left|\dfrac{dA}{A}\right|} = \frac{1}{\left|1 - A\beta\right|} \qquad \qquad \ldots (8.7)$$

S is the sensitivity of gain of the amplifier. Thus, in closed-loop operation the fractional change in amplifier gain is reduced from that in open-loop operation in the case of negative feedback. This is because in negative feedback $\left|1 - A\beta\right| > 1$. If the loop gain magnitude $\left|A\beta\right| \gg 1$, then

$$A_f \cong \frac{A}{-A\beta} \qquad \text{(neglecting 1 in comparison with } A\beta)$$

$$= -\frac{1}{\beta}.$$

The closed-loop gain in this case depends upon the feedback factor only.

8.2.1 IMPROVEMENT OF STABILITY

Let $R = \dfrac{1}{1 - A\beta}$. Then, $A_f = R.A$.

In negative feedback, $\left|R\right| < 1$ since $\left|1 - A\beta\right| > 1$.

So, any undesired change in the open-loop gain of the amplifier is multiplied by $R(<1)$ in its effect on the closed-loop gain. But, since $\left|R\right| < 1$, this undesired change in A_f is reduced compared with the original change in A. This improves the stability of amplifier gain.

8.2.2 AMPLIFIER DISTORTION REDUCTION

There can be three types of distortion in an amplifier. These are:

   (i) Frequency distortion,
  (ii) Phase distortion, and
 (iii) Nonlinear distortion.

The gain of the amplifier varies, in general, with the signal frequency. Thus, $A \equiv A(\omega)$, where ω = signal frequency in radian. So, different frequency components of the signal are amplified by different amounts by the amplifier. This introduces frequency distortion.

The gain of an amplifier is, in general, a complex quantity. It has a magnitude and a phase angle. Thus, $A(\omega) = \left|A(\omega)\right| < \varphi(\omega)$. $\varphi(\omega)$ is the phase angle which is a function of signal frequency, ω. Phase distortion occurs when the relative phases of different frequency components of the signal at the output changes from that at the input of the amplifier. This occurs because the phase shift $\varphi(\omega)$ of the signal produced by the amplifier is a function of signal frequency.

Now, if negative feedback is applied such that $|A\beta| >> 1$, then $A_f = -1/\beta$. Moreover, if the feedback network contains no reactive element then β is real and A_f is frequency independent. In this case, frequency and phase distortions are substantially reduced.

Nonlinear distortion arises from the nonlinearity of the device and gives rise to additional frequency components at the output which are not present at the input. This distortion arises due to the nonlinearity in the amplifier characteristics. We assume this distortion to be small and the principle of superposition to be valid.

Let V_s = input signal voltage,

V_{os} = output signal voltage in absence of feedback

D_0 = distortion at the output in absence of feedback.

Then, total output voltage in absence of feedback is

$$V_0 = V_{os} + D_0$$

and $V_{os} = AV_s$.

Here, A is the open-loop gain.

Let D_f = distortion in presence of feedback. Then, due to feedback the distortion changes by $A\beta D_f$ in making a round trip passage through the loop in the steady state. So,

$$D_f = D_0 + A\beta D_f$$

or, $$D_f = \frac{D_0}{1 - A\beta} \qquad \qquad \ldots (8.8)$$

For negative feedback, $|1 - A\beta| > 1$ and hence $D_f < D_0$. Negative feedback, thus, reduces nonlinear distortion. The gain of the amplifier is also reduced at the same time to a value $\frac{A}{(1 - A\beta)}$. But, gain reduction can be overcome by changing circuit parameters.

8.2.3 EFFECT ON AMPLIFIER BANDWIDTH

The gain bandwidth product of an amplifier is a constant. This can be seen from the analysis of a RC coupled amplifier in the audio frequency region or a tuned amplifier in the radio frequency region. Now,

$$|A| \times BW = \text{constant} = |A_f| \times BW_f \qquad \qquad \ldots (8.9)$$

where BW = amplifier bandwidth without feedback, BW_f = amplifier bandwidth with feedback.

$\therefore$ $$BW_f = \frac{|A|}{|A_f|} BW = |1 - A\beta| BW \cdot \qquad \qquad \ldots (8.10)$$

Since, with negative feedback $|1 - A\beta| > 1$, $BW_f > BW$.

8.2.4 EFFECT OF NEGATIVE FEEDBACK ON INPUT AND OUTPUT IMPEDANCES

In negative feedback, when the feedback signal (voltage or current) is introduced in series with the input signal, the feedback voltage opposes the signal voltage. Hence, the net input voltage is reduced. This causes a reduction of input current (I_i). The input resistance of the amplifier with feedback is defined as,

$$R_{if} = \frac{V_s}{I_i}.$$

where V_s = input signal voltage and I_i = input signal current. Since I_i is reduced, R_{if} is increased in presence of negative series feedback. The schematic circuit diagrams of voltage series feedback and current series feedback is shown in Figs. 8.2 and 8.4 respectively.

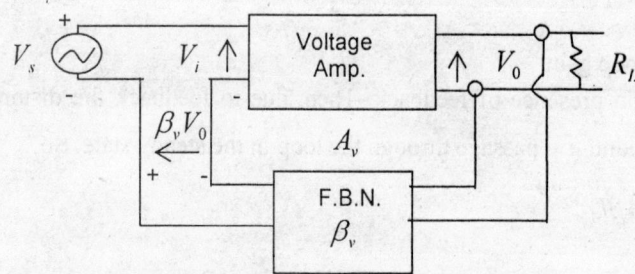

Fig. 8.2. Voltage series feedback.

Here, V_0 = output voltage,

V_i = Effective input voltage

β_v = feedback factor

R_L = load resistance

F.B.N. = Feedback network.

From, Fig. 8.2, we can write

$$V_i = V_s + \beta_v V_0$$

$$= V_s + \beta_v A_v V_i \qquad\qquad [\because V_0 = A_v V_i]$$

$$\therefore \qquad V_i(1 - A_v \beta_v) = V_s \qquad\qquad\qquad \dots(8.11)$$

$$R_{if} = \frac{V_s}{I_i} = \frac{V_i}{I_i}(1 - A_v \beta_v) = R_i(1 - A_v \beta_v) \qquad\qquad \dots(8.12)$$

where $R_i = \dfrac{V_i}{I_i}$ = input impedance without feedback.

In negative feedback, $|1 - A_v\beta_v| > 1$. Therefore, $R_{if} > R_i$. We should remember that in negative feedback, the feedback voltage $\beta_v V_0$ is in phase opposition with V_s, i.e., $\beta_v V_0$ is itself negative in Fig. 8.2.

We now calculate the output impedance (R_{of}) with feedback in voltage series feedback circuit. The corresponding ac equivalent output circuit is shown in Fig. 8.3.

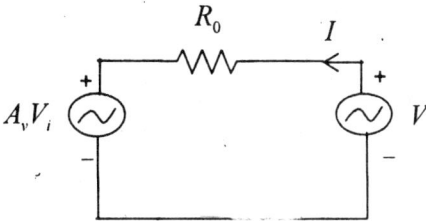

Fig. 8.3. AC equivalent output circuit of Fig. 8.2 with $V_s = 0$ and $R_L = \infty$.

To measure output impedance with feedback, set source voltage $V_s = 0$ and remove R_L. Place a test voltage source (V) at the output terminals which drives a current I.

Then,

$$R_{of} = \frac{V}{I} \qquad \qquad \ldots (8.13)$$

From Figures 8.2 and 8.3, we write

$$V = A_v V_i + R_0 I \qquad \qquad \ldots (8.14).$$

with $V_s = 0, V_i = \beta_v V_0 = \beta_v.V$ $\qquad [\because V_0 = V]$.

From (8.14), we can write

$$V = A_v \beta_v V + R_0 I$$

or, $\qquad V(1 - A_v \beta_v) = R_0 I$

or, $\qquad \dfrac{V}{I} = \dfrac{R_0}{1 - A_v\beta_v}$

$\therefore \qquad R_{of} = \dfrac{R_0}{1 - A_v\beta_v} \qquad \qquad \ldots (8.15)$

Since $|1 - A_v\beta_v| > 1$ in negative feedback, we get $R_{of} < R_0$ in voltage series feedback.

In current series feedback, the feedback voltage is proportional to output current.

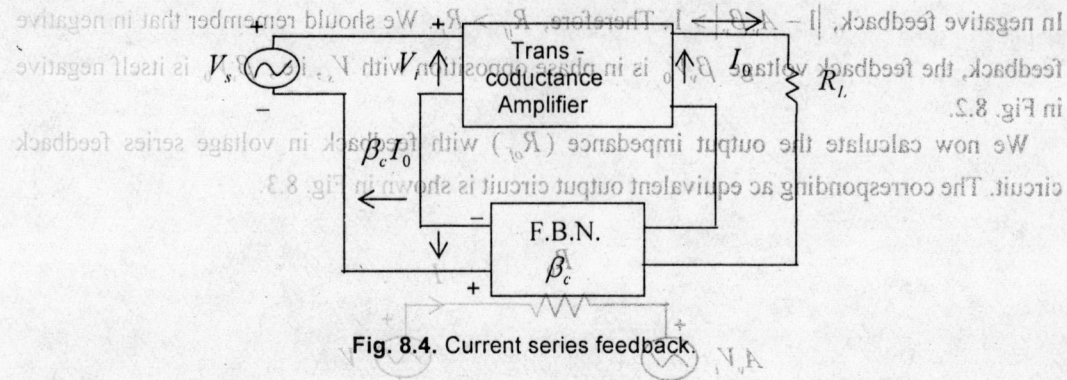

Fig. 8.4. Current series feedback.

Here, β_c = feedback factor

A_c = open-loop gain of transconductance amplifier

I_0 = output current

V_f = feedback voltage

$= \beta_c I_0$

$I_0 = A_c . V_i$

From fig. 8.4, we write

$$V_i = V_s + V_f = V_s + \beta_c I_0$$

$$= V_s + \beta_c . A_c V_i \qquad (8.16)$$

$$\therefore \quad V_i (1 - A_c \beta_c) = V_s \qquad (8.17)$$

$$\therefore \quad R_{if} = \frac{V_s}{I_i} = \frac{V_i}{I_i}(1 - A_c \beta_c)$$

$$= R_i (1 - A_c \beta_c) \qquad (8.18)$$

For negative feedback, $|1 - A_c \beta_c| \geqslant 1$. Hence, $R_{if} > R_i$.

We now calculate the output impedance in current series feedback. To do this, remove R_L and make $V_s = 0$ in Fig. 8.4. Then, apply a test voltage V at the output terminals. If this voltage source drives a current I then output impedance with feedback is

$$R_{of} = \frac{V}{I} \qquad \dots (8.19)$$

The equivalent circuit for the measurement of R_{of} in current series feedback is shown in Fig. 8.5.

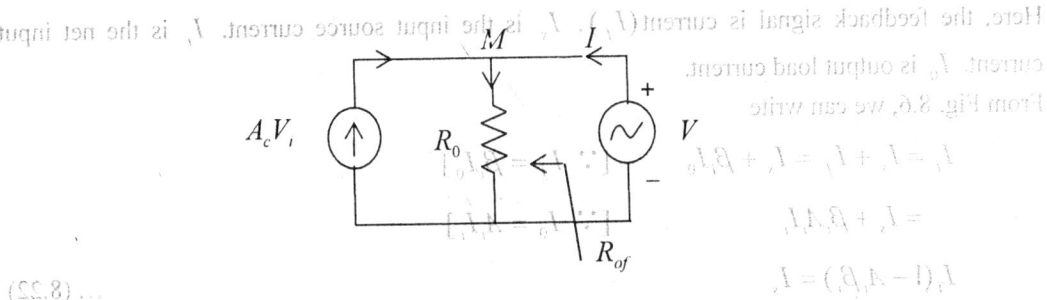

Fig. 8.5. AC equivalent output circuit of Fig. 8.4 for output impedance measurement.

Since, $V_s = 0$, $V_i = V_f = \beta_c I_0$ [where I_0 is the load current in Fig. 8.4]

When there is no feedback, $V_i = 0$. The only voltage source is V and it drives a current I through R_0 so that $R_0 = \dfrac{V}{I}$. This is in absence of feedback.

In presence of feedback, applying Kirchhoff's current law at the point M in Fig. 8.4 we get,

$$I + A_c V_i = \frac{V}{R_0} \qquad \qquad \qquad \qquad \dots (8.20)$$

or, $\qquad I(1 - A_c \beta_c) = \dfrac{V}{R_0} \qquad [\because V_i = \beta_c I_0 \text{ and } I_0 = -I]$

$$\therefore \qquad \frac{V}{I} = R_0(1 - A_c \beta_c)$$

$$\therefore \qquad R_{of} = \frac{V}{I} = R_0(1 - A_c \beta_c) \qquad \qquad \dots (8.21)$$

Since $|1 - A_c \beta_c| > 1$ in negative feedback, we have $R_{of} > R_0$ in current series feedback.

Now, we consider current shunt feedback. The circuit is shown in Fig. 8.6.

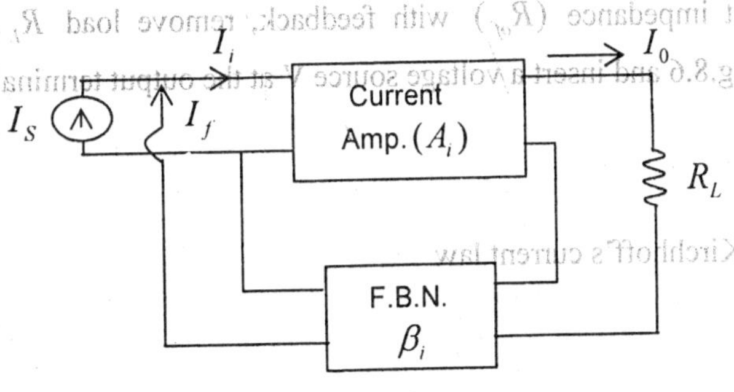

Fig. 8.6. Current shunt feedback.

Here, the feedback signal is current (I_f). I_s is the input source current. I_i is the net input current. I_0 is output load current.

From Fig. 8.6, we can write

$$I_i = I_s + I_f = I_s + \beta_i I_0 \qquad [\because I_f = \beta_i I_0]$$

$$= I_s + \beta_i A_i I_i \qquad [\because I_0 = A_i I_i]$$

$$I_i(1 - A_i \beta_i) = I_s \qquad \qquad \dots (8.22)$$

The input impedance with feedback is

$$R_{if} = \frac{V_i}{I_s} = \frac{V_i}{I_i} \cdot \frac{1}{1 - A_i \beta_i}$$

$$= \frac{R_i}{1 - A_i \beta_i} \qquad \qquad \dots (8.23)$$

Here, V_i is the amplifier input voltage and $R_i = \dfrac{V_i}{I_i}$ is the input impedance without feedback.

Since $|1 - A_i \beta_i| > 1$ in negative feedback, $R_{if} < R_i$ in current shunt feedback.

We now calculate the output impedance of the current shunt feedback circuit. The ac equivalent circuit is shown in Fig. 8.7.

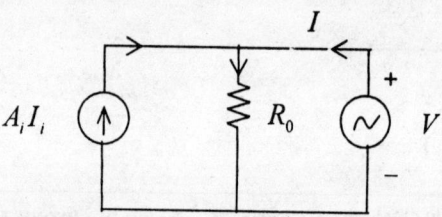

Fig. 8.7. Ac equivalent output circuit of Fig. 8.6.

To determine the output impedance (R_{of}) with feedback, remove load R_L, remove source current $(I_s = 0)$ from Fig.8.6 and insert a voltage source V at the output terminals. Now,

$$R_{of} = \frac{V}{I} \qquad \qquad \dots (8.24)$$

From Fig. 8.7 we get by Kirchhoff's current law

$$I + A_i I_i = \frac{V}{R_0} \qquad \qquad \dots (8.25)$$

where R_0 = output impedance without feedback. But, with $I_s = 0$ we get $I_i = I_f = \beta_i I_0$ (from Fig. 8.6).

$$\therefore \qquad I + A_i \beta_i I_0 = \frac{V}{R_0} \qquad\qquad \text{[from (8.25)]}$$

or, $\qquad I(1 - A_i \beta_i) = \dfrac{V}{R_0} \qquad\qquad [\because I_0 = -I\,]$

or, $\qquad \dfrac{V}{I} = R_0(1 - A_i \beta_i)$

$$\therefore \qquad R_{of} = R_0(1 - A_i \beta_i) \qquad\qquad\qquad \dots (8.26)$$

Since $|1 - A_i \beta_i| > 1$ in negative feedback, we get $R_{of} > R_0$ in current shunt feedback.

We now calculate the input and output impedances in voltage shunt feedback. The voltage shunt feedback circuit is shown in Fig. 8.8.

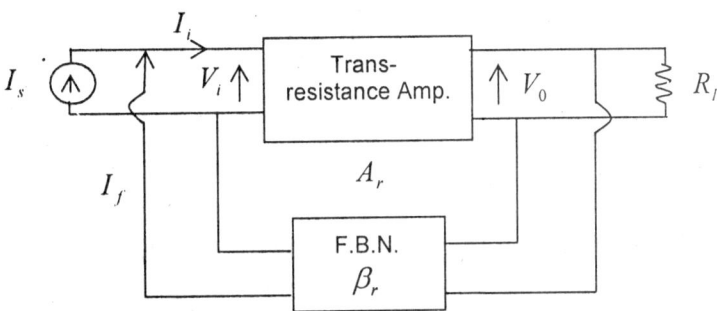

Fig. 8.8. Voltage shunt feedback.

The feedback signal is current which is proportional to output voltage. Here,

$$I_i = I_s + I_f \qquad\qquad\qquad \dots (8.27)$$

$$I_f = \beta_r V_0 \qquad\qquad\qquad \dots (8.28)$$

$$V_0 = A_r I_i \qquad\qquad\qquad \dots (8.29)$$

where A_r = open-loop gain of the transresistance amplifier. β_r = feedback factor. Now,

$$I_i = I_s + \beta_r V_0$$

$$= I_s + \beta_r . A_r I_i$$

or, $\qquad I_i(1 - A_r \beta_r) = I_s \qquad\qquad\qquad \dots (8.30)$

Input impedance with feedback is

$$R_{if} = \frac{V_i}{I_s} = \frac{V_i}{I_i} \cdot \frac{1}{1 - A_r \beta_r}$$

$$= \frac{R_i}{1 - A_r \beta_r} \qquad \qquad \qquad \dots (8.31)$$

In negative feedback, $\left| 1 - A_r \beta_r \right| > 1$ so that $R_{if} < R_i$.

For output impedance calculation in voltage shunt feedback circuit, we consider the following ac equivalent output circuit as shown in Fig. 8.9.

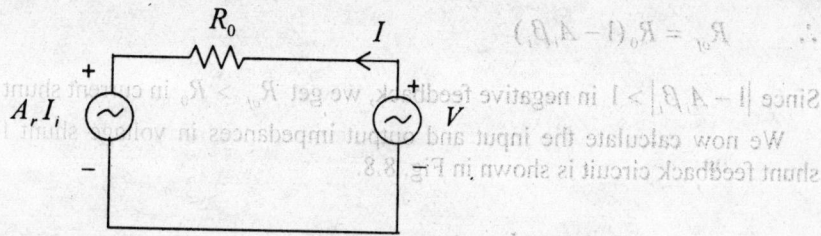

Fig. 8.9. AC equivalent output circuit of Fig. 8.8.

By applying Kirchhoff's voltage law, we get from Fig. 8.8

$$V = A_r I_i + R_0 I \qquad \qquad \dots (8.32)$$

where R_0 = output impedance without feedback. For output impedance measurement, we set $I_s = 0$ and remove R_L. Then, insert a test voltage source V at the output which drives a current I. Now, with $I_s = 0$ we have

$$I_i = I_f = \beta_r V_0$$

$$= \beta_r V \qquad \qquad \qquad [\because V_0 = V]$$

$$\therefore \quad V = A_r \beta_r V + R_0 I$$

or, $$V(1 - A_r \beta_r) = R_0 I$$

$$\therefore \quad \frac{V}{I} = \frac{R_0}{1 - A_r \beta_r}$$

$$\therefore \quad R_{of} = \frac{R_0}{1 - A_r \beta_r} \qquad \qquad \dots (8.33)$$

In negative feedback, $\left| 1 - A_r \beta_r \right| > 1$. Hence, $R_{of} < R_0$ in voltage shunt feedback.

SOLVED PROBLEMS

8.1   In a CE amplifier with voltage series feedback, calculate the gain with feedback when the amplifier has an open-loop voltage gain equal to 100 and the feedback circuit has a feedback factor 0.005.

Solution

Here, $A_v = -100, \beta_v = 0.005$

$\therefore \quad A_v \beta_v = -0.5$

Now, $1 - A_v \beta_v = 1 + 0.5 = 1.5$

$\therefore$ Closed-loop gain $A_f = \dfrac{A_v}{1 - A_v \beta_v} = \dfrac{-100}{1.5} = -66.6$.

8.2   If the above CE amplifier as mentioned in problem-1 has a bandwidth of 10 kHz under open-loop condition, what will be the bandwidth when the feedback loop is closed?

Solution

Here, open-loop bandwidth = 10 kHz

$1 - A_v \beta_v = 1 + 0.5 = 1.5$

$\therefore$ Closed-loop bandwidth = 10 kHz $\times (1 - A_v \beta_v)$

$= 10$ kHz $\times 1.5$

$= 15$ kHz

8.3   A CE voltage amplifier has an open-loop voltage gain equal to -300 which produces a harmonic distortion of 10%. Find out the value of the feedback factor if you want to reduce the distortion to 1%.

Solution

The distortion with feedback is

$$D_f = \frac{D_0}{1 - A_v \beta_v}$$

where D_0 is the distortion without feedback, A_v is the open-loop voltage gain and β_v is the feedback factor.

$D_0 = 10\% = 0.1$

$D_f = 1\% = 0.01$

Then,

$$1 - A_v \beta_v = \frac{D_0}{D_f}$$

or,

$$-A_v \beta_v = \frac{D_0}{D_f} - 1$$

$$\therefore \quad \beta_v = \frac{1}{-A_v}\left(\frac{D_0}{D_f} - 1\right)$$

$$= \frac{1}{300}\left(\frac{0.1}{0.01} - 1\right)$$

$$= \frac{9}{300} = \frac{3}{100} = 0.03$$

8.4 A CE voltage amplifier with a open-loop voltage gain equal to -200 and an input resistance of $1\,\text{k}\Omega$. You are asked to raise the input resistance equal to $5\,\text{k}\Omega$. What kind of feedback you will use? What is the value of feedback factor required for this circuit?

Solution
The feedback to be used for raising the input resistance is the negative voltage series feedback.
In this case, the input resistance with feedback is

$$R_{if} = R_i(1 - A_v \beta_v)$$

where R_i = input resistance without feedback

   A_v = open-loop voltage gain

   β_v = feedback factor.

Here, $R_{if} = 5\,\text{k}\Omega$

   $R_i = 1\,\text{k}\Omega$

   $A_v = -200$

Then, $1 - A_v \beta_v = \dfrac{R_{if}}{R_i}$.

or, $\beta_v = \dfrac{1}{-A_v}\left(\dfrac{R_{if}}{R_i} - 1\right)$

$$= \frac{1}{200}\left(\frac{5}{1} - 1\right) = \frac{4}{200} = \frac{1}{50} = 0.02$$

8.5   A CE voltage amplifier with negative feedback produces an output voltage with an input voltage of 100 millivolt. When operated without feedback this amplifier produces the same output voltage with an input voltage of 10 mv. Find out the magnitude of open-loop voltage gain of the amplifier when the feedback factor is 0.05. Also, calculate the output voltage of the amplifier.

Solution

Input voltage with feedback = 100 mv = 0.1 volt.

Gain with feedback, $A_{vf} = -\dfrac{V_0}{0.1} = -10V_0$,

where V_0 = output voltage in volts.

Under open-loop condition, gain

$$A_v = -\dfrac{V_0}{0.01}$$
$$= -100V_0. \quad [\because 10 \text{ mV} = 0.01 \text{ V}]$$

Again,

$$A_{vf} = \dfrac{A_v}{1 - A_v\beta_v}$$

$\therefore \qquad -10V_0 = \dfrac{-100V_0}{1 - A_v\beta_v}$

$\therefore \qquad 1 - A_v\beta_v = 10$

or, $\qquad -A_v = \dfrac{9}{\beta_v} = \dfrac{9}{0.05} = 180 \quad [\because \beta_v = 0.05]$

Again,

$$A_v = -100V_0$$

$\therefore \qquad V_0 = \dfrac{A_v}{-100} = \dfrac{-180}{-100} \quad = 1.8 \text{ volt.}$

PROBLEMS

8.1   A voltage amplifier has a gain of magnitude 150 and the output is phase shifted by 180° from the input. Calculate the gain of this amplifier when a negative series feedback is applied with a feedback factor of 0.01.

[**Ans.** -60]

8.2 A voltage amplifier has a mid-frequency gain of 200 and a bandwidth of 40 kHz. Its bandwidth is increased to 100 kHz by applying negative voltage series feedback. Find out the mid-frequency voltage gain with feedback. (Given that the lower half power frequency is very small compared with the upper half power frequency).

[**Ans.** 80 kHz]

8.3 A voltage amplifier with negative series feedback has a closed-loop gain of -100 and an open-loop gain of -200. In open-loop condition, the gain varies from -180 to -220 due to parameter variation. What will be the range of gain variation when feedback is applied?

[**Ans.** -95 to -105]

8.4 In a current amplifier using negative current shunt feedback, the input current is 1 mA and the output current is 2 mA. Under open-loop condition, the output current is 5 mA. The amplifier has a phase reversal of $180°$ between the input and the output. Calculate the feedback factor.

[**Ans.** 0.3]

QUESTIONS

8.1 What do you mean by 'feedback' in amplifiers? Derive an expression for closed-loop gain of an amplifier with feedback in terms of open-loop gain and the feedback factor.

8.2 What are the different types of feedback? Define negative feedback. What are the advantages of negative feedback? How do the input and output impedances of a voltage amplifier change with feedback?

8.3 How does the gain of an amplifier get stabilized through negative feedback? Discuss the effect of negative feedback on the nonlinear distortion of an amplifier.

8.4 Calculate the output impedance of a current amplifier with feedback in terms of its current gain, feedback factor and the output impedance without feedback. Show that the input impedance of a current amplifier with negative feedback is reduced from its value without feedback.

240 Analog and Digital Electronics

9.2.1 PHASE-SHIFT OSCILLATOR

Transistorized circuit

9

Oscillators

9.1 DEFINITION AND CLASSIFICATION

Electronic oscillators are devices which convert dc power to ac power without the application of any signal at the input. They generate periodic signals at the output. Depending upon the nature of the output waveform they produce, oscillators can be divided into two groups:

(i) sinusoidal, and
(ii) relaxation oscillators.

Sinusoidal oscillators generate sinusoidal waveforms at the output while relaxation oscillators generate non-sinusoidal (square or rectangular) waveforms at the output. Multivibrators are examples of nonsinusoidal oscillators.

Depending upon the magnitude of the oscillator frequency, oscillators can be classified as audio frequency oscillators and radio frequency oscillators. Typical oscillation frequency of an audio frequency oscillator can range up to 100 kHz. For RF oscillators, the oscillation frequency can range up to several MHz.

9.1.1 BARKHAUSEN CRITERION

This criterion states that an amplifier with positive feedback will oscillate in the steady state if the loop gain is unity. Mathematically, $A\beta = 1$, where $A =$ open-loop gain and $\beta =$ feedback factor. Thus, the magnitude of loop gain is unity and the loop phase shift is zero.
The gain of a feedback amplifier is

$$A_f = \frac{A}{1 - A\beta}$$

For steady state oscillation, $A\beta = 1$. So, $A_f = \infty$. This means that there is an output signal when the input signal is zero.

9.2 AUDIO FREQUENCY OSCILLATORS

In this section, we will discuss phase-shift oscillator and Wien-Bridge oscillator in detail.

9.2.1 PHASE-SHIFT OSCILLATOR

Transistorized circuit

This circuit uses a single CE amplifier with a resistive load which introduces a phase shift of $180°$ between the input and the output. Another $180°$ phase shift is introduced by 3 R-C sections so that the total phase shift becomes $180° + 180° = 360°$ (which is equivalent to $0°$). Thus, the second part of Barkhausen criterion which demands that loop phase shift must be zero, is satisfied. The circuit diagram of the transistorized phase-shift oscillator is shown in Fig. 9.1.

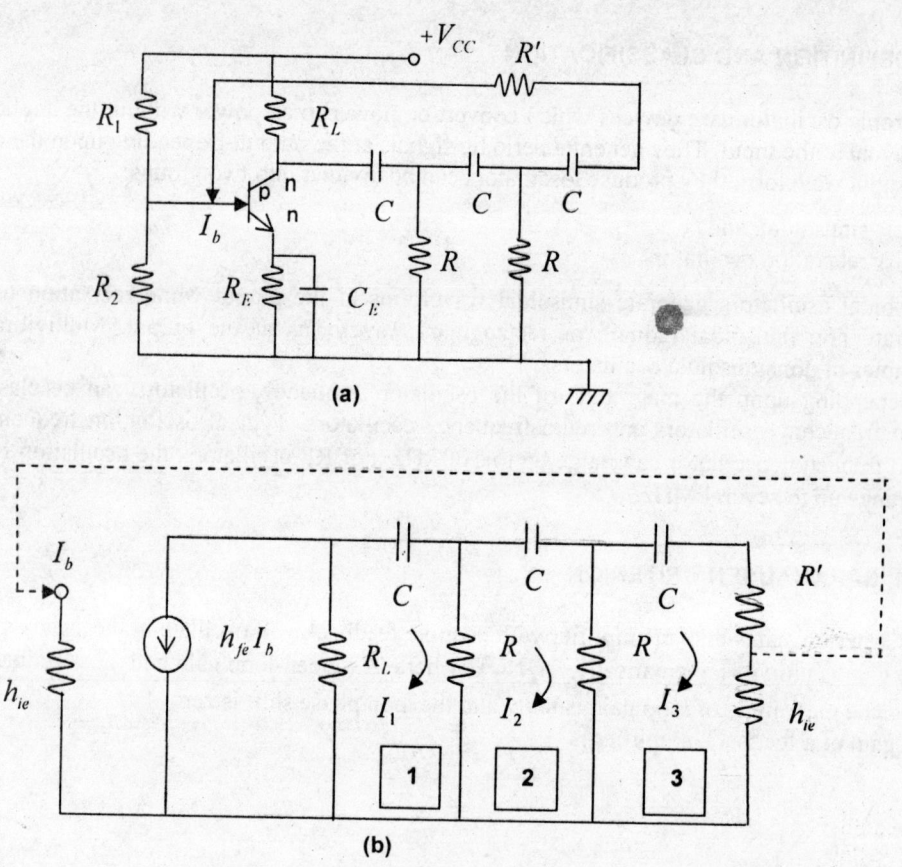

Fig. 9.1. (a) Circuit diagram and (b) AC equivalent circuit of the phase-shift oscillator.

The input resistance h_{ie} appears in series with the resistance R' in Fig. 9.1(a). We choose R' such that $R' + hie = R$.

Why do we use 3 R-C sections?

Each R-C section introduces a phase shift less than $90°$. So, to achieve a phase shift of $180°$ we require at least 3 R-C sections. Three R-C sections are chosen to be identical so that each R-C section introduces a phase-shift of $60°$. Oscillation will be possible even if we take 3 different R-C sections. In this case, the R-C sections will introduce different phase-shift but the total phase-shift will be $180°$. For simplicity in analysis, we choose 3 identical R-C sections.

The circuit uses voltage shunt feedback. The feedback signal is current which is proportional to output voltage. The transfer gain

$$A_r = \frac{X_o}{X_i} \qquad \qquad ... (9.1)$$

and feedback factor $\beta_r = \dfrac{X_f}{X_o}$ $\qquad \qquad ... (9.2)$

where X_o = output signal

X_i = input signal

X_f = feedback signal

Now, $A_r \beta_r = \dfrac{X_f}{X_i}$ $\qquad \qquad ... (9.3)$

Here, $X_o = V_o$, the output voltage

$X_i = I_b$, the input current

$X_f = I_3$, the feedback current.

According to Barkhausen criterion, we must have

$$A_r \beta_r = 1 \qquad \qquad ... (9.4)$$

i.e., $I_3 = I_b$ $\qquad \qquad ... (9.5)$

Let I_1, I_2, I_3 be the currents flowing through the loops 1, 2, and 3 respectively. Applying Kirchhoff's voltage law to the loops we get:

Loop-1:

$$R_L(-h_{fe}I_b - I_1) = (I_1 - I_2)R + \frac{I_1}{j\omega C}$$

or, $-h_{fe}R_L I_b = I_1(R + R_L + \dfrac{1}{j\omega C}) - RI_2$ $\qquad \qquad ... (9.6)$

Loop-2:

$$R(I_1 - I_2) = R(I_2 - I_3) + \frac{I_2}{j\omega C}$$

or, $RI_1 = I_2(2 + \dfrac{1}{j\omega C}) - RI_3$

$\therefore \quad I_1 = I_2(2 + \dfrac{1}{j\omega CR}) - I_3$ ... (9.7)

Loop-3:

$$R(I_2 - I_3) = (R' + h_{ie})I_3 + \frac{I_3}{j\omega C}$$

or, $\quad RI_2 = I_3(2R + \dfrac{1}{j\omega C})$ [since $R' + h_{ie} = R$]

$\therefore \quad I_2 = I_3(2 + \dfrac{1}{j\omega CR})$ ... (9.8)

From (9.6) we get

$$-h_{fe}R_L I_b = \left(R + R_L + \frac{1}{j\omega C}\right)\left[I_2\left(2 + \frac{1}{j\omega CR}\right) - I_3\right] - RI_2$$ [using eqn. (9.7)]

$$= \left(R + R_L + \frac{1}{j\omega C}\right)\left[I_3\left(2 + \frac{1}{j\omega CR}\right)^2 - I_3\right] - RI_3\left(2 + \frac{1}{j\omega CR}\right)$$ [using eqn. (9.8)]

Since, $I_3 = I_b$, we write

$$-h_{fe}R_L = \left(R + R_L + \frac{1}{j\omega C}\right)\left[4 - \frac{1}{\omega^2 C^2 R^2} + \frac{4}{j\omega CR} - 1\right] - R\left(2 + \frac{1}{j\omega CR}\right)$$

$$= (R + R_L)\left(3 - \frac{1}{\omega^2 C^2 R^2}\right) - \frac{4}{\omega^2 C^2 R} + \frac{1}{j\omega C}\left(\frac{1}{\omega^2 C^2 R}\right)$$

$$+ \frac{4}{j\omega CR}(R + R_L) - 2R - \frac{1}{j\omega C}$$ ... (9.9)

Equating imaginary parts from both sides of (9.9), we get

$$0 = -\frac{1}{\omega C}\left(3 - \frac{1}{\omega^2 C^2 R^2}\right) - \frac{4(R + R_L)}{\omega CR} + \frac{1}{\omega C}$$

or, $\quad \left(\dfrac{1}{\omega^2 C^2 R^2} - 3\right) - 4(1 + K) + 1 = 0$ where $K = \dfrac{R_L}{R}$

$$\frac{1}{\omega^2 C^2 R^2} = 3 + 4 + 4K - 1 = 6 + 4K$$

$$\therefore \qquad \omega = \frac{1}{RC} \frac{1}{\sqrt{6+4K}} \qquad\qquad \dots (9.10)$$

This condition holds good at a particular frequency which is the oscillation frequency ω_0.

$$\therefore \qquad \omega_0 = \frac{1}{RC} \frac{1}{\sqrt{6+4K}} \qquad\qquad \dots (9.11)$$

Equating real parts from eqn. (9.9), we get

$$-h_{fe}R_L = (R+R_L)\left(3 - \frac{1}{\omega_0^2 C^2 R^2}\right) - \frac{4}{\omega_0^2 C^2 R} - 2R$$

$$= (R+R_L)(3-6-4K) - \frac{4R}{\omega_0^2 C^2 R^2} - 2R$$

$$= -(R+R_L)(3+4K) - 4R(6+4K) - 2R \qquad [\text{using } (9.11)]$$

$$\therefore \qquad h_{fe} = \left(1 + \frac{1}{K}\right)(3+4K) + \frac{4}{K}(6+4K) + \frac{2}{K} \qquad [\text{ dividing by } -R_L]$$

$$= 3 + 4K + \frac{3}{K} + 4 + \frac{24}{K} + 16 + \frac{2}{K}$$

$$= 23 + 4K + \frac{29}{K} \qquad\qquad \dots (9.12)$$

For oscillation to start, we must have $A_r \beta_r > 1$ initially, i.e.,

$$h_{fe} > (23 + 4K + \frac{29}{K}) \qquad\qquad \dots (9.13)$$

So, the transistor must have a h_{fe} value specified by eqn. (9.13) in order that oscillation may occur in the circuit.

Is there any minimum value of h_{fe} below which oscillation can not occur ?

For this, we see that $\dfrac{dh_{fe}}{dK} = 4 - \dfrac{29}{K^2}$. Equating $\dfrac{dh_{fe}}{dK} = 0$, we get $K = \dfrac{\sqrt{29}}{2} = 2.7$ Now,

$\dfrac{d^2 h_{fe}}{dK^2} = +\dfrac{58}{K^3} = $ a positive quantity. Hence, h_{fe} is a minimum for $K = 2.7$. Then,

$h_{fe}\big|_{min} = 23 + 8(2.7) = 44.6$. Thus, the transistor must have $h_{fe} > 44.6$ for oscillation to take place. Transistors with $h_{fe} < 44.6$ can not oscillate in this phase shift oscillator circuit.

9.2.2 WIEN-BRIDGE OSCILLATOR

This oscillator uses a two-stage CE amplifier. Each CE amplifier provides a phase-shift of $180°$ between the input and the output. So, two-stage amplifier will produce a total phase shift of $180° + 180° = 360°$. The feedback network is in the form of an electrical bridge which is known as Wien-bridge. The feedback network introduces zero phase shift in the steady state oscillation. The circuit diagram of this oscillator is shown in Fig.9.2.

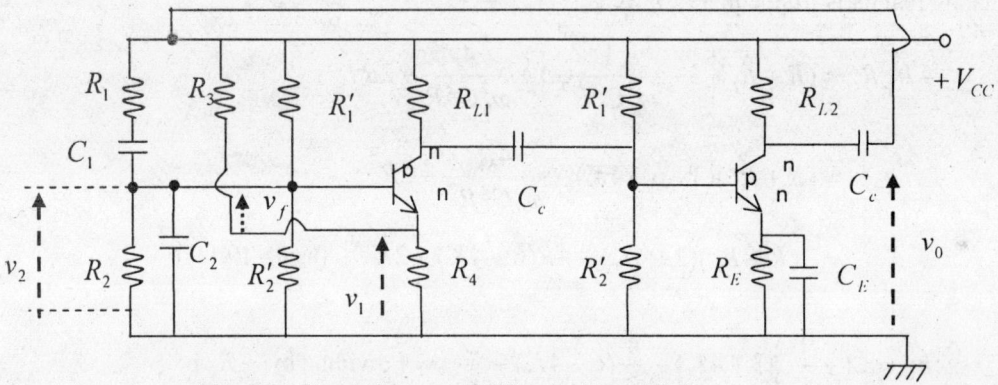

Fig. 9.2. Circuit diagram of Wien-bridge oscillator.

The output voltage is v_0 and the feedback voltage is v_f. The circuit uses voltage series feedback. The four arms of the Wien-bridge are R_1, C_1 series combination, R_2, C_2 parallel combination, R_3 and R_4. The output voltage of the bridge is v_f. From the voltage divider principle, we write

$$\frac{v_2}{v_0} = \frac{R_2 \parallel \dfrac{1}{j\omega C_2}}{\left(R_1 + \dfrac{1}{j\omega C_1}\right) + R_2 \parallel \dfrac{1}{j\omega C_2}}$$

where the sign '$\parallel$' indicates parallel combination of.

$$\therefore \quad \frac{v_2}{v_0} = \frac{R_2/j\omega C_2}{\left(R_2 + \dfrac{1}{j\omega C_2}\right)\left[R_1 + \dfrac{1}{j\omega C_1} + \dfrac{R_2/j\omega C_2}{R_2 + \dfrac{1}{j\omega C_2}}\right]}$$

$$= \frac{R_2/j\omega C_2}{\left[R_1 R_2 - \dfrac{1}{\omega^2 C_1 C_2}\right] + \dfrac{1}{j\omega}\left(\dfrac{R_1}{C_2} + \dfrac{R_2}{C_1}\right) + \left(R_2/j\omega C_2\right)}$$

$$= \frac{R_2}{j\omega C_2 \left(R_1 R_2 - \dfrac{1}{\omega^2 C_1 C_2} \right) + C_2 \left(\dfrac{R_1}{C_2} + \dfrac{R_2}{C_1} \right) + R_2}$$

$$= \frac{1}{j\left(\omega C_2 R_1 - \dfrac{1}{\omega C_1 R_2} \right) + \left(\dfrac{R_1}{R_2} + \dfrac{C_2}{C_1} \right) + 1} \qquad \qquad \text{...(9.14)}$$

Again, since V_0 appears across R_3, R_4 series combination we can write,

$$\frac{v_1}{v_0} = \frac{R_4}{R_3 + R_4} \qquad \qquad \text{...(9.15)}$$

Now, from Fig. 9.2, $v_f = v_2 - v_1$. Let the overall voltage gain of the two-stage amplifier be A_V.

Since the load resistances are R_{L1} and R_{L2}, A_V is real. The feedback factor $\beta_V = \dfrac{v_f}{v_0}$.

According to Barkhausen criterion, $A_V \beta_V = 1$. Then, since A_V is real, β_V must be real. But,

$\beta_V = \dfrac{v_2 - v_1}{v_0}$. Again, $\dfrac{v_1}{v_0}$ is real. So, $\dfrac{v_2}{v_0}$ must also be real. Then, from equation (9.14), we must

have, after equating the imaginary part to zero,

$$\omega_0 C_2 R_1 - \frac{1}{\omega_0 C_1 R_2} = 0 \qquad \qquad \text{...(9.16)}$$

Here, we have put ω_0 in place of ω since this condition holds good at the steady state of oscillation and the oscillation frequency is ω_0.

$$\therefore \qquad \omega_0 = \frac{1}{\sqrt{R_1 R_2 C_1 C_2}} \qquad \qquad \text{...(9.17)}$$

If we make $R_1 = R_2 = R$ (say) and $C_1 = C_2 = C$ (say), then

$$\omega_0 = \frac{1}{RC} \qquad \qquad \text{...(9.18)}$$

which is the expression for oscillation frequency of the Wien bridge oscillator. Now, if the oscillation is desired under the balanced condition of the bridge then $V_f = 0$. So, $\beta_V = 0$. Then,

$A_V \beta_V = 1$ leads to $A_V = \dfrac{1}{\beta_V} = \infty$. This means that oscillation under the balanced condition of

the bridge is not possible.

From eqn. (9.14), putting $\omega = \omega_0$, $R_1 = R_2 = R$ and $C_1 = C_2 = C$ we get $\dfrac{v_2}{v_0} = \dfrac{1}{3}$. Obviously,

under balanced condition of the bridge, $\dfrac{v_1}{v_0} = \dfrac{v_2}{v_0} = \dfrac{1}{3}$. To overcome this infinite gain

requirement problem under the balanced condition of the bridge, we set $\dfrac{v_1}{v_0} = \dfrac{1}{3} - \dfrac{1}{\delta}$ where

$\delta \gg 3$. Now, the bridge is slightly unbalanced. Now, $\beta_V = \dfrac{v_2}{v_0} - \dfrac{v_1}{v_0} = \dfrac{1}{3} - \left(\dfrac{1}{3} - \dfrac{1}{\delta} \right) = \dfrac{1}{\delta}$.

$$\therefore \qquad A_V \beta_V = A_V \cdot \dfrac{1}{\delta} = 1$$

or, $\qquad A_V = \delta$. $\hspace{5cm}$ ... (9.19)

This is the gain required for two-stage amplifier for steady state oscillation. Now, let us see how to find the value of δ. For this, we note that under balanced condition of the bridge

$$\dfrac{v_1}{v_0} = \dfrac{1}{3}$$

or, $\qquad \dfrac{R_4}{R_3 + R_4} = \dfrac{1}{3}$ $\quad$ [from eqn. (9.15)]

$$\therefore \qquad R_3 = 2R_4 \hspace{5cm} ... (9.20)$$

If R_3 is made a little larger than $2R_4$, then oscillation will occur in the unbalanced condition of the bridge and the value of A_V will be determined by the deviation as given in (9.19).

AMPLITUDE STABILIZATION

Oscillation amplitude grows from a very small noise voltage present in the circuit. Since initially, $A_V \beta_V > 1$, this small signal will grow in amplitude as it makes round-trip passage in the loop. Then, **how can we stabilize this growing oscillation amplitude?**

For this, we insert a tungsten filament lamp with a positive temperature coefficient in place of R_4. As the oscillation amplitude builds up, v_0 increases and hence current through R_4 increases. This produces heating in the resistance R_4 which makes R_4 increase. R_4 produces a negative current feedback. Increase in R_4 produces a corresponding increase in negative feedback. This reduces gain of the amplifier and the oscillation amplitude is stabilized.

9.3 RADIO-FREQUENCY OSCILLATORS

In this section, we will describe Hartley, Colpitts , Clapp and tuned collector oscillators.

9.3.1 GENERALIZED ANALYSIS OF HARTLEY, COLPITTS AND CLAPP OSCILLATORS

A general circuit diagram of these radio frequency oscillators is shown in Fig. 9.3. Here, RFC is a radio frequency choke which blocks the rf signal from passing to the ground through the supply V_{CC}. C_H is the Hartley capacitor which blocks the dc voltage from passing to ground through the inductance L_2 (Z_2). The feedback is voltage series feedback.

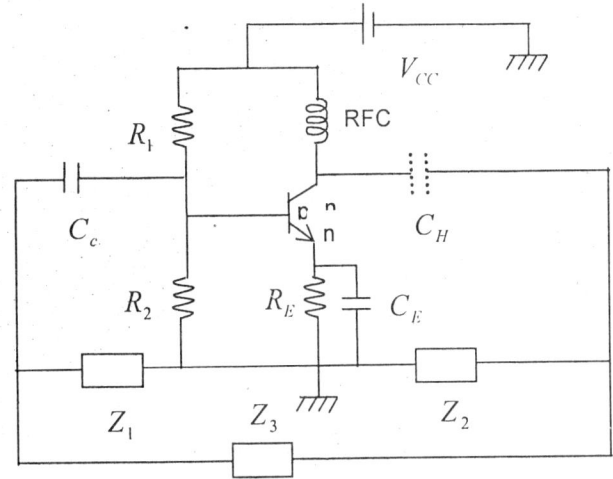

Fig. 9.3. (a) Generalized circuit diagram.

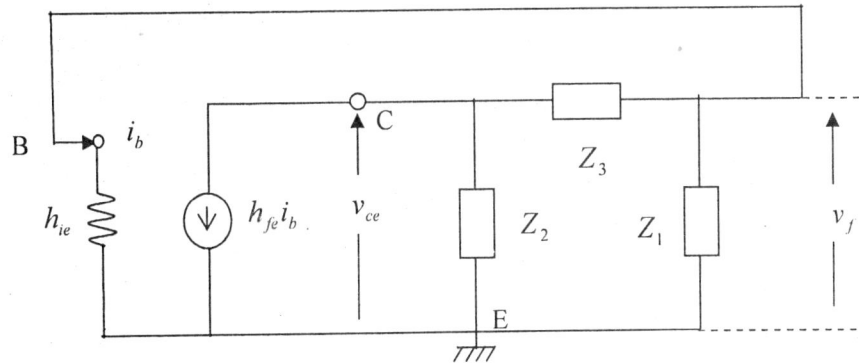

Fig. 9.3. (b) Ac equivalent circuit.

The load of the CE amplifier is

$$Z_L = Z_2 \| (Z_3 + Z_1 \| h_{ie})$$

$$= Z_2 \| \left(Z_3 + \frac{Z_1 h_{ie}}{Z_1 + h_{ie}} \right)$$

$$= \frac{Z_2 \left(Z_3 + \dfrac{Z_1 h_{ie}}{Z_1 + h_{ie}} \right)}{Z_2 + Z_3 + \dfrac{Z_1 h_{ie}}{Z_1 + h_{ie}}}$$

$$= \frac{Z_2 \left[Z_3 (Z_1 + h_{ie}) + Z_1 h_{ie} \right]}{(Z_2 + Z_3)(Z_1 + h_{ie}) + Z_1 h_{ie}} \qquad \ldots (9.21)$$

Here, Z_1, Z_2, Z_3 are impedances as shown in Fig. 9.3 (b). h_{ie} is the input resistance of the transistor amplifier. Neglecting the small h_{re} and h_{oe} parameters of the transistor, the voltage gain of the CE amplifier is

$$A_V = -\frac{h_{fe}}{h_{ie}} Z_L \qquad \ldots (9.22)$$

where h_{fe} is the short circuit current gain of the transistor amplifier. The feedback voltage is calculated as

$$v_f = \frac{Z_1 \| h_{ie}}{Z_3 + Z_1 \| h_{ie}} v_{ce}$$

$$= \frac{Z_1 h_{ie}}{Z_3 (Z_1 + h_{ie}) + Z_1 h_{ie}} v_{ce} \qquad \ldots (9.23)$$

where v_{ce} is the ac collector voltage with respect to the emitter. Then, the feedback factor

$$\beta_V = \frac{v_f}{v_{ce}} = \frac{Z_1 h_{ie}}{Z_3 (Z_1 + h_{ie}) + Z_1 h_{ie}} \qquad \ldots (9.24)$$

Then, $$A_V \beta_V = -\frac{h_{fe}}{h_{ie}} \cdot \frac{Z_2 \left[Z_3 (Z_1 + h_{ie}) + Z_1 h_{ie} \right]}{\left[(Z_2 + Z_3)(Z_1 + h_{ie}) + Z_1 h_{ie} \right]} \cdot \frac{Z_1 h_{ie}}{\left[Z_3 (Z_1 + h_{ie}) + Z_1 h_{ie} \right]}$$

$$= -\frac{h_{fe} Z_1 Z_2}{(Z_2 + Z_3)(Z_1 + h_{ie}) + Z_1 h_{ie}}$$

$$= -\frac{h_{fe} Z_1 Z_2}{h_{ie}(Z_1 + Z_2 + Z_3) + Z_1 (Z_2 + Z_3)} \qquad \ldots (9.25)$$

We choose Z_1, Z_2 and Z_3 as pure reactances. Let $Z_1 = jX_1$, $Z_2 = jX_2$ and $Z_3 = jX_3$. Then, equation (9.25) can be written as

$$A_V \beta_V = -\frac{h_{fe} jX_1 \cdot jX_2}{j(X_1 + X_2 + X_3) h_{ie} + jX_1 (jX_2 + jX_3)}$$

$$= \frac{h_{fe} X_1 \cdot X_2}{j(X_1 + X_2 + X_3)h_{ie} - X_1(X_2 + X_3)} \qquad \dots (9.26)$$

According to Barkhausen criterion, $A_V \beta_V = 1$. From equation (9.26) we get

$$h_{fe} X_1 X_2 = j(X_1 + X_2 + X_3)h_{ie} - X_1(X_2 + X_3) \qquad \dots (9.27)$$

Equating imaginary parts from both sides of equation (9.27) we get

$$X_1 + X_2 + X_3 = 0 \qquad \dots (9.28)$$

Equating real parts from both sides of equation (9.27), we get

$$h_{fe} X_1 X_2 = -X_1(X_2 + X_3) \qquad \dots (9.29)$$

From (9.29), we write $h_{fe} = -\dfrac{(X_2 + X_3)}{X_2}$

$$= \frac{X_1}{X_2} \quad \text{[using eqn.(9.28)]} \qquad \dots (9.30)$$

9.3.1.1 COLPITTS OSCILLATOR

Here, Z_1 and Z_2 are impedances of two capacitors (C_1 and C_2) while Z_3 is the impedance of a

pure inductance (L). Then, $Z_1 = \dfrac{1}{j\omega C_1} = -\dfrac{j}{\omega C_1}$, $Z_2 = \dfrac{1}{j\omega C_2} = -\dfrac{j}{\omega C_2}$ and $Z_3 = j\omega L$.

Equation (9.28) yields

$$-\left(\frac{1}{\omega C_1} + \frac{1}{\omega C_2}\right) + \omega L = 0$$

or, $\quad \omega L = \dfrac{1}{\omega}\left(\dfrac{1}{C_1} + \dfrac{1}{C_2}\right)$

$$\therefore \quad \omega = \frac{1}{\sqrt{L\left(\dfrac{C_1 C_2}{C_1 + C_2}\right)}} \qquad \dots (9.31)$$

This is the expression for oscillation frequency, ω_0.

$$\therefore \quad \omega_0 = \frac{1}{\sqrt{L\left(\dfrac{C_1 C_2}{C_1 + C_2}\right)}} \qquad \dots (9.32)$$

From equation (9.30), we get

$$h_{fe} = \left(-\frac{1}{\omega C_1} \right) \cdot (-\omega C_2) = \frac{C_2}{C_1} \qquad \qquad \dots (9.33)$$

For starting oscillation, $h_{fe} > \dfrac{C_2}{C_1}$.

9.3.1.2 HARTLEY OSCILLATOR

In Hartley oscillator, Z_1 and Z_2 are the impedances of two pure inductances L_1 and L_2 respectively while Z_3 is the impedance of a pure capacitance. Then, $Z_1 = j\omega L_1$, $Z_2 = j\omega L_2$ and $Z_3 = \dfrac{1}{j\omega C} = -\dfrac{j}{\omega C}$.

From equation (9.28), putting $\omega = \omega_0$, where ω_0 is the oscillation frequency, we get

$$\omega_0 \left(L_1 + L_2 \right) + \left(-\frac{1}{\omega_0 C} \right) = 0$$

$$\therefore \qquad \omega_0 = \frac{1}{\sqrt{C\left(L_1 + L_2 \right)}} \qquad \qquad \dots (9.34)$$

From equation (9.30), we write $h_{fe} = \dfrac{\omega L_1}{\omega L_2} = \dfrac{L_1}{L_2}$ $\qquad \qquad \dots (9.35)$

For the start of oscillation in Hartley oscillator, h_{fe} should be greater than $\dfrac{L_1}{L_2}$.

9.3.1.3 CLAPP OSCILLATOR

It is a modified Colpitts oscillator. Here, a capacitor C_3 is placed in series with L. So,

$$Z_3 = jX_3 = j\left(\omega L - \frac{1}{\omega C_3} \right), \ Z_1 = jX_1 = \frac{1}{j\omega C_1} \ \text{and} \ Z_2 = jX_2 = \frac{1}{j\omega C_2}.$$

Now, since $\quad X_1 + X_2 + X_3 = 0$, we get

$$-\frac{1}{\omega_0 C_1} - \frac{1}{\omega_0 C_2} + \left(\omega_0 L - \frac{1}{\omega_0 C_3} \right) = 0 \ , \text{where} \ \omega_0 = \text{oscillation frequency.}$$

$$\therefore \qquad -\frac{1}{\omega_0} \left(\frac{1}{C_1} + \frac{1}{C_2} + \frac{1}{C_3} \right) + \omega_0 L = 0$$

$$\therefore \quad \omega_0 = \frac{1}{\sqrt{LC_{eq}}} \quad \text{where} \quad \frac{1}{C_{eq}} = \left(\frac{1}{C_1} + \frac{1}{C_2} + \frac{1}{C_3} \right) \qquad \ldots (9.36)$$

Again, $\quad h_{fe} = \dfrac{X_1}{X_2} = \dfrac{C_2}{C_1}$ $\qquad \ldots (9.37)$

Now, C_1 and C_2 are much larger than C_3, in practice. So, $\dfrac{1}{C_{eq}} \approx \dfrac{1}{C_3}$. Hence, ω_0 is primarily

controlled by varying C_3. Again, since C_1 is large, $\dfrac{1}{\omega_0 C_1}$ is small. So, the loading effect of h_{ie}

which appears in parallel with C_1 will be very small.

9.3.2 TUNED COLLECTOR OSCILLATOR

The circuit uses a CE amplifier with a tuned circuit as the load. The CE amplifier introduces a phase shift of $180°$ between the input and output while the mutual inductance providing feedback introduces another $180°$ phase shift. The total phase shift is $180° + 180° = 360°$ which is equivalent to $0°$. Thus, a part of the Barkhausen criterion is satisfied. The circuit diagram is shown in Fig. 9.4.

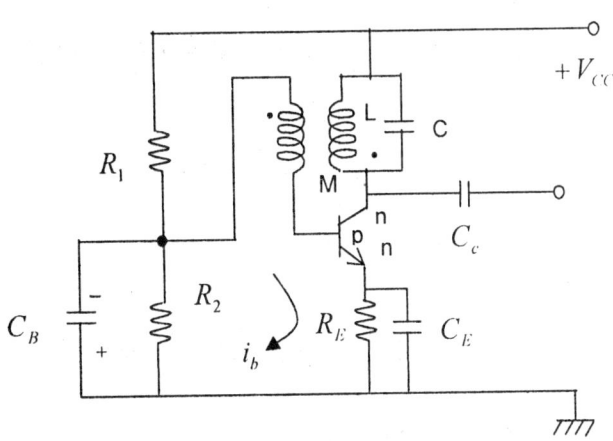

Fig. 9.4. Circuit diagram of tuned collector oscillator.

R_E, C_E provides self bias to the CE amplifier. R_1, R_2 provide constant bias to the CE amplifier. C_C is the output coupling capacitor which blocks the dc and passes the ac to the output. L and C form a parallel resonant circuit connected with the collector. The presence of dots (•) with the mutual inductance indicates a phase shift of π-radian. The ac equivalent circuit is shown in Fig. 9.5.

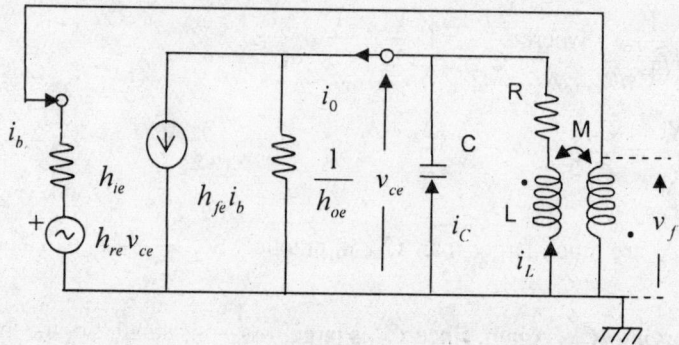

Fig. 9.5. AC equivalent circuit of tuned collector oscillator.

Here, h_{ie} is the input impedance, h_{oe} is the output impedance, h_{re} is the reverse voltage transfer ratio, h_{fe} is the forward current gain of the transistor. v_{ce} is the collector voltage and i_o is output current of the CE amplifier. i_b is input base current. R is the resistance of the inductive coil having self inductance L. i_L is the current through inductance L while i_C is the current through capacitor C. v_f is the feedback voltage.

Then, $\quad v_f = -\dfrac{d\varphi}{dt}$ ... (9.38)

where φ = magnetic flux linked with the secondary of mutual inductance.

Now, $\quad \varphi = M \cdot i_L$ ... (9.39)

and $\quad v_f = -j\omega M \cdot i_L$ ... (9.40)

Again, $i_L = -\dfrac{v_{ce}}{R + j\omega L}$ ... (9.41)

Feedback factor $\beta_v = \dfrac{v_f}{v_{ce}} = -\dfrac{j\omega M\, i_L}{v_{ce}} = \dfrac{j\omega M}{R + j\omega L}$ ... (9.42)

Here, M is itself negative due to a phase difference of π-radian between the primary and secondary of the mutual inductance. Now, the open-loop gain of the voltage amplifier is

$$A_v = -\frac{h_{fe} Z_L}{h_{ie} + Z_L \Delta^h} \qquad \text{... (9.43)}$$

here $\quad Z_L = (R + j\omega L) \| \dfrac{1}{j\omega C} = \dfrac{(R + j\omega L)\dfrac{1}{j\omega C}}{(R + j\omega L) + \dfrac{1}{j\omega C}} = \dfrac{R + j\omega L}{1 - \omega^2 LC + j\omega CR}$ ... (9.44)

and $\Delta^h = h_{ie}h_{oe} - h_{fe}\dot{h}_{re}$.

Now, $A_v\beta_v = -\dfrac{h_{fe}Z_L}{h_{ie} + Z_L\Delta^h} \cdot \dfrac{j\omega M}{R + j\omega L}$

$$= \frac{-h_{fe}(R + j\omega L)}{1 - \omega^2 LC + j\omega CR} \cdot \frac{j\omega M}{(R + j\omega L)} \cdot \frac{1}{h_{ie} + \dfrac{(R + j\omega L)\Delta^h}{1 - \omega^2 LC + j\omega CR}}$$

$$= \frac{j\omega|M|h_{fe}}{(1 - \omega^2 LC)h_{ie} + j\omega CRh_{ie} + (R + j\omega L)\Delta^h} \qquad \ldots(9.45)$$

since $M = -|M|$.

Now, since according to Barkhausen criterion, $A_v\beta_v = 1$, we get from equation (9.45),

$$j\omega|M|h_{fe} = (1 - \omega^2 LC)h_{ie} + R\Delta^h + j\omega(RCh_{ie} + L\Delta^h) \qquad \ldots(9.46)$$

Equating real parts from both sides of (9.46) we get

$$(1 - \omega_0^2 LC)h_{ie} + R\Delta^h = 0 \qquad \ldots(9.47)$$

where ω_0 is the oscillation frequency at which equation (9.47) holds good. Hence, from equation (9.47) we write

$$1 - \omega_0^2 LC = -\frac{R\Delta^h}{h_{ie}}$$

$\therefore$ $\omega_0 = \dfrac{1}{\sqrt{LC}}\sqrt{1 + \dfrac{R\Delta^h}{h_{ie}}}$ $\ldots(9.48)$

If h_{re} and h_{oe} are neglected due to their small values, we get

$$\omega_0 = \frac{1}{\sqrt{LC}} \qquad \ldots(9.49)$$

Equating imaginary parts from equation (9.46), we write

$$\omega|M|h_{fe} = \omega(RCh_{ie} + L\Delta^h)$$

$\therefore$ $h_{fe} = \dfrac{RC}{|M|}\left(h_{ie} + \dfrac{L}{CR}\Delta^h\right)$

$$= \frac{RC}{|M|}(h_{ie} + R_l\Delta^h) \qquad \ldots(9.50)$$

where $R_t = L/RC$ is the dynamic resistance of the tuned circuit. The transistor must have a current gain (h_{fe}) greater than that specified by equation (9.50) for oscillation to start.

AMPLITUDE STABILIZATION

Oscillation starts from the very small noise voltage present in the circuit. This small signal is amplified due to the presence of loop gain ($A_v \beta_v > 1$). So, the output signal (and also the feedback signal) grow in amplitude. The question is how to stabilize this oscillation amplitude to a constant value.

For this, let us consider the action of the R_2, C_B circuit at the input (Fig. 9.4). During the positive half cycle of input ac, the capacitor C_B charges to its peak value. As soon as the input ac swings down from its peak value, C_B discharges through R_2. This discharge tends to make the base of the npn transistor less positive with respect to the emitter. This, in turn, reduces the effective base bias (V_{BE}). So, the operating point of the transistor is driven towards the nonlinear region of the $V_{BE} - I_C$ transfer characteristics of the transistor as shown in Fig. 9.6.

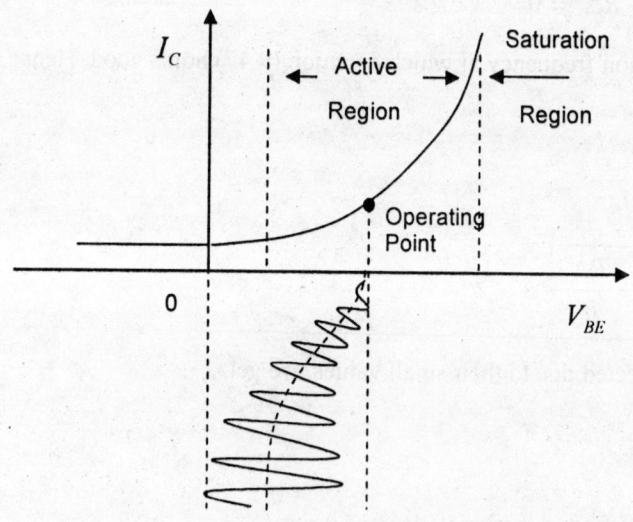

Fig. 9.6. Growth of oscillation in a tuned collector oscillator.

In effect, the gain of the amplifier is reduced which reduces the value of value of $A_v \beta_v$ (which was initially > 1) gradually. Ultimately a stage comes when $A_v \beta_v = 1$ and no more growth of oscillation takes place. The oscillation amplitude is stabilized. The growth of oscillations and its stabilization is shown in Fig. 9.6.

9.4 ASTABLE MULTIVIBRATOR

The term 'astable' means no stable state. Sometimes it is also called free-runing multivibrator. It generates pulses or square waves. A schematic circuit diagram of the transistorized astable multivibrator is shown in Fig. 9.7.

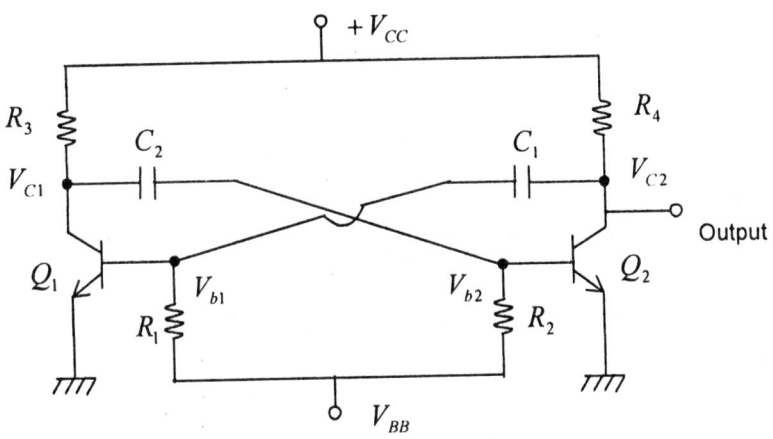

Fig. 9.7. Circuit diagram of an astable multivibrator.

The output of Q_2 is fed back to the input of Q_1 through capacitor C_1. Similarly, the output of Q_1 is fed back to the input of Q_2 through capacitor C_2 .

OPERATION

The transistors Q_1 and Q_2 are not perfectly identical in practice. So, when the power supply V_{CC} is applied, Q_1 and Q_2 conduct differently. Let us suppose that Q_1 conducts more than Q_2. The collector voltage (V_{C1}) of Q_1 falls rapidly. This negative-going change in V_{C1} appears as a negative voltage across R_2. The forward bias of Q_2 is reduced. So, current through Q_2 decreases. This causes the collector voltage (V_{C2}) of Q_2 to increase. This rapid change (increase) in V_{C2} appears across R_1 as a positive voltage. This leads to an increase of forward bias of Q_1 . So current through Q_1 increases further and V_{C1} decreases further. This fall in V_{C1} reduces the forward bias of Q_2 . So, V_{C2} increases again. This is a cumulative process. Eventually, Q_1 saturates and Q_2 is cut-off. Then, $V_{C1} \cong 0$ volt and $V_{C2} \cong V_{CC}$. The output of the circuit is V_{CC} . Now, at one end of C_1 the voltage is V_{CC} while the other end of C_1 is connected to ground through the low resistance path of the conducting emitter-base junction of Q_1. So, C_1 charges up with voltage V_{CC} .

Since, V_{C1} has reduced from V_{CC} to 0 volt, the net change V_{C1} is $0 - V_{CC} = -V_{CC}$. This change in V_{C1} appears at the input (base) of Q_2. Thus, the base voltage of Q_2 is $V_{b2} = -V_{CC}$.

Let us now look at the condition of capacitor C_2. This is shown in Fig. 9.8.

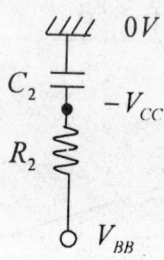

Fig. 9.8. Discharging of capacitor C_2.

At one end of C_2, the voltage (V_{C1}) is $0V$ while on the other end of C_2, the voltage is $V_{b2} = -V_{CC}$. Then, C_2 begins to discharge through R_2 until its potential V_{b2} attains $0V$ (approximately). Then, C_2 begins to charge up with V_{BB} and V_{b2} rises. Q_2 begins to conduct and its collector current increases. This leads to a decrease in collector voltage V_{c2}. This falling V_{c2} is equivalent to a negative change in the base voltage V_{b1} of Q_1. So, the forward bias of Q_1 is reduced and Q_1 comes out of saturation. V_{C1} increases until Q_1 is cut-off. This makes $V_{C1} \cong V_{CC}$.

The increase of V_{C1} from 0 volt to V_{CC} produces a forward bias change of $(V_{CC} - 0 =) V_{CC}$ of Q_2. So, Q_2 saturates and its collector voltage $V_{c2} = 0$ volt.

At the end of C_2, the voltage is $V_{C1} = V_{CC}$ while at the other end there is a low resistance path of the emitter-base junction of Q_2 going to ground. So, C_2 begins to charge up with V_{CC}. The process repeats and a full cycle is complete. The output can be taken from the collectors of Q_1 or Q_2. The waveforms appearing at the collector and base points of Q_1 and Q_2 are shown in Fig. 9.9.

ANALYSIS

When Q_1 is cut-off, its collector voltage rises according to the equation

$$V_{C1} = V_{CC}\left(1 - e^{-t/R_3 C_2}\right) \qquad \ldots (9.51)$$

Since the capacitor C_2 charges through R_3 as shown in Fig. 9.10.

Thus, $V_{b2} = V_{BB}\left(1 - e^{-t/R_2C_2}\right) - V_{CC}e^{-t/R_2C_2}$ ... (9.52)

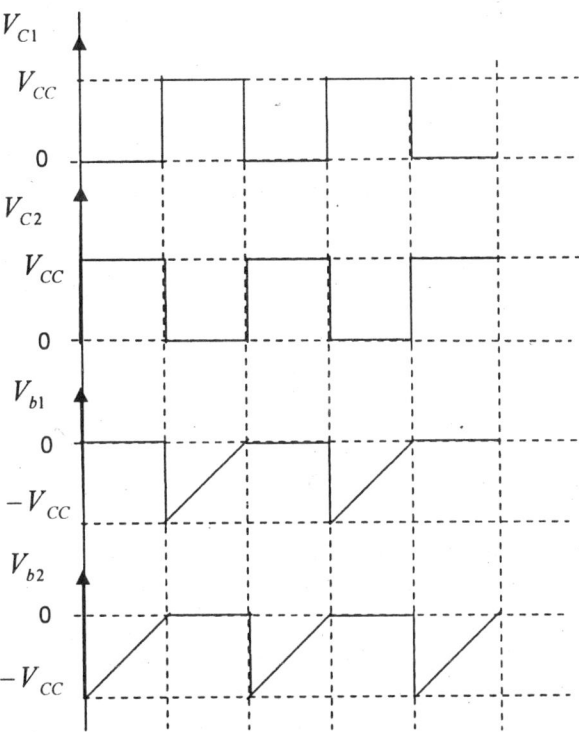

Fig. 9.9. Typical waveforms at the collector and base terminals of Q_1 and Q_2.

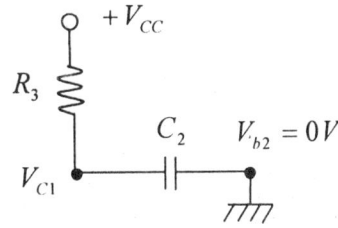

Fig. 9.10. Charging process of C_2.

When Q_1 is saturated, $V_{C1} = 0V$ and $V_{b2} = -V_{CC}$. Rise in V_{b2} is a combination of two process – charging with V_{BB} and discharging from $-V_{CC}$ voltages. The circuit is shown in Fig. 9.11.

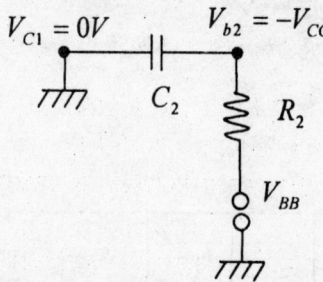

Fig. 9.11. Charging and discharging of capacitor C_2.

At $t = 0, V_{b2} = -V_{CC}$. Q_2 remains in the cut-off state until V_{b2} reaches 0 volt during the charging action of C_2. Let $T_2 = $ off time interval of Q_2 . Then, from (9.52) we write

$$0 = V_{BB}(1 - e^{-T_2/R_2C_2}) - V_{CC}e^{-T_2/R_2C_2}$$

or, $\quad (V_{CC} + V_{BB})e^{-T_2/R_2C_2} = V_{BB}$

$\therefore \qquad T_2 = -R_2C_2 \ln\left(\dfrac{V_{CC}}{V_{BB} + V_{CC}}\right)$

$$\qquad\quad = R_2C_2 \ln\left(1 + \dfrac{V_{CC}}{V_{BB}}\right) \qquad\qquad \ldots (9.53)$$

When Q_2 becomes just cut-off, its collector voltage V_{C2} rises according to the equation

$$V_{C2} = V_{CC}\left(1 - e^{-t/R_4C_1}\right) \qquad\qquad \ldots (9.54)$$

This is the charging equation of capacitor C_1 as shown in Fig. 9.12.

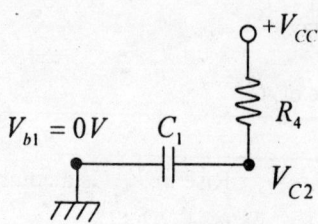

Fig. 9.12. Charging circuit of capacitor C_1.

When Q_2 is saturated, $V_{C2} \cong 0V$ and $V_{b1} \cong -V_{CC}$.

Rise in V_{b1} is a combination of two processes, viz., discharging of C_1 from $-V_{CC}$ and charging of C_1 with V_{BB}. This is shown in Fig. 9.13.

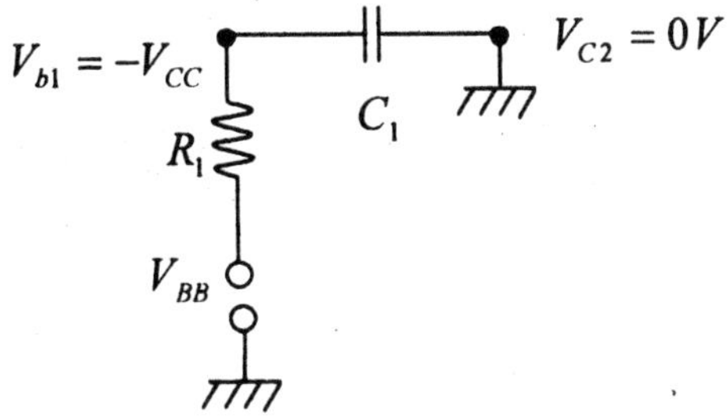

Fig. 9.13. Charging and discharging of capacitor C_1.

Thus, $V_{b1} = -V_{CC} e^{-\frac{t}{R_1 C_1}} + V_{BB}\left(1 - e^{-\frac{t}{R_1 C_1}}\right)$ ... (9.55)

Q_1 remains cut-off until V_{b1} reaches $0V$ (approximately).

Let T_1 = off time interval of Q_1. Then, from (9.55) we write

$$0 = -V_{CC} e^{-\frac{T_1}{R_1 C_1}} + V_{BB}\left(1 - e^{-\frac{T_1}{R_1 C_1}}\right)$$

or, $\left(V_{CC} + V_{BB}\right)e^{-\frac{T_1}{R_1 C_1}} = V_{BB}$

$\therefore$ $T_1 = -R_1 C_1 \ln\left(\dfrac{V_{BB}}{V_{CC} + V_{BB}}\right) = R_1 C_1 \ln\left(1 + \dfrac{V_{CC}}{V_{BB}}\right)$ ... (9.56)

Therefore, the time period of the waveform is

$$T = T_1 + T_2 = \left(R_1 C_1 + R_2 C_2\right)\ln\left(1 + \frac{V_{CC}}{V_{BB}}\right)$$ ... (9.57)

If $V_{BB} = V_{CC}$ then $T = \left(R_1 C_1 + R_2 C_2\right)\ln 2$.

For symmetrical operation, $T_1 = T_2$ which implies $R_1 C_1 = R_2 C_2$. Under this condition, we get square wave output. If $T_1 \gg T_2$, or, $T_2 \gg T_1$ we get a pulse waveform at the outputs of transistors Q_1 and Q_2 respectively.

SOLVED PROBLEMS

9.1 In a phase-shift oscillator using an npn transistor amplifier in the CE mode, three identical R-C sections are used to provide voltage shunt feedback. The amplifier uses a load resistance of $1\,\text{k}\Omega$. Calculate the oscillation frequency of the circuit when $R = 1\,\text{k}\Omega$ and $C = 0.1\,\mu\text{F}$. Calculate the minimum value of h_{fe} parameter of the transistor in order that oscillation can take place in this circuit. If the CE amplifier has a short circuit current gain of 50, decide whether it can oscillate in the above circuit.

Solution

Here, load resistance $R_L = 1\,\text{k}\Omega$.

Feedback network resistance $R = 1\,\text{k}\Omega$.

Feedback network capacitance $C = 0.1\,\mu\text{F}$.

$\therefore$ Oscillation frequency, $f_0 = \dfrac{\omega_0}{2\pi} = \dfrac{1}{2\pi} \dfrac{1}{RC} \dfrac{1}{\sqrt{6 + 4\left(\dfrac{R_L}{R}\right)}}$

$$= \dfrac{1}{2\pi} \dfrac{1}{10^3 \cdot 10^{-7}} \dfrac{1}{\sqrt{6 + 4 \times 1}}$$

$$= \dfrac{1}{2\pi} \dfrac{10^4}{\sqrt{10}} = \dfrac{\sqrt{10}}{2\pi}\,\text{kHz} \quad = 503 \text{ Hz.}$$

For oscillation to occur,

$$h_{fe} > \left(23 + 4K + \dfrac{29}{K}\right) \quad \text{where } K = \dfrac{R_L}{R}$$

Now, $K = \dfrac{1}{1} = 1.$

$\therefore \qquad 23 + 4K + \dfrac{29}{K} = 23 + 4 \cdot 1 + \dfrac{29}{1} = 56$

$\therefore \qquad h_{fe}\big|_{\min} = 56.$

The short circuit current gain of the CE amplifier $= h_{fe}$.

If $h_{fe} = 50$, then it is less than $h_{fe}\big|_{\min}$.

So, this transistor amplifier can not oscillate in the above circuit.

9.2 A Wien-bridge oscillator has equal resistances and equal capacitances in the bridge. The value of resistance is $2\,\text{k}\Omega$ and the value of capacitance is $0.047\,\mu\text{F}$. At what frequency, will this oscillator oscillate?

Solution

The oscillation frequency, $f_0 = \dfrac{\omega_0}{2\pi} = \dfrac{1}{2\pi} \dfrac{1}{RC}$

Here, $R = 2\,k\Omega = 2\cdot10^3\,\Omega$.

$C = 0.047\,\mu F = 0.047 \times 10^{-6}\,F = 47 \times 10^{-9}\,F.$

$\therefore$ Oscillation frequency, $f_0 = \dfrac{1}{2\pi \cdot 2\cdot10^3 \cdot 47 \cdot 10^{-9}}$

$$= \dfrac{10^6}{4\pi \cdot 47} = 1.69\,\text{kHz}.$$

9.3   In a Colpitts oscillator the feedback network has an inductance of $0.1\,mH$ and two equal capacitances each of value $0.01\,\mu F$. Calculate the oscillation frequency of this oscillator.

Solution

The oscillation frequency, $f = \dfrac{\omega}{2\pi} = \dfrac{1}{2\pi} \dfrac{1}{\sqrt{L\left(\dfrac{C_1 C_2}{C_1 + C_2}\right)}}$

Here, $L = 10^{-4}\,H.$
$\qquad C_1 = C_2 = C = 10^{-8}\,F,$

$\dfrac{C_1 C_2}{C_1 + C_2} = \dfrac{1}{2}C = \dfrac{1}{2}\cdot10^{-8}\,F,$

$f = \dfrac{1}{2\pi} \dfrac{1}{\sqrt{10^{-4} \cdot \dfrac{1}{2}\cdot10^{-8}}} = \dfrac{\sqrt{2}}{2\pi}\cdot10^6$

$\qquad = 225\,\text{kHz}.$

9.4   A transistorized astable multivibrator has identical RC sections for the coupling network with $R = 10\,k\Omega$ and $C = 0.1\,\mu F$. The common ends of the coupling resistances (R) is connected with the supply. Calculate the frequency of the square wave collector output of this circuit.

Solution
The time period, $T = (R_1 C_1 + R_2 C_2)\ln 2$.

$\qquad R_1 = R_2 = R = 10^4\,\Omega,$
$\qquad C_1 = C_2 = C = 10^{-7}\,F,$

$$\therefore \qquad T = 2RC \cdot \ln 2 = 2 \cdot 10^4 \cdot 10^{-7} \cdot \ln 2$$
$$= 2 \cdot \ln 2 \cdot 10^{-3} \text{ sec.}$$

$\therefore$ Frequency of oscillation, $f = \dfrac{1}{T} = \dfrac{10^3}{2\ln 2}$

$$= 721 \text{ Hz.}$$

PROBLEMS

9.1 A tuned collector transistorized oscillator with a capacitor $500\,\text{pF}$ at the tank circuit oscillates at a frequency of a $1\,\text{MHz}$. Change the circuit parameters such that it oscillates at frequency of $2\,\text{MHz}$.

[Cf. B.U. 2002]

Hint: Oscillator frequency, $f_0 = \dfrac{1}{2\pi} \dfrac{1}{\sqrt{LC}}$

$$\therefore \qquad L = \frac{1}{4\pi^2 C f_0^2}$$

change L, $\dfrac{L_2}{L_1} = \dfrac{f_{01}^2}{f_{02}^2} = \left(\dfrac{1}{2}\right)^2 = \dfrac{1}{4}$

So, inductance should be reduced to $\dfrac{1}{4}$ of its initial value.

9.2 A Colpitts oscillator has an inductive coil of inductance $50\,\mu\text{H}$. A $100\,\text{pF}$ capacitor is connected at the amplifier input and a $300\,\text{pF}$ capacitor is connected across the output. Find the frequency of oscillation and the minimum gain of the amplifier required for sustained oscillation.

[cf. B.U. 2000]

Hint: $f = \dfrac{1}{2\pi} \Bigg/ \sqrt{L\left(\dfrac{C_1 C_2}{C_1 + C_2}\right)}$

Here, $C_1 = 100\,\text{pF} = 10^{-10}\,\text{F}$

$\qquad C_2 = 300\,\text{pF} = 3 \times 10^{-10}$

$\qquad L = 50 \times 10^{-6}\,\text{H}$

$\therefore \qquad f = 2.599\,\text{MHz.}$

Minimum gain, $h_{fe}\big|_{\min} = \dfrac{C_2}{C_1} = 3$.

9.3 A transistorized Wien bridge oscillator has a feedback network consisting of a series combination of $1\,k\Omega$ resistance and $0.1\,\mu F$ capacitance. The parallel combination is also formed by identical components. Calculate the oscillation frequency of the circuit.

[**Ans.** $1.59\,kHz$]

9.4 A transistorized shift oscillator has three identical RC sections as the feedback network. The value of resistance is $1\,K\Omega$ and the value of capacitance is $0.047\,\mu F$. The amplifier with voltage shunt feedback has a load resistance of $1\,k\Omega$. Calculate the oscillation frequency. Can this circuit produce oscillation if the transistor has a h_{fe} equal to 30?

[**Ans.** $1.07\ kHz$, No]

QUESTIONS

9.1 Explain with the help of a circuit diagram the principle of operation of Wien-bridge oscillator. Find out an expression for the frequency of oscillation of this oscillator. How do you stabilize the oscillation amplitude in this circuit?

9.2 Draw a neat circuit diagram of a transistorized phase-shift oscillator. Derive an expression for the oscillation frequency of this oscillator. Why do you use three RC sections in the feedback network of this oscillator? Is there any minimum value h_{fe} of the transistor below which oscillation in this circuit is not possible?

9.3 Sketch the circuit of a tuned collector oscillator. Find out an expression for oscillation frequency of this oscillator in terms of circuit parameters. How can you stabilize the oscillation amplitude in this circuit?

9.4 Draw the circuit diagram of a Hartley oscillator. Find out the conditions of steady state oscillation in this circuit.

9.5 Explain briefly the principle of operation of the Colpitts oscillator. Derive an expression for oscillation frequency of this oscillator. What is the minimum value of h_{fe} of the transistor used in this circuit for oscillation to take place. How can you modify this circuit to construct a Clapp oscillator ?

9.6 With the help of a circuit diagram, explain the operation of an astable multivibrator. Derive an expression for the time period of the waveform generated by this circuit.

$\boxed{\textbf{10}}$

Rectifier

10.1 DEFINITION

A rectifier is an electronic device which converts alternating current into unidirectional current. It is used to generate dc power from ac power with the help of filters and voltage regulators. The filter removes the ripple content of the rectifier output and the voltage regulator maintains the output voltage constant.

Depending upon the nature of output waveform of the rectifier, rectifiers can be divided into two groups:

(i)   half-wave rectifier

(ii)   full-wave rectifier.

10.2 HALF-WAVE RECTIFIER

Since diodes conduct in one direction, we have to use diodes in the design of rectifiers. A half-wave rectifier uses a single diode (D) as shown in Fig. 10.1. R_L is the load resistance.

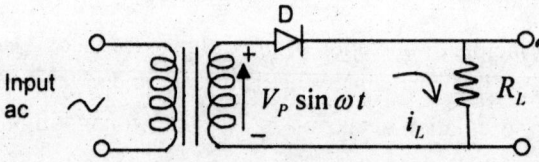

Fig. 10.1. A half-wave rectifier.

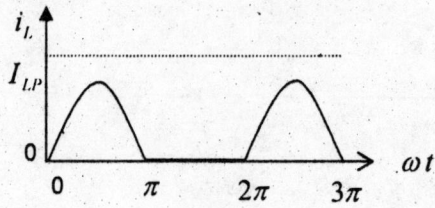

Fig. 10.2. Load current waveform of the half-wave rectifier.

If R_d be the internal resistance of the diode under forward bias, then total resistance of the circuit during conduction is $(R_L + R_d)$. Let $V_{in}(t) = V_p \sin \omega t$ be the input ac voltage appearing at the secondary of the transformer. Here, V_p is the peak input voltage and ω is the angular frequency of the input ac. The load current during conduction is

$$i_L(t) = \frac{V_{in}(t)}{R_L + R_d} = \frac{V_p}{R_L + R_d} \sin \omega t$$

$$= I_{LP} \sin \omega t \qquad \qquad \text{...(10.1)}$$

where $I_{LP} = \dfrac{V_p}{R_L + R_d}$ is the peak value of load current. The load current waveform can be represented as

$$i_L(t) = I_{LP} \sin \omega t \qquad \text{for } 0 \le \omega t \le \pi$$

$$i_L(t) = 0 \qquad \text{for } \pi \le \omega t \le 2\pi$$

This waveform is shown in Fig. 10.2.

We now make a Fourier analysis of $i_L(t)$. We write

$$i_L(t) = a_0 + \sum_{n=1}^{\infty} (a_n \cos n\omega t + b_n \sin n\omega t) \qquad \qquad \text{...(10.2)}$$

where a_0, a_n, b_n (for $n = 1$ to ∞) are coefficients to be determined. Now,

$$a_0 = \frac{1}{2\pi} \int_0^{2\pi} i_L(t) d\omega t$$

$$= \frac{1}{2\pi} \int_0^{\pi} I_{LP} \sin \omega t \, d\omega t$$

$$= \frac{I_{LP}}{2\pi} \left[-\cos \omega t \right]_0^{\pi}$$

$$= \frac{I_{LP}}{2\pi} 2 = \frac{I_{LP}}{\pi} \qquad \qquad \text{...(10.3)}$$

$$a_n = \frac{1}{\pi} \int_0^{2\pi} i_L(t) \cos n\omega t \, d\omega t$$

$$= \frac{I_{LP}}{\pi} \int_0^{\pi} \sin \omega t \, \cos n\omega t \, d\omega t$$

$$= \frac{I_{LP}}{2\pi} \int_0^\pi [\sin(n+1)\omega t - \sin(n-1)\omega t] \, d\omega t$$

$$= \frac{I_{LP}}{2\pi} \left[-\frac{\cos(n+1)\omega t}{(n+1)} + \frac{\cos(n-1)\omega t}{(n-1)} \right]_0^\pi$$

Now, $a_n = 0$ for odd n.

$$a_n = \frac{I_{LP}}{2\pi} \left[\frac{1}{(n+1)} - \frac{1}{(n-1)} + \frac{1}{(n+1)} - \frac{1}{(n-1)} \right] \text{ for even } n$$

$$= \frac{I_{LP}}{2\pi} 2 \left[\frac{1}{n+1} - \frac{1}{n-1} \right]$$

$$= \frac{I_{LP}}{\pi} \frac{(-2)}{(n^2 - 1)} \qquad \qquad \dots (10.4)$$

Again, $b_n = \frac{1}{\pi} \int_0^{2\pi} i_L(t) \sin n\omega t \, d\omega t$

$$= \frac{1}{\pi} I_{LP} \int_0^\pi \sin \omega t \sin n\omega t \, d\omega t$$

Taking $n = 1$,

$$b_1 = \frac{I_{LP}}{\pi} \int_0^\pi \sin^2 \omega t \, d\omega t$$

$$= \frac{I_{LP}}{2\pi} \int_0^\pi (1 - \cos 2\omega t) \, d\omega t$$

$$= \frac{I_{LP}}{2\pi} \pi \qquad \text{since } \int_0^\pi \cos 2\omega t \, d\omega t = 0$$

$$= \frac{I_{LP}}{2} \qquad \qquad \dots (10.5)$$

For $n > 1$,

$$b_n = \frac{I_{LP}}{2\pi} \int_0^\pi [\cos(n-1)\omega t - \cos(n+1)\omega t] \, d\omega t$$

$$= 0$$

$$\therefore \qquad i_L(t) = \frac{I_{LP}}{\pi} + \frac{I_{LP}}{2}\sin\omega t - 2\frac{I_{LP}}{\pi}\sum_{n=2,4,6..}\frac{1}{(n^2-1)}\cos n\omega t \qquad \ldots(10.6)$$

Thus, the output current of the half-wave rectifier contains dc term, fundamental frequency component ω and alternate even harmonics.

The root mean square (rms) value of a sinusoidal voltage (or current) defined as the dc voltage (or current) that produces the same amount of heat as produced by the sinusoidal voltage (or current) over a complete cycle in a given resistance.

Let the sinusoidal current be represented as

$$i(t) = I_p\sin\omega t \qquad \ldots(10.7)$$

where I_p is the peak value and ω is the angular frequency of the ac. Then,

$$I_{rms}^2.R.T = \int_0^T i^2(t)R\,dt$$

$$\therefore \qquad I_{rms}^2 = \frac{1}{T}\int_0^T i^2(t)\,dt$$

$$= \frac{1}{2\pi}\int_0^T i^2(t)\,d\omega t$$

$$= \frac{I_P^2}{2\pi}\int_0^{2\pi}\sin^2\omega t\,d\omega t$$

$$= \frac{I_P^2}{2\pi}.\frac{1}{2}\int_0^{2\pi}(1-\cos 2\omega t)\,d\omega t$$

$$= \frac{I_P^2}{2\pi}.\frac{1}{2}.2\pi$$

$$= \frac{I_P^2}{2}$$

$$\therefore \qquad I_{rms} = \frac{I_P}{\sqrt{2}} \qquad \ldots(10.8)$$

The rms value of load current of a half-wave rectifier can be calculated in two ways. The first method is,

$$I_{rms}^2 = \frac{1}{T}\int_0^T i_L^2(t)\,dt$$

$$= \frac{1}{2\pi} \int_0^{2\pi} i_L^2(t) \, d\omega t$$

$$= \frac{1}{2\pi} \int_0^{\pi} I_{LP}^2 \sin^2 \omega t \, d\omega t + \frac{1}{2\pi} \int_{\pi}^{2\pi} 0 \, d\omega t$$

$$= \frac{I_{LP}^2}{2\pi} \cdot \frac{1}{2} \int_0^{\pi} (1 - \cos 2\omega t) \, d\omega t$$

$$= \frac{I_{LP}^2}{2\pi} \cdot \frac{1}{2} \cdot \pi \quad \text{(since } \int_0^{\pi} \cos 2\omega t \, d\omega t = 0\text{)}$$

$$= \frac{I_{LP}^2}{4}$$

$$\therefore \quad I_{rms} = \frac{I_{LP}}{2} \qquad \qquad \qquad \qquad \dots (10.9)$$

The second method of calculation of I_{rms} is as follows:

The rms value of a composite wave (made up of harmonics) is the square root of the sum of the mean square values of individual components. From equation (10.6), we can write

$$I_{rms}^2 = \left(\frac{I_{LP}}{\pi}\right)^2 + \left(\frac{I_{LP}}{2}\right)^2 \cdot \frac{1}{2} + \left(\frac{2I_{LP}}{\pi}\right)^2 \cdot \frac{1}{2} \sum_{n=2,4,6..} \frac{1}{(n^2-1)^2}$$

$$= \left(\frac{I_{LP}}{\pi}\right)^2 + \frac{I_{LP}^2}{8} + \frac{2I_{LP}^2}{\pi^2} \cdot \left(\frac{\pi^2-8}{16}\right) \qquad \text{(since, } \sum_{n=2,4,6..} \frac{1}{(n^2-1)^2} = \frac{\pi^2-8}{16}\text{)}$$

$$= \frac{I_{LP}^2}{\pi^2} + \frac{I_{LP}^2}{8} + \frac{I_{LP}^2}{8} - \frac{I_{LP}^2}{\pi^2} \quad = \frac{I_{LP}^2}{4}$$

$$\therefore \quad I_{rms} = \frac{I_{LP}}{2} \qquad \qquad \qquad \qquad \dots (10.10)$$

RIPPLE FACTOR (γ)

It gives a measure of the lack of smoothness of the output waveform. The higher the fluctuation present in the waveform, the higher is the ripple factor.

It is defined as the ratio of the rms value of the ac components of the load current to the dc component of the load current.

Now,

$$I_{rms}^2 \Big|_{ac} = I_{rms}^2 - I_{dc}^2 \qquad \qquad \qquad \dots (10.11)$$

Equation (10.11) follows from the fact that the heat generated by the ac components of load current is equal to the heat generated by the composite load current minus the heat generated by the dc current component, the resistance being the same.

Now,

$$I_{rms}^2\Big|_{ac} = \frac{I_{LP}^2}{4} - \frac{I_{LP}^2}{\pi^2} = I_{LP}^2 \frac{(\pi^2 - 4)}{4\pi^2} \qquad \qquad \dots (10.12)$$

Ripple factor,

$$\gamma = \frac{I_{rms}\big|_{ac}}{I_{dc}} = \frac{I_{LP}}{2\pi}.\sqrt{\pi^2 - 4}.\frac{\pi}{I_{LP}}$$

$$= \sqrt{\frac{\pi^2}{4} - 1} = 1.21$$

The dc power output is given by

$$P_{dc} = I_{dc}^2 R_L = \frac{I_{LP}^2}{\pi^2} R_L \qquad \qquad \dots (10.13)$$

where R_L is the load resistance. Total power drawn from the source is

$$P_{in} = I_{rms}^2 (R_L + R_d) \qquad \qquad \dots (10.14)$$

Ac power to dc power conversion efficiency is

$$\eta = \frac{P_{dc}}{P_{in}} = \frac{I_{dc}^2 R_L}{I_{rms}^2 (R_L + R_d)}$$

$$= \frac{I_{LP}^2}{\pi^2} . \frac{4}{I_{LP}^2} . \frac{R_L}{R_L + R_d}$$

$$= \frac{4}{\pi^2} . \frac{1}{1 + R_d/R_L}$$

$$= \frac{0.406}{1 + R_d/R_L} \qquad \qquad \dots (10.15)$$

If $R_d \ll R_L$, then $\eta \to 40.6\%$. Peak inverse voltage applied to the diode is the full secondary peak voltage. This peak inverse voltage must be less than the breakdown voltage of the semiconductor diode.

10.3 FULL-WAVE RECTIFIER

A schematic circuit diagram of the full-wave rectifier is shown in Fig. 10.3. The load current waveform is shown in Fig. 10.4.

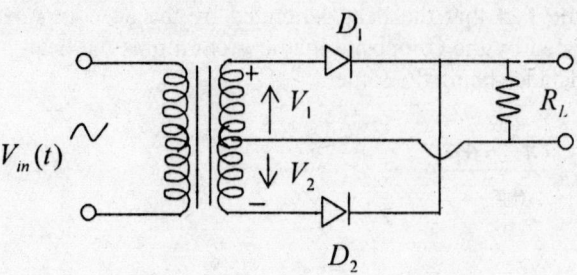

Fig. 10.3. Full-wave rectifier circuit.

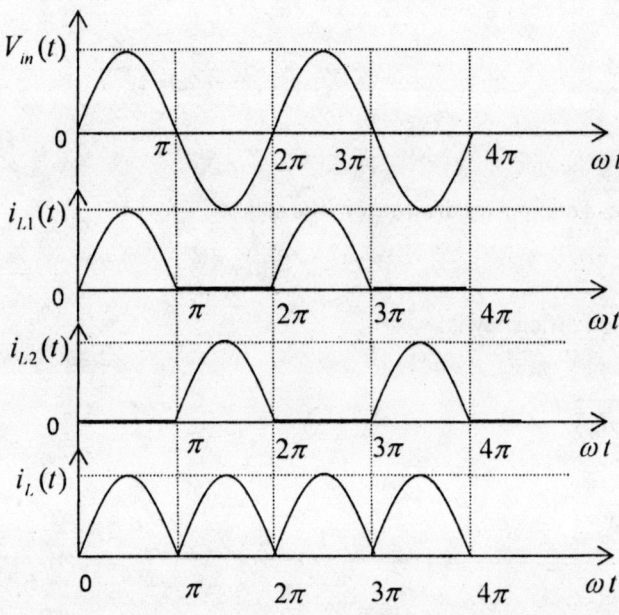

Fig. 10.4. Waveforms of input ac $\left(V_{in}(t)\right)$ and load currents $i_{l.1}(t)$, $i_{l.2}(t)$ and $i_{l.}(t)$.

The transformer in Fig. 10.3 is center-trapped. The voltages appearing at the inputs to diodes D_1 and D_2 at any instant of time can be represented as

$$V_1(t) = V_P \sin \omega t \qquad \qquad \ldots (10.16)$$

and $V_2(t) = V_P \sin(\omega t + \pi)$ $\qquad \ldots (10.17)$

where V_P is the peak ac voltage input to each diode. Diode D_1 conducts when the P-side is positive with respect to the N-side. Now, diode D_2 is non-conducting. During the next half cycle diode D_2 conducts and diode D_1 remains non-conducting. The diodes D_1 and D_2 are assumed to be identical.

The load current due to the conduction of D_1 is written as

$$i_{L1}(t) = \frac{V_1(t)}{R_L + R_d} = \frac{V_P}{R_L + R_d}\sin\omega t$$

$$= I_{LP}\sin\omega t \qquad \text{for } 0 \le \omega t \le \pi .$$

where $I_{LP} = \dfrac{V_P}{R_L + R_d}$ is the peak value of load current due to the conduction of diode D_1. R_L is the load resistance and R_d is the internal resistance of the semiconductor diode during conduction. Now,

$$i_{L1}(t) = 0 \text{ for } \pi \le \omega t \le 2\pi ,$$ since diode D_1 is non-conducting during this period.

Similarly, the load current due to the conduction of diode D_2 is

$$i_{L2}(t) = 0 \text{ for } 0 \le \omega t \le \pi$$

$$= I_{LP}\sin(\omega t + \pi) \quad \text{for } \pi \le \omega t \le 2\pi$$

Making a Fourier analysis of $i_{L1}(t)$ and $i_{L2}(t)$ as before, we get

$$i_{L1}(t) = \frac{I_{LP}}{\pi}\left[1 + \frac{\pi}{2}\sin\omega t - 2\sum_{n=2,4,6...}\frac{1}{(n^2-1)}\cos n\omega t\right] \qquad \text{... (10.18)}$$

and

$$i_{L2}(t) = \frac{I_{LP}}{\pi}\left[1 - \frac{\pi}{2}\sin\omega t - 2\sum_{n=2,4,6...}\frac{1}{(n^2-1)}\cos n\omega t\right] \qquad \text{... (10.19)}$$

Total load current is given by

$$i_L(t) = i_{L1}(t) + i_{L2}(t)$$

$$= \frac{2I_{LP}}{\pi}\left[1 - 2\sum_{n=2,4,6...}\frac{1}{(n^2-1)}\cos n\omega t\right] \qquad \text{... (10.20)}$$

Equation (10.20) shows that the lowest frequency of the ripple component is 2ω in a full wave rectifier. In half-wave rectifier, this lowest frequency of the ripple was ω.

The definitions of rms value of load current and the ripple factor remain the same as mentioned in the case of half-wave rectifier. The dc component of the load current is

$$I_{dc} = \frac{1}{2\pi}\int_0^{2\pi} i_L(t)\,d\omega t$$

$$= \frac{1}{2\pi}I_{LP}\left[\int_0^{\pi}\sin\omega t\,d\omega t + \int_{\pi}^{2\pi}\sin(\omega t + \pi)\,d\omega t\right]$$

$$= \frac{I_{LP}}{2\pi} \left[-\cos \omega t \right]_0^\pi + \frac{I_{LP}}{2\pi} \left[\cos \omega t \right]_\pi^{2\pi}$$

$$= \frac{I_{LP}}{2\pi}(2+2) = \frac{2I_{LP}}{\pi} \qquad \qquad \dots (10.21)$$

This is also seen from Equation (10.20). Now, the rms value of load current (I_{rms}) can be calculated as

$$I_{rms}^2 = \frac{1}{2\pi} \int_0^{2\pi} i_L^2(t) \, d\omega t$$

$$= \frac{I_{LP}^2}{2\pi} \left[\int_0^\pi \sin^2 \omega t \, d\omega t + \int_\pi^{2\pi} \sin^2(\omega t + \pi) \, d\omega t \right]$$

$$= \frac{I_{LP}^2}{2\pi} \left[\int_0^\pi \sin^2 \omega t \, d\omega t + \int_\pi^{2\pi} \sin^2 \omega t \, d\omega t \right]$$

$$= \frac{I_{LP}^2}{2\pi} \int_0^{2\pi} \sin^2 \omega t \, d\omega t$$

$$= \frac{I_{LP}^2}{2\pi} \cdot \frac{1}{2} \cdot \int_0^{2\pi} (1 - \cos 2\omega t) \, d\omega t$$

$$= \frac{I_{LP}^2}{2\pi} \cdot \frac{1}{2} \cdot 2\pi$$

$$= \frac{I_{LP}^2}{2}$$

$$\therefore \quad I_{rms} = \frac{I_{LP}}{\sqrt{2}} \qquad \qquad \dots (10.22a)$$

Alternatively, I_{rms} can be calculated from (10.20) as

$$I_{rms}^2 = \left(\frac{2I_{LP}}{\pi} \right)^2 + \frac{1}{2} \left(\frac{2I_{LP}}{\pi} \right)^2 \cdot 2^2 \cdot \sum_{n=2,4,6\dots} \frac{1}{(n^2-1)^2}$$

$$= \left(\frac{2I_{LP}}{\pi} \right)^2 \left[1 + 2 \cdot \frac{\pi^2 - 8}{16} \right]$$

$$= \left(\frac{2I_{LP}}{\pi} \right)^2 \left[1 + \frac{\pi^2 - 8}{8} \right]$$

$$= \frac{4I_{LP}^2}{\pi^2}\left[\frac{\pi^2}{8}\right] = \frac{I_{LP}^2}{2}$$

$$\therefore \qquad I_{rms} = \frac{I_{LP}}{\sqrt{2}} \qquad \qquad \ldots (10.22b)$$

Now,

$$I_{rms}^2\bigg|_{ac} = I_{rms}^2 - I_{dc}^2 = \frac{I_{LP}^2}{2} - \frac{4}{\pi^2}I_{LP}^2 = \frac{I_{LP}^2}{2\pi^2}(\pi^2 - 8)$$

The ripple factor (γ) is given by,

$$\gamma = \frac{I_{rms}\big|_{ac}}{I_{dc}} = \frac{I_{LP}}{\pi\sqrt{2}}\sqrt{\pi^2 - 8}\,\frac{\pi}{2I_{LP}}$$

$$= \frac{\sqrt{\pi^2 - 8}}{2\sqrt{2}} = \sqrt{\frac{\pi^2}{8} - 1}$$

$$= 0.48$$

The input power delivered to the rectifier is

$$P_{in} = I_{rms}^2(R_L + R_d) \qquad \qquad \ldots (10.23)$$

The dc power delivered to the load R_L is

$$P_{dc} = I_{dc}^2 R_L \qquad \qquad \ldots (10.24)$$

The ac power to dc power conversion efficiency is

$$\eta = \frac{P_{dc}}{P_{in}} = \frac{I_{dc}^2 R_L}{I_{rms}^2(R_L + R_d)}$$

$$= \left[\frac{2I_{LP}}{\pi} \cdot \frac{\sqrt{2}}{I_{LP}}\right]^2 \cdot \frac{R_L}{R_L + R_d}$$

$$= \frac{8}{\pi^2} \cdot \frac{R_L}{R_L + R_d} \qquad \qquad \ldots (10.25)$$

If $R_d << R_L$, $\eta \rightarrow \dfrac{8}{\pi^2} = 81.2\%$.

Peak inverse voltage appearing across any diode is the full secondary voltage $2V_P$. This voltage must be less than the semiconductor diode breakdown voltage.

10.4 BRIDGE RECTIFIER

This is a kind of full-wave rectifier, which uses four diodes. These diodes form the four arms of an electrical bridge (such as Wheatstone bridge) and hence the name bridge rectifier. Two diodes conduct at a time during one half cycle of the input ac. The conducting diode pair form the opposite arms of the bridge. The circuit diagram of the bridge rectifier is shown in Fig. 10.5.

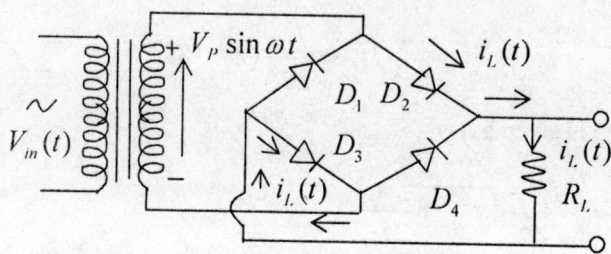

Fig. 9.5. Circuit diagram of the bridge rectifier.

Here, D_1, D_2, D_3 and D_4 are four diodes assumed to be identical. R_L is the load resistance. D_2 and D_3 conduct when upper end of the secondary is positive as shown in Fig. 10.5. D_1 and D_4 are non-conducting now. Similarly, D_1 and D_4 conduct when the polarity of the secondary voltage changes. This time D_2 and D_3 remain non-conducting. Load current flows through R_L in the same direction during the entire cycle of input ac. Since the diode internal resistance (R_d) is small, the dc component of load current is

$$I_{dc} = \frac{2I_{LP}}{\pi} \quad\quad\quad \ldots (10.26)$$

where I_{LP} is the peak load current which is almost the same as in the full-wave circuit.

One advantage of the bridge rectifier is that the full secondary voltage in bridge rectifier is one half the full secondary voltage in a full-wave circuit for the same dc voltage output. As a result, the number of secondary turns required in the bridge rectifier is one half the secondary turns in the full-wave circuit for a given input ac voltage. No center-tapping of transformer is required in this case.

Another advantage is that the peak inverse voltage (PIV) appearing across any diode during non-conducting period is the full secondary voltage. But, full secondary voltage in a bridge rectifier is one half that of the full-wave circuit. So, PIV in bridge rectifier is one half the PIV in a full-wave rectifier.

10.5 FULL-WAVE RECTIFIER WITH A CAPACITOR FILTER

How do we obtain filtering action from the capacitor?

The answer is that the capacitor stores energy during the conduction period of the diode and delivers this energy to the load when the falling input voltage makes the diode non-conducting. This minimizes load current variation and hence the ripple content of the output decreased.

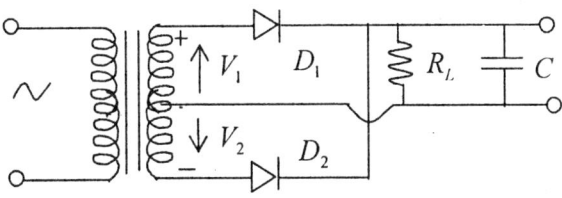

Fig. 9.6. Full-wave rectifier with capacitor filter.

Due to the charging and discharging of the capacitor (C), a positive dc voltage appears at the N-side (or cathode) of the diodes D_1 and D_2. So, the diodes conduct only when its P-side is more positive than the N-side. As a result, the diodes conduct for less than half cycle. The output waveform of load voltage $V_L(t)$ is shown in Fig. 10.7.

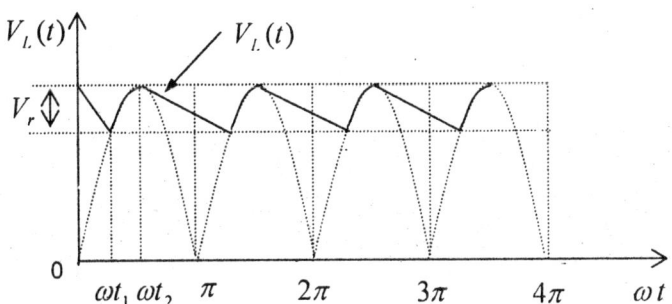

Fig. 10.7. Load voltage waveform of the full-wave rectifier with capacitor filter.

Let $V_1(t) = V_p \sin \omega t$ ... (10.27)

and $V_2(t) = V_p \sin(\omega t - \pi)$ ... (10.28)

be the input ac voltages applied to diodes D_1 and D_2 respectively. V_p is the peak input voltage. V_2 lags behind V_1 by π radian.

Let t_1 = cut-in time of D_1 when D_1 just begins to conduct and t_2 = cut-out time of D_1 when D_1 just becomes non-conducting.

Also let i_R = current through load R_L during the conduction of D_1 and i_c = current through capacitor C during the conduction of D_1. Total diode (D_1) current,

$$i = i_R + i_c$$

$$= \frac{V_P}{R_L} \sin \omega t + C \frac{dv_L(t)}{dt}$$

$$= \frac{V_P}{R_L} \sin \omega t + \omega C V_P \cos \omega t \qquad \qquad \dots (10.29)$$

where $v_L(t)$ is the load voltage developed across R_L. Now, substitute

$$\frac{V_P}{R_L} = I_P \cos \psi \qquad \qquad \dots (10.30)$$

and $\qquad \omega C V_P = I_P \sin \psi \qquad \qquad \dots (10.31)$

Then, $i = I_P \sin(\omega t + \psi) \qquad \qquad \dots (10.32)$

where

$$I_P = \sqrt{\left(\frac{V_P}{R_L}\right)^2 + (\omega C V_P)^2}$$

$$= V_P \left(\frac{1}{R_L^2} + \omega^2 C^2\right)^{1/2} \qquad \qquad \dots (10.33)$$

and $\qquad \tan \psi = \omega C R_L$

that is, $\psi = \tan^{-1}(\omega C R_L) \qquad \qquad \dots (10.34)$

Now, at $t = t_2$, $i = 0$. From (10.32)

$$\omega t_2 + \psi = \pi$$

$$\omega t_2 = \pi - \psi \qquad \qquad \dots (10.35)$$

Since V_1 and V_2 differ in phase by π radian, the extinction angle of diode D_2 is given by

$$\omega t_2' + \psi = 2\pi$$

$$\therefore \qquad \omega t_2' = 2\pi - \psi \qquad \qquad \dots (10.36)$$

Here, t_2' = cut-out time of diode D_2 when D_2 becomes just non-conducting.

The capacitor C discharges through the load R_L according to the relation

$$i_R = A e^{-\frac{t}{CR_L}} \qquad \qquad \dots (10.37)$$

where A = constant, i_R is the current through the load R_L.

During charging of the capacitor C the load current follows the input voltage and hence

$$i_R = \frac{V_P}{R_L} \sin \omega t \qquad \qquad \ldots (10.38)$$

with diode D_1 conducting. At the cut-out time t_2 of D_1, conduction ends and extinction begins. So, at $t = t_2$, both (10.37) and (10.38) must hold good. Then,

$$\frac{V_P}{R_L} \sin \omega t_2 = A e^{-\frac{t_2}{CR_L}}$$

$$\therefore \qquad A = \frac{V_P}{R_L} \sin \omega t_2 e^{\frac{t_2}{CR_L}} \qquad \qquad \ldots (10.39)$$

Diode D_2 starts conduction at $t = t_1'$ when the falling capacitor voltages becomes equal to the rising input voltage (V_2) of D_2.

Then, $i_R.R_L = V_P \sin(\omega t_1' - \pi)$

where i_R is given by (10.37).

$$\therefore \qquad A R_L e^{-\frac{t_1'}{CR_L}} = V_P \sin(\omega t_1' - \pi)$$

$$\therefore \qquad V_P \sin \omega t_2 e^{\frac{t_2}{CR_L}} .e^{-\frac{t_1'}{CR_k}} = V_P \sin(\omega t_1' - \pi) \qquad \text{using (10.39)}$$

$$\therefore \qquad \sin(\omega t_1' - \pi) = \sin \omega t_2 e^{\frac{t_2}{CR_L}} .e^{-\frac{t_1'}{CR_L}}$$

$$\therefore \qquad \omega t_1' = \pi + \sin^{-1} \left[\sin \omega t_2 .e^{\frac{(t_2 - t_1')}{CR_L}} \right] \qquad \qquad \ldots (10.40)$$

since t_2 is known from (10.35), t_1' can be found by solving (10.40).

Cut-in time (t_1) of diode D_1 differs from cut-in time (t_1') of D_2 by π radian. Thus,

$$\omega t_1 = \omega t_1' - \pi \qquad \qquad \ldots (10.41)$$

Thus, cut-in and cut-out times of D_1 and D_2 are calculated.

CALCULATION OF RIPPLE FACTOR OF FULL-WAVE RECTIFIER WITH CAPACITOR FILTER

We assume the waveform of the load voltage $\left(v_L(t) \right)$ of this rectifier as piece-wise linear. For a large value of C, $\omega CR_L \gg \pi$ and $\psi \to \frac{\pi}{2}$ [from(10.34)]. Then $\omega t_2 \to \frac{\pi}{2}$ [from(10.35)] and at this time, $v_L(t) = V_P \sin \dfrac{\pi}{2} = V_P$. So, the capacitance C is charged to the peak voltage V_P at the beginning of discharge. Since $CR_L >> (t_2 - t_1')$, we get $\omega t_1' = \pi + \omega t_2$ [using (10.40)]. So, $\omega t_1' \to 3\dfrac{\pi}{2}$. From (10.41), $\omega t_1 \to \dfrac{\pi}{2}$ in this case. The waveform of load voltage $v_L(t)$ under piece-wise linear approximation is shown in Fig. 10.8.

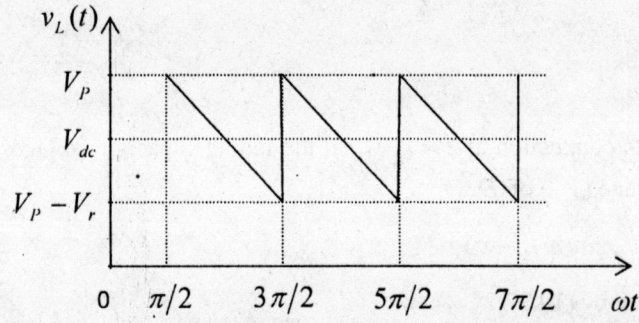

Fig. 10.8. Load voltage variation under piece-wise linear approximation.

Load voltage during the discharge of capacitor C is given by

$$v_L(t) = i_R . R_L$$

$$= \frac{V_P}{R_L} . R_L e^{-\frac{t}{CR_L}}$$

$$\approx V_P (1 - \frac{t}{CR_L}) \quad \text{since } CR_L >> \frac{\pi}{\omega}$$

Rate of fall of load voltage $= \dfrac{V_P}{CR_L}$.

Fall of load voltage over the time interval $\dfrac{\pi}{\omega}$ is $\dfrac{V_P}{CR_L} . \dfrac{\pi}{\omega}$ and this is the ripple voltage amplitude (peak to peak) V_r.

$$\therefore \quad V_r = \frac{\pi}{\omega CR_L} . V_P \qquad \qquad \ldots (10.42)$$

The load voltage can be represented as

$$v_L(t) = V_P - \frac{V_r}{\pi} \left(\omega t - \frac{\pi}{2} \right) \qquad \text{for } \frac{\pi}{2} \le \omega t \le \frac{3\pi}{2}$$

$$= \left(V_P + \frac{V_r}{2}\right) - \frac{V_r}{\pi}\omega t \qquad \qquad \ldots (10.43)$$

and

$$v_L(t) = V_P - \frac{V_r}{\pi}\left(\omega t - \frac{3\pi}{2}\right) \quad \text{for} \quad \frac{3\pi}{2} \le \omega t \le \frac{5\pi}{2}$$

$$= \left(V_P + \frac{3V_r}{2}\right) - \frac{V_r}{\pi}\omega t \qquad \qquad \ldots (10.44)$$

Now,

$$V_{rms}^2 = \frac{1}{2\pi}\int_{\pi/2}^{5\pi/2} v_L^2(t)\, d\omega t$$

$$= \frac{1}{2\pi}\left[\int_{\pi/2}^{3\pi/2}\left(V_P + \frac{V_r}{2} - \frac{V_r}{\pi}\omega t\right)^2 d\omega t + \int_{3\pi/2}^{5\pi/2}\left(V_P + \frac{3V_r}{2} - \frac{V_r}{2}\omega t\right)^2 d\omega t\right]$$

$$= \frac{1}{2\pi}\left[\int_{\pi/2}^{3\pi/2}\left\{\left(V_P + \frac{V_r}{2}\right)^2 + \frac{V_r^2}{\pi^2}\omega^2 t^2 - \frac{2V_r}{\pi}\left(V_P + \frac{V_r}{2}\right)\omega t\right\} d\omega t\right.$$

$$\left. + \int_{3\pi/2}^{5\pi/2}\left\{\left(V_P + \frac{3V_r}{2}\right)^2 + \frac{V_r^2}{\pi^2}\omega^2 t^2 - \frac{2V_r}{\pi}\left(V_P + \frac{3V_r}{2}\right)\omega t\right\} d\omega t\right]$$

$$= \frac{1}{2\pi}\left[\left(V_P + \frac{V_r}{2}\right)^2 .\pi + \frac{V_r^2}{\pi^2}.\frac{1}{3}\pi^3\left(\frac{27}{8} - \frac{1}{8}\right) - \frac{2V_r}{\pi}\left(V_P + \frac{V_r}{2}\right).\frac{\pi^2}{2}.\left(\frac{9}{4} - \frac{1}{4}\right)\right]$$

$$+ \frac{1}{2\pi}\left[\left(V_P + \frac{3V_r}{2}\right)^2 .\pi + \frac{V_r^2}{\pi^2}.\frac{\pi^3}{3}\left(\frac{125}{8} - \frac{27}{8}\right) - \frac{2V_r}{\pi}\left(V_P + \frac{3V_r}{2}\right).\frac{\pi^2}{2}.\left(\frac{25}{4} - \frac{9}{4}\right)\right]$$

$$= \left[\frac{V_P^2}{2} + \frac{V_r^2}{8} + \frac{V_P V_r}{2} + \frac{13}{24}V_r^2 - V_P V_r - \frac{V_r^2}{2}\right]$$

$$+ \left[\frac{V_P^2}{2} + \frac{9V_r^2}{8} + \frac{3V_P V_r}{2} + \frac{49}{24}V_r^2 - 2V_P V_r - 3V_r^2\right]$$

$$= V_P^2 - V_P V_r + \frac{V_r^2}{3} \qquad \qquad \ldots (10.45)$$

where V_{rms} is the rms load voltage. The dc voltage output is

$$V_{dc} = \frac{1}{2\pi} \int_{\pi/2}^{5\pi/2} v_L(t)\, d\omega t$$

$$= \frac{1}{2\pi}\left[\int_{\pi/2}^{3\pi/2}\left(V_P + \frac{V_r}{2} - \frac{V_r}{\pi}\omega t \right) d\omega t + \int_{3\pi/2}^{5\pi/2}\left(V_P + \frac{3V_r}{2} - \frac{V_r}{\pi}\omega t \right) d\omega t \right]$$

$$= \frac{1}{2\pi}\left[V_P.\pi + \frac{V_r}{2}.\pi - \frac{V_r}{\pi}.\frac{\pi^2}{2}\left(\frac{9}{4} - \frac{1}{4}\right) \right] + \frac{1}{2\pi}\left[V_P.\pi + \frac{3V_r}{2}.\pi - \frac{V_r}{\pi}.\frac{\pi^2}{2}\left(\frac{25}{4} - \frac{9}{4}\right) \right]$$

$$= \frac{V_P}{2} + \frac{V_r}{4} - \frac{V_r}{2} + \frac{V_P}{2} + \frac{3V_r}{4} - V_r$$

$$= V_P - \frac{V_r}{2} \qquad \qquad \dots (10.46)$$

The rms value of ac components of the load voltage is $v_{rms}\big|_{ac}$ where

$$v_{rms}\big|_{ac} = V_{rms}^2 - V_{dc}^2$$

$$= V_P^2 - V_P V_r + \frac{V_r^2}{3} - \left(V_P - \frac{V_r}{2} \right)^2 \qquad = \frac{V_r^2}{3} - \frac{V_r^2}{4} = \frac{V_r^2}{12}$$

$$\therefore \qquad v_{rms} = \frac{V_r}{2\sqrt{3}} \qquad \qquad \dots (10.47)$$

Ripple factor,

$$\gamma = \frac{v_{rms}\big|_{ac}}{V_{dc}}$$

$$= \frac{V_r}{2\sqrt{3}} . \frac{1}{V_P - \dfrac{V_r}{2}}$$

$$\approx \frac{V_r}{2\sqrt{3}\, V_P} \qquad \left(\text{since } \frac{V_r}{2} << V_P \right)$$

$$= \frac{1}{2\sqrt{3}} . \frac{\pi}{\omega C R_L} \qquad \text{[using (9.42)]}$$

$$= \frac{1}{4\sqrt{3}\, C R_L f} \qquad \qquad \dots (10.48)$$

where $f = \dfrac{\omega}{2\pi}$ is the frequency of input ac.

There is a minimum value of C above which (9.48) holds good. Below this minimum value of C, the piece-wise linear approximation is not valid.

For the half-wave rectifier with capacitor filter, we get

$$V_r = \frac{2\pi}{\omega CR_L} V_P \qquad\qquad\qquad \dots (10.49)$$

and

$$\gamma = \frac{1}{2\sqrt{3}} \left(\frac{V_r}{V_P} \right) = \frac{1}{2\sqrt{3}} \cdot \frac{2\pi}{\omega CR_L}$$

$$= \frac{1}{2\sqrt{3}\, fCR_L} \qquad\qquad\qquad \dots (10.50)$$

The assumption $\dfrac{V_r}{2} << V_P$ indicates that in full-wave rectifier with capacitor filter, we have

$$\frac{1}{2} \cdot \frac{\pi}{\omega CR_L} \cdot V_P << V_P$$

$$\therefore \quad \omega CR_L >> \frac{\pi}{2} \qquad\qquad\qquad \dots (10.51)$$

10.6 π-SECTION FILTER

Thus far we have considered a single capacitor at the output. Now, we add another inductor-capacitor (LC') filter with the previous capacitor C so that they form a π-section. The $C-L-C'$ combination looks like a greek letter π in structure and hence the name π filter. The π-filter is shown in Fig. 10.9. The reactance of the inductance L has a magnitude ωL. So, it provides a high impedance at high frequencies. Hence, the high frequency component of the ripple faces high impedance in passing through inductance L and the corresponding ripple current does not pass through the R_L-C' combination. There is another effect of filtering in this circuit.

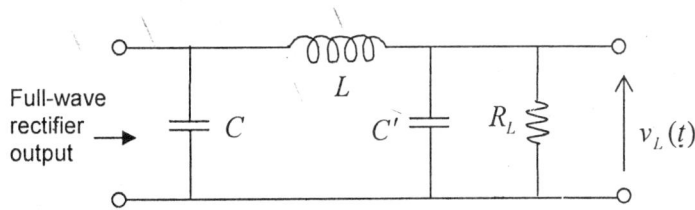

Fig. 10.9. π-section filter connected with full-wave rectifier.

The residual ripple that appear across $R_L - C'$ combination, is again shunted by the low reactance $\left(\dfrac{1}{\omega C'}\right)$ and the high frequency components of the ripple are again bypassed. Hence, the ripple suppression effect of this π-filter is better than that of a single capacitor filter.

10.7 IC VOLTAGE REGULATOR

Voltage regulators are commercially available in the form of integrated circuits (IC). Three types of IC regulators are known. They are

    (i)   Fixed positive voltage regulator (IC 7800 series).
    (ii)   Fixed negative voltage regulator (IC 7900 series).
    (iii)   Variable voltage regulator (IC-LM317).

IC 7805 gives an output voltage of $+5$ V. Similarly, IC 7812 gives an output voltage of $+12$ V. IC 7906 gives an output voltage of -6 V. IC 7915 gives an output voltage of -15 V.

These ICs have three terminals – terminal 1 is input, terminal 2 is ground and terminal 3 is output. A schematic view of the IC is given in Fig. 10.10(a) and 10.10(b).

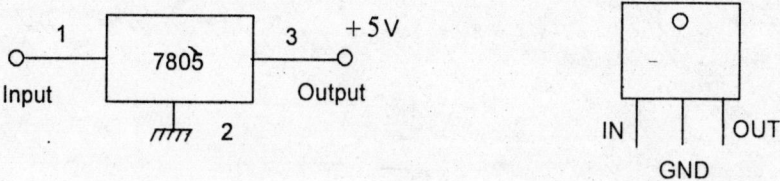

Fig. 10.10. (a) Circuit diagram of IC regulator.   Fig. 10.10. (b) A view of the IC regulator.

The output of a full-wave rectifier with a capacitor filter is fed to the input terminal of the IC regulator.

SOLVED PROBLEMS

10.1   The secondary voltage(peak) of a center-tapped transformer is $\pm 12 \sin 100\pi t$ volts. A full-wave rectifier with a capacitor filter has $C = 100$ µF and $R_L = 1$ kΩ. Calculate the peak to peak ripple voltage of this rectifier.

Solution
From (10.42),

$$V_r = \frac{\pi}{\omega C R_L} . V_P$$

Here, $V_P = 12$ volts, $\omega = 100\pi$, $C = 100$ µF, $R_L = 10^3$ Ω.

$$\therefore \quad V_r = \frac{\pi.12}{100\pi.100 \times 10^{-6}.10^3}$$

$$= \frac{12}{10} = 1.2 \text{ volt.}$$

10.2 A half-wave rectifier with a capacitor filter is fed with the ac voltage(peak) of $12\sin 100\pi t$. The capacitor and the load resistance are $C = 100\,\mu F$ and $R_L = 1\,k\Omega$ respectively. Find the value of peak to peak ripple voltage.

Solution
From (10.49),

$$V_r = \frac{2\pi}{\omega CR_L}.V_p.$$

Here, $V_p = 12$ volt, $\omega = 100\pi$ radian, $C = 100\ \mu F = 10^{-4}$ F, $R_L = 10^3\ \Omega$.

$$\therefore \quad V_r = \frac{2\pi \times 12}{100\pi \times 10^{-4} \times 10^3} = \frac{24}{10} = 2.4 \text{ volt.}$$

10.3 In problem 10.1, calculate the value of ripple factor.

Solution
From, (10.48),

$$\gamma = \frac{1}{4\sqrt{3}\,CR_L f}.$$

Here, $C = 100\,\mu F = 10^{-4} F$, $R_L = 10^3 \Omega$ $f = 50$ Hz.

$$\therefore \quad \gamma = \frac{1}{4\sqrt{3} \times 10^{-4} \times 10^3 \times 50} = \frac{1}{20\sqrt{3}} = 0.0288.$$

10.4 In problem 10.2, calculate the value of ripple factor.

Solution
From (10.50),

$$\gamma = \frac{1}{2\sqrt{3}\,CR_L f}$$

$$= \frac{1}{2\sqrt{3} \times 10^{-4} \times 10^3 \times 50} = \frac{1}{10\sqrt{3}}$$

$$= 0.0577$$

10.5 In problem 1, calculate the dc voltage output of the full-wave rectifier with capacitor filter.

Solution
From (10.46),

$$V_{dc} = V_P - \frac{V_r}{2}.$$

Here, $V_P = 12$ volt and $V_r = 1.2$ volt (from solution of problem 1).

$$\therefore \quad V_{dc} = 12 - \frac{1.2}{2} = 12 - 0.6 = 11.4 \text{ volt}.$$

10.6 In problem 2, calculate the dc voltage output of the half-wave rectifier with capacitor filter.

Solution
From, (10.46),

$$V_{dc} = V_P - \frac{V_r}{2}.$$

Here, $V_P = 12$ volt and $V_r = 2.4$ volt (from solution of problem 10.2).

$$V_{dc} = 12 - \frac{2.4}{2} = 12 - 1.2 = 10.8 \text{ volt}.$$

PROBLEMS

10.1 The diode in a half wave rectifier circuit has a forward resistance of 50 Ω. The rectifier supplies power to a load of 1 kΩ from a 200 volt rms source. Calculate the dc load current, rms load current and the output power of this rectifier.

[**Ans.** 85.7 mA, 134.7 mA, 7.34 W]

Hints : $V_p = 200\sqrt{2}$ V, $R_d = 50$ Ω, $R_L = 1$ kΩ.

$$I_{LP} = \frac{V_p}{R_L + R_d} = \frac{200\sqrt{2}}{1050} \text{ A}; \ I_{dc} = \frac{200\sqrt{2}}{1050\pi} = 85.7 \text{ mA}.$$

$$I_{rms} = \frac{I_{LP}}{2} = \frac{100\sqrt{2}}{1050} \text{ A} = 134.7 \text{ mA}.$$

$$P_{dc} = I_{dc}^2 R_L = \frac{(200)^2 \times 2}{(1050\pi)^2} 10^3 \text{ W} = 7.34 \text{ W}.$$

10.2 A half wave rectifier has a diode with forward resistance of 100 Ω and a load resistance of 1200 Ω. It is supplied with a ac voltage of 280 volt peak. Calculate the rms and average value of load current. Also calculate the dc output power, ac input power and the efficiency of the rectifier.

[**Ans.** 107.7 mA, 68.5 mA, 5.64 W, 15.07 W, 37.47 %]

Hints: $V_P = 280$ V, $R_d = 100$ Ω, $R_L = 1200$ Ω.

$$I_{LP} = \frac{V_P}{R_L + R_d} \; ; I_{dc} = \frac{I_{LP}}{\pi} = \frac{280}{1300\pi} \text{ A} = 68.5 \text{ mA.}$$

$$I_{rms} = \frac{I_{LP}}{2} = \frac{280}{1300 \times 2} = \frac{14}{130} \text{ A} = 107.7 \text{ mA.}$$

$$P_{dc} = I_{dc}^2 R_L = \frac{(280)^2 \times 1200}{(1300\pi)^2} \text{ W} = 5.64 \text{ W.}$$

$$P_{in} = I_{rms}^2 (R_L + R_d) = 15.07 \text{ W}$$

$$\eta = \frac{P_{dc}}{P_{in}} = \frac{0.406}{1 + \dfrac{100}{1200}} = \frac{12}{13} \times 0.406 = 37.47 \text{ %.}$$

10.3 A full wave rectifier is designed with identical diodes each having a forward resistance of 200 Ω. It delivers power to a load of 1kΩ. The input supply is 240 volt rms ac. Calculate the dc load current, output power and the efficiency of this rectifier.

[**Ans.** 90 mA, 8.1 W, 67.5%]

Hints: $V_P = \dfrac{240\sqrt{2}}{2} = 120\sqrt{2}$ V.

$$I_{LP} = \frac{V_P}{R_L + R_d} = \frac{120\sqrt{2}}{1200} = \frac{\sqrt{2}}{10} \text{ A}$$

$$I_{dc} = \frac{2I_{LP}}{\pi} = \frac{2\sqrt{2}}{10\pi} \text{ A} = 90 \text{ mA.}$$

$$P_{dc} = I_{dc}^2 R_L = \frac{8}{100\pi^2} 10^3 \text{ W} = 8.1 \text{ W.}$$

$$P_{in} = I_{rms}^2 (R_L + R_d)$$

$$\eta = \frac{P_{dc}}{P_{in}} = \frac{8}{\pi^2} \frac{1}{1 + \dfrac{R_d}{R_L}} = \frac{8}{\pi^2} \times \frac{5}{6} = 67.5 \text{ %.}$$

QUESTIONS

10.1 What is meant by the term 'Rectifier'? Draw a neat circuit diagram of a half-wave rectifier and explain its operation. Derive expressions for ripple factor and conversion efficiency of this filter.

10.2 Define the term 'ripple factor' of a rectifier. Explain the operation of a full-wave rectifier with the help of a neat circuit diagram. Derive expressions for dc voltage output, ripple factor and conversion efficiency of this filter.

10.3 Draw a circuit diagram of a full-wave rectifier with a capacitor filter at the output. Explain qualitatively how ripple is suppressed in this rectifier-filter. Derive expressions for cut-in and cut-out angles of the diodes in terms of circuit parameters.

10.4 In a full-wave rectifier with a capacitor filter, derive expressions for dc load voltage, rms load voltage and ripple factor using piece-wise linear approximation. What is the limitation of this approximation?

10.5 Explain how a π-section filter yields a better suppression of ripple at the output. Write a short note on IC regulator.

Modulation, Demodulation, Radar and Television

11.1 NECESSITY OF MODULATION

Voice can not be transmitted typically beyond several tens of meter in free space. This is because sound gets attenuated, reflected and scattered in the medium which result in energy dissipation of the sound wave. In order to achieve long and very long distance communication, the electrical signal corresponding to voice, video or data is made to modulate a high frequency carrier electromagnetic wave and the modulated carrier wave is radiated by an antenna into free space, which travels with the speed of light. The electrical signal corresponding to voice, video or data can not be radiated itself without modulation since the size of the radiating antenna must have a dimension of the order of wavelength of the radiation. Any attempt to radiate the electrical signal corresponding to voice (say) directly will require an antenna having a length ~tens of kilometer. Superimposing information on the high frequency carrier through modulation will reduce the antenna size dramatically. Signal transmission in a particular direction can be possible by using a higher frequency of the carrier electromagnetic wave in modulation. Broadcasting would not be possible without modulation. The high speed of message transmission would not be possible without modulation of a high frequency electromagnetic wave. Modulation makes a frequency translation of the message signal which enables us to avoid interference between several message signals propagating simultaneously through the same channel. This is done by proper carrier frequency selection. In optical communication, the modulation of a lightwave by the information signal gives immunity to the transmitted information against electromagnetic interference.

11.2 MODULATION

Modulation is the process which involves two waves- one is the modulating signal which represents the message and the other is the carrier wave. The frequency of the carrier wave is much greater than the frequency of the modulating wave. The carrier wave has three parameters viz. amplitude, frequency and phase. Modulation is the process in which one of these three characteristics is changed in accordance with the instantaneous value of the modulating wave.

Since there are three characteristic parameters of a carrier wave, there can be three types of modulation. These are

1. amplitude modulation

2. frequency modulation, and

3. phase modulation

The modulation discussed here is analog modulation since the modulating signal is a continuously varying signal. There can be another type of modulation known as digital modulation which will be discussed later.

11.3 AMPLITUDE MODULATION (AM)

In AM, the amplitude of the carrier wave is varied in accordance with the instantaneous value of the modulating signal. The instantaneous amplitude deviation from the unmodulated value is proportional to the instantaneous value of the modulating signal.

Let $\quad v_C(t) = V_C \cos \omega_C t$ ... (11.1)

be the carrier wave and

$$v_m(t) = V_m \cos \omega_m t \qquad \text{... (11.2)}$$

be the modulating signal. Then, if $V_{am}(t)$ be the amplitude of the AM signal, we write

$$\left[V_{am}(t) - V_C\right] \propto v_m(t)$$

$$\therefore \quad V_{am}(t) = V_C + k v_m(t) \qquad \text{... (11.3)}$$

where k is a constant of proportionality. The AM wave can now be represented as

$$v_{am}(t) = V_{am}(t) \cos \omega_C t$$
$$= \left[V_C + k v_m(t)\right] \cos \omega_C t$$
$$= V_C \left[1 + m \cos \omega_m t\right] \cos \omega_C t \qquad \text{... (11.4)}$$

where $m = \dfrac{kV_m}{V_C}$ is known as amplitude modulation index. A schematic plot of the AM wave is shown in Fig. 11.1.

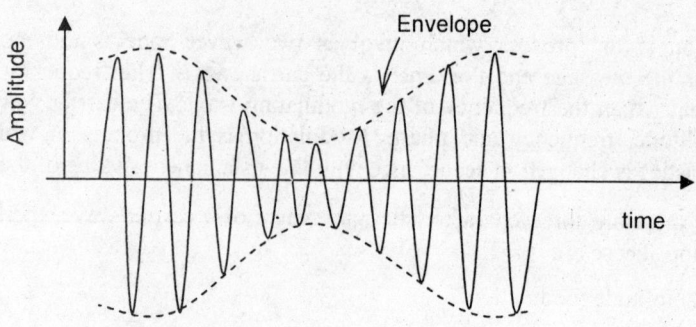

Fig. 11.1 Amplitude Modulation

The maximum and minimum values of the amplitude of the AM signal are

$$V_{am}(t)\big|_{max} = V_C(1+m) \qquad \ldots (11.5)$$

$$V_{am}(t)\big|_{min} = V_C(1-m) \qquad \ldots (11.6)$$

Then, $$\dfrac{V_{am}(t)\big|_{max} - V_{am}(t)\big|_{min}}{V_{am}(t)\big|_{max} + V_{am}(t)\big|_{min}} = \dfrac{2mV_C}{V_C} = m \qquad \ldots (11.7)$$

Thus, by measuring the minimum values of the amplitude of the AM wave the AM index (m) can be measured.

Now,

$$v_{am}(t) = V_C\big(1 + m\cos\omega_m t\big)\cos\omega_C t$$

$$= V_C\cos\omega_C t + \frac{mV_C}{2}\cos(\omega_C + \omega_m)t + \frac{mV_C}{2}\cos(\omega_C - \omega_m)t$$

↓	↓	↓
Carrier	Upper sideband	Lower sideband

The separation of the AM signal is shown in Fig. 11.2

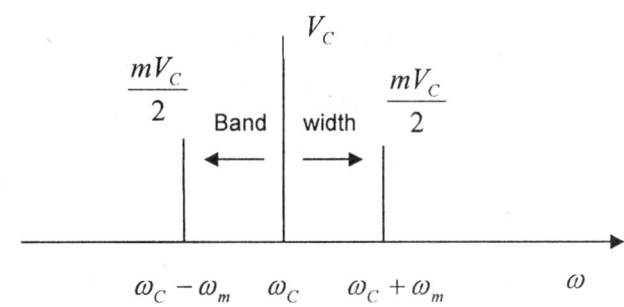

Fig. 11.2 Amplitude spectrum of the AM signal.

There is the carrier and two sidebands, viz., upper sideband (USB) and lower sideband (LSB). The carrier amplitude is the same as its unmodulated value V_C while the sidebands have an amplitude $\dfrac{mV_C}{2}$ each. USB is located at a frequency $\omega_C + \omega_m$ and the LSB is located at a frequency $\omega_C - \omega_m$ in the spectrum.

Now,   Power $\propto$ (Amplitude)^2.

Then,   carrier power $P_C \propto V_C^2$

USB power, $P_{USB} \propto \dfrac{m^2}{4} V_C^2$

LSB power, $P_{LSB} \propto \dfrac{m^2}{4} V_C^2$

$\therefore \qquad P_{USB} = P_{LSB} = \dfrac{m^2}{4} P_C .$

Total sideband power, $P_{SB} = P_{USB} + P_{LSB} = \dfrac{m^2}{2} P_C$ ... (11.8)

For a 100% modulated AM wave, $m = 1$ and $P_{SB} = \dfrac{P_C}{2}$. Then, with $m = 1$, total AM signal power

$$P_T = P_C + P_{SB} = P_C + \frac{P_C}{2} = \frac{3}{2} P_C \qquad\qquad ... (11.9)$$

Thus, for a 100% modulated AM wave the total signal power is three times the total sideband power.

In AM, the carrier convey no information. Useful information is in the sidebands. The bandwidth of the AM wave is $2\omega_m$. If the carrier is suppressed, we get double sideband suppressed carrier (DSBSC) signal. Again, both sidebands contain the same information. So, one sideband can be suppressed and one can be transmitted. This results in single sideband (SSB) transmission. Bandwidth is saved by half in SSB. But, at the same time half of the useful power is gone.

11.3.1 SQUARE-LAW MODULATOR

It utilizes the square-law I-V characteristics of a p-n junction diode. A schematic circuit diagram is shown in Fig. 11.3.

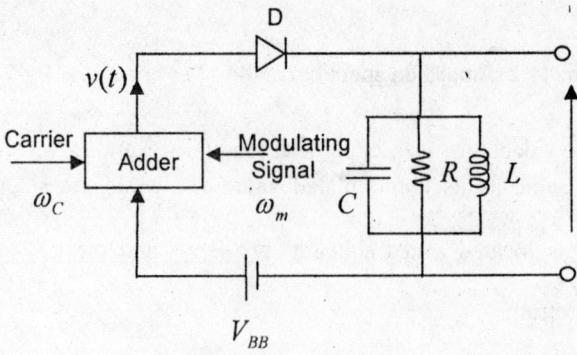

Fig. 11.3. Schematic circuit diagram of a square-law modulator.

Here, RLC forms a tank circuit resonant at the carrier frequency, ω_C. A typical current-voltage characteristics of the diode showing the square-law region is shown in Fig. 11.3.

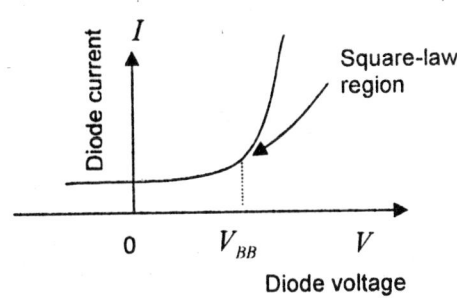

Fig. 11.4. $I - V$ characteristics of a pn-junction diode.

Let $\quad v_C(t) = V_C \cos\omega_C t$ be the carrier wave and $v_m(t) = V_m \cos\omega_m t$ be the modulating signal. Also, let V_{BB} be the dc bias voltage required for square-law operation of the diode.
Total voltage applied to the diode is

$$v(t) = v_C(t) + v_m(t) + V_{BB} \qquad \text{... (11.10)}$$

$v(t)$ is the output voltage of the adder adding the carrier, modulating signal and the dc bias voltage. Let the square-law characteristics be represented as

$$i = a_1 v(t) + a_2 v^2(t) \qquad \text{... (11.11)}$$

where i is the diode current, and a_1 and a_2 are constant coefficients. Substituting (11.10) in (11.11) we get

$$i = a_1 \left[V_C \cos\omega_C t + V_m \cos\omega_m t + V_{BB} \right]$$

$$+ \left[V_{BB}^2 + \frac{V_C^2}{2}(\cos 2\omega_C t + 1) + \frac{V_m^2}{2}(1 + \cos 2\omega_m t) \right.$$

$$+ 2V_{BB}V_C \cos\omega_C t + 2V_{BB}V_m \cos\omega_m t$$

$$\left. + V_C V_m \cos(\omega_C - \omega_m)t + V_C V_m \cos(\omega_C + \omega_m)t \right] \qquad \text{... (11.12)}$$

Thus, there are frequency components ω_C, ω_m, $2\omega_C$, $2\omega_m$, $\omega_C \pm \omega_m$ and dc present in the output current. The LCR tuned circuit has a bandpass response around ω_C with a bandwidth of $2\omega_m$. The only ω_C, and $\omega_C \pm \omega_m$ frequency components will be present at the output voltage $v_0(t)$ developed across the tuned circuit. Now, neglecting frequencies other than ω_C, and $\omega_C \pm \omega_m$ from (11.12), we get

$$v_0(t) = V_C(a_1 + 2a_2 V_{BB})\cos\omega_C t + a_2 V_C V_m 2\cos\omega_C t \cdot \cos\omega_m t$$

$$= V_C(a_1 + 2a_2 V_{BB})[1 + m\cos\omega_m t]\cos\omega_C t \qquad \ldots (11.13)$$

where $m = \dfrac{2a_2 V_m}{a_1 + 2a_2 V_{BB}}$ is the AM index. Equation (11.13) represents an AM signal.

11.3.2 AM SIGNAL GENERATION IN CLASS-C AMPLIFIER

A schematic circuit diagram of an amplitude modulated class-C amplifier is shown in Fig. 11.5. Here, the carrier wave is inductively coupled to a resonant circuit ($L_1 C_1$) which applies the carrier to the input of a CE amplifier. The $R_B C_B$ circuit provides necessary bias for class-C operation. The modulating signal, $v_m(t)$, applied to the collector of the transistor through a transformer T and hence it is called a collector modulator. The dc supply V_{CC} is also added with $v_m(t)$ and applied to the collector of the transistor. The RFC is a radio frequency choke which prevents the AM signal passing through the modulating signal source. In class-C operation, collector current flows in pulses for very short durations. Thus, the collector injects energy pulses into the output tank circuit ($L_2 C_2$) which tends to oscillate at its resonance frequency ω_C. This tank circuit oscillation is analogous with a pendulum driven by short impulses. The capacitor C_S provides a short circuit for the rf signal to ground.

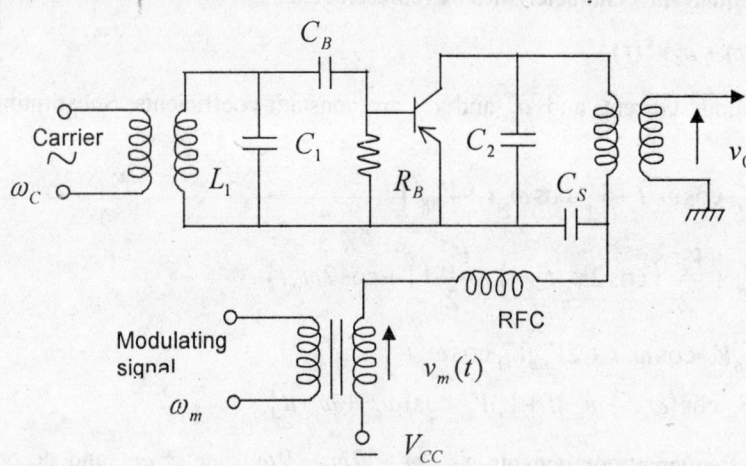

Fig. 11.5. A schematic circuit diagram of an amplitude- modulated class-C amplifier.

The output tank circuit ($L_2 C_2$) oscillation is of the form $V_l \cos\omega_C t$. The collector voltage (v_{ce}) at any time is given by

$$v_{ce} = V_{CC} + V_m \cos\omega_m t + V_l \cos\omega_C t \qquad \ldots (11.14)$$

where, $v_m(t) = V_m \cos\omega_m t$ is the modulating signal voltage. Equation (11.14) is written as

$$v_{ce} = V_{CC}(1 + m\cos\omega_m t) + V_t \cos\omega_c t \qquad \qquad \ldots (11.15)$$

where $m = \dfrac{V_m}{V_{CC}}$. The input carrier amplitude is adjusted to drive the transistor just into saturation

during the peak of the carrier cycle. Then, since $v_{ce} = 0$ at saturation, we have

$$V_t = V_{CC}(1 + m\cos\omega_m t) \qquad \qquad \ldots (11.16)$$

since $\cos\omega_c t = -1$ at saturation.

Now, $\quad v_{ce} = V_{CC}(1 + m\cos\omega_m t)(1 + \cos\omega_c t) \qquad \qquad \ldots (11.17)$

The average collector voltage over one carrier cycle is

$$\overline{v_{ce}} = \frac{1}{2\pi}\int_0^{2\pi} v_{ce}\, d\omega_c t = V_{CC}(1 + m\cos\omega_m t) \qquad \qquad \ldots (11.18)$$

The term $\cos\omega_m t$ does not vary appreciably over one carrier cycle and hence taken as constant in the integration. The output voltage $v_0(t)$ is the alternating part of v_{ce}.

Then, $\quad v_0(t) = v_{ce} - \overline{v_{ce}} = V_{CC}(1 + m\cos\omega_m t)\cos\omega_c t \qquad \qquad \ldots (11.19)$

This is the equation of an AM signal.

11.4 AM DETECTORS

In this section, we will describe the detectors and a receiver for amplitude modulated signals.

11.4.1 SQUARE-LAW DIODE DETECTOR

Detection is the process which extracts the modulating signal from the modulated wave. The amplitude modulated wave is represented by the voltage equation

$$v_{AM}(t) = V_C(1 + m\cos\omega_m t)\cos\omega_c t \qquad \qquad \ldots (11.20)$$

where m is the AM index. The circuit diagram of the square-law diode detector is shown in Fig.11.6.

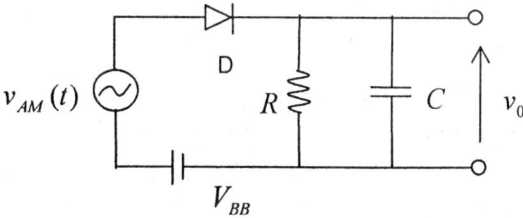

Fig. 11.6. Circuit diagram of the square-law diode detector.

The RC circuit acts as a low pass filter. A typical I-V characteristic is shown in Fig. 11.7.

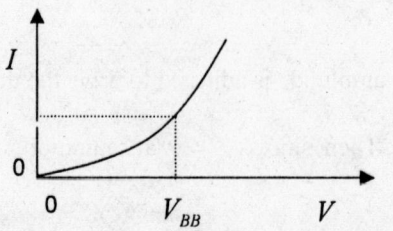

Fig. 11.7. Typical $I - V$ characteristics of the junction diode.

Let the square-law characteristics of the diode be represented by the equation

$$I = b_1 v(t) + b_2 v^2(t) \qquad \qquad \text{...(11.21)}$$

where b_1 and b_2 are constant coefficients. The diode input voltage

$$v(t) = V_{BB} + v_{AM}(t) \qquad \qquad \text{...(11.22)}$$

Simplifying eqn. (11.21), we get

$$
\begin{aligned}
I = b_1 & \Bigg[V_{BB} + V_C \cos\omega_C t + \frac{mV_C}{2}\cos(\omega_C + \omega_m)t \\
& + \frac{mV_C}{2}\cos(\omega_C - \omega_m)t \Bigg] + b_2 \Bigg[V_{BB}^2 + \frac{V_C^2}{2}\left(1 + \frac{m^2}{2}\right) \\
& + \frac{V_C^2}{2}\Bigg\{ \frac{m^2}{2}\cos 2\omega_m t + 2m\cos\omega_m t + \left(1 + \frac{m^2}{2}\right)\cos 2\omega_C t \\
& + m\cos(2\omega_C + \omega_m)t + m\cos(2\omega_C - \omega_m)t \\
& + \frac{m^2}{4}\cos 2(\omega_C - \omega_m)t + \frac{m^2}{4}\cos 2(\omega_C + \omega_m)t \Bigg\} + 2V_{BB}V_C \cdot \\
& \times \Bigg\{ \cos\omega_C t + \frac{m}{2}\cos(\omega_C + \omega_m)t + \frac{m}{2}\cos(\omega_C - \omega_m)t \Bigg\} \Bigg]
\end{aligned}
\qquad \text{...(11.23)}
$$

The high frequency components ω_C, $(\omega_C \pm \omega_m)$, $2\omega_C$, $2(\omega_C \pm \omega_m)$ and $(2\omega_C \pm \omega_m)$ are bypassed by the capacitor C. The low frequency components are dc, signal component (ω_m) and second harmonic distortion ($2\omega_m$). The signal component of output current is

$$I_m = b_2 V_C^2 m\cos\omega_m t \qquad \qquad \text{...(11.24)}$$

while the second harmonic component is

$$I_2 = b_2 \frac{m^2}{4} V_C^2 \cos 2\omega_m t \qquad\qquad \dots (11.25)$$

The second harmonic distortion is

$$D = \frac{\text{Amplitude of } I_2}{\text{Amplitude of } I_m} = \frac{m}{4} \qquad\qquad \dots (11.26)$$

To keep second harmonic distortion below 10%, we must keep $m < 40\%$. For a 100% modulated AM wave, $D = \dfrac{100}{4} = 25\%$. So, this detector is unsuitable for 100% modulated AM wave. Here, we have considered single tone modulation. For, multiple tone modulation, intermodulation distortion will appear.

11.4.2 ENVELOPE DECTETOR

Input voltage is assumed to be high so that the diode operates in the linear region of its I-V characteristics. The circuit diagram of the envelope detector is shown in Fig. 11.8

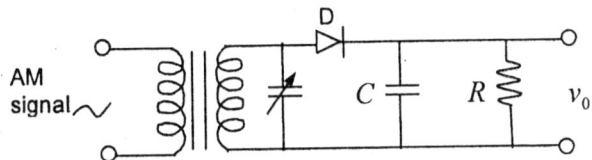

Fig. 11.8. Circuit diagram of the envelope detector.

The output capacitance C must be selected so that $\dfrac{1}{\omega_c C} \ll R$ and $\dfrac{1}{\omega_m C} \gg R$, where ω_c = angular frequency of the carrier wave and ω_m = angular frequency of the modulating signal. These conditions ensure that carrier frequency components are bypassed by C while modulating signal components produce a voltage drop at the output.

The diode conducts during the positive half cycle of the input rf signal. The capacitor C charges with a voltage equal to the input voltage less the diode drop. When the input signal decreases from its peak, the capacitor discharges through the resistor R. When the input signal rises, the condenser voltage also rises. If the time constant RC is chosen properly, the capacitor voltage will tend to follow the modulation envelope and the output voltage will become a replica of the modulating signal. Thus, AM signal is detected. Since the output voltage is proportional to the envelope of the input signal, it is called envelope detector.

CHOICE OF THE TIME CONSTANT (RC)

If the time constant (RC) becomes very small, the condenser discharge curve tends to be vertical. On the other hand, if the value of RC is very high, the discharge curve tends to be horizontal resulting in clippimg of the modulation envelope. So, we desire the time constant to be large enough subject to the condition that clipping does not occur.

The envelope voltage is expressed as

$$V(t) = V_C\left(1 + m\cos\omega_m t\right) \qquad \ldots (11.27)$$

$$\frac{dV(t)}{dt} = -V_C\, m\omega_m \sin\omega_m t \qquad \ldots (11.28)$$

At the beginning of discharge, the condenser is charged at the envelope voltage $V(t)$. So, the rate of discharge of the condenser is described as

$$\frac{dv_C(t)}{dt} = \frac{d}{dt}\left[V\,e^{-t/RC}\right]$$

$$= -\frac{v_C(t)}{RC} \qquad \ldots (11.29)$$

where V is the condenser voltage at the start of discharge as given by (11.27). $V(t)$ is taken as constant over one carrier frequency cycle. $v_C(t)$ is the condenser voltage at time t which is taken equal to $V(t)$ over one carrier cycle. Then,

$$\frac{dv_C(t)}{dt} = -\frac{V(t)}{RC}$$

$$= -\frac{V_C\left(1 + m\cos\omega_m t\right)}{RC} \qquad \ldots (11.30)$$

In order that the discharge curve follows the modulation envelope, we must have

$$\left|\frac{dv_C(t)}{dt}\right| = \left|\frac{dV(t)}{dt}\right| \qquad \ldots (11.31)$$

Using equation (11.28) and (11.30), we get

$$\frac{1}{RC} = \frac{m\omega_m \sin\omega_m t}{1 + m\cos\omega_m t} \qquad \ldots (11.32)$$

Now, the right side at (11.32) is maximum when

$$\frac{d}{dt}\left[\frac{m\sin\omega_m t}{1 + m\cos\omega_m t}\right] = 0 \qquad \ldots (11.33)$$

i.e., $\qquad \cos\omega_m t = -m \qquad \ldots (11.34)$

$$\therefore \qquad \frac{1}{RC} \geq \frac{m\omega_m}{\sqrt{1-m^2}} \qquad \ldots (11.35)$$

The greater than sign indicates that no clipping should occur. If $m \to 1$, then $RC \to 0$. But, very small value of RC indicates that carrier frequency components are not filtered out at the output. Hence, envelope detection is not suitable for 100% AM.

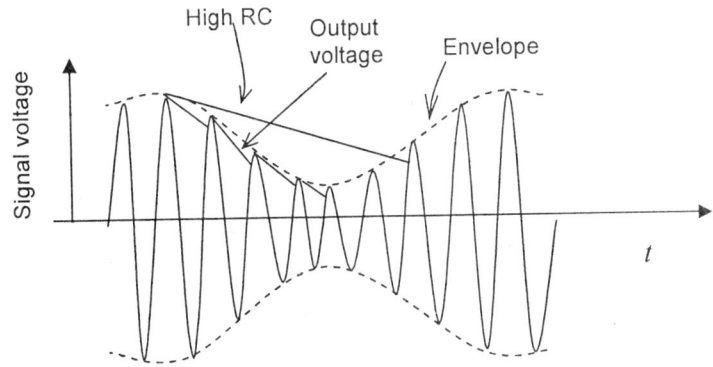

Fig. 11.9. Effect of the value of RC on the envelope detection process.

DETECTION EFFICIENCY

It is the ratio of the average load voltage to the peak rf voltage at the input. The effective diode voltage

$$v_d = V_C(1 + m\cos\omega_m t)\cos\omega_c t - V_0 \qquad \qquad \text{... (11.36)}$$

where V_0 is the average load voltage which is assumed to be constant over an rf cycle. The diode conducts when $v_d > 0$. Let the diode conduct from $\omega_c t = -\theta_1$ to $\omega_c t = \theta_1$ of an rf cycle. The rf cycle peaks at $\omega_c t = 0$. Now, the average diode current over a carrier frequency cycle is

$$I_{av} = \frac{1}{2\pi}\int_{-\pi}^{\pi} i_d(t)\, d\omega_c t$$

where $i(t)$ is the diode current. Also, $i_d(t) = \dfrac{v_d}{r_d}$, where r_d is the forward diode resistance.

Then,

$$I_{av} = \frac{1}{2\pi}\int_{-\theta_1}^{\theta_1} \frac{V_C(1 + m\cos\omega_m t)\cos\omega_c t - V_0}{r_d}\, d\omega_c t$$

$$= \frac{1}{\pi r_d}\left[V_C(1 + m\cos\omega_m t)\sin\theta_1 - V_0\theta_1\right] \qquad \text{... (11.37)}$$

since the dc load voltage remains nearly constant during discharge, we can take

$$V_0 = V_C(1 + m\cos\omega_m t)\cos\theta_1 \qquad \qquad \text{... (11.38)}$$

Here, $\cos\omega_m t$ remains almost constant over a carrier cycle since $\omega_c \gg \omega_m$. Now, the average load voltage is calculated as

$$V_0 = I_{av}R = \frac{R}{\pi r_d}V_C(1 + m\cos\omega_m t)\left[\sin\theta_1 - \theta_1\cos\theta_1\right] \qquad \text{... (11.39)}$$

using equation (11.37) and (11.38). From equation (11.38) and (11.39), we write

$$\cos\theta_1 = \frac{R}{\pi r_d}\left[\sin\theta_1 - \theta_1\cos\theta_1\right] \qquad \therefore \qquad \frac{\pi r_d}{R} = \tan\theta_1 - \theta_1$$

$$\therefore \qquad \theta_1 = \left(\frac{3\pi r_d}{R}\right)^{\frac{1}{3}} \qquad\qquad\qquad \dots (11.40)$$

The detector efficiency

$$\eta = \frac{V_0}{V_C\left(1 + m\cos\omega_m t\right)} = \cos\theta_1$$

$$\simeq 1 - \frac{\theta_1^2}{2} = 1 - \frac{1}{2}\left(\frac{3\pi r_d}{R}\right)^{\frac{2}{3}} \qquad\qquad \dots (11.41)$$

If the time constant is not selected properly, there will be distortion of the detected signal. Since, the diode I-V characteristic is not perfectly linear, there can be some distortion of the detected signal also.

11.4.3 SUPERHETERODYNE RADIO RECEIVER

Superheterodyne radio receiver was designed to overcome the following problems of the tuned radio frequency receiver:
- High gain at high frequencies may lead to spurious oscillations in rf amplifiers.
- The Q-value of the tuned circuits do not increase with the signal frequency adequately so that the effective bandwidth of the tuned amplifiers, $\Delta f = f_0/Q$, increases with frequency. As a result, adjacent channels are picked up at high frequencies producing interference in desired channels.

A schematic circuit diagram of the superheterodyne receiver is in Fig. 11.10

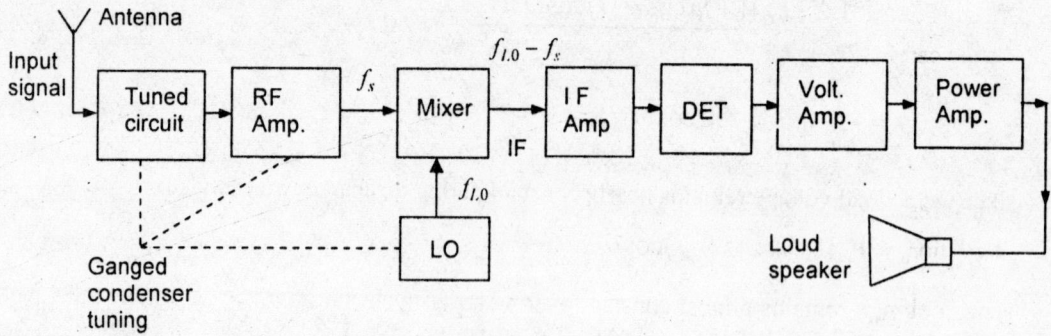

Fig. 11.10. Schematic circuit diagram of the superheterodyne radio receiver.

The desired radio channel is selected by the tuned circuit and amplified by the tuned rf amplifier. The adjacent channels are rejected by the tuned circuits which are tuned to the desired channel carrier frequency. The received signal is AM whose carrier frequency lies in the ranged 540 kHz-1650 kHz for medium waveband.

The mixer is a nonlinear device like bipolar transistor or FET which has a square-law input-output characteristics. The mixer produces multiplication of the amplified signal (f_s) and the local oscillator signal (f_{LO}) and produces a difference frequency ($f_{LO} - f_s$) signal. This difference frequency signal is known as the intermediate frequency (IF) signal. The tuned circuit, rf amplifier and the local oscillator are tuned by using a ganged condenser so that the intermediate frequency, f_{IF}, is held constant. This IF has a value of 455 kHz. The local oscillator is a rf oscillator like Hartley oscillator. Every channel is downconverted to the constant IF signal. The IF contains the same amplitude modulation as the input signal. The IF signal is then detected in envelope diode detector which produces a replica of the amplitude modulation at its output. The detected signal is amplified by voltage and power amplifier and then fed to the loud speaker.

As an example, if $f_s = 800$ kHz, we must have $f_{LO} = f_s + f_{IF} = 800 + 455 = 1255$ kHz. The LO frequency (f_{LO}) should not be selected as ($f_s - f_{IF}$) since it makes the ratio of maximum to minimum capacitance variation of the ganged condenser quite large. Taking $f_{LO} = f_s + f_{IF}$ makes the ganged capacitance variation ratio small.

There is another problem called image frequency rejection. Suppose a signal of frequency $f_s + 2f_{IF}$ appears at the input of the receiver along with the desired channel f_s. Then, this undesired channel will produce the same IF since $(f_s + 2f_{IF}) - f_{LO} = f_{IF}$. Such signal will not be suppressed at the intermediate stage and are called image signal. This image signal is rejected by the tuned circuit and the rf amplifier at input of the superheterodyne radio receiver.

11.5 FREQUENCY MODULATION

Frequency modulation is the process in which the frequency of a high frequency carrier wave is changed in accordance with the instantaneous value of the modulating wave.

Let $\qquad v_C = V_C \cos \omega_C t$ ... (11.42)

be the carrier wave and let

$$v_m = V_m \cos \omega_m t \qquad\qquad\qquad ... (11.43)$$

be the modulating signal. ω_C and ω_m are the frequencies of the carrier wave and the modulating signal respectively. $\omega_C \gg \omega_m$. If $\omega(t)$ be the instantaneous frequency of the FM signal then, by definition,

$$\left[\omega(t) - \omega_C\right] \propto v_m(t)$$

$\therefore \qquad \omega(t) = \omega_C + K_f v_m(t)$ ... (11.44)

where K_f is the constant of proportionality. Equation (11.44) can be recast as

$$\omega(t) = \omega_C + \Delta \cos \omega_m t \qquad \qquad \ldots (11.45)$$

where $\Delta = K_f V_m$ = peak frequency deviation in radian.

The phase of the FM signal is

$$\phi(t) = \int_0^t \omega(t)\, dt = \omega_C t + \frac{\Delta}{\omega_m} \sin \omega_m t$$

$$= \omega_C t + m_f \sin \omega_m t \qquad \qquad \ldots (11.46)$$

where $m_f = \dfrac{\Delta}{\omega_m}$ is the index of FM.

The FM signal can now be represented as

$$v_{FM}(t) = V_C \cos \phi(t)$$

$$= V_C \cos(\omega_C t + m_f \sin \omega_m t)$$

$$= V_C \cos \omega_C t \cos(m_f \sin \omega_m t) - V_C \sin \omega_C t \sin(m_f \sin \omega_m t) \qquad \ldots (11.47)$$

Now, $\quad \cos(m_f \sin \omega_m t) = J_0(m_f) + 2 \displaystyle\sum_{n=1}^{\infty} J_{2n}(m_f) \cos 2n\omega_m t \qquad \qquad \ldots (11.48)$

and $\quad \sin(m_f \sin \omega_m t) = 2 \displaystyle\sum_{n=1}^{\infty} J_{2n-1}(m_f) \sin(2n-1)\omega_m t \qquad \qquad \ldots (11.49)$

Here, $J_n(m_f)$ is the Bessel function of the first kind of order n and argument m_f.
Now eqn. (11.47) can be recast as

$$v_{FM}(t) = V_C \left[J_0(m_f) \cos \omega_C t + \sum_{n=1}^{\infty} J_{2n}(m_f)\{\cos(\omega_C + 2n\omega_m)t + \cos(\omega_C - 2n\omega_m)t\} \right.$$

$$\left. - \sum_{k=1}^{\infty} J_{2k-1}(m_f)\{\cos(\omega_C - (2k-1)\omega_m)t - \cos(\omega_C + (2k-1)\omega_m)t\} \right] \quad \ldots (11.50)$$

Equation (11.50) indicates that the FM signal spectrum has the carrier as well as infinite number of sidebands. The carrier amplitude is proportional to $J_0(m_f)$ while the n-th order sideband amplitude is proportional to $J_n(m_f)$.

Theoretically, the bandwidth of the FM signal seems to be infinite. But, in practice, the sideband amplitude $J_n(m_f)$ becomes insignificant for $n > (1 + m_f)$. The sidebands have a spacing of $f_m \left(= \dfrac{\omega_m}{2\pi} \right)$ in the frequency scale. The thumb rule which gives the bandwidth of the M signal is

$$B = 2(1 + m_f) f_m \text{ in Hz} \qquad \qquad \ldots (11.51).$$

This is known as Carson's rule. This is the bandwidth required in order to have a faithful reproduction of an FM signal without any appreciable loss of information. Equation (11.51) can be written as

$$B = 2(f_m + \delta) \qquad\qquad \dots (11.52)$$

where $\delta = f_m m_f$

$$= \frac{\omega_m m_f}{2\pi} = \frac{\Delta}{2\pi}$$

δ is the peak frequency deviation in Hz.

If $m_f \gg 1$, $B = 2\delta$.

If $m_f \ll 1$, $B = 2f_m$. So, when the FM index is small enough, the FM signal bandwidth is only $2f_m$. There exists only one upper sideband and one lower sideband effectively. The FM is said to be a narrow band FM. It is similar with the AM signal relating to bandwidth. The two sidebands in narrow band FM differ in phase by π radian in the case of cosine modulation. But, the two sidebands in the AM signal are in phase in cosine modulation.

Distinction between AM and FM

(i)   In AM there are three frequencies in the spectrum and its bandwidth is $2f_m$, where f_m is the modulating signal frequency. But, in FM there are many sidebands in general and its bandwidth is greater than $2f_m$ except in the special case of $m_f \ll 1$.

(ii)   In AM, the sideband amplitude is proportional to AM index. But, in FM, the sideband amplitude is proportional to $J_n(m_f)$, where $J_n(m_f)$ is the Bessel function of the first kind of order n and argument m_f.

(iii)   In AM, the carrier power remains unchanged due to modulation. But, in FM the carrier power changes due to modulation and it is proportional to $J_0^2(m_f)$.

(iv)   In AM, the total signal power increases due to modulation. But, in FM, the total power remains constant before and after modulation.

(v)   For certain values of FM index (m_f), the FM carrier power can be zero when the total power is in the sidebands. But, in AM, the carrier power is constant.

(vi)   The phase relationship between the corresponding sidebands in AM and FM are different.

The fact that the total power in an FM signal is equal to the unmodulated carrier power follows from the identity.

$$J_0^2(m_f) + 2\sum_{n=1}^{\infty} J_n^2(m_f) = 1 \qquad\qquad \dots (11.53)$$

11.5.1 FM GENERATION

An FM signal can be generated by varying the reactance of an active device by the application of a voltage and this reactance variation is used to change the frequency of a tuned oscillator. The corresponding modulator is called a reactance modulator. The circuit diagram of a reactance modulator using a JFET is shown in Fig. 11.11.

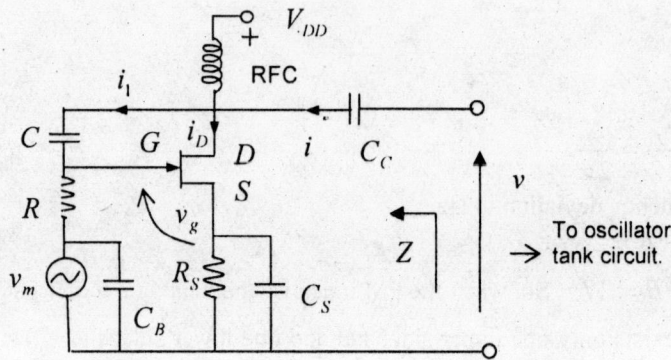

Fig. 11.11. Schematic circuit diagram of a JFET reactance modulator.

The radio frequency choke blocks the passage of rf current through the dc supply source, V_{DD}. C_B is the bypass capacitor which bypasses the rf signal to ground so that the rf signal does not pass through the modulating signal source. R_S, C_S combination provides self bias to the junction field effect transistor. C_C is the coupling capacitor.

Let us now find the impedance presented by this circuit to the tank circuit of the oscillator. For this, let us apply a rf voltage v at the output which produces an output current i. Then, the impedance presented by this circuit is

$$Z = \frac{v}{i} \qquad \qquad \dots (11.54)$$

Now, $i = i_D + i_1$ $\qquad \qquad \dots (11.55)$

where i_D is drain current and i_1 is the current through the capacitor C.

The gate voltage measured with respect to the source is

$$v_g = \frac{vR}{R + \dfrac{1}{j\omega C}} \qquad \qquad \dots (11.56)$$

Again, $i_D = g_m v_g$

$$= \frac{g_m R v}{R - \dfrac{j}{\omega C}} \qquad \qquad \dots (11.57)$$

where g_m is the transconductance of the JFET.

Now, $i_1 = \dfrac{v}{R + \dfrac{1}{j\omega C}} = \dfrac{v}{R - \dfrac{j}{\omega C}}$ ... (11.58)

From (11.55) we write

$$i = \dfrac{v}{R - \dfrac{j}{\omega C}}(1 + g_m R)$$ ... (11.59)

Then, $Z = \dfrac{R - \dfrac{j}{\omega C}}{1 + g_m R}$ ... (11.60)

using eqn. (11.54).

Let $\dfrac{1}{\omega C} \gg R$. Then,

$$Z = -\dfrac{j}{\omega C(1 + g_m R)} = \dfrac{1}{j\omega C(1 + g_m R)}$$

$$= \dfrac{1}{j\omega C_{eq}}$$ ... (11.61)

where $C_{eq} = C(1 + g_m R)$ is the equivalent capacitance presented by the circuit of Fig. 11.11. It is known that g_m varies linearly with the gate voltage and hence with v_m. The resonant frequency of the oscillator tank circuit is

$$\omega = \dfrac{1}{\sqrt{L(C_0 + C_{eq})}}$$ ... (11.62)

where L is the inductance and C_0 is the capacitance of the tank circuit. C_{eq} is the change in capacitance produced by the circuit.

Now,

$$\omega = \dfrac{1}{\sqrt{LC_0}}\left(1 + \dfrac{C_{eq}}{C_0}\right)^{-\frac{1}{2}}$$

$$\cong \omega_0\left(1 - \dfrac{C_{eq}}{2C_0}\right)$$

$$= \omega_0 - \dfrac{\omega_0}{2} \cdot \dfrac{C}{C_0}(1 + g_m R)$$ ... (11.63)

Equation (11.63) shows that the oscillator frequency ω varies linearly with g_m and hence with v_m. This yields frequency modulation of the carrier generated by the oscillator.

11.5.2 FOSTER-SEELY DISCRIMINATOR

Discriminator is a device which produces an output voltage proportional to the frequency deviation of the input signal from a nominal frequency called as the crossover frequency of the discriminator. This device demodulates an FM signal. Two types of FM demodulators are common – (i) slope detector and (ii) FM discriminator. In slope detection, the carrier frequency of the FM signal falls at one side of the tuned circuit response curve and there occurs a linear voltage variation with the frequency of the FM signal. In the FM discriminator, dc cancellation occur at the output and the response has a wide range of linearity. The FM discriminator is known as Foster-Seely discriminator after the name of its inventor.

The circuit diagram of the Foster-Seely discriminator is shown in Fig. 11.12 below. RFC is the radio frequency choke which blocks the rf from passing to the output.

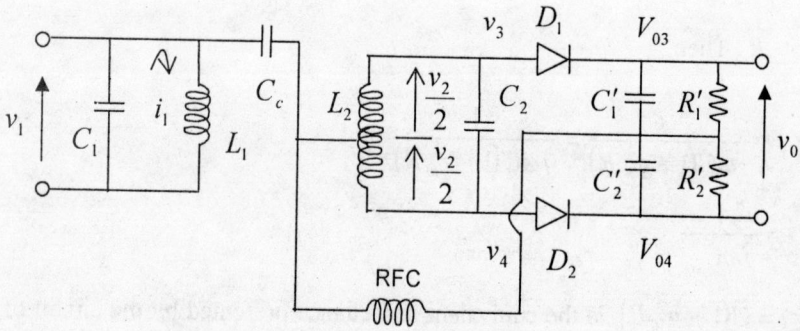

Fig. 11.12. Schematic circuit diagram of the Foster-Seely discriminator.

The simplified ac equivalent circuit of the discriminator is shown in Fig. 11.13.

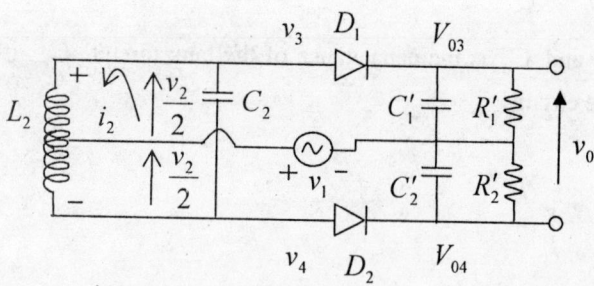

Fig. 11.13. simplified AC equivalent circuit of Foster-Seely discriminator

Let the input FM signal be v_1 which is applied to the primary tuned circuit $(L_1 C_1)$. If i_1 be the current through the inductance L_1 then

$$v_1 \cong j\omega L_1 i_1 \qquad \qquad \dots (11.64)$$

$$\therefore \qquad i_1 = \frac{v_1}{j\omega L_1} \qquad\qquad\qquad \dots (11.65)$$

Here, we have neglected the coil resistance R_1 of L_1.

From the circuit of Fig. 11.12, we write

$$v_0 = V_{03} - V_{04} \qquad\qquad\qquad \dots (11.66)$$

where V_{03} and V_{04} are voltages at the output of diodes D_1 and D_4 respectively. v_0 is the discriminator output voltage.

Assuming linear diode detectors D_1 and D_2, we write

$$V_{03} = k|V_3| \qquad\qquad\qquad \dots (11.67)$$

and $\qquad V_{04} = k|V_4| \qquad\qquad\qquad \dots (11.68)$

where k is the constant of the detector. v_3 and v_4 are diode input voltages for D_1 and D_2 respectively. Now,

$$v_3 = v_1 + \frac{v_2}{2} \qquad\qquad\qquad \dots (11.69)$$

and $\qquad v_4 = v_1 - \frac{v_2}{2} \qquad\qquad\qquad \dots (11.70)$

v_2 is the full voltage of the centre-tapped secondary circuit.

The voltage equation of the secondary tuned circuit $(L_2 C_2 R_2)$ is

$$R_2 i_2 + j\omega L_2 i_2 + \frac{1}{j\omega C_2} i_2 = M \frac{di_1}{dt} \qquad\qquad\qquad \dots (11.71)$$

where i_2 is the rf current in the secondary tuned circuit, M is the mutual inductance of the coils and ω is the input signal frequency in radian. The resonance frequency of the secondary tank circuit is ω_0 where

$$\omega_0^2 = \frac{1}{L_2 C_2} \qquad\qquad\qquad \dots (11.72)$$

ω_0 coincides with the FM carrier frequency. Now, eqn. (11.71) can be simplified as

$$i_2 \left[R_2 + j\omega L_2 \left(1 - \frac{\omega_0^2}{\omega^2} \right) \right] = j\omega M i_1$$

or, $\qquad R_2 i_2 [1 + jx] = j\omega M i_1 \qquad\qquad\qquad \dots (11.73)$

where $x = \left(\dfrac{\omega}{\omega_0} - \dfrac{\omega_0}{\omega} \right) Q_2$ and $Q_2 = \dfrac{\omega_0 L_2}{R_2}$ is the Q-factor of the secondary tuned circuit.

Substituting for i_1 from (11.65) in (11.73), we get

$$i_2 = \frac{Mv_1}{L_1} \frac{1}{R_2(1 + jx)} \qquad \qquad \qquad \ldots (11.74)$$

Now, $\quad v_2 \approx j\omega L_2 i_2$

$$\cong j\frac{Q_2}{L_1} Mv_1 \frac{1}{1 + jx}$$

putting $\dfrac{\omega L_2}{R_2} \cong Q_2$. Then,

$$v_2 = \frac{j\alpha}{1 + jx} v_1 \qquad \qquad \qquad \ldots (11.75)$$

where $\alpha = \dfrac{MQ_2}{L_1}$.

Now, $\quad v_3 = v_1 + \dfrac{1}{2} \dfrac{j\alpha}{1 + jx} v_1$

$$= \frac{v_1}{2} \frac{[2 + j(2x + \alpha)]}{1 + jx} \qquad \qquad \qquad \ldots (11.76)$$

and $\quad v_4 = v_1 - \dfrac{1}{2} \dfrac{j\alpha}{1 + jx} v_1$

$$= \frac{v_1}{2} \frac{[2 + j(2x - \alpha)]}{1 + jx} \qquad \qquad \qquad \ldots (11.77)$$

$\therefore \qquad v_0 = k\left(|v_3| - |v_4| \right)$

$$= \frac{k}{2} |v_1| f(x) \qquad \qquad \qquad \ldots (11.78)$$

where $\quad f(x) = \dfrac{\sqrt{4 + (2x + \alpha)^2} - \sqrt{4 + (2x - \alpha)^2}}{\sqrt{1 + x^2}} \qquad \qquad \ldots (11.79)$

The plot of the function $f(x)$ with x is shown in Fig. 11.14 with α as a parameter. The plot is nearly linear over a certain range about the origin $x = 0$ (i.e., $\omega = \omega_0$). The discriminator operation should be confined over this linear region.

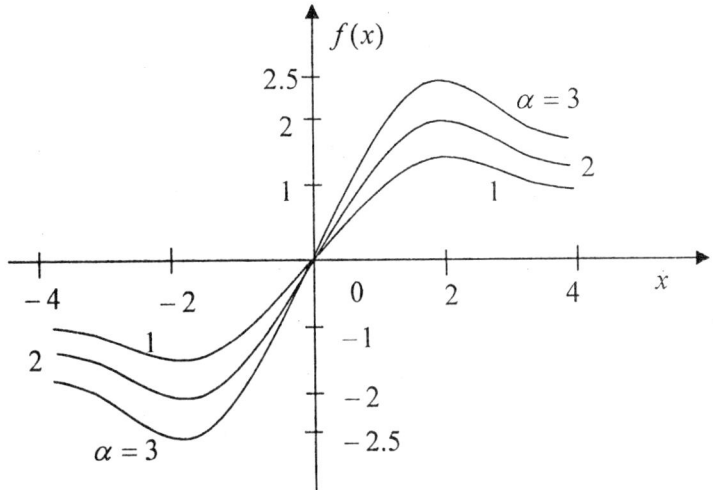

Fig. 11.14. Plot of the function $f(x)$ with x using α as a parameter.

11.6 RADAR

Radar is an acronym for RAdio Detection And Ranging. It is a device which can measure the range, velocity, altitude and even the direction of a target by sending pulses or CW signals. The target reflects the incident pulses or CW signal which is processed in a radar receiver and the desired measurement is made. In the case of pulsed radar, the time of transit of the pulse in forward and backward motion is measured and the range of the target is measured. In Doppler radar, the change in frequency of the reflected pulse from the moving target is measured which gives a measure of the target velocity. The radar has various kinds of uses such as in

(i)   Navigation – positions of land, mountain, ocean, river, buildings, forests etc. can be detected from a flying aircraft with the help of a radar. In ship, information of other ships in the ocean, presence of buoys, distance of land etc. can be obtained from a radar.

(ii)   Military – radar is used in bombing planes, guiding missiles, detection of submarines and attacking hostile aircrafts. This detection process is not affected by poor visibility or darkness.

(iii)   Weather forecasting – presence of cloud, location of cyclones and storms, turbulence in upper atmosphere can be detected by a radar. Radars are vital for air traffic control. The landing of an aircraft is precisely controlled by the radar.

A schematic block diagram of a pulsed radar is shown in Fig. 11.15. The pulse generator in Fig. 11.15 generates rf pulses of short duration. Duplexer is a switch which connects the pulse generator with the antenna during transmission. The same antenna is used for transmission and reception. During reception, the duplexer disconnects the transmitter from the antenna but connects the antenna with the receiver. The receiver must be a low noise, wideband receiver to handle short pulses. The output is displayed on a cathode ray tube (CRT) indicator. The CRT sweep and the pulse generator are synchronized.

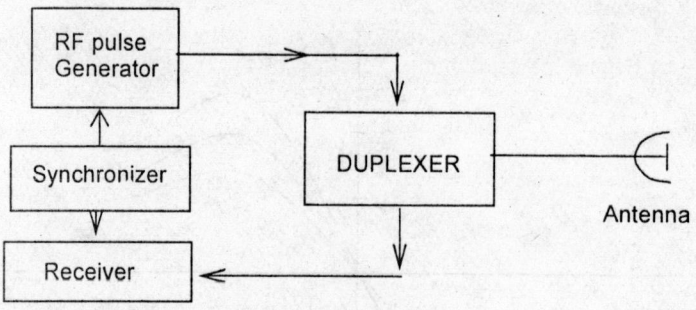

Fig. 11.15. Schematic block diagram of a pulsed radar.

Depending upon the operating frequency of the radar, different bands have been designated. These are UHF, L, S, C, X, Ku, K, Ka etc. The ultra high frequency (UHF) band extends from 300 MHz-1 GHz, L-band extends from 1 GHz-1.5 GHz, S-band ranges from 1.5 GHz-3.9 GHz, C-band extends from 3.9 GHz-8 GHz, X-band ranges from 8 GHz-12.4 GHz, Ku band extends from 12.4 GHz-18 GHz, K-band ranges from 18 GHz-26.5 GHz and Ka band has a range from 26.5 GHz-40 GHz. There are radars which operate at higher bands in the mm-wave region also.

11.6.1 RANGE EQUATION OF RADAR

Let us derive an expression for the calculation of maximum range of the target which can be detected by the radar.

Let P_t be the peak power of the transmitter pulse which is fed to the antenna. The antenna is assumed to be isotropic. The power density of the radiation at a distance r from the antenna is

$$P' = \frac{P_t}{4\pi r^2} \qquad \qquad \dots (11.80)$$

But, the radar antenna is directional in practice. The power gain (A_p) of a directional antenna is defined as the ratio of the power radiated by an isotropic antenna to the power radiated by the directional antenna in the direction of maximum radiation in order to produce a given field at a given distance. So, the actual power density at a distance r from the directional antenna will be greater than P' and will be multiplied by A_p.

∴ The power density at the target is

$$P'' = A_p \cdot P' \qquad \qquad \dots (11.81)$$

Now, how much power will be intercepted by the target from the incident power will depend upon the effective area of the target which is known as radar cross section (S). Let us now define radar cross section. It is such an area that if the power contained in this area (S) of the incident wavefront were radiated by an isotropic antenna placed on the target, the field produced at the receiving antenna will be the same as the actual field produced by the echo signal.

The power intercepted by the target is then

$$P = P'' \cdot S = A_p S P' \qquad \qquad \dots (11.82)$$

The target is assumed to be an isotropic radiator.

The power density of the radiation received by the receiving antenna is

$$P_r = \frac{P}{4\pi r^2} = \frac{A_p S P_t}{\left(4\pi r^2\right)^2} \qquad \qquad \dots (11.83)$$

using eqn. (11.82) and (11.80).

The receiving antenna intercepts a fraction of the incident power. The capture area of the antenna is

$$A_0 = kA \qquad \qquad \dots (11.84)$$

where k is a constant which depends upon the structure of the antenna and A is the geometrical mouth area or, aperture area of the antenna. The received power is

$$P_R = P_r \cdot A_0$$

$$= \frac{A_p S A_0 P_t}{\left(4\pi r^2\right)^2} \qquad \qquad \dots (11.85)$$

The gain of the receiving antenna is

$$G_R = 4\pi \frac{A_0}{\lambda^2} \qquad \qquad \dots (11.86)$$

where λ is the wavelength of the radiation. If we use the same antenna for transmission and reception then $G_R = A_p$. Under this condition, eqn. (11.85) can be recast as

$$P_R = \frac{S A_0^2 P_t}{4\pi r^4 \lambda^2} \qquad \qquad \dots (11.87)$$

using (11.86). If P_{min} be the minimum power detectable by the receiver then the range r becomes a maximum, say, r_{max}. Equation (11.87) can then be written as

$$P_{min} = \frac{S A_0^2 P_t}{4\pi r_{max}^4 \lambda^2} \qquad \qquad \dots (11.88)$$

Again, substituting for A_0 in terms of G_R from (11.86) we get from (11.88)

$$P_{min} = \frac{S A_p^2 \lambda^2 P_t}{(4\pi)^3 r_{max}^4} \qquad \qquad \dots (11.89)$$

The maximum range is then given by

$$r_{max} = \left[\frac{S A_0^2 P_t}{4\pi \lambda^2 P_{min}} \right]^{1/4} \qquad \qquad \dots (11.90)$$

This is the range equation of radar.

The implications of the range equation are the following:

(i) $r_{max} \propto P_t^{1/4}$. Thus, to double the maximum range the transmitter power has to be increased 16 times.

(ii) $r_{max} \propto \dfrac{1}{\sqrt{\lambda}}$. So, reducing the wavelength of the radiation, or, equivalently increasing the frequency of radar, r_{max} can be increased. Again, the antenna beam width $\propto \dfrac{\lambda}{D}$, where D = diameter of the parabolic antenna. So, reducing λ we get a narrow beam width. Narrow beam makes it possible to discriminate between adjacent multiple targets. But, it takes a long time to search a wide portion of the space for targets.

(iii) $r_{max} \propto P_{min}^{-1/4}$. If the gain of the receiver is increased, P_{min} can be reduced. But, at the same time the receiver becomes more likely to be affected by interference.

Now, we consider an interesting question. **Why are the ground-hopping military aircrafts not detected by radars?**

It is known that military planes fly very close to ground just grazing the buildings of city. The reason is that due to the finite conductivity of the ground, there occur some reflection of the radiated energy of the antenna from the ground. This produces an interference pattern in the antenna radiation. As a result, the lowest lobe of the antenna is a few degree above the horizontal plane. So, there is a zone just above the ground where there is no antenna lobe. This is the null zone. If the military aircraft fly close to ground and falls in this null zone it will not be intercepted by any antenna lobe and hence will not be detected by the radar until it is very close to the radar. By that time, the military operation (e.g. bombing) will be carried out. This is shown in Fig. 11.16.

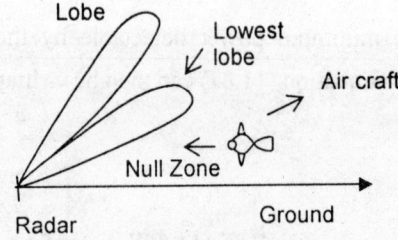

Fig. 11.16. Military air-craft flying in the null zone and undetected by radar.

There is a minimum value of radar range also. This follows from the fact that the receiver is disconnected from the antenna during the transmission of the pulse. If the reflected echo signal comes back during this period of disconnection, it will not be received. If the pulse width is 1 μsec. then it corresponds to a signal path of $\left(10^{-6} \times 3 \times 10^8\right) m = 300\,m$. So, any target lying within a distance of $(300/2 =) 150\,m$ will not be detected by the radar.

11.6.2 MOVING TARGET INDICATOR RADAR

Static objects like mountains, forests, buildings etc. produce reflections of the radar pulse which produce permanent echoes on the radar screen. This is known as ground clutter. If the desired echo originating from a moving target is weak then it will be difficult to detect it in presence of ground clutter. One way to detect the moving target echo on the radar screen is to suppress the ground clutter.

Since the objects like mountains, forests, buildings etc. are stationary, echoes resulting from them will have the some phase with respect to the transmitted pulse for successive pulses. But, when the target is moving, the phase of the echo changes from pulse to pulse continuously since the distance of the target from the radar is changing continuously. This continuous change in phase of the echo signal gives rise to a change in frequency of the echo. Suppose that the target moves a distance ΔD in time Δt. This produces a phase change $\Delta \psi$ of the echo during its to and fro motion where

$$\Delta \psi = \frac{2\pi}{\lambda} \cdot 2 \Delta D \qquad \qquad \dots (11.91)$$

Here, λ is the wavelength of the transmitted pulse. This gives rise to a frequency change of the echo signal from that of the transmitted pulse equal to Δf where

$$\Delta f = \frac{1}{2\pi} \frac{\Delta \psi}{\Delta t} = \frac{2}{\lambda} V_r \qquad \qquad \dots (11.92)$$

Here, V_r is the radial velocity of the target. This change in frequency of the reflected signal from a moving target is known as Doppler effect. This fact that the frequency and hence the phase of the echo changes from pulse to pulse while the phase of the echo from stationary objects remains fixed with respect to the transmitted pulse for successive pulses, gives the idea of ground clutter suppression. Consider two successive echoes. If they are subtracted then they cancel for stationary objects. But, for a moving target they will not cancel, in general, due to different phases.

11.6.3 CW RADAR

The CW radar transmits CW sign is from the transmitter. It can

(i)   detect the presence of a moving target, and
(ii)   measure speeds of moving targets.

The moving target produces a Doppler shift in frequency of the reflected signal. The signals at the transmitter frequency and the Doppler shifted frequency are mixed at the receiver to generate a difference frequency signal. The magnitude of this difference frequency gives a measurement of radial velocity of the target. A schematic circuit of the CW radar is shown in Fig. 11.17.

CW radar is limited in maximum transmitted power. It can not measure the range of the target since it can not realize which particular cycle of transmitted oscillation is represented by a received cycle of oscillation. So, the time difference between transmitted and received cycles can not be determined. It can be used to detect moving mass of people, speeding vehicles, speeds of aircraft etc. It has no minimum limit on range because the receiver is connected with the antenna at all times.

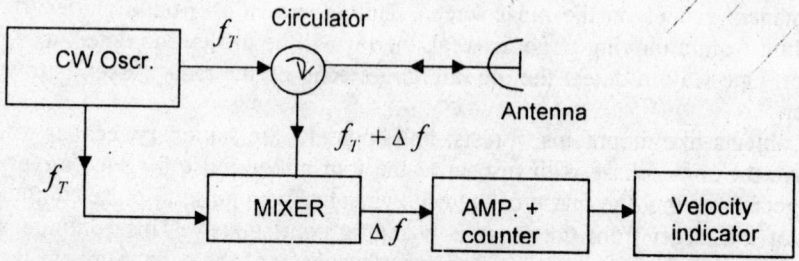

Fig. 11.17. Schematic circuit diagram of a CW radar.

11.6.4 FM RADAR

In FM radar, the transmitter frequency is modulated. The modulation can be by a sinusoidal signal, a sawtooth wave or any other waveform. The transmitted signal takes sometime to reach the target and come back to the receiver. During this time interval, the transmitter frequency has changed considerably. Now, if the reflected wave and the transmitter signal at any instant of time are mixed in a mixer, a difference frequency signal is generated. The magnitude of difference frequency is proportional to the time taken by the transmitted signal to make a round trip passage. So, by measuring the difference frequency, we get an estimation of range of the target. The target can be fixed or in motion. However, the Doppler shift in frequency due to target motion can be neglected in comparison with the large frequency shift produced by frequency modulation. The main application of FM radar is as FM altimeter. The height of an aircraft above ground can be measured by this FM radar. The incapability of a CW radar for range measurement is overcome by FM radar. It is advantageous to pulsed radar in the sense that it can measure short heights since there is no limitation on minimum range. A block diagram of the FM radar is shown in Fig. 11.18.

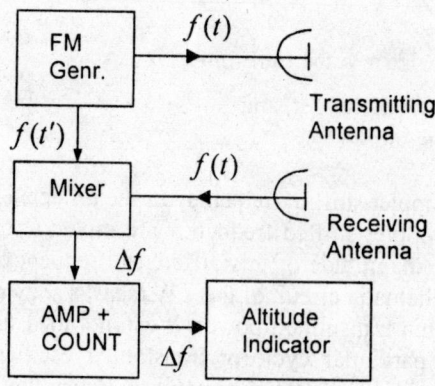

Fig. 11.18. Block diagram of an FM radar.

11.7 TELEVISION

The word 'Television' means 'to see at a distance'. Thus, what is happening at a distant point can be visualized sitting at a different location. In television there are two basic informations – viz., picture and sound. Now, how can we transmit picture and sound from one place to another? For this we must convert both the picture and the sound into their corresponding electrical signals. Brightness information of the picture is converted into an electrical signal by using photoelectric effect. In photoelectric effect photo (i.e., light) is converted into an electrical signal. A static picture has a spatial variation of brightness. But, moving picture will have both spatial and time variations of brightness. The brightness variations of all points of a picture can not be transmitted simultaneously because it will require so many channels. To avoid this problem, a picture is sequentially scanned and the brightness information of a particular point is converted into the electrical signal at a given time. The picture produces an optical image on a photosensitive surface resulting in photoelectric emission. The photoelectric surface is scanned by an electron beam horizontally from left to right and vertically from top to bottom. The electron beam is cut-off during horizontal and vertical fly back. In the reception process, a similar scanning of the picture tube is done by an electron beam which makes the picture on the TV receiver screen. The scanning process in transmission and reception are synchronized.

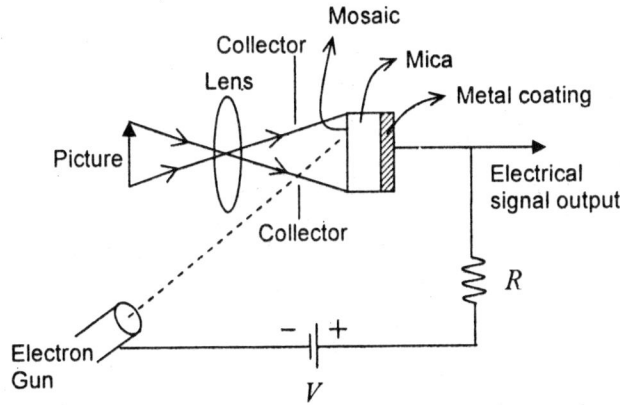

Fig. 11.19. Schematic diagram of Iconoscope.

A basic device which converts picture information into an electrical signal in Iconoscope as shown in Fig. 11.19. On one side of a mica sheet photo-sensitive compound of caesium-silver in the from of globules is deposited. While on the other side, a metal film is coated. Mica is a dielectric. So, each photo-sensitive globule with the help of the metal film and the dielectric in between form a tiny capacitor. An image of the picture (object) is focused on the mosaic by a lens. Photo-electrons are emitted from the photo-sensitive surface of the mosaic by the incident light. The photo-electrons are collected by a collector electrode. A positive charge distribution appears on the mosaic which is proportional to the spatial light intensity variation. When this mosaic is scanned by an electron beam the charge on the photo-sensitive surface is neutralized. This produces a current to flow through the tiny capacitors and a voltage is developed across the resistance R. Thus, the brightness information of the picture is converted into an electrical signal.

11.7.1 SCANNING

To convert the picture information into an equivalent electrical signal, the image of the picture is scanned. To cover the full area of the image, the scanning is done horizontally from left to right. At the end of each line, the electron beam is brought back to its left end position. This retrace or fly back of the electron beam from right end to left end is blanked so that the retrace is not visible in the TV receiver. Blanking is done precisely by transmitting blanking pulses. After each horizontal line, the beam comes down a little vertically and performs another horizontal scanning from left to right. The beam then fly back to the left end but at a little lower position vertically. In this way, both horizontal and vertical scanning of the image of the picture is done.

The picture is scanned very fast in order to generate a sensation of continuity in the picture, otherwise flicker will occur. To avoid the problem of flicker, the picture is scanned at a rate of 50 Hz. The sensation of continuity originates from the persistence of vision effect of our eyes. A moving picture can be taken as stationary during one vertical (top to bottom) scan period.

If $\delta\theta$ be the minimum resolvable angle produced at the eye of the spectator by two consecutive lines of scanning, then

$$D \cdot \delta\theta = \text{vertical distance between two successive lines} = d\,(\text{say})$$

where D = distance of the viewer from the TV screen and $\delta\theta$ is in radian.

If H be the height of the TV screen and N be the number of white (or black) lines on the screen then

$$N = \frac{H}{d} = \frac{H}{D(\delta\theta)}$$

Typical value of $\delta\theta \approx \left(\dfrac{1}{60}\right)^{\circ}$. It depends upon the resolution of our eyes. To get an idea of the total number of lines, black or white, on the screen, let us take $D = 6$ times H. Then,

$$N = \frac{1}{6} \cdot \frac{1}{\dfrac{1}{60} \cdot \dfrac{\pi}{180}} = \frac{1800}{\pi} = 573 \text{ lines.}$$

11.7.2 INTERLACED SCANNING

The International Radio Consultative committee (CCIR) is a regulatory body of television broadcasting. Indian TV transmission is according to CCIR-B standard. In CCIR-B system, there will be 625 lines of scanning per picture. If the picture is scanned 50 times per second then the horizontal line scanning frequency is $50 \times 625 = 31250$ Hz. This will require a video bandwidth of $\simeq 10\,\text{MHz}$. To reduce this bandwidth requirement, the concept of interlaced scanning has been proposed. In interlaced scanning, shown in Fig. 11.20, the picture is scanned in two steps – each step uses $\frac{625}{2} = 312\frac{1}{2}$ lines and two steps of scanning are interleaved. Each step of $312\frac{1}{2}$ line scanning is called a field. So, two fields constitute a picture frame. The first $312\frac{1}{2}$ lines constitute the odd field and the remaining $312\frac{1}{2}$ lines form the even field. The first field ends at the middle of the 313 th line. From this point, the beam goes up from bottom centre to top centre of the screen and begins the first $1/2$ line of the second (even) field. This bottom to top fly back of the beam takes some time which is equivalent to some horizontal line periods. A field takes $1/50$ seconds to complete while a picture frame takes $\frac{1}{50} + \frac{1}{50} = \frac{1}{25}$ seconds for scanning. In the even field, the scanned lines are placed in between two consecutive lines of the odd field. This is interleaving of fields. So, in effect, 625 lines are scanned in $1/25$ seconds of time in this interlaced scanning. This corresponds to a line frequency of 15625 Hz. This reduces the bandwidth requirement to 5 MHz. In interlaced scanning the field frequency is 50 Hz but the picture frame frequency is 25 Hz.

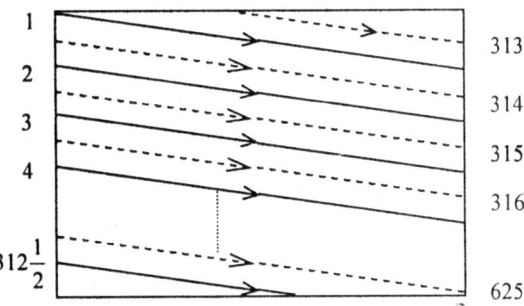

Fig. 11.20. Interlaced scanning. Solid lines correspond to odd field and dotted lines correspond to even field.

11.7.3 BLANKING PULSE

Blanking pulses are transmitted from the TV transmitter to cut-off the electron beam in the picture tube during horizontal flyback from the right end of the line to its left end position. The same blanking is done during the vertical flyback from the bottom in one field to the top. The blanking pulses modulate the picture carrier.

The time taken by the scanning beam to complete one horizontal line is $t_H = 1/15615 = 64$ µs.

The horizontal blanking period is 19% of the line period i.e.,

$$t_{HB} = 0.19 \times 64 = 12 \text{ µs.}$$

The vertical blanking period is equal to 20 horizontal line periods, i.e.,

$$t_{VB} = 20 \times 64 = 1280 \text{ µs.}$$

11.7.4 SYNC PULSE

The synchronization pulses are transmitted during the blanking period in order to synchronize the scanning operation in the transmitter and the receiver. These sync pulses make no effect on the picture because during their transmission the beam in the picture tube is cut-off. The width of sync pulse used for horizontal line synchronization is 4.7 µs.

In order to synchronize the field scanning transitions, vertical sync pulses are transmitted on the picture carrier. The field sync pulses appear at the end of the $312\frac{1}{2}$ th line and at the end of 625 th line.

11.7.5 VESTIGIAL SIDEBAND (VSB) MODULATION

The video signal has a bandwidth of 5 MHz in CCIR-B system. The video signal frequency ranges from dc to a maximum of 5 MHz. This video signal modulates the amplitude of a high frequency carrier in the VHF (very high frequency) or UHF (ultra high frequency) region. If we use double sideband with carrier (DSB+C) transmission, the bandwidth of the video AM signal will be 10 MHz. In order to reduce this wide bandwidth requirement, a special type of amplitude modulation is used which is called vestigial sideband modulation (VSB). The term 'vestige' means 'a portion'. In VSB, one sideband is fully transmitted and a small portion of the other sideband is transmitted. This is because the filter response, in practice, is not perfectly sharp at the edges. In order to accommodate a slope in the filter response, the vestige of the suppressed sideband is transmitted. Through VSB, about one half the video bandwidth is saved. The picture carrier is also transmitted along with the VSB so that the resultant signal is called VSB+C (C means carrier) signal. In VSB, the sideband selection filter is given a slope of 0.5 MHz on both sides of its response in CCIR-B system. The fully transmitted upper sideband has a bandwidth of $(5 + 0.5 =) 5.5$ MHz. In the suppressed lower sideband, frequencies down to 0.75 MHz from the picture carrier are fully transmitted so that considering the effect of filter slope the bandwidth of the lower sideband partially transmitted is $(0.75 + 0.5 =) 1.25$ MHz.

The CCIR-B system uses negative amplitude modulation. The negative modulation indicates that higher amplitude of the video carrier corresponds to a decrease in brightness level and a smaller carrier amplitude of the video signal corresponds to an increase in brightness level of the picture. In CCIR-B system, 75% of carrier amplitude corresponds to black level, 100% amplitude of the carrier corresponds to blacker than black level whereas 10% carrier amplitude corresponds to white level of the picture.

11.7.6 SOUND MODULATION IN TV

The sound (including voice and music) is used to modulate the frequency of the sound carrier. The sound carrier is positioned a distance of 5.5 MHz above the picture carrier in CCIR-B system. The peak frequency deviation of the FM signal is 50 kHz. Frequency modulation (FM) is chosen in order to get noise immunity in sound reproduction. The effect of noise is much less in FM system compared with the AM system. The higher frequencies in the audio signal have less power and the effect of noise is pronounced in the high frequency region in FM. So, signal-to-noise power ratio degrades at higher frequencies in FM. In order to overcome this problem, the high frequency components of the audio signal are relatively boosted by passing the audio signal through a RC high pass filter. This is known as pre-emphasis. The time constant of the RC filter is 50 μs in CCIR-B system. The channel characteristics is shown in Fig. 11.21 in CCIR-B system. A guard band of 0.25 MHz is allowed between adjacent channels to avoid any overlap between these channels. So, the total width of the TV channel in CCIR-B system is $(0.5 + 0.75 + 5 + 0.5 + 0.25 =) 7$ MHz.

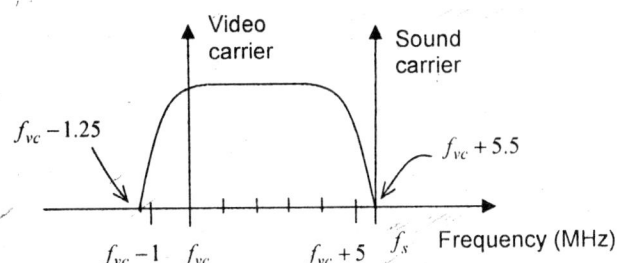

Fig. 11.21. CCIR-B system TV channel characteristics. f_{vc} = video carrier frequency, f_s = sound carrier frequency.

11.7.7 FREQUENCY ALLOCATION TO TV CHANNELS

The TV channels are distributed in the VHF and UHF frequency bands. The VHF band ranges from 30 MHz-300 MHz region while UHF spans over the range 300 MHz-3 GHz region. The VHF region has three bands, viz., band-I, band-II and band-III. Band-I has 4 - channels each of bandwidth 7 MHz and ranges from 41 MHz to 68 MHz. channel-1 spanning over 41 MHz- 47 MHz is used for services other than TV. So, in band-I there are three TV channels- channel-2, channel-3 and channel-4. Band-II ranges from 88 MHz-108 MHz is used for FM broadcasting. Band-III ranging from 174 MHz- 230 MHz has 8 TV channels each of 7 MHz bandwidth. These channels are channel 5, 6, 7, 8, 9, 10, 11 and 12. In the UHF range there are band-IV and band-V. Band-IV has a frequency range 470 MHz- 582 MHz while band-V has a frequency range 606 MHz- 790 MHz.

The TV signal in the VHF and UHF bands propagate from transmitter to receiver by the line of sight mode of propagation. There is a question, why FM is not used for video signal modulation? One reason is it requires much wider bandwidth. The other reason is that when the same TV signal reaches the TV receiver along different paths they produce interference at the receiver due to various delays in various paths. This results in ghost images in the receiver. The repetition frequency of the ghosts varies with the variation of the brightness of the picture from one area to another.

11.7.8 COLOUR TV

There are three primary colours, viz., red (R), green (G), and blue (B). Here, an electron gun produces three electron beams – one beam excites the red phosphor dot, the second excites the green phosphor dot while the third excites the blue phosphor dot. The screen consists of phosphor dot triads (a combination of three dots red, green and blue). These individual dot, are so small and so close that they act as one composite color unit. The scanning mechanism in color TV is also a three-beam scanning system. A color dot corresponding to a particular colour is excited by the corresponding electron beam. The colour information of the picture elements are transformed into corresponding electrical signals which is called as the chrominance signal. The electrical signals corresponding to red, green and blue colours have their characteristic amplitude and phase. The colour signals are combined to produce

    i) a luminance signal (Y) which represents the brightness of the picture, and

    ii) two difference signals-(R-Y) and (B-Y).

These are called chrominance signals. The chrominance signals are transmitted with the help of a colour subcarrier inside the composite video signal bandwidth. The luminance signal is also transmitted.

SOLVED PROBLEMS

11.1 An AM broadcast transmitter transmits 1 kW power whose carrier power is 800 W. calculate the modulation index.

Solution

The transmitter power is

$$P_t = P_C\left(1 + \frac{m^2}{2}\right)$$

where P_C = carrier power and m = AM index.

Here, $P_t = 1\,\text{kW}$, $P_C = 800\ \text{W} = 0.8\,\text{kW}$

$\therefore \qquad 1 + \dfrac{m^2}{2} = \dfrac{1}{0.8}$

or, $\qquad \dfrac{m^2}{2} = \dfrac{1 - 0.8}{0.8} = \dfrac{1}{4}$

$\therefore \qquad m = \dfrac{1}{\sqrt{2}} = 0.707 = 70.7\%.$

11.2 A AM broadcast transmitter has an antenna currents of 5 Amp and 5.5 Amp, respectively, when only carrier and then the AM signal is transmitted. Calculate the AM index.

Solution

Carrier power, $P_C = i_C^2 R_L$

and AM signal power, $P_t = i_t^2 R_L$

Where i_C and i_t are antenna currents corresponds to carrier transmission and signal transmission respectively. R_L is the load resistance to which the antenna delivers current. Here, $i_C = 5\ \text{A}$ and $i_t = 5.5\ \text{A}$.

$\therefore \qquad \dfrac{P_t}{P_C} = \left(\dfrac{i_t}{i_C}\right)^2$

or, $1 + \dfrac{m^2}{2} = \left(\dfrac{5.5}{5}\right)^2$

or, $m^2 = 2(1.21 - 1)$

$\therefore$ $m = 64\%$.

11.3 The envelope voltages of a signal varies from 3 volts to 1 volts. Calculate the AM index.

Solution

Here, $V_{max} = 3$ V

$V_{min} = 1$ V

$\therefore$ $m = \dfrac{V_{max} - V_{min}}{V_{max} + V_{min}}$

$= \dfrac{3 - 1}{3 + 1} = \dfrac{2}{4} = \dfrac{1}{2} = 0.5$.

11.4 An FM transmitter has a modulation sensitivity of 20 kHz/volts. What is the peak frequency deviation for sinusoidal modulation of amplitude 0.5 V? If the modulating signal frequency is 5 kHz, calculate the FM index.

Solution

Modulation sensitivity, $S = 20$ kHz/V

Peak frequency deviation, $\delta = S \cdot (0.5 \text{ V})$

$= 10$ kHz.

Again, $\delta = m_f \, f_m$ where $m_f = $ FM index.

$\therefore$ $m_f = \dfrac{\delta}{f_m} = \dfrac{10 \text{ kHz}}{5 \text{ kHz}} = 2$.

11.5 An FM signal is described by the voltage equation

$$v_{FM}(t) = 5 \cos(10^6 \pi t + 2 \sin 10^4 \pi t) \text{ volt}$$

How much is the carrier amplitude and carrier frequency? How much is the FM index and modulation frequency? Calculate the peak frequency deviation.

Solution

Comparing the given equation with the standard FM signal equation

$$v_{FM}(t) = V_C \cos(\omega_C t + m_f \sin \omega_m t)$$

we get

$V_C = 5$ V

$$\omega_C = 10^6 \pi \ \text{radian}$$

$$m_f = 2$$

$$\omega_m = 10^4 \pi \ \text{radian}$$

$$\therefore \ f_C = \frac{\omega_C}{2\pi} = 500 \, \text{kHz}$$

$$f_m = \frac{\omega_m}{2\pi} = \frac{10^4 \pi}{2\pi} = 5 \, \text{kHz}.$$

Peak frequency deviation. $\delta = m_f f_m = 2 \times 5 \, \text{kHz} = 10 \, \text{kHz}$

11.6 The transmitter power of a pulsed radar is $500 \, \text{kw}$. The radar cross section is $10 \, \text{m}^2$ and the capture area of the receiving antenna is $1 \, \text{m}^2$. The minimum detectable power of the receiver is $5 \times 10^{-13} \, \text{W}$ and the operating frequency is $10 \, \text{GHz}$. Calculate the maximum range of the radar.

Solution

Here, $\ \ P_t = 500 \, \text{kW} = 5 \times 10^5 \, \text{W}$

$$S = 10 \, \text{m}^2$$

$$A_0 = 1 \, \text{m}^2, \ P_{\text{min}} = 5 \times 10^{-13} \, \text{W}$$

$$f = 10 \, \text{GHz} = 10^{10} \, \text{Hz}$$

$$\lambda = \frac{c}{f} = \frac{3 \times 10^8}{10^{10}} = 0.03 \, \text{m}$$

From equation (11. 90) we get

$$r_{\text{max}} = \left[\frac{10 \cdot (1)^2 \cdot 5 \cdot (10)^5}{4\pi(.03)^2 \times 5 \times 10^{-13}} \right]^{\frac{1}{4}}$$

$$= \left[\frac{10^{23}}{36\pi} \right]^{\frac{1}{4}} = [8.842]^{\frac{1}{4}} . 10^5 \, \text{m}$$

$$= 1.724 \times 10^2 \, \text{km} = 172.4 \, \text{km}.$$

11.7 A pulsed radar transmit pulses of $2 \, \mu s$ duration. What is the minimum range of the radar?

Solution

Pulse duration, $T_D = 2 \, \mu s = 2 \times 10^{-6} \, \text{sec}.$

Distance travelled in time T_D is

$$R_m = T_D \cdot c = 2 \times 10^{-6} \cdot 3 \times 10^8 \, \text{m} = 600 \ \text{m}.$$

$\therefore$ Minimum range of the radar $= \dfrac{R_m}{2} = \dfrac{600}{2} = 300$ m.

11.8 A pulsed radar operates at a frequency of 3 GHz. An aircraft moves with a velocity of 1000 km/hour radially towards the radar. Calculate the change in frequency of the echo resulting from the aircraft.

Solution

Here, $f = 3$ GHz.

Operating wavelength $\lambda = \dfrac{c}{f} = \dfrac{3 \times 10^8}{3 \times 10^9} = 0.1$ m.

Radial velocity of the aircraft is

$$V_r = \dfrac{10^6}{60 \times 60} \text{ m/sec.}$$

$\therefore$ Change in frequency of the echo is

$$\Delta f = 2\dfrac{V_r}{\lambda} = 2 \times \dfrac{10^6}{60 \times 60 \times 0.1} \text{ s}^{-1}$$

$$= \dfrac{10^6}{180} \text{ Hz} = 5.55 \text{ kHz.}$$

11.9 A pulsed radar operating at 1 GHz transmits 1000 pulses per second. Calculate the first two blind speeds for this radar.

Solution

Here, the operating frequency, $f = 1 \text{ GHz} = 10^9$ Hz.

$\therefore$ The operating wavelength, $\lambda = \dfrac{c}{f} = \dfrac{3 \times 10^8}{10^9}$ m $= 0.3$ m.

Pulse repetition frequency $= 1000$ /sec.

$\therefore$ Pulse period, $T = \dfrac{1}{1000}$ sec. $= 10^{-3}$ sec.

The target moves a distance of $V_r.T$ meter between two successive pulses, where V_r = radial velocity of the target in meter.

The difference in signal path between two successive pulses is

$$\Delta d = 2V_r T$$

This difference in signal path produces a phase difference between two successive pulses which is

$$\Delta \phi = \dfrac{2\pi}{\lambda}.\Delta d \ .$$

Blind corresponds to the condition $\Delta\phi = 2n\pi$ where, $n = 1, 2, 3, \dots$ etc.

For the first blind speed, $n = 1$. Then, $\dfrac{2\pi}{\lambda} . 2V_r T = 2\pi$

$$\therefore \qquad V_r = \frac{\lambda}{2T} = \frac{0.3}{2} \cdot 10^3 = 150 \text{ m/s} = 540 \text{ km/s}.$$

For the second blind speed, $n = 2$. This second blind speed is given by $2 \times 540 = 1080$ km/s.

PROBLEMS

11.1 The modulation depth of a AM signal is 50%. The carrier power is $1\,\text{kW}$. Calculate the AM signal power.

[**Ans.** $1.125\,\text{kW}$]

11.2 A AM transmitter transmits a AM signal with a power of $600\,\text{W}$. The AM index is 60%. How much is the carrier power?

[**Ans.** $508.4\,\text{W}$]

11.3 The antenna current in a AM transmitter is $4\,\text{Amp}$ with 40% modulation depth. What will be the antenna current when only the carrier is transmitted?

[**Ans.** $3.85\,\text{A}$]

11.4 The maximum envelope voltage of a AM signal is 2 volts and the modulation depth is 60%. Calculate the minimum envelope voltage.

[**Ans.** $0.5\,\text{V}$]

11.5 The peak frequency deviation of an FM signal is $20\,\text{kHz}$. The FM index is 4. What is the modulation frequency? If the modulating signal voltage amplitude is 2 volts, what is the modulation sensitivity?

[**Ans.** $5\,\text{kHz}$, $10\,\text{kHz/V}$]

11.6 A pulsed radar with a transmitter power of $400\,\text{kW}$ operates at a frequency of $1\,\text{GHz}$. The radar cross section is $10\,\text{m}^2$ and the minimum detectable signal power is $2 \times 10^{-13}\,\text{W}$. The same antenna is used for transmission and reception with a gain of 1000. Calculate the maximum range of the radar.

[**Ans.** $173.5\,\text{km}$]

Hints: use equation (11.89).

11.7 An aircraft flying with a uniform radial velocity produces a change in frequency of a pulsed radar echo of $4\,\text{kHz}$. The radar operates at a frequency of $2\,\text{GHz}$. Calculate the radial velocity of the aircraft.

[**Ans.** $1080\,\text{km/hour}$]

QUESTIONS

11.1 What do you mean by 'modulation' of a wave? What is the necessity of modulation? What are the different kinds of analog modulation?

11.2 What is meant by the term 'amplitude modulation'? Find out a voltage equation for a AM signal starting from the definition. Show the spectrum of the AM signal. Show that the AM signal power is $3/2$ times the carrier power for 100% modulation.

11.3 Describe, with necessary circuit diagram, the operation of a square-law modulator.

11.4 Explain the method of AM signal generation in a class-C amplifier. Develop the necessary theory.

11.5 Give the principle of AM signal detection in a square-law diode detector. Can you detect a 100% modulated AM signal in this detector? Give reasons for your answer.

11.6 Explain briefly the principle of operation of an envelope detector. Derive an expression for the optimum time constant of this detector. Find out an expression for detection efficiency.

11.7 What is frequency modulation of a wave? Starting from the definition, derive a voltage equation for the FM wave. Find out the spectrum of the FM wave.

11.8 What is Carson's rule in connection with the FM signal. Distinguish between AM and FM waves. Show that the signal power does not change due to frequency modulation.

11.9 Decribe how FM signal is generated in a reactance modulator. Why is this circuit called a reactance modulator?

11.10 Give a neat circuit diagram of the Foster-Seely FM discriminator. Derive an expression for the output voltage of this discriminator and plot a schematic response of this discriminator.

11.11 What do you mean by the term 'radar'? Mention some uses of the radar. Give a circuit diagram of a pulsed radar.

11.12 Derive an expression for the maximum range of the pulsed radar. What are the implications of the range equation? Why are the ground-hopping military aircrafts not detected by a radar? Is there any minimum value of range of a pulsed radar?

11.13 Explain the principle of a moving target indicator radar. What is ground clutter? How is the Doppler effect associated with a moving target?

11.14 How can you measure the velocity of a target by a CW radar? Can a CW radar be used for the measurement of range of a target? Explain.
What is an FM altimeter? Describe, with a block diagram, the principle of operation of the FM altimeter.

11.15 What is a Television? Explain, briefly, how does a TV system operate. Describe the operation of an Iconoscope.

11.16 What is the scanning of a picture? What is interlaced scanning? Why is it necessary?

11.17 What is the necessity of blanking pulses in a TV? What is the reason of transmission of sync pulse in a TV? What is CCIR-B system. Describe the features of the CCIR-B TV system.

11.18 What is VSB+C modulation? Why is it used instead of other modulation formats in a TV? How do you transmit sound in a CCIR-B TV system? Show the channel characteristics of a CCIR-B TV system. Why is the video signal transmitted through AM and not FM in CCIR-B TV?

11.19 Describe the frequency allocation to TV channels in VHF and UHF bands. Explain, in short, the principle of coloured picture transmission in a colour TV.

$$\boxed{12}$$

Propagation of Radio Waves

12.1 CLASSIFICATION

There are five different ways by which a radio wave can propagate from a transmitter to a receiver. These are (i) ground wave (ii) sky wave (iii) space wave (iv) tropospheric wave and (v) artificially scattered wave via satellite.

12.2 GROUND WAVE

The transmitting and receiving antenna in this case are close to Earth's surface. The waves launched are vertically polarized. Otherwise, in case of horizontal polarization the horizontal component of the electric field of the wave would be short-circuited by the Earth. The ground wave induces charge on the Earth surface and these charges travel with the electromagnetic wave. This constitutes a ground current due to non-zero conductivity of the Earth surface. This ground current causes a loss of energy of the wave. So, long distance communication is not possible by ground wave. Ground wave attenuation increases with increase of the signal frequency.

12.3 SKY WAVE

Sky wave is the mode of electromagnetic wave propagation through ionosphere. Ionosphere is the region of Earth's atmosphere extending from a height of 50 km to several thousand kilometers above ground. Ionospheric propagation was discovered when Marconi succeeded in making transatlantic communication using radio waves in 1901. Kennely and Heaviside proposed that there exists a reflecting layer of radio waves in the upper atmosphere of earth. This layer contains ionized molecules of gases. This ionized layer is known as Ionosphere and also as Kennely-Heaviside layer after the name of its proposer. Long distance communication is possible through reflection of radio waves by the ionosphere.

The intensity of solar radiation increases with the increase of altitude above the ground. But, the pressure and the density of the gas molecules decrease with the increase of height above the Earth's surface. This leads to a maximum of ion density at some altitude above the ground. This maximum of ionization occurs at a height of about 300 kms above the earth's surface. Since the radiation intensity varies at different hours of day and night, the ionization density also varies at different times.

The ionosphere can be divided into layers of nearly constant ionization densities which are called D, E and F layers. The F layer is the layer of maximum ionization density which splits into two layers F_1 and F_2 during the day time. A long-term average of the ionization density as a function of altitude is shown in Fig. 12.1. The D layer is present during the day time only. Some layers appear in the E layer temporarily, which are called sporadic E layers.

The ionized gas molecules contain electrons and ions. These electrons and ions form a plasma. This plasma has a natural frequency of oscillation about the equilibrium position.

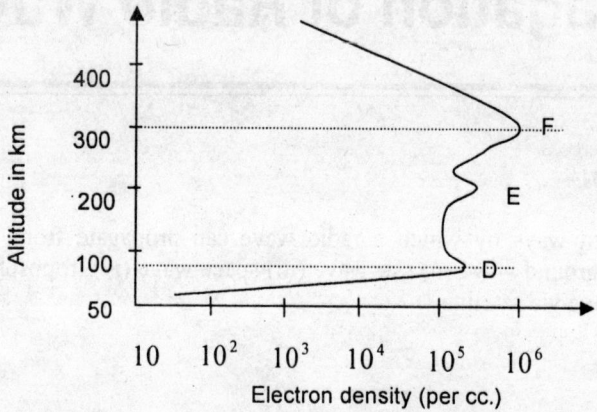

Fig. 12.1. Variation of ionospheric electron density with altitude.

Since ions are heavy, they may be assumed to be at rest and the electrons are supposed to execute oscillation. The natural oscillation frequency of the plasma is written as

$$\omega_p = \left(\frac{Ne^2}{m\varepsilon_0}\right)^{\frac{1}{2}} \qquad \ldots (12.1)$$

where e is the electronic charge, m is the electron mass, N is the electron density and ε_0 is the absolute permittivity of vacuum. We will use MKSQ system of units throughout. In presence of an external oscillatory electric field, ionospheric plasma will oscillate at the electromagnetic field frequency passing through it. The electron current density

$$\vec{J} = nq\vec{v} \qquad \ldots (12.2)$$

where n is the electronic density in the ionospheric plasma, q is the magnitude of electronic charge and $\vec{v}$ is the electron velocity in presence of the electromagnetic field. We assume sinusoidal oscillations of electrons with a small amplitude of oscillation in the ionospheric plasma. Then, we can write

$$n = N + \text{Re}\left[n_0 e^{j\omega t}\right] \qquad \ldots (12.3)$$

and $\qquad \vec{v} = \text{Re}\left[\vec{V}_0 e^{j\omega t}\right] \qquad \ldots (12.4)$

where ' Re ' stands for 'real part of' . N is the average electron number density. The velocity v has no drift component since the passing electromagnetic field is purely oscillatory. Then, from (12.2) we get -

$$\vec{J} = Nq \operatorname{Re}\left[\vec{V_0} e^{j\omega t}\right] \qquad \dots (12.5)$$

The product of two small quantities $n_0 V_0$ is neglected. Neglecting the effect of any external magnetic field, the equation of motion of a plasma electron due to the oscillatory electric field $\vec{E}$ is written as

$$m\frac{d\vec{v}}{dt} = q\vec{E} \qquad \dots (12.6)$$

where m is the mass of the electron . Therefore,

$$m(j\omega)\vec{v} = q\vec{E}$$

$$\therefore \quad \vec{v} = \frac{q\vec{E}}{j\omega m} \qquad \dots (12.7)$$

From Maxwell's electromagnetic field equation,

$$\vec{\nabla} \times \vec{H} = \varepsilon_0 \frac{\partial \vec{E}}{\partial t} + \vec{J} \qquad \dots (12.8)$$

where $\vec{H}$ is the magnetic field vector of the passing electromagnetic wave and ε_0 is the absolute permittivity of vacuum. Let

$$\vec{E} = \operatorname{Re}\left[\vec{E_0} e^{j\omega t}\right] \qquad \dots (12.9)$$

be the oscillatory electric field. From equation (12.8), we write

$$\vec{\nabla} \times \vec{H} = j\omega\varepsilon_0 \vec{E} + Nq\vec{v}$$

$$= j\omega\varepsilon_0 \vec{E} + \frac{Nq^2}{j\omega m}\vec{E}$$

$$= j\omega\varepsilon_0 \left[1 - \frac{Nq^2}{\omega^2 m\varepsilon_0}\right]\vec{E} \qquad \dots (12.10)$$

If ε be the equivalent absolute dielectric permittivity of the plasma, then

$$\vec{\nabla} \times \vec{H} = j\omega\varepsilon \vec{E} \qquad \dots (12.11)$$

since $\vec{J} = 0$ for a dielectric.

Comparing (12.10) and (12.11) we get

$$\varepsilon = \varepsilon_0 \left[1 - \frac{Nq^2}{m\varepsilon_0 \omega^2} \right] \qquad \dots (12.12)$$

If electron-electron collision and electron-ion collision in the plasma are taken into account, the plasma is found to possess a dielectric permittivity

$$\varepsilon = \varepsilon_0 \left[1 - \frac{Nq^2}{m\varepsilon_0 (\upsilon^2 + \omega^2)} \right] \qquad \dots (12.13)$$

and a conductivity

$$\sigma = \frac{Nq^2 \upsilon}{m(\upsilon^2 + \omega^2)} \qquad \dots (12.14)$$

where υ is the collision frequency.

For high frequency radio waves, $\omega \gg \upsilon$ and

$$\varepsilon = \varepsilon_0 \left[1 - \frac{Nq^2}{m\varepsilon_0 \omega^2} \right] \qquad \dots (12.15)$$

and $\sigma \approx 0$.

The radio wave is attenuated if the plasma has a non-zero conductivity. Refractive index of the ionospheric layer is given by

$$n = \frac{c}{v_p} = \sqrt{\frac{\mu\varepsilon}{\mu_0 \varepsilon_0}} \approx \sqrt{\frac{\varepsilon}{\varepsilon_0}} \qquad \dots (12.16)$$

where c = vacuum velocity of radio waves and v_p = phase velocity of the radio wave in the plasma. $\mu \approx \mu_0$ where μ is the permeability of the medium and μ_0 is the same for the vacuum. From equation (12.12) we get

$$n = \left[1 - \frac{Nq^2}{m\varepsilon_0 \omega^2} \right]^{\frac{1}{2}} \qquad \dots (12.17)$$

The refractive index of the ionosphere depends upon the electron density N and signal frequency ω.

12.3.1 REFRACTION OF RADIO WAVES IN THE IONOSPHERE

Suppose that the radio wave is incident at an angle φ_i in the lower edge of the ionosphere. It undergoes a continuous refraction in the ionosphere. This is shown in Fig. 12.2.

Let φ_r be the angle made by the path of the radio wave with the normal to the ionospheric layer of electron density N. By Snell's law of refraction

$$\frac{Sin\varphi_i}{Sin\varphi_r} = n \qquad \qquad \ldots (12.18)$$

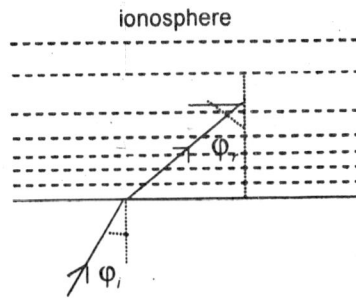

Fig. 12.2. Refraction of radio waves in the ionosphere.

Taking $q = 1.59 \times 10^{-19}$ Coulomb, $m = 9 \times 10^{-31}$ kg, and $\varepsilon_0 = \dfrac{10^{-9}}{36\pi}$ F/m, we get

$$n = \left[1 - \frac{81N}{f^2} \right]^{\frac{1}{2}} \qquad \qquad \ldots (12.19)$$

where N = electron density in m^{-3} and f is in Hz. Equation (12.19) is also correct if N is in per c.c. and f is in kHz.

When $\varphi_r = \pi/2$, the radio wave travels tangentially to the layer. If ' N' ' be the electron density of a layer where $\varphi_r = \pi/2$, we get

$$\sin\varphi_i = \sqrt{1 - \frac{81N'}{f^2}}$$

or, $$N' = \frac{f^2 \cos^2\varphi_i}{81} \qquad \qquad \ldots (12.20)$$

For $N > N'$, total internal reflection of the radio wave occurs. N' is maximum when $\varphi_i = 0$, i.e., for vertical incidence.

12.3.2 CRITICAL FREQUENCY

The critical frequency of an ionospheric layer is the frequency of the radio wave which is just reflected back for vertical incidence. Taking $\varphi_i = 0$, the critical frequency, f_{cr}, is given by

$$f_{cr} = \sqrt{81N} \qquad \qquad \ldots (12.21)$$

where N is the electron density of the layer.

12.3.3 MAXIMUM USABLE FREQUENCY (MUF)

The maximum frequency of the radio wave that can be reflected back by an ionospheric layer for a given distance of transmission is called the maximum usable frequency. This will depend upon the transmitter receiver separation. Then,

$$MUF = \frac{\sqrt{81N}}{\cos\varphi_i}$$

$$= f_{cr}\sec\varphi_i \qquad \qquad \text{... (12.22)}$$

using equation (12.20). For a certain ionospheric layer at a height of h above the earth's surface, the maximum value of φ_i is

$$\varphi_i\big|_{max} = \sin^{-1}\left[\frac{r}{r+h}\right] \qquad \qquad \text{... (12.23)}$$

This is shown in Fig. 12.3. This corresponds to a maximum transmitter-receiver separation on the Earth surface. Then,

$$MUF = f_{cr}\sec\varphi_i\big|_{max} \qquad \qquad \text{... (12.24)}$$

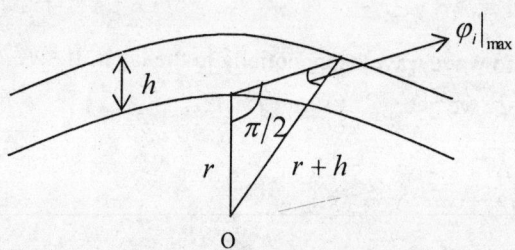

Fig. 12.3. Maximum angle of incidence to find MUF.

12.3.4 VIRTUAL HEIGHT

The radio wave follows a curved path in the ionosphere due to continuous refraction. The path of the radio ray is straight up to the lower edge of the ionosphere and then it bends in the ionosphere and finally gets reflected back. If the incident and reflected ray paths are extrapolated in the ionosphere they meet at a point from where the radio wave appears to be reflected. The height of this virtual point of reflection measured from the ground is called virtual height. In Fig. 12.4, GQ is the virtual height.

Typical values of virtual height $h \sim 300$ kms in winter day for signal frequency $f \sim 4-10$ MHz, and $h \sim 400-700$ kms in summer day for $f \sim 5-7$ MHz.

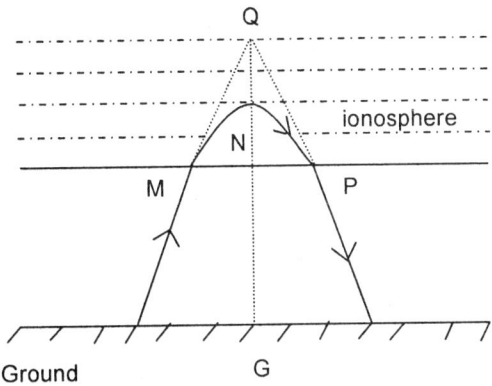

Fig. 12.4. Virtual height of the reflected radio wave

12.3.5 SKIP DISTANCE

The point where a radio wave is reflected back to ground from the ionosphere depends upon the angle of incidence at the lower edge of the ionosphere. At very large angles of incidence, the radio wave suffers reflection from a lower ionospheric layer and the wave returns to ground at a long distance from the transmitter. The angle of incidence is greater than the critical angle in this case. As the angle of incidence is made smaller gradually, the wave returns closer and closer to the transmitter. This is shown in Fig. 12.5 by rays 1, 2, and 3. If the angle of incidence is made less than the critical angle, the radio wave penetrates the ionosphere and never comes back to Earth . Rays 4 and 5 in Fig. 12.5 are examples. When a ray (ray 6 in Fig. 12.5) is incident at the critical angle, the ray travels horizontally in the ionosphere up to certain distance and finally returns to ground.

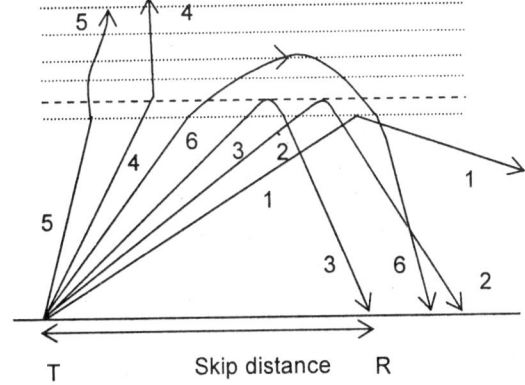

Fig. 12.5. Explanation of skip distance. T = Transmitter, R = Receiver.

The distance from the transmitter within which a radio wave of given frequency fails to be reflected back is the skip distance for that frequency.

The critical angle of any layer at radio frequency f is given by

$$\varphi_c = \sin^{-1}(n) = \sin^{-1}\left[1 - \frac{81N}{f^2}\right]^{\frac{1}{2}} \qquad \qquad \dots (12.25)$$

where N the is electron density of the ionospheric layer in per m^3 and f is in Hz.

12.4 SPACE WAVE

Space waves travel from the transmitting antenna to the receiving antenna along a straight line. This is called a direct ray. This mode of propagation is called line-of-sight propagation. The maximum distance of the transmitting antenna and the receiving antenna in space wave propagation is limited by the curvature of Earth. The signal frequencies can be greater than the maximum frequency that can be reflected by the ionosphere. The wave from the transmitter can also be reflected by the ground and the reflected wave reaches the receiving antenna along with the direct wave. This produces interference. In the TV receiver this interference resulting from large objects nearby is called as 'ghost'. In Fig. 12.6, T is the transmitting antenna and R is receiving antenna.

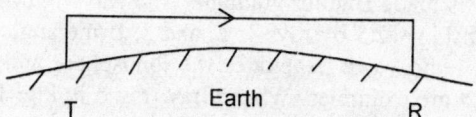

Fig. 12.6. Space wave propagation

12.5 TROPOSPHERIC PROPAGATION

Troposphere is the region of the atmosphere extending up to 10 km above the Earth surface. Normally, the temperature in the troposphere decreases at a rate of $6.5°$ C/km. Radio waves can be reflected, diffracted or scattered in this region. The most interesting propagation that can take place in the troposphere is called 'ducting'. It may so happen sometimes in the troposphere that the temperature increases with height over a certain height instead of normal decrease of temperature with height. This is called temperature inversion. This leads to a rapid reduction of refraction index with height.

Microwaves are bent continuously by this region of troposphere and finally reaches the ground. Again, the ground reflects this wave which undergoes further refraction in the troposphere and returns to ground again. In this way, propagation can occur over long distances ~1000 km or more. This mode of propagation is called 'ducting'. Tropospheric duct propagation is shown in Fig. 12.7.

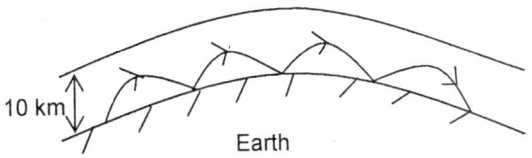

Fig. 12.7. Ducting in the troposphere.

12.6 ARTIFICIALLY SCATTERED WAVES

At very high frequencies exceeding 100 MHz, ionospheric bending of radio waves due to refraction becomes very small. Microwaves can be transmitted from a Earth station to a geostationary satellite which can retransmit the signal to another distantly located Earth station as if a reflector (here satellite) has been placed in the sky. The approximate height of the satellite above ground is 36000 km. These satellites move around the earth with the same angular velocity as that of the Earth. These satellites are placed on equatorial orbits so that the satellite experiences equal attraction from both the hemispheres of the earth. Otherwise, the satellite will drift from its orbit. Obviously, these satellites appear as stationary over a fixed point on the equator and are known as geostationary satellites. Scattering of microwaves by geostationary satellites is a way which makes global communication possible. However, with the advent of optical fibers, lasers and photodiodes, ultra long distance communication is being realized at present.

SOLVED PROBLEMS

12.1 Calculate the critical frequency of an ionospheric layer having an electron density of 2.5×10^5 /cm^3.

Solution

$$f_{cr} = \sqrt{81N} \text{ kHz} \qquad \text{where } N \text{ is the electron density in /cm}^3.$$

$$= \left[81 \times 2.5 \times 10^5\right]^{\frac{1}{2}} \text{ kHz}$$

$$= 4.5 \text{ MHz}$$

12.2 Examine whether a radio wave of frequency 1 MHz can be reflected by the ionosphere having a maximum electron density of $3.6 \times 10^5 / cm^3$.

Solution
The critical frequency

$$f_{cr} = \sqrt{81N} \text{ kHz} \qquad \text{where } N \text{ is in /cm}^3$$

$$= \left[81 \times 3.6 \times 10^5\right]^{\frac{1}{2}} \text{kHz}$$

$$= 5.4 \text{ MHz}$$

Hence, the radio wave of frequency 1 MHz can be reflected by the ionosphere.

12.3 The transmitting and receiving antenna on the Earth surface are separated by a distance of 1500 km and communication between them occur by HF radio waves through ionospheric reflection. The critical frequency of the ionosphere is 7 MHz for the reflecting layer at a height of 300 km above the ground. Assuming single–hop propagation, calculate the maximum usable frequency (MUF) for the given system.

Solution

$$MUF = f_{cr} \sec \varphi_i,$$

where φ_i = angle of incidence on the ionosphere.

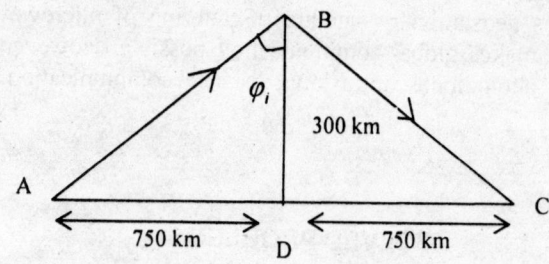

Fig. 12.8

From Fig. 12.8,

$$\sec \varphi_i = \frac{AB}{BD}$$

$$= \frac{\sqrt{750^2 + 300^2}}{300}$$

$$= \frac{\sqrt{(7.5)^2 + 9}}{3}$$

$$= 2.6925.$$

$$\therefore \quad MUF = 7 \times 2.6925$$

$$= 18.84 \text{ MHz}.$$

PROBLEMS

12.1 A radio signal of frequency 1MHz is incident on the inospheric layer at an angle $30°$ and travel horizontally to the layer. Calculate the electron density of the layer.

Hints: From eqn. (12.20), $N' = \dfrac{f^2 \cos^2 \phi_i}{81}$

Here, $f = 10^6$ Hz, $\phi_i = 30°$. So, $N' = 9.259 \times 10^9 /m^3$.

12.2 An ionospheric layer has an electron density of 5×10^4 / c.c. Calculate the refractive index of the layer for a radio signal of frequency 3 MHz.

Hints: From eqn. (20.19), $n = \left[1 - \dfrac{81N}{f^2}\right]^{1/2}$. Here, $N = 5 \times 10^4 /c.c. = 5 \times 10^{10} /m^3$

$f = 3 \times 10^6$ Hz., $n = 0.741$

12.3 What is the critical frequency of an ionospheric layer having an electron density 9×10^4 / c.c.?

Hints: $f_{er} = \sqrt{81N}$. Here, $N = 9 \times 10^4 /c.c. = 9 \times 10^{10} /m^3$

$f_{er} = 2.7$ MHz.

12.4 An ionospheric layer has an electron density $10^4 /c.c.$ Calculate the value of the critical angle for this layer when a radio signal of frequency 2 MHz is incident upon it.

Hints: $\sin \phi_C = \left[1 - \dfrac{81N}{f^2}\right]^{1/2} = \left[1 - \dfrac{0.81}{4}\right]^{1/2} = \sin^{-1}(0.893) = 63.25°$.

Here, $N = 10^4 /c.c. = 10^{10} /m^3$, $f = 2 \times 10^6$ Hz.

QUESTIONS

12.1 What are the different modes of propagation of a radio wave? Describe how a high frequency radio wave is reflected from the ionosphere.

12.2 Explain what is meant by 'ionosphere'. Derive an expression for the refractive index of an ionosphere layer of electron electron density N and signal frequency ω.

12.3 Define the following terms:
(a) Critical frequency (b) MUF (c) Virtual height (d) Skip distance

12.4 What is Tropospheric 'ducting'? What do you mean by artificially scattering of microwaves? Explain space wave propagation.

12.5 A short wave radio signal of frequency f Hz is incident at the F_1 layer of the ionosphere having an electron density N/m^3. Find out an expression for the refractive index of the layer for this radio wave.

Hence explain how this radio wave is reflected back to the Earth by the ionosphere.

Operational Amplifier

13.1 BASIC IDEA AND PROPERTIES

Operational amplifier, which is abbreviated as OP Amp, is a high gain, wideband amplifier. The frequency response of this amplifier can be changed by applying feedback. It can perform linear operations like addition, subtraction, differentiation, integration etc. It can also perform nonlinear operations such as logarithmic and antilogarithmic operations, generation of sinusoidal and non-sinusoidal waveforms etc. The characteristics of an ideal OP-Amp are the following:

- Infinite voltage gain
- Infinite bandwidth
- Infinite input resistance
- Zero output resistance
- Perfectly balanced- i.e., if same voltage is applied to both the inputs, the output voltage becomes zero.
- The device characteristics is insensitive to temperature.

DEFINITIONS

(i) INPUT OFFSET VOLTAGE

It is the dc voltage which should be applied between the input terminals of the OP-Amp in order to make the output voltage of the OP-Amp equal to zero.

(ii) OUTPUT OFFSET VOLTAGE

It is the dc output voltage of an OP-Amp when the two input terminals of the OP-Amp are grounded.

(iii) INPUT OFFSET CURRENT

It I_1 and I_2 be the dc currents entering the two input terminals of the OP-Amp which make the OP-Amp balanced (i.e., gives zero output voltage) then the current difference $|I_1 - I_2|$ is called the input offset current.

(iv) SLEW RATE

The operational amplifier can not respond to a very rapidly varying input voltage. This is because the internal capacitances of the OP-Amp require some time to charge. As a result, the OP-Amp out-put can not follow the rapidly changing input voltage immediately. The maximum possible value of the rate of change of the output voltage is called the slew rate of the OP-Amp. If the amplitude and frequency of the input signal results in an output which exceeds the slew rate, distortion of the output signal will occur.

Let the output signal voltage of the OP-Amp be given by

$$v_0(t) = V_m \sin \omega t \qquad \qquad \dots (13.1)$$

Then, $\dfrac{dv_0(t)}{dt} = V_m \omega \cos \omega t \qquad \qquad \dots (13.2)$

Here, V_m = voltage amplitude and ω is the radian frequency of the signal. Now,

$$\left. \frac{dv_0(t)}{dt} \right|_{max} = V_m \cdot \omega = S \text{ (by definition)}$$

where S is the slew rate.

Thus, $S = 2\pi f V_m \qquad \qquad \dots (13.3)$

Equation (13.3) defines the relation between slew rate, signal frequency and maximum possible voltage amplitude of the OP-Amp.

13.1.1 VIRTUAL GROUND

A circuit diagram of the operational amplifier is shown in Fig. 13.1 where the non-inverting terminal is grounded.

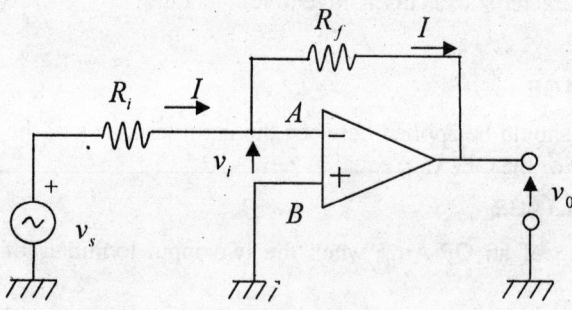

Fig. 13.1. OP-Amp circuit explaining virtual ground.

In Fig. 13.1, the resistance R_f provides negative voltage shunt feedback. The input current I does not enter the OP-Amp due to its infinite input resistance and instead flows through R_f. So, there is no potential drop between the input terminals A and B. The potentials of the points A and B are the same resulting in $v_i = 0$. Normally, when two points in a circuit having a potential difference are short circuited, a current flows through this short. Here, the points A and

B represent a virtual short circuit ($v_i = 0$) but no current flows into this short circuit. For this reason, it is called a virtual ground of the OP-Amp.

13.1.2 COMMON MODE REJECTION

In a practical OP-Amp, the output signal amplitude not only depends upon the difference of the voltages applied to the two input terminals but also depends upon the average level of the two input signals. Thus, if v_1 and v_2 be signal voltages applied to the inverting and non-inverting terminals of an OP-Amp, then the difference signal is $v_d = (v_2 - v_1)$ and the average signal is $v_c = \left(\dfrac{v_1 + v_2}{2}\right)$. v_c is called the common mode signal.

Let A be the voltage gain from the inverting input to the output when the non-inverting input is grounded. Similarly, let A' be the voltage gain from the non-inverting input to the output when the inverting input is grounded. Then, the output voltage of the OP-Amp can be expressed as

$$v_0 = Av_1 + A'v_2 \qquad \qquad \text{... (13.4)}$$

Now, $\quad v_1 = v_c - \dfrac{1}{2}v_d \qquad \qquad \text{... (13.5)}$

and $\quad v_2 = v_c + \dfrac{1}{2}v_d \qquad \qquad \text{... (13.6)}$

Substituting (13.5) and (13.6) in (13.4), we get

$$v_0 = \frac{1}{2}v_d(A' - A) + v_c(A + A')$$

$$= A_D v_d + A_c v_c \qquad \qquad \text{... (13.7)}$$

where $\quad A_D = \dfrac{1}{2}(A' - A)$ and $A_c = (A + A')$.

A_D is called difference mode gain and A_c is called the common mode gain of the OP-Amp.

Common mode rejection ratio (CMRR) is defined as the ratio of difference mode gain to the common mode gain of the OP-Amp. Denoting the CMRR by ρ we get

$$\rho = \frac{A_D}{A_c} \qquad \qquad \text{... (13.8)}$$

Now, eqn. (13.7) can be written as

$$v_0 = A_D v_d + \frac{1}{\rho}A_D v_c$$

$$= A_D v_d \left[1 + \frac{1}{\rho}\frac{v_c}{v_d}\right] \qquad \qquad \text{... (13.9)}$$

13.2 APPLICATIONS

In this section, we will first describe linear applications of OP-Amp. Then, we will describe some nonlinear applications.

13.2.1 DIFFERENCE AMPLIFIER AND SUBTRACTOR

Let us consider the circuit shown in Fig 13.2.

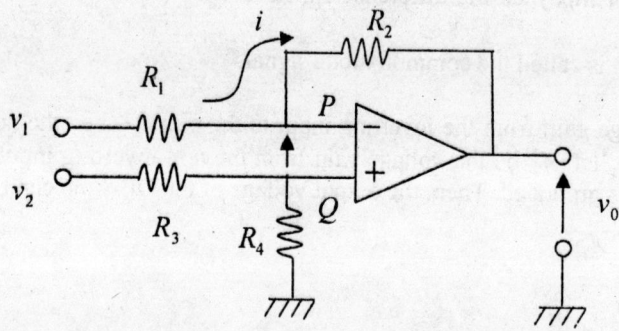

Fig. 13.2. OP-Amp acting as difference amplifier.

Due to the existence of virtual ground at the input of OP-Amp, we have $v_P = v_Q$.

Now, $$v_Q = \frac{v_2 R_4}{R_3 + R_4} \qquad \qquad \dots (13.10)$$

Again, $$i = \frac{v_1 - v_P}{R_1} = \frac{v_P - v_0}{R_2}$$

$$\therefore \quad \left[v_1 - \frac{v_2 R_4}{R_3 + R_4} \right] \frac{1}{R_1} = \left[\frac{v_2 R_4}{R_3 + R_4} - v_0 \right] \frac{1}{R_2}$$

$$\therefore \quad v_0 = \frac{v_2 R_4}{R_3 + R_4} - \frac{R_2}{R_1} \left[v_1 - \frac{v_2 R_4}{R_3 + R_4} \right]$$

$$= \frac{v_2 R_4}{R_3 + R_4} \left(1 + \frac{R_2}{R_1} \right) - \frac{R_2}{R_1} v_1 \qquad \qquad \dots (13.11)$$

Now, if $\dfrac{R_2}{R_1} = \dfrac{R_4}{R_3}$ then

$$v_0 = \frac{v_2 R_4}{R_3 + R_4} \left(1 + \frac{R_4}{R_3} \right) - \frac{R_2}{R_1} v_1$$

$$= \frac{R_4}{R_3} v_2 - \frac{R_2}{R_1} v_1$$

$$= \frac{R_2}{R_1} (v_2 - v_1) \qquad \qquad \text{... (13.12)}$$

Equation (13.12) shows that the difference signal $(v_2 - v_1)$ is amplified (R_2/R_1) times by this circuit. As a special case, if we take $R_1 = R_2$ then

$$v_0 = v_2 - v_1 \qquad \qquad \text{... (13.13)}$$

Thus, this circuit acts as a subtractor.

13.2.2 INVERTING AMPLIFIER

Consider the OP-Amp circuit as shown in Fig. 13.3.

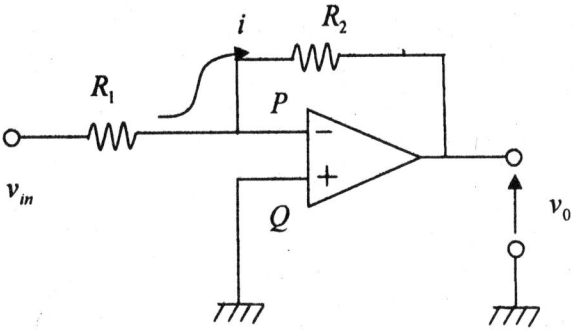

Fig. 13.3. Circuit diagram of the inverting amplifier.

Here, $v_Q = 0$.

$\therefore$ $v_P = v_Q = 0$.

Now, $\dfrac{v_{in} - v_P}{R_1} = i = \dfrac{v_P - v_0}{R_2}$

$\therefore$ $\dfrac{v_{in}}{R_1} = -\dfrac{v_0}{R_2}$

$\therefore$ Voltage gain, $A_v = \dfrac{v_0}{v_{in}} = -\dfrac{R_2}{R_1}$ $\qquad \qquad \text{... (13.14)}$

there is a phase difference of π radian between the input and output of this amplifier. This is the feature of the inverting amplifier.

13.2.3 NON-INVERTING AMPLIFIER

The OP-Amp based non-inverting amplifier circuit is shown in Fig. 13.4.

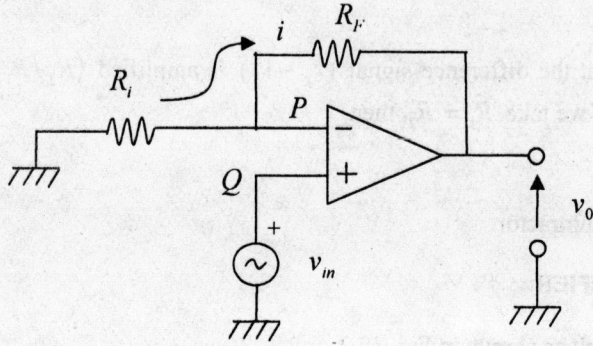

Fig. 13.4. Circuit diagram of the Non-inverting amplifier.

Here, $v_Q = v_{in}$. Also, $v_P = v_Q$.

Current, $i = \dfrac{0 - v_P}{R_i} = -\dfrac{v_P}{R_i} = -\dfrac{v_{in}}{R_i}$ ... (13.15)

Again, $i = \dfrac{v_P - v_0}{R_F} = \dfrac{v_{in} - v_0}{R_F}$ ... (13.16)

Equating (13.15) and (13.16), we get

$$-\frac{v_{in}}{R_i} = \frac{v_{in} - v_0}{R_F}$$

$\therefore \qquad v_0 = \left(1 + \dfrac{R_F}{R_i}\right) v_{in}$ ... (13.17)

$\therefore$ Voltage gain $A_v = \dfrac{v_0}{v_{in}} = 1 + \dfrac{R_F}{R_i}$ ... (13.18)

Thus, there is no phase inversion in this amplifier.

13.2.4 INVERTING ADDER

A circuit of the inverting adder is shown in Fig. 13.5.

Here, $v_Q = 0$. So, $v_P = v_Q = 0$.

Now, $i = \dfrac{(v_1 - v_P)}{R_1} + \dfrac{(v_2 - v_P)}{R_2} + \dfrac{(v_3 - v_P)}{R_3}$

$$= \frac{v_1}{R_1} + \frac{v_2}{R_2} + \frac{v_3}{R_3} \qquad \qquad \dots (13.19)$$

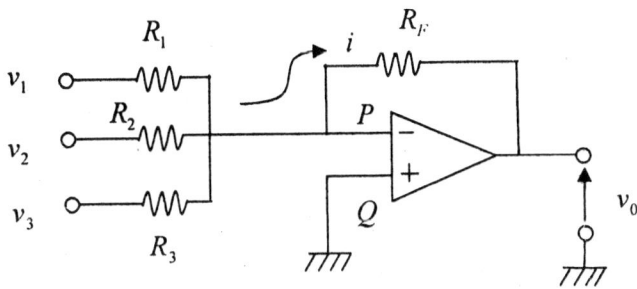

Fig. 13.5. Inverting adder circuit.

Again, $i = \dfrac{v_P - v_0}{R_F} = -\dfrac{v_0}{R_F}$ $\qquad \qquad \dots (13.20)$

Equating (13.19) and (13.20) we get

$$-\frac{v_0}{R_F} = \left(\frac{v_1}{R_1} + \frac{v_2}{R_2} + \frac{v_3}{R_3} \right)$$

$$\therefore \qquad v_0 = -R_F \left(\frac{v_1}{R_1} + \frac{v_2}{R_2} + \frac{v_3}{R_3} \right) \qquad \qquad \dots (13.21)$$

If $R_1 = R_2 = R_3 = R_F$,

then,

$$v_0 = -\left(v_1 + v_2 + v_3 \right) \qquad \qquad \dots (13.22)$$

The negative sign in eqn. (13.22) indicates phase reversal. Thus, the output voltage of this circuit is the sum of input voltages with inverted phase.

13.2.5 NON-INVERTING ADDER

Consider the circuit shown in Fig. 13.6. Here, $v_P = v_Q$. To find v_Q we must consider the contribution from two voltage sources v_1 and v_2. The contributions can be found out by applying superposition theorem. According to this theorem, the contribution of v_1 source towards v_Q is obtained by removing v_2 and the contribution of source v_2 towards v_Q is obtained by removing v_1. Then,

$$v_Q = v_1 - \frac{v_1 R_1}{R_1 + R_2} + \frac{v_2 R_1}{R_1 + R_2}$$

$$= \frac{v_1 R_2}{R_1 + R_2} + \frac{v_2 R_2}{R_1 + R_2}$$

$$= \frac{1}{R_1 + R_2}(v_1 R_2 + v_2 R_1) \qquad \qquad \dots (13.23)$$

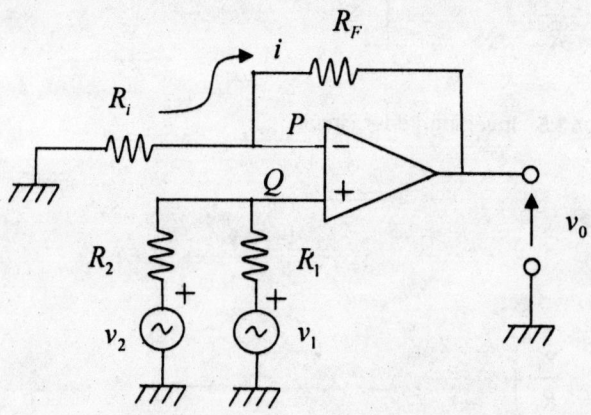

Fig. 13.6. Circuit diagram of non-inverting adder.

Now, current $i = \dfrac{0 - v_P}{R_i} = \dfrac{v_P - v_0}{R_F}$

$\therefore \qquad v_P - v_0 = -\dfrac{R_F}{R_i}v_P$

$\therefore \qquad v_0 = \left(1 + \dfrac{R_F}{R_i}\right)v_P$

$\qquad\qquad = \left(1 + \dfrac{R_F}{R_i}\right)\dfrac{1}{R_1 + R_2}(v_1 R_2 + v_2 R_1)$ (since $v_P = v_Q$) $\qquad \dots (13.24)$

Taking $R_1 = R_2$ and $R_F = R_i$ we get

$$v_0 = 2 \cdot \frac{R_1}{2R}(v_1 + v_2) = v_1 + v_2 \qquad \qquad \dots (13.25)$$

13.2.6 DIFFERENTIATOR

The circuit of the differentiator is shown in Fig.13.7.

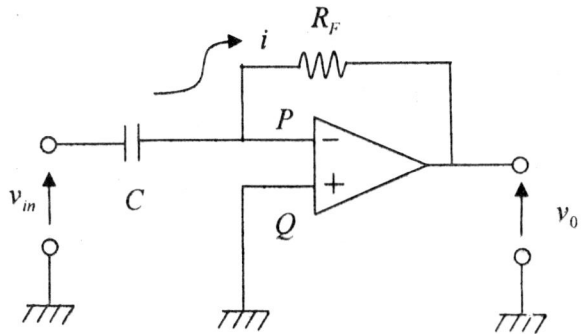

Fig. 13.7. Schematic circuit diagram of an OP-Amp differentiator.

Here, $v_Q = 0$.

$\therefore\ v_P = 0$.

Charge on the capacitor C is

$$q(t) = C[v_{in}(t) - v_P] = Cv_{in}(t).$$

Then, current $\quad i = \dfrac{dq(t)}{dt} = C\dfrac{dv_{in}(t)}{dt}$ ... (13.26)

Again, $\quad i = \dfrac{v_P - v_0}{R_F} = -\dfrac{v_0}{R_F}$ ... (13.27)

Equating (13.26) and (13.27) we get

$$v_0 = -R_F C\dfrac{d}{dt}v_{in}(t)$$ ... (13.28)

If we take $R_F C = 1$, then

$$v_0 = -\dfrac{d}{dt}v_{in}(t)$$ ... (13.29)

Thus, the output of this circuit is differentiated input except for a negative sign. To get rid of this negative sign, the output can be passed through a unity gain inverting amplifier which changes the phase by π radian. Here, the output of a sinusoidal signal is proportional to ω, where ω is the signal frequency in radian. For this reason, this circuit enhances high frequency noise components.

13.2.7 INTEGRATOR

The circuit diagram of the OP-Amp based integrator is shown in Fig. 13.8. Here, $v_Q = 0$. So, $v_p = 0$.

The current $i = \dfrac{v_{in} - v_p}{R} = \dfrac{v_{in}}{R}$ ... (13.30)

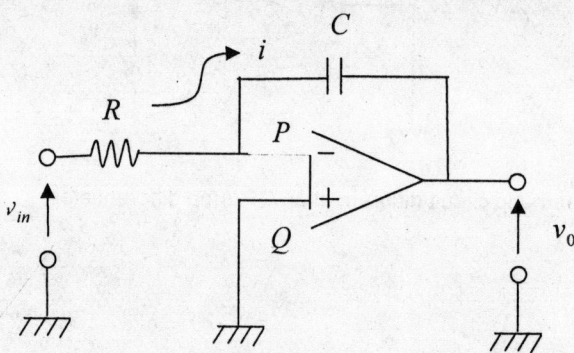

Fig. 13.8. Circuit diagram of an OP-Amp integrator.

Again, the charge on capacitor C is

$$q(t) = C(v_p - v_0) = -Cv_0$$

Then, current $i = \dfrac{dq(t)}{dt} = -C\dfrac{dv_0}{dt}$ ... (13.31)

From (13.30) and (13.31), we get

$$-C\frac{dv_0}{dt} = \frac{v_{in}}{R}$$

$\therefore \qquad v_0 = -\dfrac{1}{RC} \displaystyle\int v_{in} dt$ ... (13.32)

Taking $RC = 1$, $v_0 = -\displaystyle\int v_{in} dt$ ... (13.33)

The output voltage is the integrated version of input voltage except for a phase change of π radian. The negative sign can be eliminated by passing this output through a unity gain inverting amplifier. Giving a sinusoidal input, we get a cosine waveform at the output. Moreover, the output voltage is proportional to $\dfrac{1}{\omega}$, where ω is the signal frequency in radian. For this reason, the integrator suppresses high frequency noise.

13.2.8 ACTIVE LOW-PASS FILTER

The circuit of the low pass filter (LPF) is shown in Fig. 13.9. Here, $v_Q = 0$. So, $v_P = 0$.

Impedance of the feedback circuit is

$$Z = R_F \parallel \frac{1}{j\omega C} = \frac{R_F}{1 + j\omega CR}$$

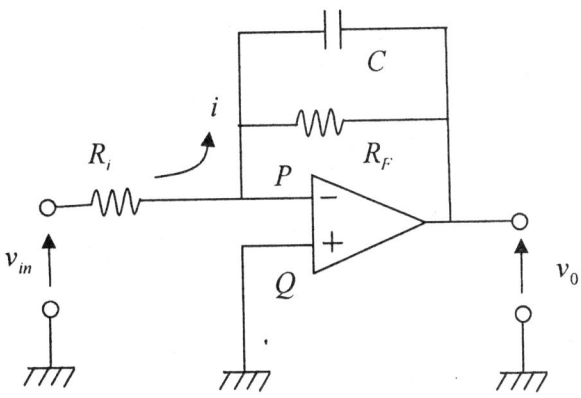

Fig. 13.9. Active low pass filter circuit.

Current, $i = \dfrac{v_{in} - 0}{R_i} = \dfrac{v_{in}}{R_i}$ ...(13.34)

Again, $i = \dfrac{v_P - v_0}{Z} = -\dfrac{v_0}{Z}$ ...(13.35)

From (13.34) and (13.35), we write

$$\frac{v_{in}}{R_i} = -\frac{v_0}{Z}$$

$$\therefore \quad \frac{v_0}{v_{in}} = -\frac{Z}{R_i} = -\frac{R_F / R_i}{1 + j\omega CR_F}$$

$\therefore$ Transfer function of the LPF is

$$H(\omega) = \frac{v_0}{v_{in}} = -\frac{R_F / R_i}{1 + j\omega CR_F} \qquad ...(13.36)$$

Then, $\quad H(0) = -\dfrac{R_F}{R_i}$

We put $\omega_C = \dfrac{1}{R_F C}$

Now, $H(\omega) = \dfrac{H(0)}{1 + j\dfrac{\omega}{\omega_C}}$ ... (13.37)

So, $|H(\omega)| = \dfrac{|H(0)|}{\sqrt{1 + \left(\dfrac{\omega}{\omega_C}\right)^2}}$ ... (13.38)

At $\omega = \omega_C$, $|H(\omega)| = \dfrac{1}{\sqrt{2}}|H(0)|$. Thus, ω_C is the cut-off radian frequency of the LPF where

the magnitude of the transfer function falls to $\dfrac{1}{\sqrt{2}}$ of its zero frequency value. The cut-off

frequency is

$$f_C = \dfrac{\omega}{2\pi} = \dfrac{1}{2\pi R_F C}$$ ... (13.39)

The phase shift produced by this filter is

$$\theta = \pi - \tan^{-1}\left(\dfrac{\omega}{\omega_C}\right)$$ ... (13.40)

13.2.9 PHASE INVERTER

The phase inverter circuit is shown in Fig. 13.10. Here, the feedback resistance R_F is made equal to the input resistance (R) connected with the inverting terminal.

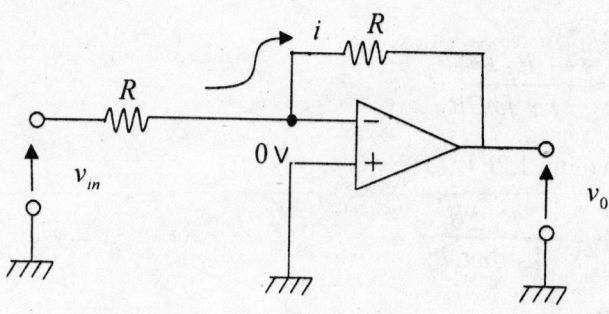

Fig. 13.10. Phase inverter circuit.

Now, current $i = \dfrac{v_{in} - 0}{R} = \dfrac{v_{in}}{R}$ ... (13.41)

Again, $i = \dfrac{0 - v_0}{R} = -\dfrac{v_0}{R}$ ... (13.42)

$\therefore$ $v_0 = -v_{in}$ ... (13.43)

Thus, output voltage is equal to the input voltage with a phase inversion. A phase change of $180°$ takes place in this circuit. This circuit is also called a sign changer circuit.

13.2.10 VOLTAGE FOLLOWER

In voltage follower, the output voltage of the device follows the input signal voltage. To realize this function in an OP-Amp circuit, the inverting terminal is directly connected with the output. This means the output voltage equal to the inverting input voltage. The signal voltage is applied to the non-inverting terminal. Due to the existence of a virtual ground, the inverting input voltage becomes equal to the signal voltage. This results in a voltage following action. The circuit diagram is shown in Fig. 13.11.

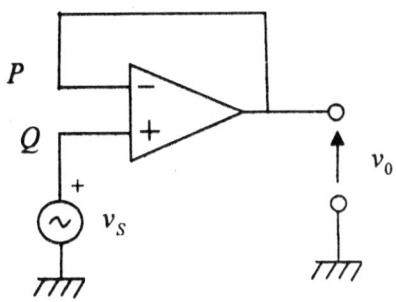

Fig. 13.11. Voltage follower circuit.

Potential at Q is $V_Q = v_S$.

Potential at P is $V_P = V_Q$.

Again, $v_0 = V_P$, since the output is connected with P.

Now, $V_P = V_Q = v_S$.

$\therefore$ $v_0 = v_S$ ... (13.44)

This circuit finds application as a impedance matching device. It can match a very high impedance circuit with a low impedance load. Voltage remains the same but possibility of impedance mismatch is eliminated. In this sense, it acts as a buffer stage.

13.2.11 VOLTAGE-TO-CURRENT CONVERTER

This circuit produces a load current which is proportional to the signal voltage. Here, the load resistance R_L acts as the feedback resistance. The circuit is shown in Fig. 13.12.

From the circuit, potential at Q is $V_Q = v_S$.

Since there is a virtual ground at the input, we have $V_P = V_Q$.

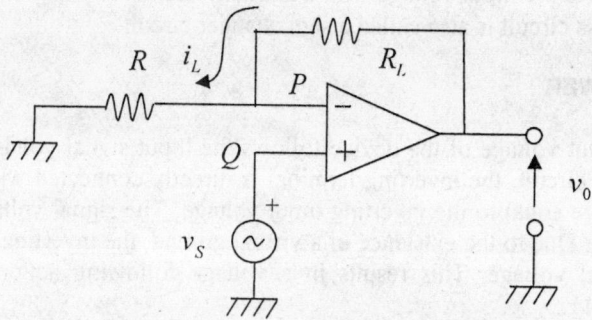

Fig. 13.12. Voltage-to-current converter circuit.

Then, load current $i_L = \dfrac{V_P - 0}{R} = \dfrac{v_S}{R}$. Therefore, $i_L \propto v_S$.

It finds application to supply current to the deflection coils in the picture tube of a television receiver.

13.2.12 CURRENT-TO-VOLTAGE CONVERTER

Here, the load resistance R_L acts as the feedback resistance of the OP-Amp. The non-inverting input is grounded so that there is a virtual ground at the inverting input of the OP-Amp. A current source I_S drives the inverting input of the OP-Amp. The converter circuit is shown in Fig.13.13.

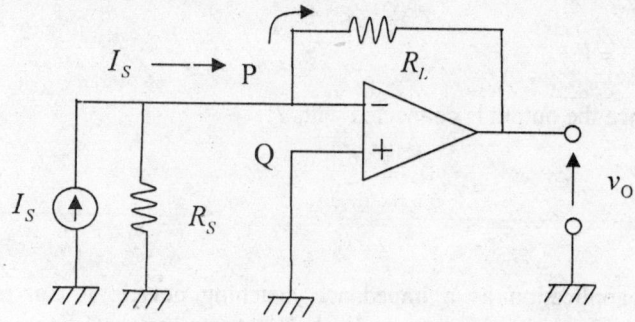

Fig. 13.13. Current-to-voltage converter circuit.

From the circuit, in Fig. 13.13, potential at Q, $V_Q = 0$.

Due to the existence of virtual ground, $V_P = V_Q$.

The current I_S flows through the load R_L. Due to high input resistance I_S does not enter the OP-Amp.

$$\therefore \qquad I_S = \frac{0 - v_0}{R_L} = -\frac{v_0}{R_L}$$

$$\therefore \qquad v_0 = -R_L I_S \qquad\qquad\qquad \ldots (13.45)$$

So, $\qquad v_0 \propto I_S$.

This circuit produces a voltage which is proportional to signal current and hence the name current-to-voltage converter. It finds application to convert the output current of a photomultiplier tube into a corresponding voltage signal.

13.2.13 ACTIVE HIGH-PASS FILTER

A schematic circuit of the active high pass filter is shown in Fig. 13.14.

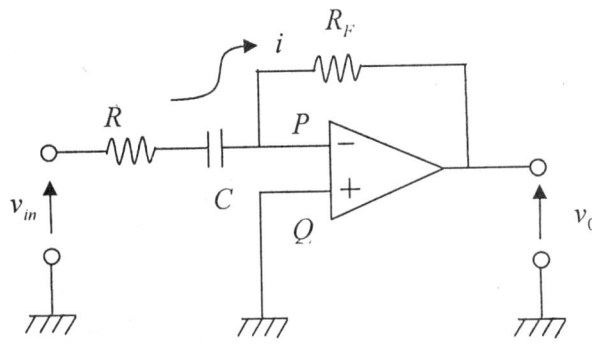

Fig. 13.14. Active high pass filter circuit.

Here, $\quad V_Q = 0$. So, $V_P = 0$.

The series combination of R and C has an impedance

$$Z = R + \frac{1}{j\omega C} = \frac{1 + j\omega CR}{j\omega C}$$

Current, $i = \dfrac{v_{in} - 0}{Z} = \dfrac{v_{im}}{Z}$ $\qquad\qquad \ldots (13.46)$

Again, $i = \dfrac{V_P - v_0}{R_F} = -\dfrac{v_0}{R_F}$ $\qquad\qquad \ldots (13.47)$

Equating (13.46) and (13.47), we get

$$-\frac{v_0}{R_F} = \frac{v_{in}}{Z}$$

$$\therefore \quad \frac{v_0}{v_{in}} = -\frac{R_F}{Z} = -\frac{j\omega C R_F}{1 + j\omega C R}$$

$$= -\frac{R_F}{R\left(1 + \dfrac{1}{j\omega C R}\right)} = -\frac{R_F/R}{1 - j\dfrac{\omega_H}{\omega}} \qquad \dots (13.48)$$

where $\omega_H = \dfrac{1}{RC}$. The transfer function of the filter is

$$H(j\omega) = \frac{v_0}{v_{in}} = -\frac{R_F/R}{1 - j\dfrac{\omega_H}{\omega}} \qquad \dots (13.49)$$

In the high frequency limit when $\omega \to \infty$, the transfer function $H(j\omega) \to -(R_F/R)$.

Now, $|H(j\omega)| = \dfrac{(R_F/R)}{\sqrt{1 + (\omega_H/\omega)^2}}$ $\qquad \dots (13.50)$

When $\omega = \omega_H$, $|H(j\omega)| = \dfrac{(R_F/R)}{\sqrt{2}}$ $\qquad \dots (13.51)$

At $\omega = \omega_H$, the magnitude of the transfer function falls to $\dfrac{1}{\sqrt{2}}$ of its high frequency value (R_F/R). If we take $R_F = R$, this high frequency limit of the transfer function becomes equal to 1. ω_H is the cut-off radian frequency of this high pass filter. The phase shift produced by this filter is

$$\theta_H = \pi + \tan^{-1}(\omega_H/\omega) \qquad \dots (13.52)$$

We now consider another active high pass filter circuit. This is shown in Fig. 13.15.

Here, $V_P = V_Q$

Again, $v_Q = \dfrac{v_{in} R}{R + \dfrac{1}{j\omega C}} = \dfrac{v_{in} \cdot j\omega C R}{1 + j\omega C R}$ $\qquad \dots (13.53)$

Now, $\dfrac{v_0 - V_P}{R_F} = i = \dfrac{V_P - 0}{R_1}$

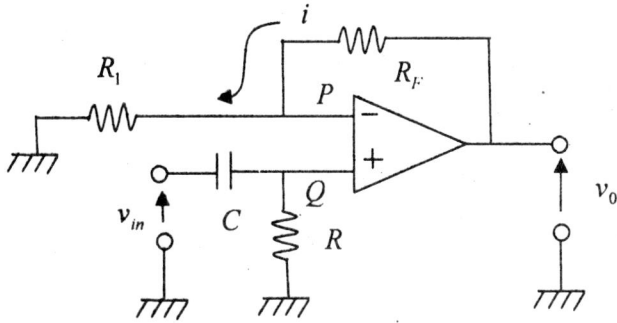

Fig. 13.15. A second circuit for active high pass filter.

$$\therefore \quad v_0 - V_P = \frac{R_F}{R_1} V_P$$

$$\therefore \quad v_0 = \left(1 + \frac{R_F}{R_1}\right) V_P$$

$$= \left(1 + \frac{R_F}{R_1}\right) \frac{j\omega CR}{1 + j\omega CR} v_{in} \qquad \qquad \dots (13.54)$$

since $V_P = V_Q$. Here, equation (13.53) has been used.

The transfer function of the filter is

$$H(j\omega) = \frac{v_0}{v_{in}} = \left(1 + \frac{R_F}{R_1}\right) \frac{1}{1 + \dfrac{1}{j\omega CR}} \qquad \qquad \dots (13.55)$$

$$= \left(1 + \frac{R_F}{R_1}\right) \frac{1}{1 - j\dfrac{\omega_H}{\omega}} \quad \text{where } \omega_H = \frac{1}{RC}.$$

$$\therefore \quad |H(j\omega)| = \left(1 + \frac{R_F}{R_1}\right) \frac{1}{\sqrt{1 + (\omega_H/\omega)^2}} \qquad \qquad \dots (13.56)$$

When ω is very large (i.e, $\omega \gg \omega_H$), $|H(j\omega)| = \left(1 + \dfrac{R_F}{R_1}\right)$.

At $\quad \omega = \omega_H$, $|H(j\omega)| = \dfrac{1}{\sqrt{2}}\left(1 + \dfrac{R_F}{R_1}\right)$.

ω_H is called the cut-off frequency of the high pass filter in radian. The phase shift produced by this filter is

$$\theta_H = \tan^{-1}(\omega_H/\omega) \qquad \qquad \dots (13.57)$$

13.2.14 LOGARITHMIC AMPLIFIER

The circuit diagram of the logarithmic amplifier using an OP-Amp is shown in Fig. 13.16.

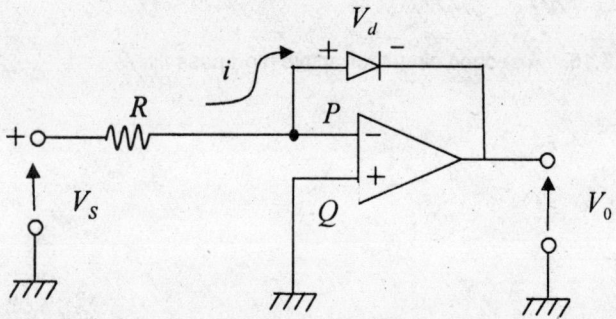

Fig. 13.16. Logarithmic amplifier.

A p-n junction diode is inserted between the inverting terminal and the output. The output voltage V_0 is proportional to the natural logarithm of the input voltage.
The diode current

$$I_d = I_0\left(e^{V_d/\eta V_T} - 1\right) \qquad \qquad \dots (13.58)$$

where I_0 = reverse saturation current of the diode,

$$V_T = \frac{kT}{q}$$

k = Boltzmann constant
T = absolute temperature
q = magnitude of electronic charge.
Under forward bias condition, $I_d \gg I_0$ since $V_d \gg V_T$.

This makes $e^{V_d/\eta V_T} \gg 1$. $\therefore \qquad I_d = I_0 e^{V_d/\eta V_T}$

or, $\ln I_d = \ln I_0 + \dfrac{V_d}{\eta V_T}$

$\therefore \qquad V_d = \eta V_T\left(\ln I_d - \ln I_0\right) \qquad \qquad \dots (13.59)$

Now, $V_P = V_Q = 0$

$\therefore \qquad i = I_d = \dfrac{V_S}{R}$ $\qquad\qquad\qquad$ ... (13.60)

where I_d = diode current and

$\qquad V_S$ = input voltage.

Again, $V_0 + V_d = V_P = 0$

$\therefore \qquad V_d = -V_0$ $\qquad\qquad\qquad\qquad$ ... (13.61)

From eqn. (13.59) and (13.61) we get

$$V_0 = -\eta V_T \left(\ln I_d - \ln I_0 \right)$$

$$= -\eta V_T \left(\ln \dfrac{V_S}{R} - \ln I_0 \right)$$

$$= -\eta V_T \left(\ln V_S - \ln V_R \right) \qquad\qquad ...(13.62)$$

where $V_R = I_0 R$. ηV_T is scale factor.

Now, $V_0 = 0$ when $V_S = V_R$. Again, when $\ln V_S = 0$, we have $V_0 = \eta V_T \ln V_R$. If V_0 is plotted against $\ln V_S$ we get a straight line with slope $-\eta V_T$ and an intercept of $(\eta V_T \ln V_R)$ with the $(\ln V_S)$ axis. The plot is shown in Fig. 13.17.

If $\ln V_R > 0$ then the straight line will pass through first quadrant. If $\ln V_R < 0$ then the straight line will pass through the 3rd quadrant. The slope is the same in both cases.

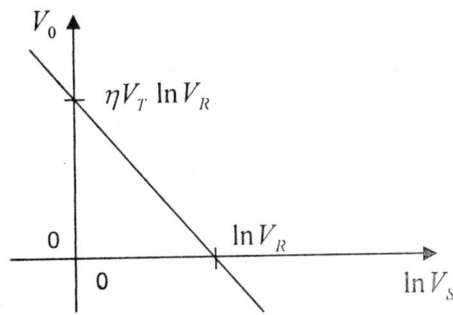

Fig. 13.17. Logarithmic amplifier characteristics.

13.2.15 Logarithmic amplifier with matched transistor pair

Here, we use a pair of identical transistors along with a pair of operational amplifiers to design a logarithmic amplifier. The effect of reverse saturation current of emitter-base diodes of the transistors is cancelled. As a result, the characteristics of the logarithmic amplifier becomes much less sensitive to temperature.

The corresponding circuit is shown in Fig. 13.18.

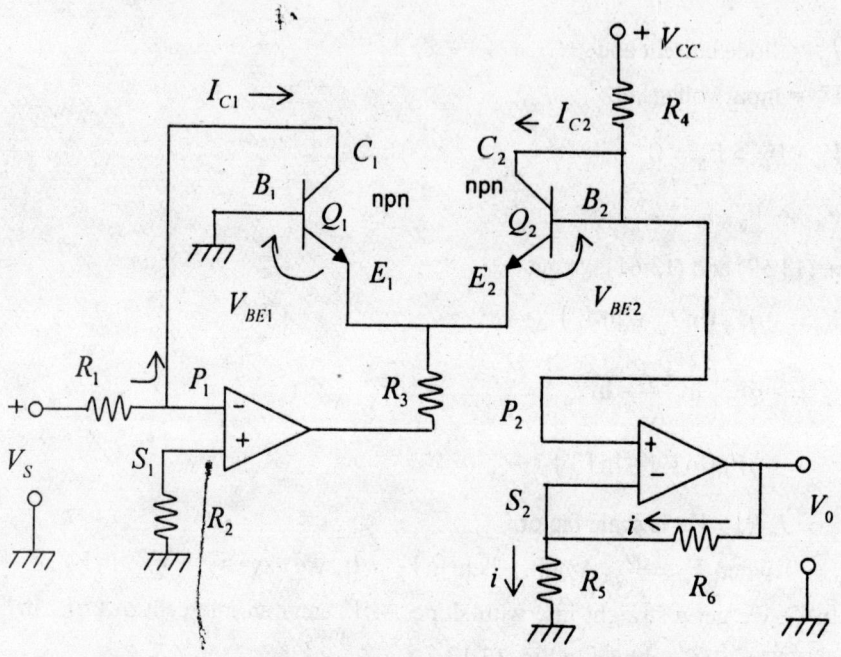

Fig. 13.18. Logarithmic amplifier circuit with a matched transistor pair.

With reference to the above circuit, the potential difference between B_2 and B_1 is

$$V_{B2} - V_{B1} = V_{BE2} - V_{BE1}$$

$$\therefore \quad V_{B2} = V_{BE2} - V_{BE1} \qquad \qquad \dots (13.63)$$

Since, $V_{B1} = 0$ (B_1 being grounded).

Again, $v_{P1} = v_{S1}$ $\qquad$ and $\qquad$ $v_{P2} = v_{S2}$.

Now, $\dfrac{V_0 - v_{S2}}{R_6} = i = \dfrac{v_{S2} - 0}{R_5}$

$$\therefore \quad V_0 = v_{S2}\left(1 + \frac{R_6}{R_5}\right)$$

$$= v_{P2}\left(1 + \frac{R_6}{R_5}\right) \qquad \qquad \dots (13.64)$$

Again, from circuit $v_{P2} = V_{B2}$.

The emitter current (I_{E1}) of transistor Q_1 is

$$I_{E1} = I_S\ e^{V_{BE1}/\eta V_T} \qquad \qquad \dots (13.65)$$

$\therefore \qquad \ln I_{E1} = \ln I_S + \dfrac{V_{BE1}}{\eta V_T}$

$\therefore \qquad V_{BE1} = \eta V_T \left(\ln I_{E1} - \ln I_S \right) \qquad \qquad \dots (13.66)$

Similarly, the emitter current (I_{E2}) of transistor Q_2

$$I_{E2} = I_S e^{V_{BE2}/\eta V_T} \qquad \qquad \dots (13.67)$$

$\therefore \qquad V_{BE2} = \eta V_T \left(\ln I_{E2} - \ln I_S \right) \qquad \qquad \dots (13.68)$

Reverse saturation current is the same for the identical transistors Q_1 and Q_2.

$\therefore \qquad V_{B2} = \eta V_T \left(\ln I_{E2} - \ln I_{E1} \right)$

$$= \eta V_T \ln \left(\dfrac{I_{E2}}{I_{E1}} \right) \qquad \qquad \dots (13.69)$$

Again, since the base currents are smaller than the collector currents, we write

$$I_{B2} \ll I_{C2} \text{ and } I_{B1} \ll I_{C1}.$$

So, $\qquad I_{E2} \cong I_{C2}$ and $I_{E1} \cong I_{C1}$

$\therefore \qquad V_{B2} = \eta V_T \ln \left(\dfrac{I_{C2}}{I_{C1}} \right) \qquad \qquad \dots (13.70)$

From circuit, $I_{C2} = \dfrac{V_{CC} - V_{B2}}{R_4}$, neglecting I_{B2} in comparison with I_{C2}.

$$\cong \dfrac{V_{CC}}{R_4} \quad \text{since } |V_{B2}| \ll V_{CC}. \qquad \qquad \dots (13.71)$$

V_{CC} is the collector supply voltage of Q_2. Due to the existence of the virtual ground at P_1 and S_1, we have $v_{P1} = v_{S1} = 0$

$\therefore \qquad I_{C1} = \dfrac{V_S - 0}{R_1} = \dfrac{V_S}{R_1} \qquad \qquad \dots (13.72)$

Here, V_S is the input voltage.

From eqns. (13.64) and (13.70) and noting that $v_{p2} = V_{B2}$, we write

$$V_0 = \left(1 + \frac{R_6}{R_5}\right) \eta V_T \ln\left(\frac{I_{C2}}{I_{C1}}\right)$$

$$= -\left(1 + \frac{R_6}{R_5}\right) \eta V_T \ln\left(\frac{I_{C1}}{I_{C2}}\right)$$

$$= -\left(1 + \frac{R_6}{R_5}\right) \eta V_T \ln\left(\frac{V_S}{R_1} \frac{R_4}{V_{CC}}\right)$$

$$= -\eta V_T \left(1 + \frac{R_6}{R_5}\right) \left[\ln V_S - \ln\left(\frac{R_1}{R_4} V_{CC}\right)\right] \qquad \ldots (13.73)$$

Thus, V_0 varies linearly with ($\ln V_S$). The slope of the straight line is $-\eta V_T \left(1 + \frac{R_6}{R_5}\right)$. The

intercept on the $\ln V_S$ axis is $\eta V_T \left(1 + \frac{R_6}{R_5}\right) \ln\left(\frac{R_1}{R_4} V_{CC}\right) = \ln V_{S0}$ while that on V_0 axis is

$V_{00} = \ln\left(\frac{R_1}{R_4} V_{CC}\right)$.

If the values of R_1, R_4 and V_{CC} are such that $\ln\left(\frac{R_1}{R_4} V_{CC}\right) > 0$ then this straight line passes

across the first quadrant. On the other hand, if $\ln\left(\frac{R_1}{R_4} V_{CC}\right) < 0$ then this straight line passes

across the third quadrant. The plot is shown schematically in Fig. 13.19.

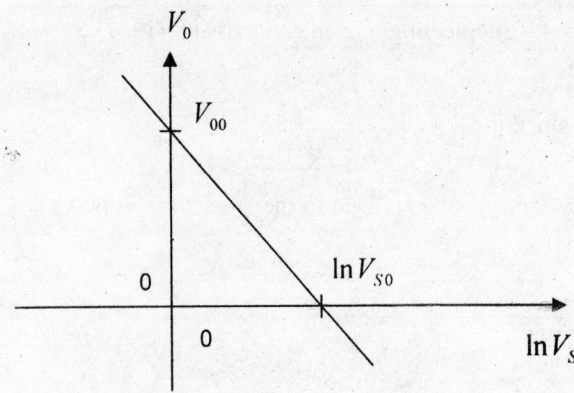

Fig. 13.19. Schematic plot of V_0 vs $\ln V_S$ characteristics in logarithmic amplifier with matched transistor pair.

13.2.16 ANTILOG AMPLIFIER

The circuit of the antilog amplifier is shown in Fig. 13.20. The output of this circuit is proportional to the antilog of the input. Here, we use two transistors Q_1 and Q_2 along with two OP-Amps.

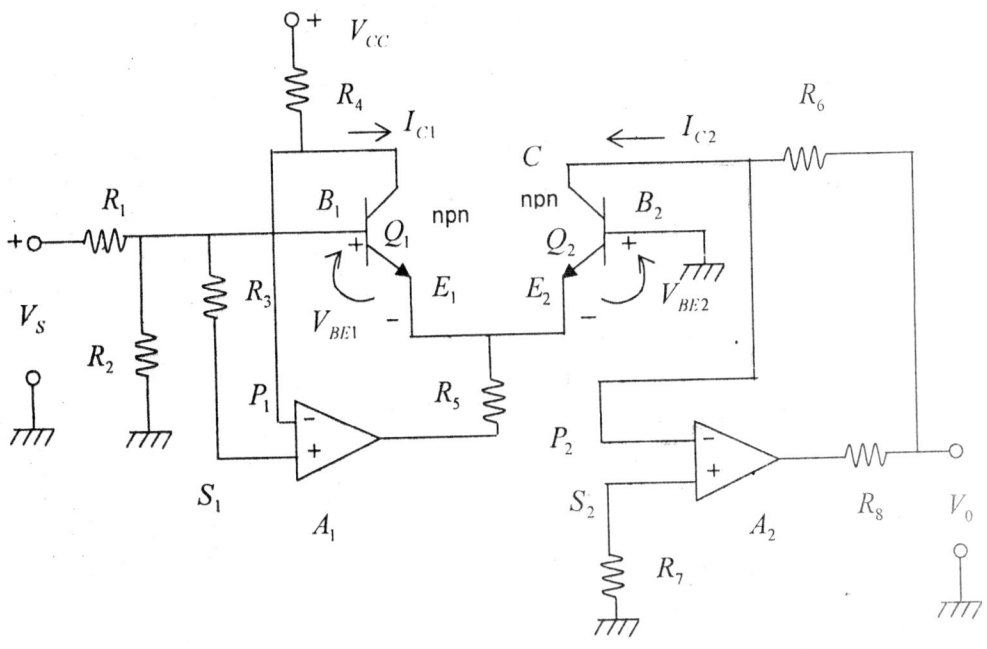

Fig. 13.20. Circuit diagram of the antilog amplifier.

Here, $v_{P2} = v_{S2} = 0$ ... (13.74)
(since there exists a virtual ground at P_2 and S_2)
Now, collector current of Q_2 is

$$I_{C2} = \frac{V_0 - v_{P2}}{R_6} = \frac{V_0}{R_6} \qquad \text{... (13.75)}$$

No current enters the P_2 terminal of OP-Amp A_2.
The voltage at the base B_1 is

$$v_{B1} = \frac{R_2 V_S}{R_1 + R_2} \qquad \text{... (13.76)}$$

The voltage at B_1 with respect to B_2 is

$$v_{B1} = V_{BE1} - V_{BE2} \qquad \text{... (13.77)}$$

The emitter current of the emitter-base diode of the transistor is given by

$$I_E = I_S e^{V_{BE}/\eta V_T} \qquad \qquad \dots (13.78)$$

where I_S = reverse saturation current.

Then, $\quad V_{BE} = \eta V_T \ln\left(\dfrac{I_E}{I_S}\right)$

$$\simeq \eta V_T \ln\left(\dfrac{I_c}{I_S}\right) \qquad \qquad \dots (13.79)$$

since $I_E \cong I_C$, the base current being small.
Then,

$$v_{B1} = \eta V_T \left[\ln\left(\dfrac{I_{C1}}{I_S}\right) - \ln\left(\dfrac{I_{C2}}{I_S}\right) \right]$$

$$= \eta V_T \ln\left(\dfrac{I_{C1}}{I_{C2}}\right) \qquad \qquad \dots (13.80)$$

Now, $I_{C1} = \dfrac{V_{CC} - v_{P1}}{R_4} = \dfrac{V_{CC} - v_{S1}}{R_4} \qquad [\text{ since } v_{P1} = v_{S1}]$

$$= \dfrac{V_{CC} - v_{B1}}{R_4} \qquad [\because v_{S1} = v_{B1}]$$

$$\simeq \dfrac{V_{CC}}{R_4} \qquad \qquad \dots (13.81)$$

since $|v_{B1}| << V_{CC}|$. Thus, I_{C1} is constant.

From eqn. (13.80) we get, using eqns. (13.75) and (13.81),

$$v_{B1} = \eta V_T \ln\left(\dfrac{V_{CC}}{R_4}\dfrac{R_6}{V_0}\right) \qquad \qquad \dots (13.82)$$

substituting for v_{B1} from eqn. (13.76) in (13.82) we write

$$\dfrac{R_2 V_S}{R_1 + R_2} = \eta V_T \ln\left(\dfrac{V_{CC}}{V_0}\dfrac{R_6}{R_4}\right)$$

$$= -\eta V_T \ln\left(\dfrac{R_4}{R_6}\dfrac{V_0}{V_{CC}}\right) \qquad \qquad \dots (13.83)$$

$$\therefore \qquad V_0 = \frac{R_6}{R_4} V_{CC} \ln^{-1}\left[-\frac{R_2 V_S}{(R_1 + R_2)\eta V_T} \right] \qquad\qquad ...(13.84)$$

Also, $\quad V_0 = \frac{R_6}{R_4} V_{CC}\, e^{-\frac{R_2 V_S}{(R_1 + R_2)\eta V_T}} \qquad\qquad\qquad ...(13.85)$

If $\qquad V_S = 0,\ V_0 = \dfrac{R_6}{R_4} V_{CC} = V_{00}$ (say)

From eqn. (13.83) we can write

$$V_S = -\eta V_T \left(1 + \frac{R_1}{R_2}\right) \ln\left(\frac{V_0}{V_{00}} \right) \qquad\qquad ...(13.86)$$

The V_S vs ($\ln V_0$) characteristic is a straight line with a slope $-\eta V_T\left(1 + \dfrac{R_1}{R_2}\right)$. The intercept on

the V_S axis is $\eta V_T\left(1 + \dfrac{R_1}{R_2}\right)\ln V_{00}$. This plot is shown in Fig.13.21.

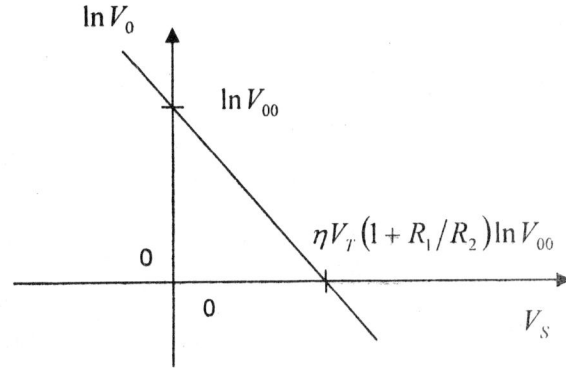

Fig. 13.21. Schematic plot of $\ln V_0$ vs. V_S of an antilog amplifier.

13.2.17 SCHMITT TRIGGER

An operational amplifier has a very high internal (open loop) gain which is ideally infinite. In absence of any feedback, if there exists a differential input of the OP-Amp the output of the OP-Amp saturates to the supply voltage ($\pm V_{SS}$). Suppose the voltage applied to the inverting terminal is greater than that of the non-inverting terminal. Then, the output saturates to $-V_{SS}$. On the other hand, when the non-inverting terminal voltage is greater than that of the inverting terminal voltage, its output saturates to $-V_{SS}$. In this way, the OP-Amp can act as a voltage comparator which can compare two voltages and gives an output voltage whose polarity indicates which of the two voltages is greater.

To design a Schmitt Trigger, we apply a positive feed back to the otherwise open-loop OP-Amp. The output voltage of the OP-Amp is divided by a potential divider and a fraction of the output voltage is fed back to the non-inverting terminal of the OP-Amp. The Schmitt Trigger circuit is shown in Fig.13.22.

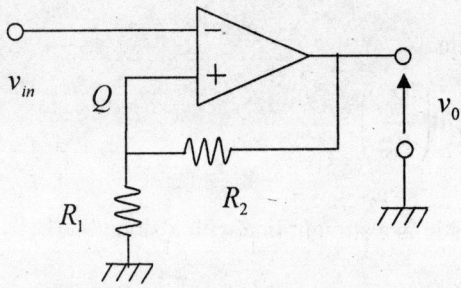

Fig. 13.22. Schmitt Trigger circuit.

The potential at Q is

$$V_Q = \frac{v_0 R_1}{R_1 + R_2} \qquad \qquad \ldots (13.87)$$

Since $\quad v_0 = \pm V_{SS}, V_Q = \pm \dfrac{R_1 V_{SS}}{R_1 + R_2} \qquad \qquad \ldots (13.88)$

When the input signal voltage v_{in} exceeds the value $\dfrac{R_1 V_{SS}}{R_1 + R_2}$, the Schmitt Trigger switches to the negative saturation $(-V_{SS})$. This trigger level is called upper trip point (UTP). Similarly, when the input signal voltage v_{in} goes below the value $-\dfrac{R_1 V_{SS}}{R_1 + R_2}$, the Schmitt Trigger switches to the positive saturation $(+V_{SS})$. This trigger level is called the lower trip point (LPT). Thus,

$$UTP = \frac{R_1}{R_1 + R_2} V_{SS} \qquad \qquad \ldots (13.89)$$

and $\quad LTP = -\dfrac{R_1}{R_1 + R_2} V_{SS} \qquad \qquad \ldots (13.90)$

The output waveform of the Schmitt Trigger is shown in Fig. 13.23.

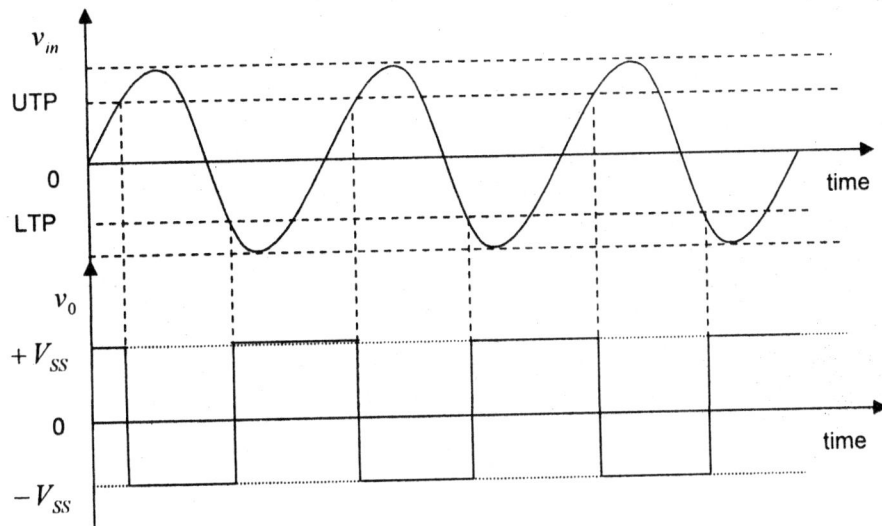

Fig. 13.23. Output waveform (v_0) of the Schmitt Trigger.

HYSTEESIS LOOP

If we observe the variation of output voltage of the Schmitt Trigger with the voltage amplitude of the input signal, we can notice that the output does not follow the same path during increasing and decreasing of the input voltage. When the input voltage increases, the output changes state at a certain value of the input voltage. Now, if the input voltage is gradually reduced, the output of the device does not go back to its previous state at that value of the input voltage. This difference in behaviour of the device produces a loop in the input voltage vs. output voltage characteristics of the Schmitt Trigger. This is called the hystersis loop. A typical hystersis loop is shown in Fig. 13.24.

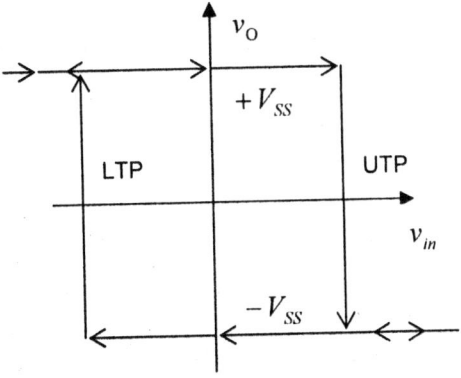

Fig. 13.24. Hysteresis loop of the Schmitt Trigger.

When the OP-Amp supply voltages ($\pm V_{SS}$) have same magnitudes, the loop should be symmetric as shown in Fig. 13.24. The loop can be made asymmetric by applying a reference voltage with the end of the resistance R_1 (instead of grounding R_1).

13.2.18 ASTABLE MULTIVIBRATOR

The term 'astable' means 'no stable state'. The term 'multivibrator' means many vibrations, i.e., many frequencies are present at its output. A multivibrator is a device which generates non-sinusoidal oscillations.

The astable multivibrator circuit using an operational amplifier is shown in Fig. 13.25.

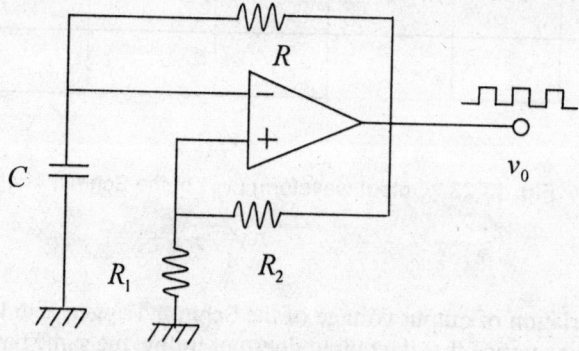

Fig. 13.25. Astable multivibrator circuit.

To understand the operation of the circuit, suppose that the RC network is removed from the circuit for the time being. Then, we have an OP-Amp with positive feedback applied to the non-inverting terminal. The feedback factor $\beta = \dfrac{R_1}{R_1 + R_2}$. When a voltage is applied to the inverting terminal of the OP-Amp, the device will switch to $+V_{SS}$ or $-V_{SS}$ depending upon which input is higher. $\pm V_{SS}$ are the supply voltages of the OP-Amp. This operation is similar to that of a Schmitt Trigger. The device will give an output of $+V_{SS}$ when the inverting input voltage just goes below the lower trip point (LPT). This is shown in Fig.13.23. Thus, we get a state of the device defined by $v_0 = V_{SS}$ and $v_{in} = LPT = -\beta V_{SS}$.

Now, connect the RC network as shown in Fig. 13.26. The RC network has two voltages applied to it – one is V_{SS} at the end of R and another is $-\beta V_{SS}$ at the junction point of C and R. This is shown in Fig.13.26. The capacitor C thus charges up by the voltage V_{SS} and discharges by the negative voltage $-\beta V_{SS}$ simultaneously.

The capacitor voltage at any time t is given by

$$v_C(t) = V_{SS}\left(1 - e^{-t/RC}\right) - \beta V_{SS}e^{-t/RC}$$

$$\dots (13.91)$$

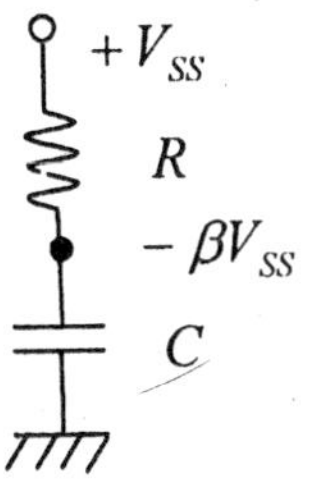

Fig. 13.26. Charging and discharging of the capacitor.

At $t = 0$, $v_C(0) = -\beta V_{SS}$ and at $t = \alpha$, $v_C(t = \infty) = V_{SS}$.

The inverting terminal voltage rises from a value $-\beta V_{SS}$ to a value $+\beta V_{SS}$ (UTP) during charging. When the inverting terminal voltage crosses the value $+\beta V_{SS}$, the OP-Amp switches to the negative saturation voltage $-V_{SS}$ at its output. Let T_1 be the time of charging of the capacitor to $+\beta V_{SS}$. Then,

$$\beta V_{SS} = V_{SS}\left(1 - e^{-T_1/RC}\right) - \beta V_{SS}e^{-T_1/RC}$$

$$= V_{SS} - (1 + \beta)V_{SS}e^{-T_1/RC} \qquad \ldots (13.92)$$

$$\therefore \qquad T_1 = RC \ln\left(\frac{1+\beta}{1-\beta}\right) \qquad \ldots (13.93)$$

This is one half of the complete cycle of operation.

In the remaining half cycle, the capacitor voltage charges from $+\beta V_{SS}$ (UTP) to $-\beta V_{SS}$ (LTP). At the beginning of this half cycle, the output voltage of the OP-Amp is at the negative saturation level $-V_{SS}$.

The capacitor voltage is now described by the equation

$$v_C(t) = -V_{SS}\left(1 - e^{-t/RC}\right) + \beta V_{SS}e^{-t/RC} \qquad \ldots (13.94)$$

At $t = 0$, $v_C(0) = \beta V_{SS}$ and at $t = \infty$, $v_C(t = \infty) = -V_{SS}$.

The corresponding circuit is shown below in Fig. 13.27.

Let T_2 be the time taken by the capacitor to change its voltage from $+\beta V_{SS}$ to $-\beta V_{SS}$. Then,

$$-\beta V_{SS} = -V_{SS}\left(1 - e^{-T_2/RC}\right) + \beta V_{SS}e^{-T_2/RC}$$

$$= -V_{SS} + (1 + \beta)V_{SS}e^{-T_2/RC} \qquad \ldots (13.95)$$

$$-V_{SS}$$
$$R$$
$$\beta V_{SS}$$
$$C$$

Fig. 13.27. Capacitor charging and discharging in the second half cycle.

$$\therefore \quad 1 - \beta = (1 + \beta) e^{-T_2/RC}$$

$$\therefore \quad T_2 = RC \ln\left(\frac{1+\beta}{1-\beta}\right) \qquad \qquad \dots (13.96)$$

We see that $T_1 = T_2$. The waveform is a square wave.
The total time period of oscillation is

$$T = T_1 + T_2 = 2RC \ln\left(\frac{1+\beta}{1-\beta}\right)$$

$$= 2RC \ln\left(1 + \frac{2R_1}{R_2}\right) \qquad \qquad \dots (13.97)$$

The frequency of oscillation is $f = \dfrac{1}{T}$.

The capacitor voltage and output voltage waveforms are shown in Fig. 13.28.

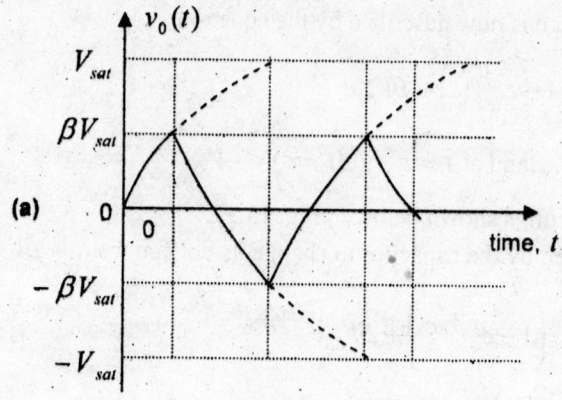

Fig. 13.28. (a) Capacitor voltage variation with time in astable multivibrator.

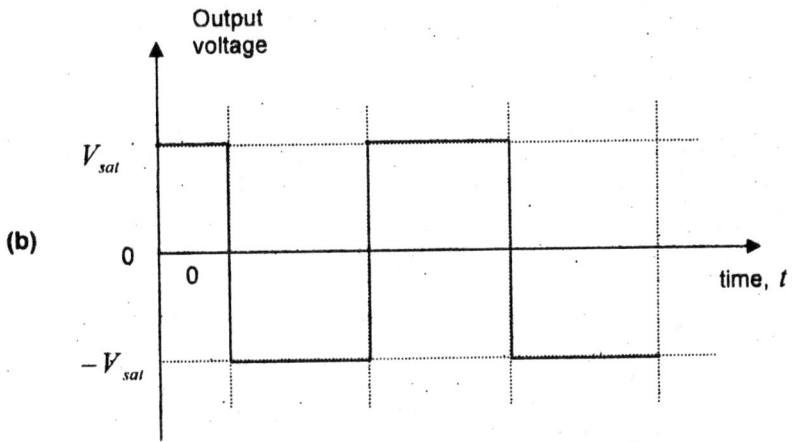

(b)

Fig.13.28. (b) Output voltage variation with time of an astable multivibrator.

13.3 DIGITAL-TO-ANALOG CONVERTER (D/A CONVERTER)

An analog quantity is one which assumes a continuous set of values over a specified range while a digital quantity assumes a discrete set of values. Any measurable quantity is analog in nature – such as pressure, temperature, time, velocity etc. A voltage which can assume any value between 0 to +10 volt in analog representation can be represented by a 4-bit binary code. In 4-bit binary code, 0-16 volt can be represented in steps of unity in a discrete manner.

Electronic systems sometimes require a digital signal at the input to be converted into analog signal in the form of voltage or current. This is done by a D/A converter.

13.3.1 BINARY- WEIGHTED DAC

The block diagram of a binary-weighted D/A converter is shown in Fig. 13.29 below. This method uses a resistance network which uses resistances having values proportional to binary weights of input digits. For a N-bit input binary word, there are N electronic switches S_0, S_1, ..., S_{N-1} which are controlled by input bits.

The smallest resistance (R) corresponds to the most significant bit (MSB) having the highest binary weight. The highest resistance ($2^{N-1}R$) corresponds to least significant bit (LSB) having the lowest binary weight.

The output voltage of the N-bit D/A converter is given as

$$V_0 = -\left(B_0 \frac{V_{ref}}{2^{N-1}R} + B_1 \frac{V_{ref}}{2^{N-2}R} + \ldots\ldots\ldots + B_{N-1} \frac{V_{ref}}{2^0 R} \right) R_f$$

$$= -\left(B_0 + 2B_1 + \ldots\ldots\ldots + 2^{N-1} B_{N-1} \right) V_{ref} \frac{R_f}{2^{N-1}R} \qquad \ldots (13.98)$$

$$= \left(B_0 + 2B_1 + \ldots\ldots\ldots + 2^{N-1} B_{N-1} \right) |V_{ref}| \frac{R_f}{2^{N-1}R} \qquad \text{when } V_{ref} = -|V_{ref}|.$$

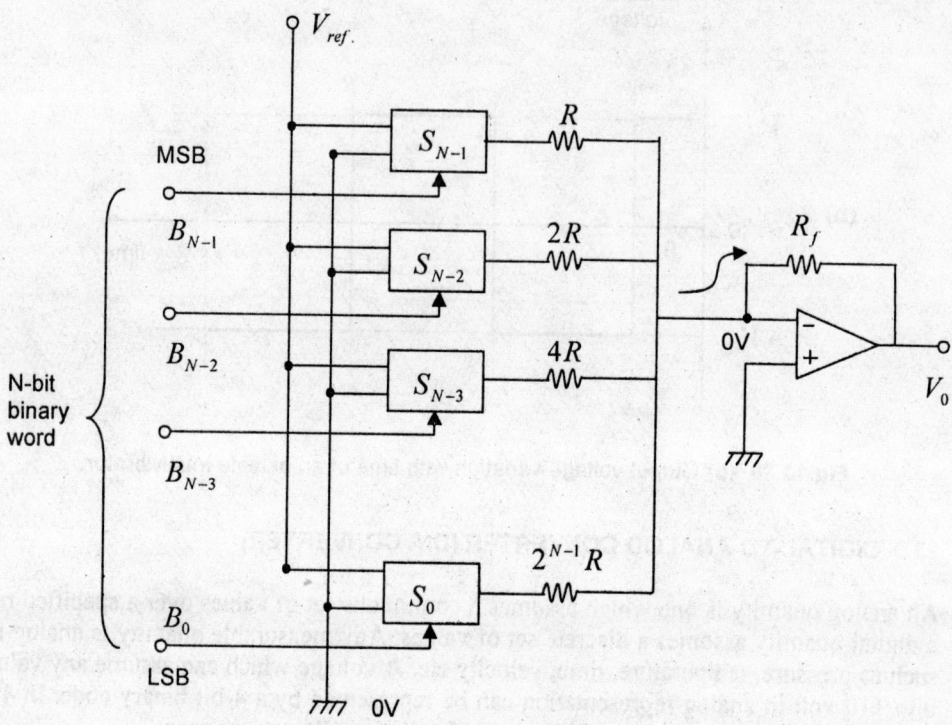

Fig. 13.29. Binary-weighted D/A converter.

The input bits B_0, B_1, ..., B_{N-1} can have values 0 or 1. When the input bit is 0, the switch connects to ground (0 V). But, when the input bit is 1, the switch connects to the reference source voltage (V_{ref}). R_f is the feedback resistance connected with the output OP-Amp acting as current-to-voltage converter.

As an example, take $R_f = \dfrac{R}{2}$. Take a 4-bit input word so that $N = 4$. Also, take $V_{ref} = -16$ V. Then,

$$V_0 = +\left(B_0 + 2B_1 + ...+ 8B_3\right)\dfrac{16}{16}$$

$$= +\left(B_0 + 2B_1 + ...+ 8B_3\right) \text{volt}. \qquad \qquad ... (13.99)$$

If the input word is 0001, then $B_0 = 1$ and $B_1 = B_2 = B_3 = 0$. Then, $V_0 = +1$ volt.

Similarly, when the input word is 0011, then $B_0 = B_1 = 1$ and $B_2 = B_3 = 0$. Now, $V_0 = 3$ volt. Thus, the circuit converts a digital word into an analog voltage.

The disadvantage of this binary weighted D/A converter is the requirement of accuracy of the resistors and the control of temperature variation of resistors required. In general, it is difficult to obtain and maintain stable, accurate resistors without increasing cost.

VOLTAGE SWITCH IN D/A CONVERTER

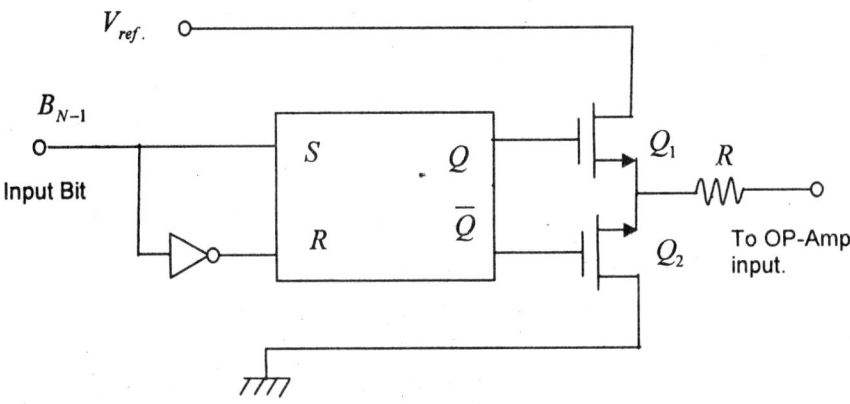

Fig. 13.30. Schematic diagram of the voltage switch in DAC.

Q_1, Q_2 are two n-channel MOSFETs. The S-R Flip-flop is driven by the input bit. When the input bit is 1, the FF is set and its output $Q = 1$ and $\overline{Q} = 0$. Here, 1 means a voltage V_{ref} and 0 means ground. A '1' on the gate of MOSFET Q_1 makes Q_1 ON. Similarly, a 0 on the gate of Q_2 makes Q_2 OFF. Thus, V_{ref} appears at the end of resistor R. This is shown in Fig. 13.30.

When the input bit is a '0', the FF is reset and $Q = 0$ with $\overline{Q} = 1$. Now, MOSFET Q_2 is ON and MOSFET Q_1 is OFF. With Q_2 ON, the resistor R is connected to ground. Thus, voltage switching is achieved.

13.3.2 R/2R LADDER D/A CONVERTER

The $R/2R$ ladder network acting as a D/A converter is shown in Fig. 13.31. N_0, N_1, N_2,.....$(N-1)$ are N number of nodes and S_0, S_1, S_2,.....S_{N-1} are N number of switches. These switches are transistor switches which are connected either to 0 V (ground) or to the reference voltage V_{ref} corresponding to '1' state. Consider (N-1)th node. Suppose all the switches S_0, S_1, S_2,..... S_{N-2} are grounded (logic 0) and only S_{N-1} switch is connected to V_{ref}. Thus, MSB is 1 and all other input bits are zero. For a 4-bit DAC, the digital input corresponds to 1000.

On the left side of any node, the resistance seen is $2R$. Below any node, the resistance is $2R$. On the right of any node the resistance seen is also $2R$. Consider node N_0 as shown in Fig. 13.32.

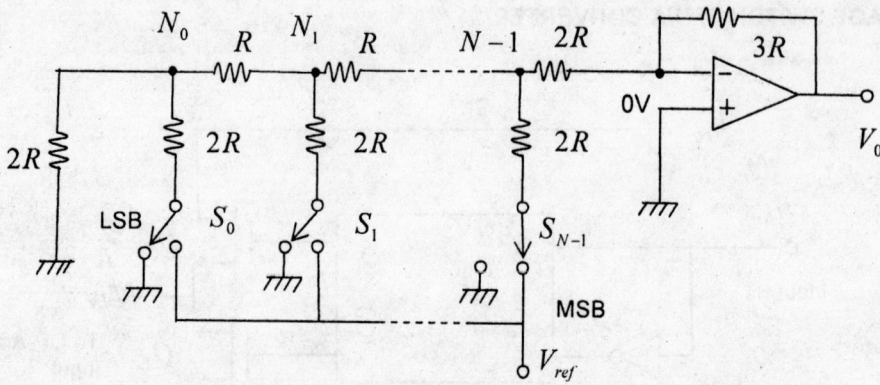

Fig. 13.31. R/2R ladder type DAC.

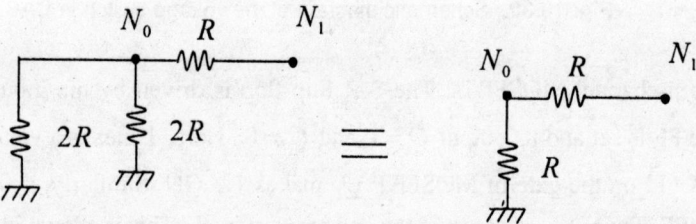

Fig. 13.32. Network simplification for node N_0.

Here, $2R \parallel 2R = R$. Similarly, for node N_1 we have the following circuit. As shown in Fig. 13.33.

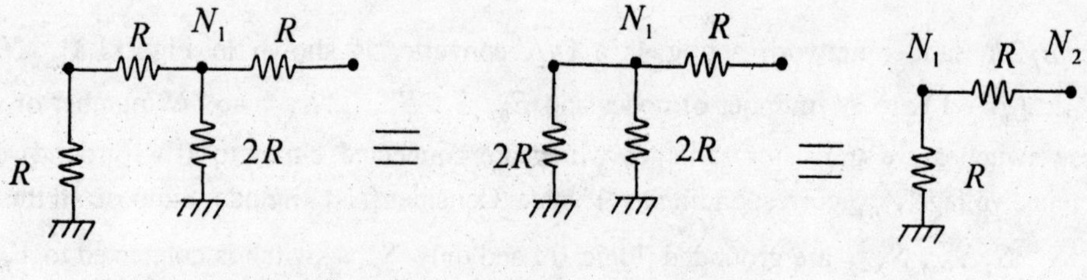

Fig. 13.33. Network simplification for node N_1.

The voltage at node $(N-1)$ is

$$V_{N-1} = \frac{V_{ref} \cdot R}{(R+2R)} = \frac{V_{ref}}{3} \qquad \ldots (13.100)$$

Then, $V_0 = -V_{N-1} \cdot \dfrac{3R}{2R} = -\dfrac{V_{ref}}{2}$ ... (13.101)

If V_{ref} is negative then $V_0 = \dfrac{|V_{ref}|}{2}$.

For a 4-bit DAC, we take $V_{ref} = -16$ V, so that $V_0 = 8$ V which corresponds to the digital signal 1000.

The equivalent circuit for ($N-1$)th node voltage calculation as shown in Fig. 13.34 is shown below.

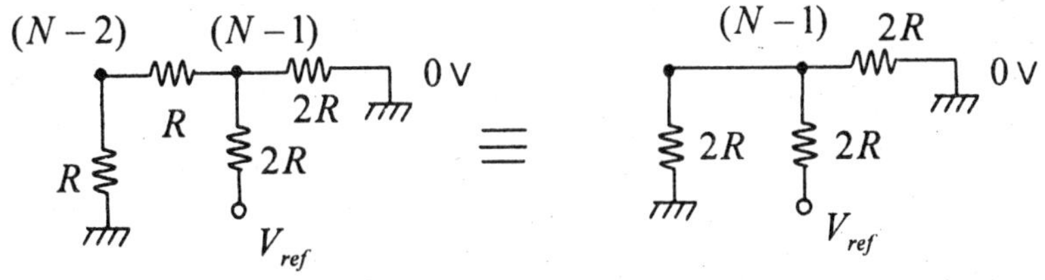

Fig. 13.34. Network simplification for node ($N-1$).

Suppose the switch S_{N-2} is connected to V_{ref} and all other switches are connected to ground. Now, the equivalent circuit is as follows (shown in Fig. 13.35).

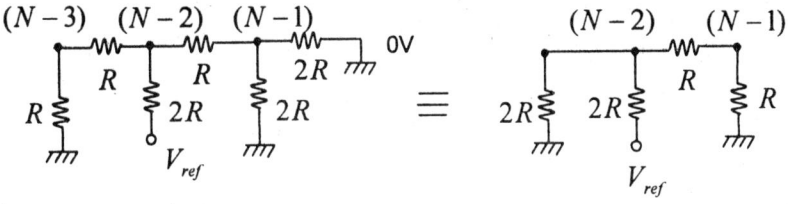

Fig. 13.35. Voltage calculation at node ($N-2$).

Then, voltage at node $(N-2)$ is

$$V_{N-2} = \frac{V_{ref}(2R \parallel 2R)}{(2R + 2R \parallel 2R)} = \frac{V_{ref} \cdot R}{(2R + R)} = \frac{V_{ref}}{3} \qquad \ldots (13.102)$$

Now, $\quad V_{N-1} = \frac{V_{N-2}R}{(R+R)} = \frac{V_{N-2}}{2} = \frac{V_{ref}}{6} \qquad \ldots (13.103)$

The output voltage is

$$V_0 = -V_{N-1} \cdot \frac{3R}{2R} = -\frac{V_{ref}}{6} \cdot \frac{3}{2} = -\frac{V_{ref}}{4} = -\frac{V_{ref}}{2^2} \qquad \ldots$$

Taking V_{ref} as a negative voltage,

$$V_0 = -\frac{|V_{ref}|}{2^2} \qquad \ldots (13.104)$$

Taking $V_{ref} = -16$ V we get $V_0 = 4$ V which corresponds to the digital signal 0100.

Similarly, if the switch S_{N-3} is connected to V_{ref} and all other switches are grounded, then the output voltage

$$V_0 = -\frac{V_{ref}}{8} = -\frac{V_{ref}}{2^3} = \frac{|V_{ref}|}{2^3} \qquad \ldots (13.105)$$

With $V_{ref} = -16$ V for a 4-bit DAC we get $V_0 = 2$ V which corresponds to the digital signal 0010.

Lastly, if S_0 is connected to V_{ref} and all other switches are grounded then we get

$$V_0 = +\frac{|V_{ref}|}{2^4} \qquad \ldots (13.106)$$

With $V_{ref} = 16$ V, we get $V_0 = 1$ V which corresponds to the digital signal 0001. Hence, we get an analog voltage corresponding to the binary input signal resulting in a D/A conversion.

The advantage of this ladder network is that it requires only two resistors R and $2R$. It does not require many precise resistances.

13.4 ANALOG-TO-DIGITAL CONVERTER (ADC)

An ADC takes an analog signal as the input and delivers a corresponding digital signal at its output.

There are four types of ADC. These are

(i)	counter type ADC	(iii)	Dual slope ADC, and
(ii)	Simultaneous type ADC	(iv)	Successive approximation type ADC.

We discuss these converters in the following subsections.

13.4.1 COUNTER-TYPE ADC

This ADC consists of a comparator, an AND gate, a binary counter and a D/A converter as shown in Fig. 13.36. An operational amplifier acts as the voltage comparator. The AND gate passes the clock pulse to the input of the binary counter when the output of the comparator goes high. The counter counts the number of clock pulses until the comparator output becomes low and makes a digital output corresponding to the number of pulses counted. The digital output of the binary counter is fed to a D/A converter which produces an analog voltage output corresponding to the binary number. This analog voltage is applied to the inverting terminal of the operational amplifier for comparison with the input analog signal.

When the input analog voltage applied to the noninverting input is higher than the D/A converter output, the comparator produces a high ('1') output and the clock pulse passes through the AND gate. For the first clock pulse, the 8-bit counter yields an output 001. The D/A converter produces a voltage equal to 1 volt. If the input signal is, as an example, 5 volt then the comparator produces a high output. Now, the second clock pulse passes through the AND gate and counted. The counter output is now 010 and the D/A converter produces a 2 volt output. Since, 5 volt > 2 volt, the comparator produces a high output. The third pulse is counted and so on. When the fifth pulse is counted, the counter output becomes 101. This is the converted digital output. Now, the comparator will have 5 volt at both inputs. So, its output will be low and the AND gate is disabled. There is no pulse to count. The counter is reset and the cycle begins again. A clear pulse resets the counter.

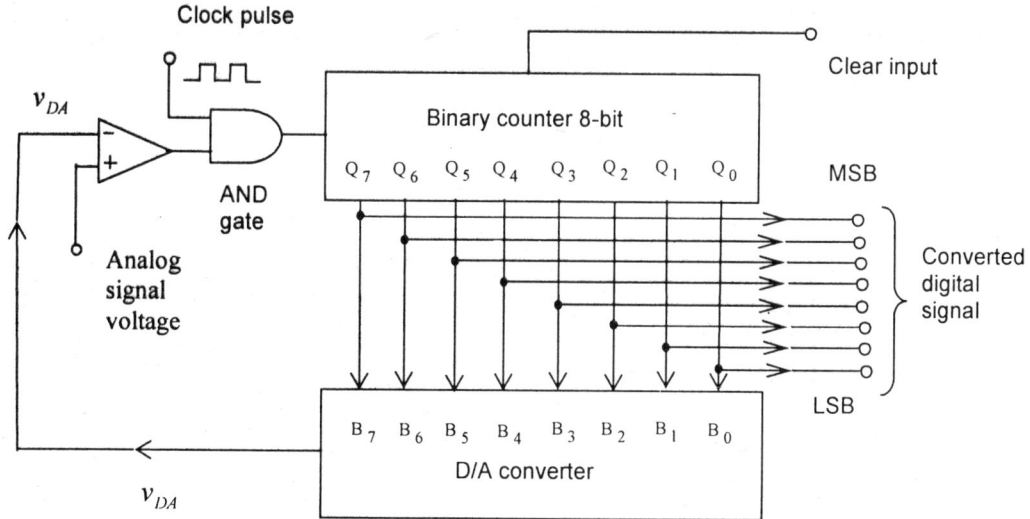

Fig. 13.36. Schematic block diagram of a counter-type ADC.

13.4.2 SIMULTANEOUS ADC

This ADC is also known as flash or parallel type ADC. Here, the analog voltage is compared with a number of reference voltages in voltage comparators and this comparison is done at the same time. This is why it is called simultaneous or parallel ADC. An 'n'-bit ADC requires $(2^n - 1)$ number of comparators. An operational amplifier can act as a voltage comparator. The outputs from $(2^n - 1)$ comparators act as $(2^n - 1)$ inputs to the priority encoder. A priority encoder is a device which produces an n-bit output code corresponding to the highest order data line when one or more input lines goes high (level 1). So, the priority encoder produces an output code corresponding to the reference voltage which is closest to the analog input voltage but below it. The reference voltages are applied to the inverting terminals of the comparator while the analog voltage is applied to non-inverting terminals. The reference voltages which are below the analog input voltage produce a high output (logical 1) of the comparators. On the other hand, the reference voltages which are equal to or above the analog input voltage produce a low output (logical 0) of the corresponding comparators. The encoder gives priority to the highest order data producing a high output of the comparator and yields a binary code leading to ADC. The circuit diagram of 4-bit ADC using 15 comparators and 16 equal resistances (R) is shown in Fig. 13.37. If V_R be the reference voltage then the potential drop across each resistor is $V_R/16$. The reference voltage levels used for comparison range from $\dfrac{15}{16}V_R$ to $\dfrac{1}{16}V_R$. This ADC is very fast. The only delay involved come from the comparator and priority encoder. The only disadvantage is that it requires a large number of comparators, $(2^n - 1)$ number for a n-bit ADC.

In Fig.13.37, C_1, C_2,......C_{15} are comparators and $Y_3 Y_2 Y_1 Y_0$ is the converted digital signal (code) corresponding to the analog input signal. As an example, suppose that the analog voltage (v_{in}) is such that $\dfrac{6V_R}{16} < v_{in} < \dfrac{7V_R}{16}$. Then, $C_1 = C_2 = C_3 = C_4 = C_5 = C_6 = 1$ and $C_7 = C_8 = C_9 = C_{10} = ... = C_{15} = 0$. Then, the highest order reference voltage corresponds to $C_6 = 1$. The priority encoder will generate a digital output corresponding to this 6-th line which is equal to 0110.

13.4.3 DUAL-SLOPE ADC

It is also known as integrating type ADC. It uses two ramp voltage of different slopes and hence its name dual-slope ADC. The circuit of the ADC is shown in Fig. 13.38. Suppose a sample of the analog voltage, v_{in}, assumed to be positive, is applied to an OP-Amp integrator. The time constant of this integrator is RC. This integrator produces a negative-going ramp voltage as shown in Fig. 13.39. The term 'ramp' means a voltage that increases or decreases linearly with time. The slope of this ramp is proportional to the magnitude of v_{in}. The negative voltage at the output of the integrator is applied to the inverting input of an OP-Amp comparator. The non-inverting terminal of the comparator is grounded. The output of the voltage comparator goes high which drives an AND gate. A clock signal is applied to the other input of the AND gate. So, the clock pulses pass through the AND gate and the number of clock pulses is counted by a binary

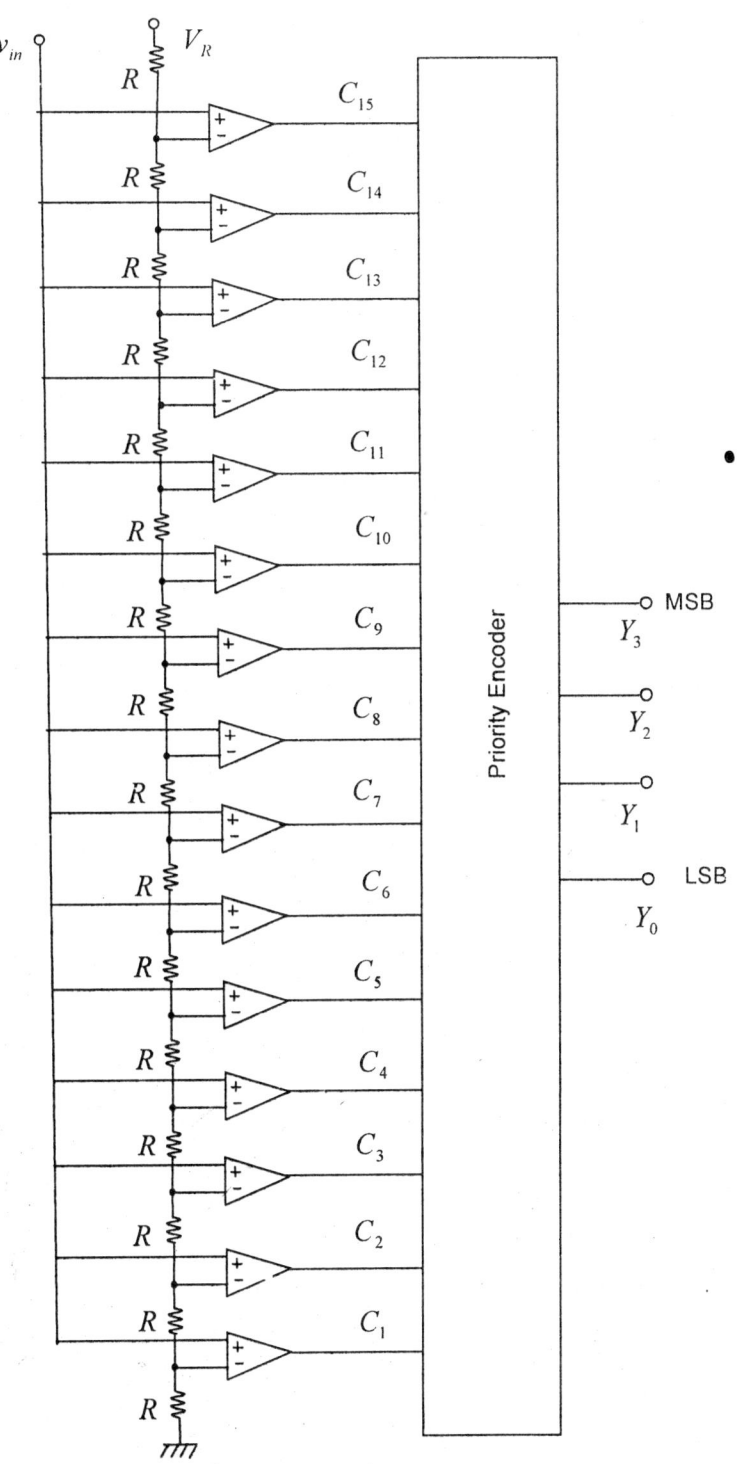

Fig. 13.37. Circuit diagram of a simultaneous 4-bit ADC.

counter. When the counter reaches a definite count, the integrator output reaches a value $-V_p$, and the counter is reset. This change of state can be used to drive a control circuit which operates a switch. This switch at the input disconnects the analog input voltage from the integrator and instead connects a reference voltage, V_R. This reference voltage is of opposite sign, i.e., if we take v_{in} positive then V_R must be negative. The comparator output runs the AND gate which passes the clock pulses through it. The counter begins to count pulses from its initial state (state 0). The counter continues to count pulses until the AND gate stops to transmit pulses. The negative reference voltage (V_R) at the integrator input now produces a positive-going ramp. This ramp starts from the value $-V_p$ reached during first integration. When this positive-going-ramp reaches a value 0 volt from its starting value $-V_p$, the OP-Amp comparator output becomes low (logical 0). The clock pulse propagation through the AND gate is now inhibited and the counter now stops. The slope of this second integration is proportional to V_R. The digital output of the counter after second integration is proportional to the analog sample voltage, v_{in}, leading to analog to digital conversion.

The current through the resistance R is

$$i = \frac{v_{in}}{R} \qquad \qquad \dots (13.107)$$

The same current charges the capacitor C. The charge accumulated in the capacitor at any time t is $q(t)$ where

$$\frac{dq(t)}{dt} = i = \frac{v_{in}}{R} \qquad \qquad \dots (13.108)$$

$$\therefore \qquad q(t) = \frac{1}{R} \int v_{in} dt \qquad \qquad \dots (13.109)$$

The output voltage of the integrator is equal to the negative of the capacitor voltage. The integrator output voltage is

$$V_0 = -\frac{q(t)}{C} = -\frac{1}{RC} \int v_{in} dt \qquad \qquad \dots (13.110)$$

If T_1 be the time of first integration then the integrator output voltage reached in time interval T_1 is $-V_p$ where

$$V_P = \frac{1}{RC} \int_0^{T_1} v_{in} dt \qquad \qquad \dots (13.111)$$

$$= \frac{v_{in}}{RC} T_1 \qquad \qquad \dots (13.112)$$

since v_{in} is a given sample of analog voltage which has a fixed value. The second integration is performed over the time interval T_1 and T_2 when the integrator output changes from $-V_p$ to 0.

$\therefore$ $\qquad$ $0 - (-V_p) =$ voltage change

$$= \frac{1}{RC} \int_{T_1}^{T_2} V_R dt$$

$\therefore$ $\qquad$ $V_P = \frac{V_R}{RC}(T_2 - T_1)$ $\qquad$ ... (13.113)

Equating (13.112) and (13.113), we get

$$\frac{v_{in}}{RC} T_1 = \frac{V_R}{RC}(T_2 - T_1)$$

$\therefore$ $\qquad$ $v_{in} = \frac{V_R}{T_1}(T_2 - T_1)$ $\qquad$ ... (13.114)

Since V_R and T_1 are constants,

$$v_{in} \propto (T_2 - T_1).$$

The digital output of the ADC is the representation of the number of pulses counted during the interval $(T_2 - T_1)$. The number of pulses counted is proportional to $(T_2 - T_1)$. So, the digital output is proportional to $(T_2 - T_1)$ and hence the digital output is proportional to v_{in}, the analog sample voltage.

The accuracy of conversion is independent of the time constant RC of the integrator. The accuracy does not also depend upon the clock frequency. The integrator reduces high frequency noise. So, the high frequency noise components present in the analog input voltage is reduced in

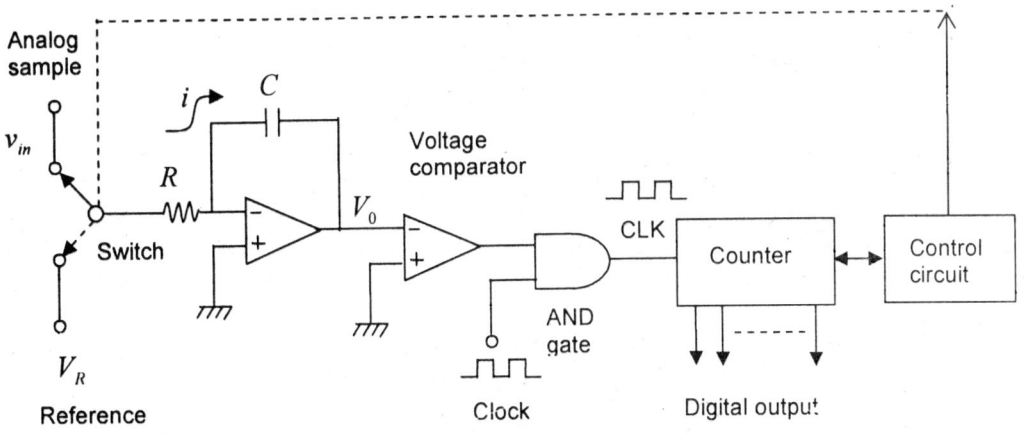

Fig. 13.38. Schematic circuit diagram of a dual-slope ADC.

this ADC. The conversion time is comparably long. The circuit is slow in speed. If finds application digital voltmeters.

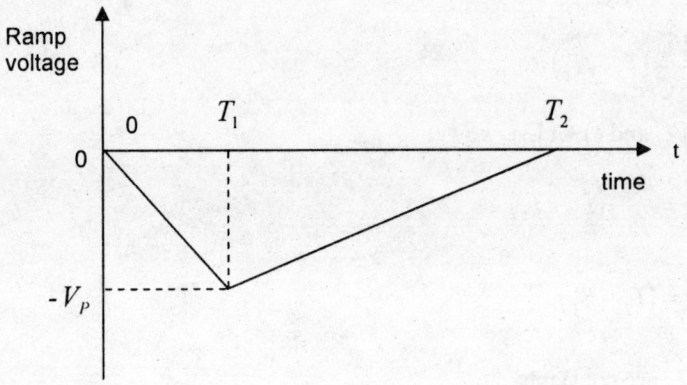

Fig. 13.39. Typical ramp voltage at the output of the integrator. V_P = maximum negative ramp voltage reached during first integration.

13.4.4 SUCCESSIVE APPROXIMATION ADC

This ADC consists of a programmer, the digital output of which is applied to a D/A converter. The DAC converts its digital input into an analog voltage. The converted analog voltage from the DAC is compared with the analog input voltage in a voltage comparator. The comparator output controls the operation of the programmer. The programmer contains a counter and a register and is driven by a clock pulse. When the first clock pulse appears, the register in the programmer is loaded with the most significant bit (MSB) equal to 1. For a 4-bit ADC, as an example, the programmer output is 1000 when the first pulse appears. The D/A converter converts this digital input (1000) into an 8 volt analog voltage. The comparator compares this 8 volt signal with the analog sample voltage. If the sample analog voltage is greater than 8 volt then the comparator output goes high (logical 1) which instructs the programmer to load the register by the next lower bit. The output of the programmer becomes now 1100. On the other hand, if the sample analog voltage is less than 8 volt then the comparator output goes low (level 0) which, in turn, instructs the programmer to shift the MSB of 1000 to the next lower position. This gives a programmer output equal to 0100. Assuming that the input analog voltage is greater than 8 volt, the DAC converts this digital signal (1100) into 12 volt analog voltage. If the sample analog voltage is greater than 12 volt then the comparator output goes high (level 1) which tells the programmer to load its register by the next lower bit which makes its output equal to 1110. On the other hand, if the sample analog voltage is less than 12 volt then the comparator output goes low (logical 0) which instructs the programmer to shift the lower bit of 1100 to the next lower bit position. This will lead to an output of 1010. This process continues until the programmer output approaches the correct value. A schematic circuit of this ADC is shown in Fig. 13.40.

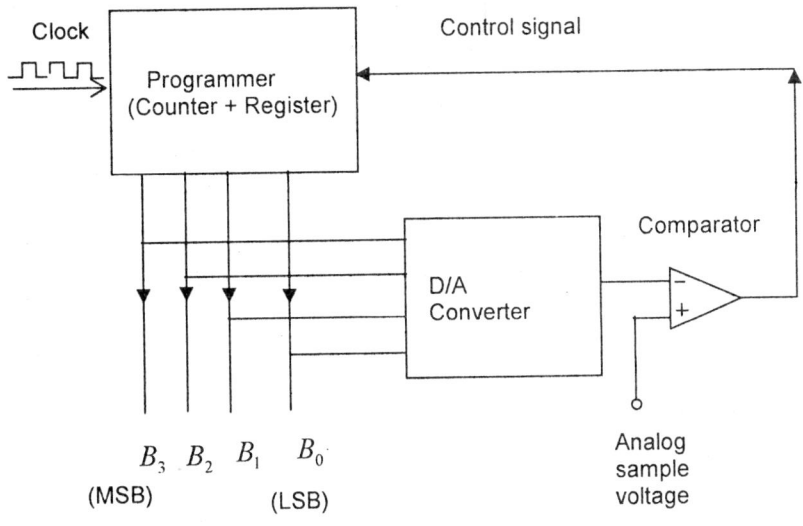

Fig. 13.40. Schematic block diagram of a 4-bit successive approximation ADC.

SOLVED PROBLEMS

13.1 The common mode gain of an amplifier is 2 and its CMMR is 100. Calculate the output of the amplifier if the two input signals are $40\,mV$ and $60\,mV$.

Solution

Here, CMRR, $\rho = 100$

Common mode gain, $A_C = 2$

$\therefore$ Difference mode gain, $A_D = \rho A_C = 200$.

Difference signal, $v_d = 60 - 40 = 20$ mV.

Common mode signal, $v_C = \dfrac{60 + 40}{2} = 50$ mV.

Now, the output voltage $V_0 = A_D v_d \left[1 + \dfrac{1}{\rho} \dfrac{v_C}{v_d} \right]$

$$= 200 \cdot 20 \left[1 + \dfrac{1}{100} \cdot \dfrac{50}{20} \right]$$

$$= 4000 \left[1 + \dfrac{1}{40} \right]$$

$$= 4100\,mV = 4.1 \text{ volt}.$$

13.2 Find the output voltage v_0 in terms of input voltages v_1 and v_2 of the following circuit.

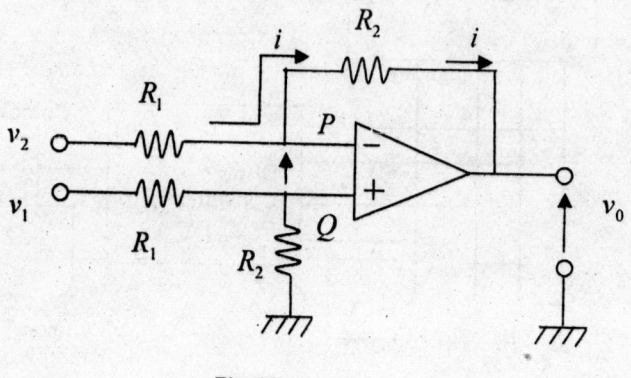

Fig.13.41

Solution

Now, potential at Q, $V_Q = \dfrac{v_1 R_2}{R_1 + R_2}$

Due to the existence of the virtual ground at input,

$$V_P = V_Q$$

Then, $i = \dfrac{v_2 - V_P}{R_1}$

Again, $i = \dfrac{V_P - v_0}{R_2}$

$\therefore \qquad \dfrac{V_P - v_0}{R_2} = \dfrac{v_2 - V_P}{R_1}$

$\therefore \qquad V_P - v_0 = \dfrac{R_2}{R_1}\left(v_2 - V_P\right)$

$\therefore \qquad v_0 = \dfrac{R_2}{R_1}\left(V_P - v_2\right) + V_P$

$\qquad = \left(1 + \dfrac{R_2}{R_1}\right)V_P - \dfrac{R_2}{R_1}v_2$

$\qquad = \dfrac{R_1 + R_2}{R_1} \cdot \dfrac{v_1 R_2}{R_1 + R_2} - \dfrac{R_2}{R_1}v_2$ [substituting the value of V_P]

$$= \frac{R_2}{R_1}(v_1 - v_2).$$

13.3 The ideal OP-Amp circuit shown below is supplied with an input sinusoidal signal of amplitude 100 mv. The output voltage amplitudes at input signal frequencies of 1 kHz and 5 kHz are V_{01} and V_{02}. Find the ratio V_{01}/V_{02}.

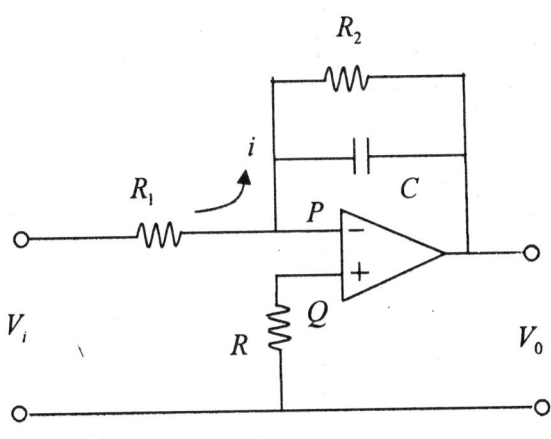

Fig.13.42

Solution

Let $\quad Z = R_2 \parallel \dfrac{1}{j\omega C} = \dfrac{R_2 \dfrac{1}{j\omega C}}{R_2 + \dfrac{1}{j\omega C}} = \dfrac{R_2}{1 + j\omega C R_2}$

Now, $\quad V_P = V_Q = 0$.

Again, $\quad \dfrac{V_i - V_P}{R_1} = i = \dfrac{V_P - V_0}{Z}$

$\therefore \quad \dfrac{V_i}{R_1} = -\dfrac{V_0}{Z}$

$\therefore \quad \dfrac{V_0}{V_i} = -\dfrac{Z}{R_1} = -\dfrac{R_2/R_1}{1 + j\omega C R_2}$

Let $\quad \omega_1 = 2\pi f_1 = 2\pi.10^3$ and $\omega_2 = 2\pi f_2 = 2\pi.5.10^3 = \pi.10^4$

Now, $\dfrac{V_{01}}{V_i} = -\dfrac{R_2/R_1}{1 + j\omega_1 CR_2}$ and $\dfrac{V_{02}}{V_i} = -\dfrac{R_2/R_1}{1 + j\omega_2 CR_2}$

$\therefore$ $\dfrac{V_{01}}{V_{02}} = -\dfrac{1 + j\omega_2 CR_2}{1 + j\omega_1 CR_2}$, i.e., $\left|\dfrac{V_{01}}{V_{02}}\right| = \left[\dfrac{1 + (\omega_2 CR_2)^2}{1 + (\omega_1 CR_2)^2}\right]^{1/2}$

When $\omega CR_2 \gg 1$, the circuit will behave as an integrator.

Then,

$$\left|\dfrac{V_{01}}{V_{02}}\right| = \dfrac{\omega_2}{\omega_1} = \dfrac{5}{1} = 5.$$

13.4 The OP-Amp circuit as shown has the following parameters. $C = .01\,\mu\text{F}$, $R_1 = 200$ kΩ, $R_2 = 2$ MΩ, $R = 2$ MΩ, saturation voltage $= \pm 10$ V.

Calculate the positive and negative threshold voltage at the inverting terminal for which the astable multivibrator will switch to the other state. Calculate the frequency of oscillation. Sketch V_{out} waveform.

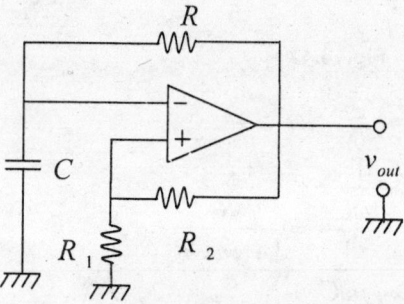

Fig.13.43

Solution

Feedback factor $\beta = \dfrac{R_1}{R_1 + R_2} = \dfrac{200 \cdot 10^3}{(200 + 2000) \times 10^3} = \dfrac{1}{11}$

Threshold voltage $= \pm \beta V_{sat} = \pm 0.909$ V.

Time period

$$T = 2RC \ln \dfrac{1 + \beta}{1 - \beta}$$

$$= 2 \cdot 2 \cdot 10^6 \cdot 0.01 \cdot 10^{-6} \ln \frac{1 + \dfrac{1}{11}}{1 - \dfrac{1}{11}}$$

$$= .04 \ln(1.2) = 7.29286 \times 10^{-3} \text{ sec.}$$

$$\therefore \quad f = \frac{1}{T} = 137 \text{ Hz.}$$

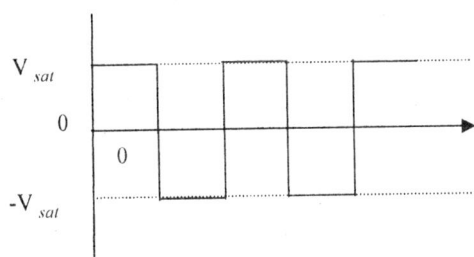

Fig.13.44

13.5 Calculate the output voltage V for $V_1 = 0.2$ volt. and $V_2 = 0.8$ volt. Assume ideal OP-Amp.

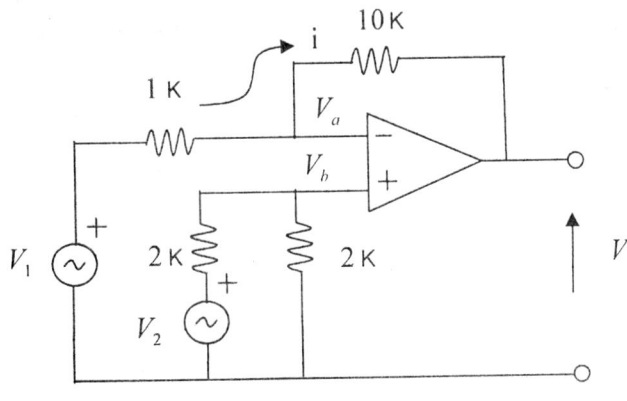

Fig.13.45

Solution

From the given circuit we can set up the following equations:

$$\frac{-V_a + V_1}{10^3} = \frac{V_a - V}{10^4} \qquad \qquad \dots (13.115)$$

$$\frac{V_2 - V_b}{2 \cdot 10^3} = \frac{V_b - 0}{2 \cdot 10^3} \qquad \qquad \ldots (13.116)$$

$$\therefore \quad 2V_b = V_2$$

i.e., $\quad V_b = \dfrac{V_2}{2} = 0.4 \text{ V}.$

Since there is virtual ground at the input, $V_a = V_b$.
From (13.115), we get

$$11V_a = 10V_1 + V$$

$$\therefore \quad V = 11V_a - 10V_1 = 4.4 - 2 = 2.4 \text{ V}.$$

13.6 Calculate the value of v_0 in the following OP-Amp circuit.

$$[\text{GATE 2001}]$$

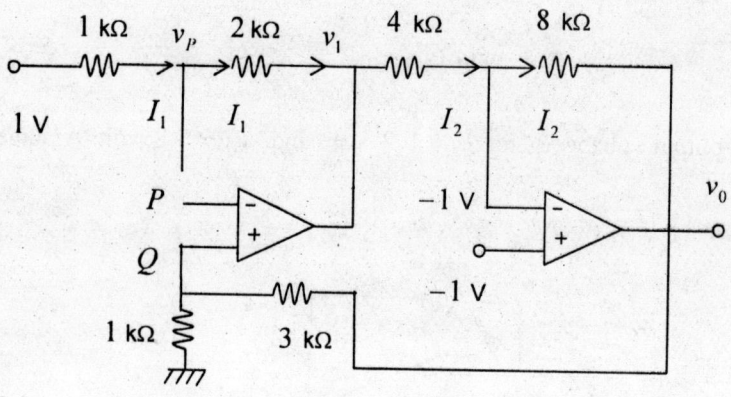

Fig.13.46

Solution

Here, potential at Q is $v_Q = \dfrac{v_0 \cdot 1}{3 + 1} = \dfrac{v_0}{4}$

Also $\quad v_P = v_Q$

$$\therefore \quad v_P = \frac{v_0}{4}$$

From circuit, $\quad I_1 = \dfrac{1 - v_P}{1} = (1 - v_P) \text{ mA}.$

Again, $v_p - v_1 = I_1 \cdot 2$

$$\therefore \qquad v_1 = v_p - 2I_1$$
$$= v_p - 2(1 - v_p)$$
$$= 3v_p - 2 = \left(3\frac{v_0}{4} - 2\right) \text{volt.}$$

Further from the second OP-Amp, we have

$$\left[v_1 - (-1)\right] = 4I_2$$
$$\therefore \qquad I_2 = \frac{1}{4}(1 + v_1) = \frac{1}{4}\left(\frac{3v_0}{4} - 1\right)$$

Lastly,

$$\left[-1 - v_0\right] = 8I_2$$

$$\therefore \qquad v_0 = -8I_2 - 1 = -8 \cdot \frac{1}{4}\left(\frac{3v_0}{4} - 1\right) - 1$$

$$= -\frac{3}{2}v_0 + 1$$

or, $\qquad \dfrac{5}{2}v_0 = 1$

$$\therefore \qquad v_0 = \frac{2}{5} = 0.4 \text{ volt.}$$

13.7 Find the transfer function of the following OP-Amp circuit. Whether this circuit behaves as a filter? If yes, then find out the cut-off frequency.

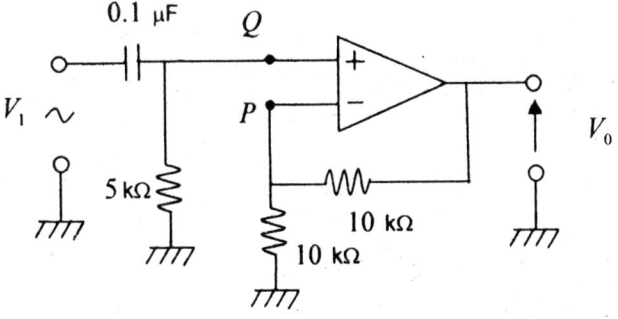

Fig.13.47

Solution

Voltage at P is $V_P = \dfrac{V_0 \cdot 10^4}{(10+10) \cdot 10^3} = \dfrac{V_0}{2}$

This is because no current enters the OP-Amp due to its infinite input resistance.

Now, $V_Q = V_P$. Let $R = 5\,k\Omega$ and $C = 0.1\,\mu F = 10^{-7}\,F$

Now, $V_Q = \dfrac{V_1 R}{R + \dfrac{1}{j\omega C}} = \dfrac{j\omega CR}{1 + j\omega CR} V_1$

$\therefore \quad \dfrac{V_0}{2} = \dfrac{j\omega CR}{1 + j\omega CR} V_1$ [since $V_Q = V_P = \dfrac{V_0}{2}$]

$\therefore$ Transfer function,

$$H(j\omega) = \frac{V_0}{V_1} = \frac{2\,j\omega CR}{1 + j\omega CR} = \frac{2}{1 - j\dfrac{\omega_C}{\omega}}$$

where $\omega_C = \dfrac{1}{RC}$. This circuit behaves like a high pass filter with a low frequency cut-off at ω_C.
The cut-off frequency

$$f_C = \frac{\omega_C}{2\pi} = \frac{1}{2\pi RC} = \frac{1}{2\pi . 5 \times 10^3 . 10^{-7}} = \frac{10^3}{\pi} = 318.3\,Hz.$$

13.8 Design an OP-Amp based circuit which produces an output V_0 given by $V_0 = 4V_1 + 6V_2$ where V_1 and V_2 are two input signals.

[B.U. 1999]

Solution

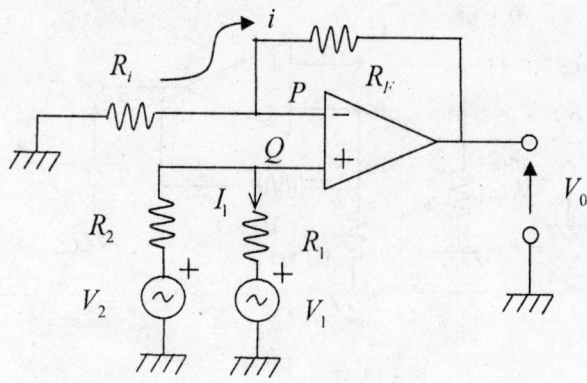

Fig.13.48

This is the required OP-Amp based circuit. We have to find out the values of resistances.
By superposition theorem, net current through R_1 is

$$I_1 = \frac{V_2}{R_1 + R_2} - \frac{V_1}{R_1 + R_2}$$

Then, potential at Q is $V_Q = V_1 + I_1 R_1 = (V_2 R_1 + V_1 R_2)\frac{1}{R_1 + R_2}$

Now,

$$V_Q = V_P$$

Again, $$\frac{0 - V_P}{R_i} = \frac{V_P - V_0}{R_F}$$

or, $$V_0 = \left(1 + \frac{R_F}{R_i}\right)V_P = \left(1 + \frac{R_F}{R_i}\right)\frac{1}{R_1 + R_2}(V_2 R_1 + V_1 R_2)$$

Since the given function is $V_0 = 4V_1 + 6V_2$, we have $\left(1 + \frac{R_F}{R_i}\right)\frac{R_2}{R_1 + R_2} = 4$ and

$$\left(1 + \frac{R_F}{R_i}\right)\frac{R_1}{R_1 + R_2} = 6$$

Dividing these two equations, we get $\dfrac{R_1}{R_2} = \dfrac{3}{2}$.

Then, $$\frac{R_1}{R_1 + R_2} = \frac{\frac{3}{2}R_2}{\frac{5}{2}R_2} = \frac{3}{5} \quad \text{and} \quad \frac{R_2}{R_1 + R_2} = \frac{R_2}{\frac{5}{2}R_2} = \frac{2}{5}.$$

Now, $$\left(1 + \frac{R_F}{R_i}\right)\cdot\frac{2}{5} = 4$$

$$\therefore \quad R_F = 9R_i .$$

If we take $R_i = 10\,\text{k}\Omega$ then $R_F = 90\,\text{k}\Omega$. Taking $R_2 = 10\,\text{k}\Omega$, $R_1 = \frac{3}{2}R_2 = 15\,\text{k}\Omega$.

This completes the design.

13.9 Design an OP-Amp based circuit to get $V_0 = 2.5V_1 + 2.5V_2 - 3V_3$, where V_1, V_2 and V_3 are input voltages.

[B.U. 2001]

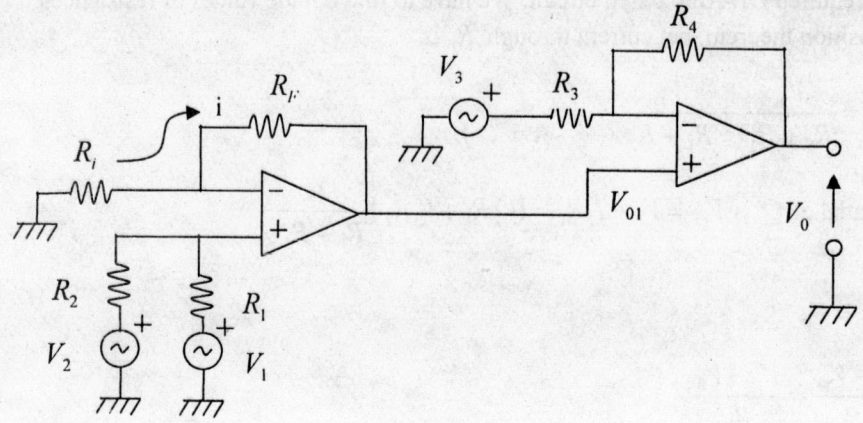

Fig.13.49

Solution
Producing in the same way as in the previous problem, we get

$$V_{01} = \left(1 + \frac{R_F}{R_i}\right) \frac{1}{R_1 + R_2} \left(V_2 R_1 + V_1 R_2\right)$$

From the second OP-Amp, we can write

$$\frac{V_3 - V_{01}}{R_3} = \frac{V_{01} - V_0}{R_4}$$

$$\therefore \qquad V_0 = V_{01}\left(1 + \frac{R_4}{R_3}\right) - \frac{R_4}{R_3} V_3$$

$$= \left(1 + \frac{R_F}{R_i}\right)\left(1 + \frac{R_4}{R_3}\right)\left(V_2 R_1 + V_1 R_2\right)\frac{1}{R_1 + R_2} - \frac{R_4}{R_3} V_3$$

Since the given function has equal coefficient for V_1 and V_2, we have $R_1 = R_2$.

Then, $\dfrac{R_1}{R_1 + R_2} = \dfrac{R_2}{R_1 + R_2} = \dfrac{1}{2}$

Comparing the coefficient of V_3, we get $\dfrac{R_4}{R_3} = 3$

Again, $\left(1 + \dfrac{R_F}{R_i}\right)\left(1 + \dfrac{R_4}{R_3}\right)\dfrac{R_1}{R_1 + R_2} = 2.5$

or, $\left(1+\dfrac{R_F}{R_i}\right)\cdot 4\cdot\dfrac{1}{2}=2.5$

$\therefore$ $\dfrac{R_F}{R_i}=\dfrac{1}{4}$

As an example, we can take $R_1=R_2=10\,k\Omega$, $R_i=40\,k\Omega$, $R_F=10\,k\Omega$ and $R_4=30\,k\Omega$.

13.10 Calculate the voltage gain of the inverting OP-Amp with open loop gain A_v, input impedance Z and feedback impedance Z' considering the loading of the amplifier.

Solution

The equivalent circuit of the inverting OP-Amp taking into account the loading effect is shown below.

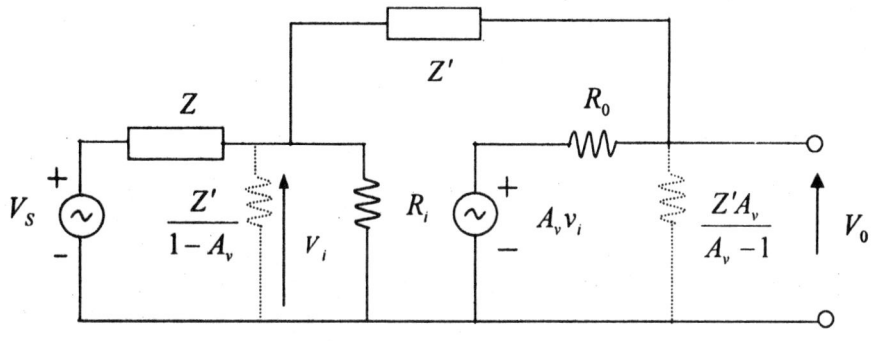

Fig.13.50

The voltage gain with feedback taking into account the effect of loading is given by

$$A_{vf}=\frac{V_0}{V_S}=\frac{V_0}{V_i}\cdot\frac{V_i}{V_S}=A_V\frac{V_i}{V_S}$$

where $A_V=\dfrac{V_0}{V_i}=$ internal voltage gain with feedback.

Now, $\dfrac{V_i}{V_S}=\dfrac{R_i\,\|\,\dfrac{Z'}{1-A_v}}{Z+R_i\,\|\,\dfrac{Z'}{1-A_v}}$

Here, the feedback effect of Z' has been transformed into an impedance $\dfrac{Z'}{1-A_V}$ at the input and

an impedance $\dfrac{Z'A_V}{A_v-1}$ at the output by Miller theorem.

Now, $\dfrac{V_i}{V_S} = \dfrac{\dfrac{R_i Z'}{1-A_V}}{Z\left(R_i + \dfrac{Z'}{1-A_V}\right) + \dfrac{R_i Z'}{1-A_V}} = \dfrac{1}{1 + \dfrac{Z}{Z'}\dfrac{1}{R_i}\left[R_i(1-A_V)+Z'\right]}$

Now, put $Y = \dfrac{1}{Z}$, $Y' = \dfrac{1}{Z'}$, $Y_i = \dfrac{1}{R_i}$

Then, $A_{vf} = A_V \cdot \dfrac{V_i}{V_S}$

$$= \dfrac{1}{\dfrac{1}{A_V} + \dfrac{Y_i Y'}{Y}\left[\dfrac{1}{Y_i}\cdot\left(\dfrac{1}{A_V}-1\right) + \dfrac{1}{Y'}\dfrac{1}{A_V}\right]}$$

$$= \dfrac{1}{-\dfrac{Y'}{Y} + \dfrac{1}{A_V}\left(1 + \dfrac{Y'}{Y} + \dfrac{Y_i}{Y}\right)}$$

$$= -\dfrac{Y}{Y' - \dfrac{1}{A_V}\left(Y + Y' + Y_i\right)}$$

In the case of ideal OP-Amp, $R_0 = 0$. Then, $\dfrac{V_0}{V_i} = A_V = \infty$ since the internal voltage gain

without feedback $A_v = \infty$. Then,

$$A_{vf} = -\dfrac{Y}{Y'} = -\dfrac{Z'}{Z}.$$

13.11 Prove that the OP-Amp circuit shown below has an output resistance R_{of} given by the relation

$$\dfrac{1}{R_{0f}} = \dfrac{1}{R_0}\left(1 - A_v\dfrac{R}{R+R'}\right) + \dfrac{1}{R+R'}$$

where A_v is the open-loop gain of the OP-Amp without loading. Given $R_i = \infty$.

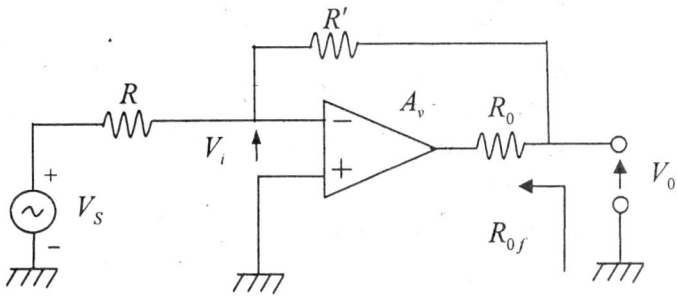

Fig.13.51

Solution

To find the output resistance R_{of}, remove V_S and put a short circuit in its place. Then, connect a test voltage source V_0 at the output which drives a current I.

Since, $R_i = \infty$ (given), no current flows at the input. The equivalent circuit is shown below.

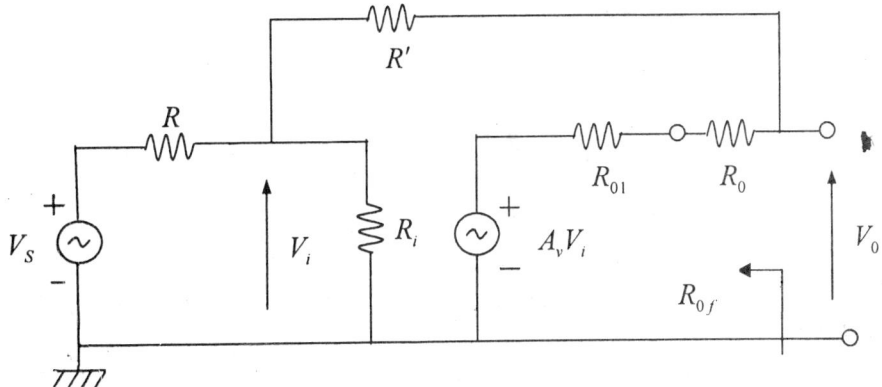

Fig.13.52

Now, $V_i = \dfrac{V_0 R}{R + R'}$

Internal voltage gain, $A_V = \dfrac{V_0}{V_i} = 1 + \dfrac{R'}{R}$

Output resistance of the OP-Amp without loading is $R_{01} \ll R_0$.

By Miller effect, the feedback resistance R' introduces a resistance $\dfrac{R'}{1-1/A_V}$ in parallel with the output. The output circuit will then look like the following.

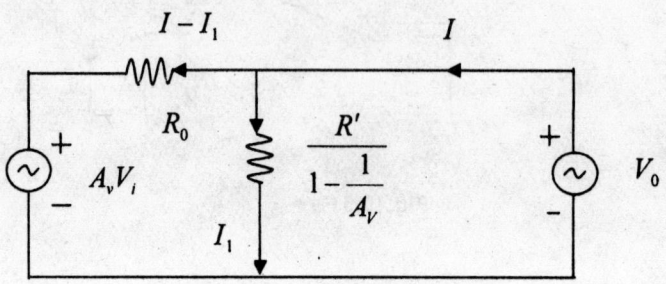

Fig.13.53

Now, $V_0 = \dfrac{R'}{1-1/A_V} \cdot I_1$

Again,

$$V_0 - A_v V_i = R_0 \left(I - I_1\right)$$

Now,

$$\left[1 - \frac{A_v}{A_V} + \frac{R_0}{R'}\left(1 - \frac{1}{A_V}\right)\right] V_0 = R_0 I$$

Since $V_i = \dfrac{V_0}{A_V}$ and $I_1 = \dfrac{V_0}{R'}\left(1 - \dfrac{1}{A_V}\right)$

$$\therefore \quad \frac{1}{R_{of}} = \frac{I}{V_0} = \frac{1}{R_0}\left[1 - A_v \frac{R}{R+R'} + \frac{R_0}{R'}\left(1 - \frac{R}{R+R'}\right)\right] \text{, putting the expression of } A_V.$$

$$= \frac{1}{R_0}\left[1 - A_v \frac{R}{R+R'}\right] + \frac{1}{R+R'}.$$

A_v is the open-loop gain of the OP-Amp without loading.

13.12 The Schmitt Trigger circuit shown below is operated with a saturation voltage of ± 15 volt. Calculate upper and lower trip points. Given $R_2 = 10$ kΩ and $R_1 = 2$ kΩ.

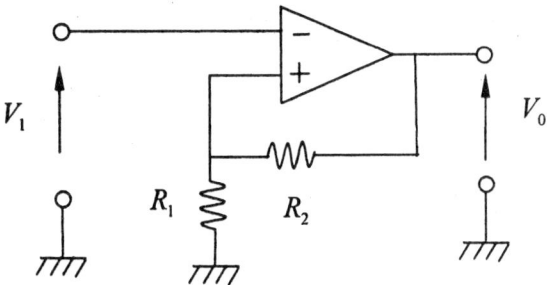

Fig.13.54

Solution

Here, saturation voltage $V_{sat} = \pm 15$ volt.

Upper trip point, UTP $= \dfrac{R_1}{R_1 + R_2} V_{sat} = \dfrac{2}{12} \times 15 = 2.5$ V.

Lower trip point, LTP $= \dfrac{R_1}{R_1 + R_2}(-15) = -\dfrac{2}{12} \times 15 = -2.5$ V.

PROBLEMS

13.1 Find the transfer function of the following OP-Amp circuit. Does this circuit behave like a filter? If yes, find out the cut-off frequency.

$$\left[\textbf{Ans. } H(j\omega) = \frac{3}{1 + j\omega \cdot 2 \times 10^{-5}}, \text{ Yes, LPF; 7.95 kHz.}\right]$$

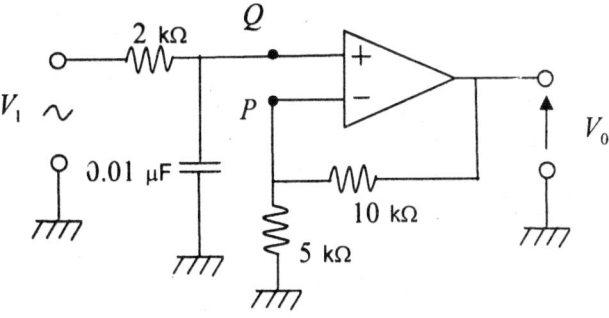

Fig.13.55

13.2 Find out the voltage gain $\dfrac{V_0}{V_1}$ of the following OP-Amp circuit.

[**Ans.** 10]

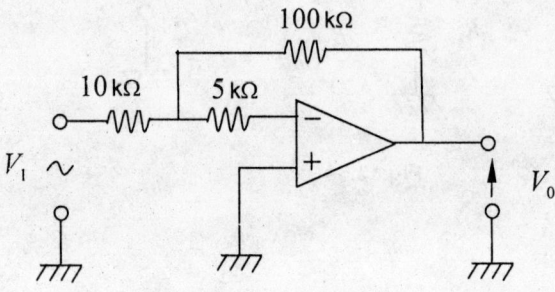

Fig.13.56

13.3 Find out $\dfrac{V_0}{V_1}$ for the following circuit.

[**Ans.** -20]

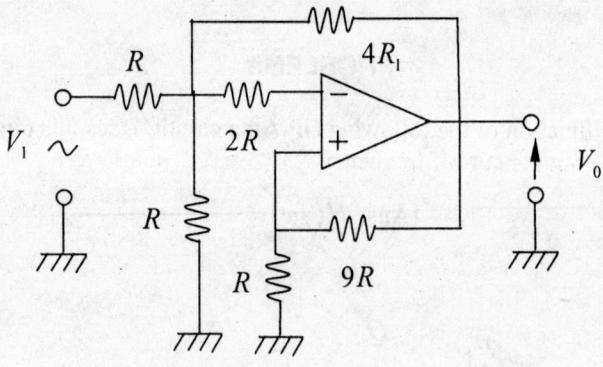

Fig.13.57

13.4 Find the transfer function $\dfrac{V_0}{V_1}$ of the following OP-Amp circuit. Show that this circuit can function as an integrator under certain condition. What is the condition?

[**Ans.** $H(j\omega) = -\dfrac{R_2}{R_1 + j\omega L}$; Yes, $\omega L \gg R_1$]

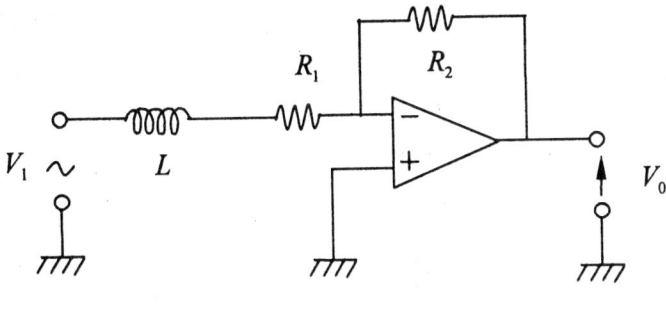

Fig.13.58

13.5 Find the voltage gain of the following OP-Amp. What will be the value of the gain if $R_1 = 0$ in this circuit?

$$[\text{Ans. } -\frac{1}{\left[\dfrac{R_3}{R_4}\left(1+\dfrac{R_1}{R_2}\right)+\dfrac{R_1}{R_2}-\dfrac{R_4}{R_3}\right]}]$$

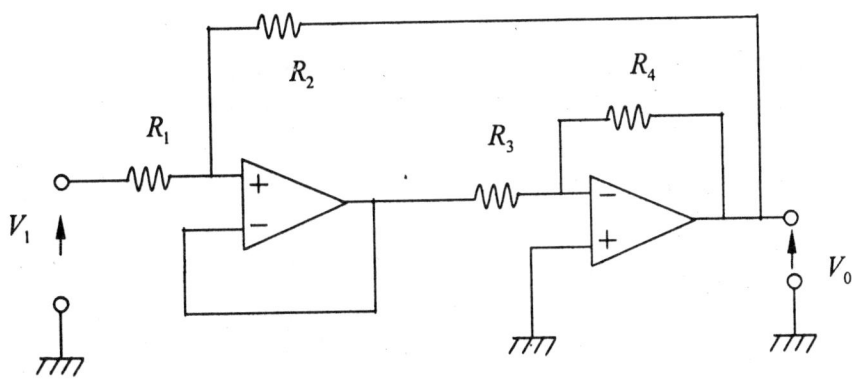

Fig.13.59

13.6 Find the oscillation frequency and the minimum value of amplifier gain of the following phase shift oscillator circuit.

$$[\text{Ans. } f_0 = \frac{1}{2\pi\sqrt{6}RC}\text{; } 29]$$

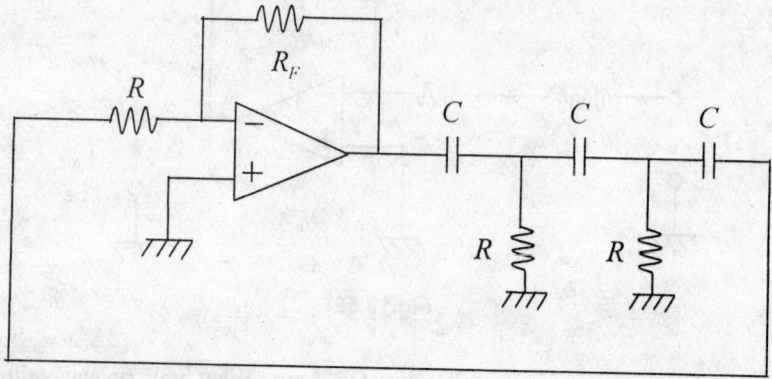

Fig.13.60

Hints: Find the feedback factor β for the network consisting of 3 C-R sections. β must introduce a phase shift of π radian. So, β = real. Equate imaginary part to zero. Get f_0. Minimum gain $A_v = \dfrac{1}{|\beta|}$.

QUESTIONS

13.1 What is an operational amplifier? What are the characteristics of an OP-Amp? Define the terms 'input off-set voltage', 'input off-set current' and 'output off-set voltage.

13.2 What do you mean by 'slew rate' of an OP-Amp? How does it depend on the signal frequency? Explain the meaning of the term 'virtual ground' in an OP-Amp.

13.3 What do you mean by CMRR of an OP-Amp? Derive an expression for output voltage of an OP-Amp in terms of difference mode gain and the CMRR.

13.4 Design a circuit using an OP-Amp for realizing the subtraction of two analog signals. Derive expressions for gain of an inverting and non-inverting amplifier.

13.5 How can you realize inverting and non-inverting addition of two analog signals by using an OP-Amp?

13.6 Analyze the performances of an OP-Amp differentiator and an integrator. What is the advantage of an integrator over a differentiator?

13.7 Draw the circuit diagrams of an OP-Amp active low pass and active high pass filter. Derive expressions for transfer functions of these filters.

13.8 How do you use an OP-Amp as a voltage follower? Explain the use of an OP-Amp as a voltage-to-current converter and current-to-voltage converter.

13.9 Mention some nonlinear applications of an OP-Amp. Give the theory of log and antilog amplifiers using OP-Amps.

13.10 Design a logarithmic amplifier using operational amplifiers and a pair of matched transistors. What is the advantage of this logarithmic amplifier over that using a diode?

13.11 What is a Schmitt Trigger? Explain its operation. What are upper and lower trip points? What is the hystersis loop exhibited by Schmitt Trigger?

13.12 What do you mean by an astable multivibrator? Give the circuit diagram of an OP-Amp-based astable multivibrator and derive an expression for the time period of oscillation of this multivibrator.

13.13 Design the operation of a binary-weighted digital-to-analog converter. What is the disadvantage of this converter?

13.14 Explain the operation of an $R/2R$ ladder type D/A converter. What is the advantage of this converter?

13.15 What is an analog-to-digital converter? What are the different types of A/D converters? Explain the operation of a counter type ADC.

13.16 Discuss, in brief, the operation of a simultaneous ADC. What is its advantage over the counter type ADC?

13.17 Describe, with necessary circuit diagram, the operation of a dual-slope ADC. Develop the theory of this ADC. What are the merits and demerits of this ADC?

13.18 What is a successive-approximation ADC? Explain, briefly, its operation.

14

Number System

14.1 NUMBER SYSTEM

A system is said to be 'digital' if its functioning involves discrete units. As an example, table, chair, cow, house, hand, etc. are discrete entities. On the other hand, analog systems involve continuously variable quantities. A good example of this analog number are rotations of the needle of a voltmeter, ammeter or ohm meter. A number 5 can be represented by a $50°$ rotation of the pointer over a calibrated scale.

Let us first consider the number systems which are used in digital electronics. There are four number systems in digital electronics. These are (i) Decimal (ii) Binary (iii) Hexadecimal and (iv) Octal.

14.1.1 DECIMAL SYSTEM

Decimal system is the oldest number system. Human beings have, in general, ten fingers and ten toes. In the early days of human civilization, human became able to count their fingers and toes, and got the idea of the decimal system somewhat unknowingly. The decimal system uses 10 symbols $0, 1, 2, 3, 4, 5, 6, 7, 8$ and 9. The base or radix of the system is 10 because it uses 10 symbols. Any number in this system can be expressed as a combination of these symbols. A positional weight is attributed to these symbols. The position of the symbol is counted from a point called the decimal point. The position just on the left of the decimal point is zeroth position while that just on the right of the decimal point is -1 position. The positional weight in decimal system is 10^K where K is the position of the digit with respect to the decimal point. Any number in decimal system can be expressed as

$$W = \sum_K d_K 10^K \qquad \qquad \dots (14.1)$$

where d_K is the digit (or symbol) at the K-th position having a positional weight 10^K. The range of K will depend upon the extension of the number. An example of a decimal number is

$$725.12 = 7 \times 10^2 + 2 \times 10^1 + 5 \times 10^0 + 1 \times 10^{-1} + 2 \times 10^{-2}$$

14.1.2 BINARY NUMBER SYSTEM

This system uses only two symbols 0 and 1. So, its base or radix is 2. The position of the symbol is counted with reference to a point called the binary point. The positional weight of the symbol is 2^K where K is the position of the digit with respect to the binary point. Any number in this system can be expressed as

$$W = \sum_K d_K 2^K \qquad \qquad \dots (14.2)$$

where d_K is the digit (which can be 0 or 1) at the K-th position with respect to the binary point and has a positional weight 2^K. This is, by far, the simplest number system. But, since it uses only two symbols for representation of a number, binary numbers are long containing a large number of digits. An example of the binary number is

$$1011.01_2 = 1 \times 2^3 + 0 \times 2^2 + 1 \times 2^1 + 1 \times 2^0 + 0 \times 2^{-1} + 1 \times 2^{-2}$$

The subscript '2' associated with a number implies that it is a binary number. This binary system is very useful in electronics. Many electronic devices have got two states. A diode is either conducting or nonconducting. A transistor can be saturated or cut-off. A switch can be ON or OFF.

14.1.3 HEXADECIMAL NUMBER SYSTEM

This number system uses 16 symbols which are ten numerals $0,1,2,3,4,5,6,7,8,9$ and six characters of English alphabet which are $A,B,C,D,E,$ and F. Since it has 16 symbols its base or radix is 16. The positional weight corresponding to K-th position is 16^K. Here, A stands for decimal 10, B for decimal 11, C for decimal 12, D for decimal 13, E for decimal 14, and F for decimal 15. Any hexadecimal number can be expressed as

$$W = \sum_K d_K 16^K \qquad \qquad \dots (14.3)$$

Where d_K is the hex symbol corresponding to K-th position. The position of the symbol is counted with respect to a point, called the hex point. An example of hex number is

$$1A8.52_{16} = 1 \times 16^2 + 10 \times 16^1 + 8 \times 16^0 + 5 \times 16^{-1} + 2 \times 16^{-2}$$

The subscript '16' associated with a number indicates that it is a hexadecimal number.

14.1.4 OCTAL NUMBER SYSTEM

This number system has 8 symbols which are $0,1,2,3,4,5,6$ and 7. Since it is based on 8 symbols, its base or radix is 8. The positional weight corresponding to K-th position is 8^K. The position of the symbol is counted with respect to a point called the octal point. The position just on the left of the octal point is 0-th position and that just on the right of the octal point is -1. Any number in octal system can be expressed as

$$W = \sum_K d_K 8^K$$

... (14.4)

where d_K is the symbol at the K-th position with respect to the octal point. As an example,

$$2517.32_8 = 2 \times 8^3 + 5 \times 8^2 + 1 \times 8^1 + 7 \times 8^0 + 3 \times 8^{-1} + 2 \times 8^{-2}$$

The subscript '8' attached with a number indicates that the number is a octal number.

14.2 INTERCONVERSION OF NUMBERS OF DIFFERENT SYSTEMS

In this section, we convert one number of a particular system into the equivalent number of another system.

14.2.1 CONVERSION OF DECIMAL NUMBER INTO BINARY NUMBER

Go on dividing the decimal number successively by 2 until the quotient is zero and record the remainders. Write down the remainders so that the last remainder forms the most significant bit (MSB). The MSB has the highest positional weight.

(i) Convert 37_{10} into binary.

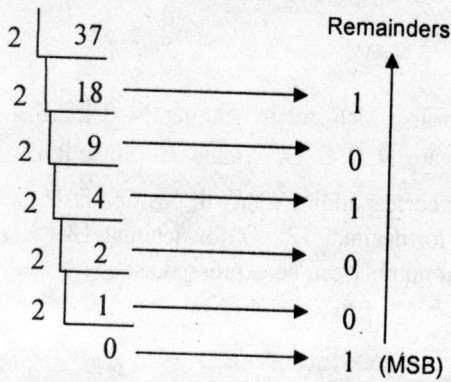

$\therefore \qquad 37_{10} = 100101_2.$

(ii) Convert 100_{10} into binary.

$\therefore \quad 100_{10} = 1100100_2$.

When the decimal number contains both integer and fraction, convert the integer into its corresponding binary number as above. Then, convert the fractional part into its binary equivalent as described below.

(iii) Convert 20.15_{10} into binary.

Here, the integer is 20. So, convert 20 into binary first.

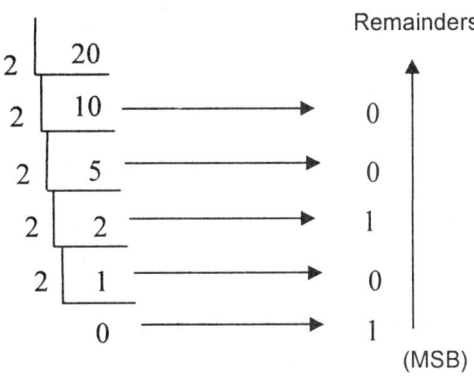

$\therefore \quad 20_{10} = 10100_2$.

Now, we have to convert the fraction 0.15_{10} into binary. Go on multiplying the decimal fraction by 2 and write down all digits on the left side of the decimal point of the product serially. Then, read down the binary fraction thus obtained.

$$
\begin{array}{llll}
0.15 \times 2 & = & & 0.30 \\
0.30 \times 2 & = & & 0.60 \\
0.60 \times 2 & = & & 1.20 \\
0.20 \times 2 & = & & 0.40 \\
0.40 \times 2 & = & & 0.80 \\
0.80 \times 2 & = & & 1.60
\end{array}
$$

$\therefore \ 0.15_{10} = 0.001001_2$

up to 6 places after the binary point.

(iii) Convert 0.125_{10} into binary.

$$
\begin{array}{llll}
0.125 \times 2 & = & & 0.25 \\
0.25 \ \times 2 & = & & 0.50 \\
0.50 \ \times 2 & = & & 1.00
\end{array}
$$

$\therefore \quad 0.125_{10} = 0.001_2$.

14.2.2 CONVERSION OF DECIMAL NUMBER INTO HEXADECIMAL NUMBER

Go on dividing the integer decimal number by 16 until the quotient is zero. Write down the remainders from bottom to top. The last remainder is the most significant bit (MSB).

(i) Convert 532.25_{10} into hexadecimal number

First convert the integer 532.

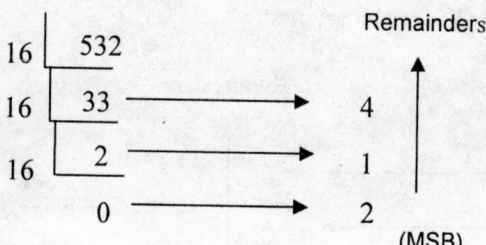

$$\therefore \quad 532_{10} = 214_{16}.$$

Now, convert 0.25_{10} into hexadecimal fraction. For this, go on multiplying by 16 and collect the digits on the left side of the decimal point. Then, read down

$$0.25 \times 16 = 4.00 \quad \downarrow$$

$$\therefore \quad 0.25_{10} = 0.4_{16}.$$

Hence, $532.25_{10} = 214.4_{16}$.

14.2.3 CONVERSION OF DECIMAL NUMBER INTO OCTAL NUMBER

If the decimal number is an integer, go on dividing by 8 successively until the quotient is zero. Write down the remainders and read up. The last remainder is the most significant bit (MSB).

(i) Convert 2080.55_{10} into octal number.

First convert the integer 2080_{10} into octal equivalent.

8	2080	Remainders	
8	260	0	
8	32	4	
8	4	0	
	0	4	(MSB)

$$\therefore \quad 2080_{10} = 4040_{8}.$$

Now convert 0.55_{10} into octal number. For this, multiply the given fraction by 8 and go on multiplying the fractions of the products by 8 and read the digits on the left side of the octal point from top to bottom. If the fraction of the product turns out to be 0 then the series terminates. Otherwise, the process has to be terminated after the desired number of digits following the octal point.

$$0.55 \times 8 = 4.40$$
$$0.40 \times 8 = 3.20$$
$$0.20 \times 8 = 1.60$$
$$0.60 \times 8 = 4.80$$
$$0.80 \times 8 = 6.40$$

$\therefore 0.55_{10} = 0.43146_8$

up to 5 places after the octal point.

$\therefore \quad 2080.55_{10} = 4040.43146_8$.

14.2.4 CONVERSION OF BINARY NUMBER INTO DECIMAL NUMBER

Multiply the bit at K-th position by 2^K and sum up all such products. This will give the required decimal number.

(i) Convert 1011.11_2 into decimal number.

$$1011.11_2 = 1 \times 2^3 + 0 \times 2^2 + 1 \times 2^1 + 1 \times 2^0 + 1 \times 2^{-1} + 1 \times 2^{-2}$$
$$= 8 + 0 + 2 + 1 + 0.5 + 0.25$$
$$= 11.75_{10}.$$

(ii) Convert 11010.011_2 into decimal number.

$$11010.011_2 = 1 \times 2^4 + 1 \times 2^3 + 0 \times 2^2 + 1 \times 2^1 + 0 \times 2^0 + 0 \times 2^{-1} + 1 \times 2^{-2} + 1 \times 2^{-3}$$
$$= 16 + 8 + 0 + 2 + 0 + 0 + 0.25 + 0.125$$
$$= 26.375_{10}.$$

14.2.5 CONVERSION OF BINARY NUMBER INTO HEXADECIMAL NUMBER

Starting at the binary point, make groups taking 4 bits successively. Then convert each group of 4 bits into its hexadecimal equivalent. To make this group, we can put zeros on the right end of the fractional part and zeros on the left end of the integer part, if required.

(i) Convert 1100101011.01_2 into hexadecimal equivalent.

Now, 1011 makes one group, then 0010 makes 2nd group and (00)11 makes the 3rd group on the left side of the binary point. Now,

$$1011_2 = B_{16}$$
$$0010_2 = 2_{16}$$
$$0011_2 = 3_{16}$$

On the right side of binary point we have .01 which is the same as .0100 . Then, $0100_2 = 4_{16}$

Thus, $\underbrace{11}_{3}\,\underbrace{0010}_{2}\,\underbrace{1011}_{B}\,\underbrace{.01}_{4} = 32B.4_{16}$

(ii)   Convert 11001111010011.1_2 into hexadecimal equivalent.

Now, $\underbrace{11}_{3}\,\underbrace{0011}_{3}\,\underbrace{1101}_{D}\,\underbrace{0011}_{3}\,\underbrace{.1000}_{8}$

Now, $0011_2 = 3_{16}$

$1101_2 = D_{16}$

$0011_2 = 3_{16}$

$(00)11_2 = 3_{16}$

$1000_2 = 8_{16}$

∴ $11\,0011\,1101\,0011 \cdot 1_2 = 33D3.8_{16}$

14.2.6 CONVERSION OF BINARY NUMBER INTO OCTAL NUMBER

Make groups by taking 3 bits in each group starting at the binary point. Then, convert each group of 3 bits into its equivalent octal number.

(i) Convert 1100111.11_2 into octal number.

Now, the grouping is $\underbrace{1}\,\underbrace{100}\,\underbrace{111}\,\underbrace{\cdot110}_2$

Now, $111_2 = 7_8$

$100_2 = 4_8$

$(00)1_2 = 1_8$

$110_2 = 6_8$

∴ $1\,100\,111\cdot11_2 = 147\cdot6_8.$

14.2.7 CONVERSION OF HEXADECIMAL NUMBER INTO DECIMAL NUMBER

Multiply each hex digit by its positional weight and then sum up all such products. This yields the decimal equivalent.

(i) Convert $1D9.48_{16}$ into its decimal equivalent.

$$1D9.48_{16} = 1\times16^2 + 13\times16^1 + 9\times16^0 + 4\times16^{-1} + 8\times16^{-2}$$
$$= 256 + 208 + 9 + 0.25 + 0.03125 \quad = 473.28125_{10}$$

(Remember D means decimal 13).

(ii) Convert $AB.84_{16}$ into its decimal equivalent.

Here, A means decimal 10 and B means decimal 11.

$$AB.84_{16} = 10 \times 16^1 + 11 \times 16^0 + 8 \times 16^{-1} + 4 \times 16^{-2}$$
$$= 160 + 11 + 0.5 + 0.015625$$
$$= 171.515625_{10}$$

14.2.8 CONVERSION OF HEXADECIMAL NUMBER INTO BINARY NUMBER

Convert each hex digit into a binary group of 4 bits. This gives the binary number.
(i) Convert $3BC4_{16}$ into its binary equivalent.

$$\begin{array}{cccc} 3 & B & C & 4 \\ 0011 & 1011 & 1100 & 0100 \end{array}$$

$\therefore \qquad 3BC4_{16} = 1110111100\ 0100_2$.

(ii) Convert $2F8.5_{16}$ into its binary equivalent.

$$\begin{array}{ccccc} 2 & F & 8 & \cdot & 5 \\ 0010 & 1111 & 1000 & \cdot & 0101 \end{array}$$

$\therefore \qquad 2F8.5_{16} = 10\,1111\,1000.0101_2$

14.2.9 CONVERSION OF HEXADECIMAL NUMBER INTO OCTAL NUMBER

The easiest method to convert a hex number into octal number is to convert first into binary and then convert binary number into octal number.

(i) Convert $1A5.4_{16}$ into octal number.

$$\begin{array}{ccccc} 1 & A & 5 & \cdot & 4 \\ 0001 & 1010 & 0101 & \cdot & 0100 \\ 6 & 4 & 5 & & 2 \end{array}$$

$\therefore \qquad 1A5.4_{16} = 110\,100\,101.01_2 = 645.2_8$.

14.2.10 CONVERSION OF OCTAL NUMBER INTO DECIMAL NUMBER

Multiply each octal digit by its positional weight and sum up the products. This yields the decimal number.

(i) Convert 7521.6_8 into its decimal equivalent.

Now, $7521.6_8 = 7 \times 8^3 + 5 \times 8^2 + 2 \times 8^1 + 1 \times 8^0 + 6 \times 8^{-1}$

$$= 3584 + 320 + 16 + 1 + 0.75$$

$$= 3921.75_{10}.$$

(ii) Convert 312.4_8 into its decimal equivalent.

Now, $312.4_8 = 3 \times 8^2 + 1 \times 8^1 + 2 \times 8^0 + 4 \times 8^{-1}$

$$= 192 + 8 + 2 + 0.5$$

$$= 202.5_{10}.$$

14.2.11 CONVERSION OF OCTAL NUMBER INTO ITS BINARY EQUIVALENT

Convert each octal digit into a group of 3 binary digits.

(i) Convert 123.7_8 into its binary equivalent.

$$
\begin{array}{ccccc}
\text{Now,} & 1 & 2 & 3 & \cdot & 7 \\
& 001 & 010 & 011 & \cdot & 111
\end{array}
$$

$\therefore \quad 123.7_8 = 1\,010\,011.111_2.$

(ii) Convert 405.6_8 into its binary equivalent.

$$
\begin{array}{ccccc}
\text{Now,} & 4 & 0 & 5 & \cdot & 6 \\
& 100 & 000 & 101 & \cdot & 110
\end{array}
$$

$\therefore \quad 405.6_8 = 100\,000\,101.11_2.$

14.2.12 CONVERSION OF OCTAL NUMBER INTO HEXADECIMAL NUMBER

We can convert the octal number into its binary equivalent first and then convert the binary number into hexadecimal number.

(i) Convert 225.5_8 into its equivalent hexadecimal number

$$
\begin{array}{ccccc}
\text{Now,} & 2 & 2 & 5 & \cdot & 5 \\
& 010 & 010 & 101 & \cdot & 1010 \\
& & 9 & 5 & & A
\end{array}
$$

First to obtain binary equivalent, we express each octal digit in a group of 3 bits. Then, make groups of 4 bits starting at the binary point. Convert each group of 4 bits into its equivalent hex digit. Note that 101 after the binary point should be written as 101(0) to make a group of 4 bits. This 1010 after the binary point is equivalent to hex digit A.

$$\therefore \quad 225.5_8 = 10010101.101_2 = 95.A_{16}.$$

14.3 BINARY ADDITION AND SUBTRACTION

The following is the binary addition table.

	0	1
0	0	1
1	1	10

The table shows that

$$0 + 0 = 0$$
$$0 + 1 = 1$$
$$1 + 0 = 1$$
$$1 + 1 = 10$$

(i) Add the following binary numbers:

$$1011_2 + 1101_2$$

Now,
$$
\begin{array}{r}
1011 \\
+ \quad 1101 \\
\hline
11000
\end{array}
$$

$\therefore \ 1011_2 + 1101_2 = 11000_2$

We should note that in the addition of $1 + 1 = 10$ the sum bit is 0 and carry bit is 1. The carry bit is taken into account in the next sum.

(ii) Subtract the following

$$1000_2 - 11_2$$

Now,
$$
\begin{array}{r}
1 \ 1 \ 10 \\
\cancel{1} \ \cancel{0} \ \cancel{0} \ 0 \\
- \qquad 1 \ 1 \\
\hline
1 \ 0 \ 1
\end{array}
$$

$\therefore \ 1000_2 - 11_2 = 101_2$

In the above subtraction, we have to subtract 1 from 0 first. To do this we must borrow a 1 from the previous (first) position. But, there is a '0' again. So, we look to the previous (i.e., 2nd) position for borrow. Again, there is a '0'. Hence, we look to the 3rd position and there is a '1'. So, borrow '1' from this position. The second position number becomes 10. We borrow a '1' from this 10 and a '1' is left at the 2nd position. The first position number is now 10 and a '1' is borrowed to zeroth position. So, a '1' is left at the first position. Now, we can subtract 1 from 10 in the zeroth position.

(iii) Subtract:

$$101010_2 - 1011_2$$

$$
\begin{array}{r}
1 \; 10 \; 1 \; 10 \; 10 \\
\cancel{1} \; \cancel{0} \; \cancel{1} \; \cancel{0} \; \cancel{1} \; 0 \\
- 1 \; 0 \; 1 \; 1 \\
\hline
1 \; 1 \; 1 \; 1 \; 1
\end{array}
$$

Now,

$$\therefore 101010_2 - 1011_2 = 11111_2$$

(iv) Subtract:

$$11001.01_2 - 1110.11_2$$

Now,

$$
\begin{array}{r}
10 \; 1 \; 10 \; 0 \; 10 \\
\cancel{1} \; \cancel{1} \; \cancel{0} \; \cancel{0} \; \cancel{1}.0 \; 1 \\
1 \; 1 \; 1 \; 0.1 \; 1 \\
\hline
1 \; 0 \; 1 \; 0.1 \; 0
\end{array}
$$

$$\therefore 11001.01_2 - 1110.11_2 = 1010.10_2$$

14.4 BINARY MULTIPLICATION AND DIVISION

The following is the binary multiplication table.

	0	1
0	0	0
1	0	1

(i) Multiply : $1011_2 \times 1100_2$

$$
\begin{array}{r}
1 \; 0 \; 1 \; 1 \\
\times \; 1 \; 1 \; 0 \; 0 \\
\hline
0 \; 0 \; 0 \; 0 \\
0 \; 0 \; 0 \; 0 \times \\
1 \; 0 \; 1 \; 1 \times \\
1 \; 0 \; 1 \; 1 \times \\
\hline
1 \; 0 \; 0 \; 0 \; 0 \; 1 \; 0 \; 0
\end{array}
$$

$$\therefore 1011_2 \times 1100_2 = 10000100_2$$

(ii) Multiply : $10.1_2 \times 11.1_2$

$$10.1$$
$$11.1$$
$$\overline{}$$
$$101$$
$$101\times$$
$$101\times$$
$$\overline{}$$
$$1000.11$$

$\therefore\ 10.1_2 \times 11.1_2 = 1000.11_2$

Binary division is similar to decimal division process.

(ii) Divide : $101.11_2 \div 10_2$

$$10 \,\big)\, 101.110 \,\big(\, 10.111$$
$$\underline{10}$$
$$11$$
$$\underline{10}$$
$$11$$
$$\underline{10}$$
$$10$$
$$\underline{10}$$

$\therefore \quad 101.11_2 \div 10_2 = 10.111_2.$

14.5 1'S COMPLEMENT AND 2'S COMPLEMENT SUBTRACTION

In section 13.3, we have discussed direct method of subtraction of binary numbers. In this section, we will discuss 1's complement method and 2's complement method of subtraction.

14.5.1 1'S COMPLEMENT METHOD

The procedure is to find 1's complement of the smaller number first. Then, add this 1's complement to the larger number. Remove the final carry after this addition and add this final carry with the result thus obtained. The final carry is known as end-around carry.

(i) Subtract by 1's complement method :

$$0111_2 - 0011_2$$

Smaller number : 0011_2

1's complement of $0011_2 : 1100_2$

larger number : 0111_2

Now,

$$
\begin{array}{r}
0\,1\,1\,1 \\
+\ 1\,1\,0\,0 \\
\hline
0\,0\,1\,1
\end{array}
$$

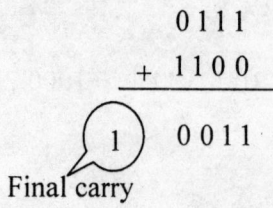

Final carry

Remove final carry and get 0011.
Add the final carry (i.e, 1) with 0011.

$$
\begin{array}{r}
0\,0\,1\,1 \\
+\quad 1 \\
\hline
0\,1\,0\,0
\end{array} \longrightarrow \text{Result}
$$

$$\therefore \quad 0111_2 - 0011_2 = 0100_2 .$$

(ii) Subtract by 1's complement method :

$$1101_2 - 1010_2$$

Smaller number : 1010_2

1's complement of $1010_2 : 0101_2$

Add 0101 with larger number :

$$
\begin{array}{r}
1\,1\,0\,1 \\
+\ 0\,1\,0\,1 \\
\hline
0\,0\,1\,0
\end{array}
$$

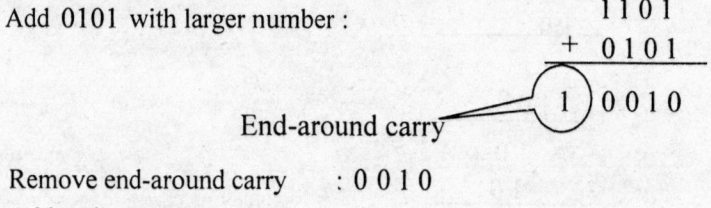

End-around carry

Remove end-around carry $: 0\,0\,1\,0$
Add end-around carry $:+\qquad 1$
                            $\overline{0\,0\,1\,1}\qquad \longrightarrow \quad \text{Result}$

$$\therefore \quad 1101_2 - 1010_2 = 0011_2 .$$

14.5.2 2'S COMPLEMENT METHOD

2's complement of a binary number B is defined as $B' = \overline{B} + 1$.

If we take 2's complement twice we get $B'' = \overline{B'} + 1$. Now, $B' = B$. We will prove it. Take an example. Let $B = 1011$.

Then, $\overline{B} = 0100$

$B' = 0100 + 1 = 0101$

$\overline{B'} = 1010$

$B'' = 1010 + 1 = 1011 = B$.

This is true for any binary number.

If we take a decimal number and change its sign twice we get back the original number.
For any decimal number A,

$$A \xrightarrow[\text{sign}]{\text{change}} -A \xrightarrow[\text{sign}]{\text{change}} -(-A) = A.$$

With the help of this analogy, 2's complement of a binary number is said to be equivalent to the negative of the decimal number.

(i) Subtract by 2's complement method

$$1110_2 - 1001_2.$$

Now, 2's complement of $1001 = 0110 + 1 = 0111$.

$$\begin{array}{r} 1\,1\,1\,0 \\ +\ 0\,1\,1\,1 \\ \hline 0\,1\,0\,1 \end{array}$$

$$\therefore\ 1110_2 - 1001_2 = 0101_2$$

Neglect final carry

The procedure is to find 2's complement of the negative smaller number first and then add it with the larger number. Neglect the final carry and get the result.

(ii) Subtract by 2's complement method $10110_2 - 10001_2.$

Now, 2's complement of $10001 = 01110 + 1 = 01111$.

$$\begin{array}{r} 1\,0\,1\,1\,0 \\ +\ 0\,1\,1\,1\,1 \\ \hline 0\,0\,1\,0\,1 \end{array}$$

$$\therefore\ 10110_2 - 10001_2 = 101_2$$

Neglect final carry

14.6 ADDITION AND SUBTRACTION OF HEX NUMBER

(i) Add the following : $1A8_{16} + 67B_{16}$

$$\begin{array}{r} 1\ A\ 8 \\ +\ 6\ 7\ B \\ \hline 8\ 2\ 3 \end{array}$$

First column : $8 + B = 8 + 11_{10} = 19_{10} = 16 + 3 = 13_{16}$

We get, sum $= 3$ and carry $= 1$.

Second column : $A + 7 + 1 = 10 + 7 + 1_{10} = 18_{10} = 16 + 2 = 12_{16}$

So, here we have sum $= 2$ and carry $= 1$.

Third column : $1 + 6 + 1 = 8_{16}$

Here, we get sum $= 8$ and carry $= 0$.

The result is 823_{16}.

(ii) Add the hex numbers : $1A7_{16} + 2C9_{16}$

$$
\begin{array}{r}
1\,A\,7 \\
+\ \ 2\,C\,9 \\
\hline
4\,7\,0
\end{array}
$$

First column : $7 + 9 = 16_{10} = 10_{16}$

This gives sum = 0 and carry = 1.
This carry will be added in the next column.

Second column : $A + C + 1 = 10 + 12 + 1_{10} = 23_{10} = 17_{16}$

So, sum = 7 and carry = 1.

Third column : $1 + 2 + 1 = 4_{16}$

Here, we get sum = 8 and carry = 0.
Result of addition is 470_{16}.

(iii) Subtract by direct method   : $67B_{16} - 1A8_{16}$

$$
\begin{array}{r}
5\,1 \\
\not{6}\,7\,B \\
-\ \ 1\,A\,8 \\
\hline
4\,D\,3
\end{array}
$$

First column : $B - 8 = 11_{10} - 8 = 3_{16}$

Second column : $7 < A$. So, borrow a '1' from the 3rd column.

Then, $17_{16} - A = 23_{10} - 10 = D_{16}$

Third column : since we have borrowed a '1' from 6, we have $5 - 1 = 4_{16}$.

Result of subtraction is 470_{16}.

14.7 SIGNED BINARY NUMBER

In signed binary number, there will be a sign (positive or negative) prefixed to this number. In the representation of such numbers, there will be two parts. One part stands for the sign of the number while the other part stands for the magnitude of this number. There are three different ways of representation of a signed binary number. These are
- Sign-magnitude representation
- 1's complement representation
- 2's complement representation.

When the sign is positive, we put a (0) before the magnitude of the binary number. This zero (0) indicates the positive sign. On the other hand, when the sign is negative, we put a '1' before the magnitude of the binary number. In a n-bit system, the n-th bit (Most significant bit) will be the sign bit. The remaining ($n-1$) bits will be the magnitude bits. The representation of a positive binary number will be same in all the three representations. Thus, $+7_{10}$ in an 8-bit system will be 00000111_2. The same number ($+7_{10}$) in a 4-bit system will be

$$0\ \underbrace{111}_{}\,_2$$

Sign bit   Magnitude bit

When the sign before the binary number is negative, there can be three different representations. In sign-magnitude number representation, we put a '1' in the extreme left bit (MSB) position which defines the sign bit. After this '1', we write the magnitude part of the binary number. Thus, -7_{10} in an 8-bit system will be represented as

$$1\ \underbrace{0000111}_{}$$

Sign bit   Magnitude bit

In a 4-bit system, -7_{10} will be represented as 1 111.

Sign bit   Magnitude bit

In 1's complement representation, we write the left most bit as '1' which indicates the sign bit and then complement the magnitude part of the binary number. This is also equivalent to represent first the same number with a positive sign. Then, we complement the whole positive number including the sign bit.

Thus, in an 8-bit system $-7_{10} \rightarrow \overline{0}\ \overline{0000111} \rightarrow 1\ 1111000$

Sign bit

In a 4-bit system, $-7_{10} \rightarrow \overline{0}\ \overline{111} \rightarrow 1\ 000$

Sign bit

In 2's complement representation, put a '1' in the extreme left position of the binary number which indicates a negative sign. Then, find the 2's complement of the magnitude part. This can also be obtained by taking 2's complement of the whole positive binary number including the sign bit.

In an 8-bit system $-7_{10} \rightarrow \overline{0}\ \overline{0000111} + 1 \rightarrow 1\ 1111001$.

In a 4-bit system, $-7_{10} \rightarrow \overline{0}\ \overline{111} + 1 \rightarrow 1\ 001$.

PROBLEMS

14.1. Convert the following binary numbers into decimal numbers:
   (i) 11011.01   (ii) 10011.10   (iii) 101010.11

[**Ans.** (i) 27.25   (ii) 19.5   (iii) 42.75]

14.2 Convert the following decimal numbers into binary numbers:
   (i) 24.5   (ii) 202.75   (iii) 12.375

[**Ans.** (i) 11000.1   (ii) 11001010.11   (iii) 1100.011]

14.3 Convert the following decimal numbers into hexadecimal numbers:
   (i) 400.25   (ii) 2748   (iii) 421.3

[**Ans.** (i) 190.4   (ii) ABC (iii) $1A5.4CC$ up to 3 places after the hex point.]

14.4 Convert the following hexadecimal numbers into binary numbers:
   (i) $FACE$ (ii) $1B7.5$ (iii) $20D.1$

[**Ans.** (i) 1111101011001110   (ii) 110110111.0101   (iii) 1000001110.0001]

14.5 Convert the following numbers:
   (i) $ABDC_{16}$ into binary
   (ii) 2_{10}^{16} into hexadecimal.

Hint: (i) convert each hex digit into equivalent binary in
$$ABDC_{16} = 1010\,1011\,1101\,1100_2$$

(ii) write $2_{10}^{16} = 1\,0000\,0000\,0000\,0000_2$
$$\therefore\ 2_{10}^{16} = 10000_{16}.$$

14.6 (i) Perform the following subtraction:
$$11001011_2 - 10011001_2$$

[**Ans.** 110010]

(ii) Represent the following numbers in an 8-bit system in 1's complement form:
   (a) -12_{10} (b) $+5_{10}$ (c) -15_{10}

Hints: (a) $+12_{10} = 00001100_2$
$$\therefore\ -12_{10} = \overline{00001100} = 11110011$$

(b) $+5_{10} = 00000101$

(c) $+15_{10} = 00001111_2$
$$\therefore\ -15_{10} = \overline{00001111} = 11110000.$$

QUESTIONS

14.1 What do you mean by a digital system? What are the different number systems in digital electronics? Compare the basic features of these number systems.

14.2 How do you convert a decimal number into binary number and vice-versa? Explain with example.

14.3 Explain the process of conversion of a decimal number into hexadecimal number and vice-versa.

14.4 How do you convert an octal number into the decimal number and a decimal number into the octal number? Explain with examples.

14.5 Describe the process of conversion of hexadecimal number into binary and octal equivalents.

14.6 What are 1's complement and 2's complement method of subtraction? Why do you call the 2's complement of a binary number equivalent to the negative of the decimal number?

15

Boolean Algebra and Combinational Logic

15.1 BOOLEAN ALGEBRA

Gorge Boole developed an algebra for the mathematical analysis of logic in 1854. This algebra is known as Boolean algebra. Boolean algebra was a philosophical subject until 1938 when C. E. Shannon applied this algebra for solving problem electrical relays.

Boolean algebra is quite different from our decimal algebra. It is also different from binary algebra. For example, $1+1=2$ in decimal algebra, $1+1=10$ in binary algebra and $1+1=1$ in Boolean algebra. Boolean algebra uses only two constants which are 0 and 1. This algebra has no fractional or negative numbers.

15.1.1 LAWS OF BOOLEAN ALGEBRA

Boolean algebra is based upon some laws.

(i) Laws of complementation

$\overline{0} = 1$

$\overline{1} = 0$

$\overline{\overline{A}} = A$, where A is a Boolean variable.

(ii) Laws of Boolean addition known as OR operation

$A + 0 = A$

$A + 1 = 1$

$A + A = A$

$A + \overline{A} = 1$

(iii) Laws of Boolean multiplication known as AND operation

$A \cdot 0 = 0$

$A \cdot 1 = A$

$A \cdot A = A$

$A \cdot \overline{A} = 0$

(iv) Commutative laws

It permits the interchange of positions of variables in OR and AND operation.

$$A + B = B + A$$
$$A \cdot B = B \cdot A$$

(v) Associative laws

These laws permit the grouping of Boolean variables in OR and AND operations.

$$A + (B + C) = (A + B) + C$$
$$A \cdot (B \cdot C) = (A \cdot B) \cdot C$$

(vi) Distributive laws

It allows factoring in Boolean expressions involving AND and OR operation.

$$A \cdot (B + C) = A \cdot B + A \cdot C$$

In addition to these laws, there are two more laws which can, however, be proved using the above laws. These are

$$A + (B \cdot C) = (A + B) \cdot (A + C)$$
$$A + (\overline{A} \cdot B) = A + B.$$

15.2 DE MORGAN'S THEOREM

The theorem states that if $A, B, C, \ldots\ldots$ are Boolean variables, then

(i) the complement of the product of a number of Boolean variables is equal to the sum of complemented variables.

Mathematically, $\overline{A \cdot B \cdot C \cdots} = \overline{A} + \overline{B} + \overline{C} + \cdots$

and

(ii) the complement of the sum of a number of Boolean variables is equal to the product of the complemented variables.

Mathematically, $\overline{A + B + C + \cdots} = \overline{A} \cdot \overline{B} \cdot \overline{C} \cdots$

15.3 KARNAUGH MAPPING

This is a useful method of simplifying combinational logic equations to the fewest possible terms. Karnaugh invented this technique of simplification and hence the name of this technique. To do this, we first discuss Minterms.

15.3.1 MINTERM

A minterm is a product of all the Boolean variables in the system where the variables can appear with or without a bar.

Consider a two variables system. If A and B are these variables, then the possible minterms are AB, $\overline{A}B$, $A\overline{B}$ and $\overline{A}\overline{B}$. Similarly, for a 3 variable system with variables A, B, and C, the minterms are ABC, $\overline{A}BC$, $A\overline{B}C$, $AB\overline{C}$, $\overline{A}\overline{B}C$, $\overline{A}B\overline{C}$, $A\overline{B}\overline{C}$ and $\overline{A}\overline{B}\overline{C}$.

To designate a minterm by a short hand notation, the procedure is as follows :

- Frame a binary word putting a '0' for a variable with a bar and a '1' for a variable without a bar. The variable 'A' is considered as the most significant bit (MSB) in this designation.

- Find the decimal equivalent of the binary word.

- Use this decimal number as the subscript of lower case 'm'

As an example, the minterm $ABCD$ has the designation is m_{15}. This is because putting a '1' for each of A, B, C, and D the binary word is 1111_2 whose decimal equivalent is 15.

15.3.2 DRAWING OF MAP AND ITS REDUCTION

If there are 'n' Boolean variables, there will be 2^n small squares which take the shape of a larger rectangle or square. As an example, if there are two variables A and B in a system there will be $2^2 = 4$ small squares which from a large square together. If there are 3 variables, there will be $2^3 = 8$ small squares. Each small square can accommodate a particular minterm. Presence of a particular minterm is indicated by putting a large '1' in the proper square. Similarly, absence of a particular minterm is indicated by large '0' in the proper square. Each square has a label which is the designation number of the concerned minterm.

15.3.2.1 TWO VARIABLE MAP

The Boolean function to be simplified by Karnaugh mapping be

$$F = \overline{AB} + A\overline{B} + AB = \sum m(0,2,3)$$

The map is shown in Fig. 15.1. A and B are placed on either side of a diagonal line with their possible values (0 and 1) along the left-most column and the top row respectively.

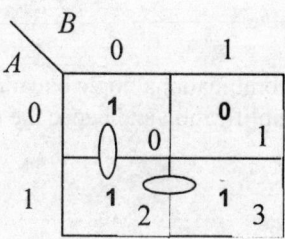

Fig. 15.1 Two variable map.

For map reduction, choose the square with least adjacencies. Here, the filled square (m_0) has only one adjacent square (m_2) filled while the other adjacent square (m_1) is empty. So, we can take m_0-square to start with. One can argue that the m_3-square has also adjacency on one side only. That is true. So, we can take m_3-square as the starting square also.

Staring from 0-th (m_0) square we connect 2nd square (m_2) with a view to form a square or rectangle in a 2^n $(n = 1, 2, 3, \ldots$etc.) square configuration. We can not link more squares because it is not possible to form any larger square or rectangle in 2^n $(n > 1)$ square configuration.

Look for the remaining squares, if any, and select one with lowest adjacencies. Here, the remaining square is the 3rd square (m_3). Starting from m_3 square connect m_2 square to form a rectangle in a 2^1 configuration. Any square can be connected more than once, if required, for this purpose. A 2^n $(n = 1, 2, 3, \ldots$etc.) square thus connected will eliminate n variables. The variables which assume different values (i.e., 0 and 1) in the minterms associated with this 2^n-square configuration will be neglected. For example, in the above two variable map, A assumes values 0 and 1 in the connected zeroth and second squares. So, A must be dropped and the simplification of this 2^1 squares gives only $\overline{B}$, since $B = 0$ in the map implies $\overline{B}$. Similarly, second and third squares connected yields only A, because B assumes values both 0 and 1 over this 2^1-square configuration. Final result will be the sum of all such simplified results.

Hence, $F = A + \overline{B}$.

15.3.2.2 THREE VARIABLE MAP

Let the function be $F = \sum m(0, 2, 4, 6)$
$$= \overline{A}\,\overline{B}\,\overline{C} + \overline{A}B\overline{C} + A\overline{B}\,\overline{C} + AB\overline{C}$$

The map is shown in Fig. 15.2.

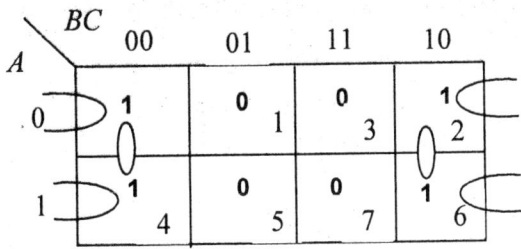

Fig. 15.2 Three variable map

Here, while labelling BC values in the top row we must remember that B and C cannot change values simultaneously. So, we have written 11 after 01 while labeling BC values. In a 3-variable map, there will be $2^3 = 8$ small squares. Only 0th, 2nd, 4th and 6th squares will be occupied containing a big '1'. All other small squares will contain a big '0'.

In simplifying the map, the map can be folded both horizontally and vertically. Making a vertical folding of the map in Fig. 15.2, we see that 0th, 2nd, 4th, and 6th squares are all adjacent squares. So, linking these squares we get a $2^2 = 4$-squares configuration. This linked 4-square configuration eliminates 2 variables. Now, looking into this 4-square combination we find that A changes values (0 and 1) and B also changes values (0 and 1). So, A and B will be dropped from the result. Again, C is 0 (zero) always over this 4-square configuration. So, we are left with $\overline{C}$.

$$\therefore \qquad F = \overline{C}.$$

15.3.2.3 FOUR VARIABLE MAP

Let us map and simplify the function :

$$F = \sum m(0,1,3,4,5,7,10,13,14,15)$$
$$= \overline{A}\,\overline{B}\,\overline{C}\,\overline{D} + \overline{A}\,\overline{B}\,\overline{C}D + \overline{A}\,\overline{B}CD + \overline{A}B\overline{C}\,\overline{D} + \overline{A}B\overline{C}D$$
$$+ \overline{A}BCD + A\overline{B}C\overline{D} + AB\overline{C}D + ABC\overline{D} + ABCD$$

There will be $2^4 = 16$ small squares in the map. Possible values of variables AB and CD are shown in the left-most column and top row of the map. One should remember that no two variables AB (or CD) can change values simultaneously. So, AB can not have values (10) after (01). Hence, after (01) it should be (11) so that only A changes from 0 to 1 and B retains its value at 1 as before. The map is shown in Fig. 15.3.

Squares $0,1,3,4,5,7,10,13,14,15$ are occupied by large '1's. That is these minterms are present in the map. All other squares contain big '0's indicating that these minterms are absent. m_{10} has only one adjacent term (m_{14}). So, starting from m_{10} square connect m_{14} square. This makes a 2^1 square configuration. We can not expand this link any more since any further linking do not produce a square or a rectangle in a 2^2-square configuration. In this 2-square configuration, B changes values while $A = 1$, $C = 1$, and $D = 0$. The result of this 2-square is $AC\overline{D}$. Then, m_4 is a square which has two adjacent terms (m_5 and m_0). So, start linking from m_4 square. m_4, m_5, m_1 and m_0 form a $2^2 = 4$-square configuration. Hence, two variables will be eliminated from this configuration. B and D changes values (0 and 1) here. So, B and D will be eliminated from the result of this 4-square. Moreover, A and C are both zero in this case and we get the term $\overline{A}\,\overline{C}$.

Again, m_3-square has two adjacent terms m_1 and m_7. So, starting from m_3-square we go on linking squares. Thus, we connect a 4-square with m_3, m_1, m_5 and m_7 squares connected. Over this 4-square, B and C changes values. So, we drop B and C from the result of this 4-square. Again, since $A = 0$ and $D = 1$ over this 4-square we get $\overline{A}D$ as the result of this 4-square.

Thus far we have not included m_{13} and m_{15} squares. So, starting from m_{13}-square and linking we get a 4-square involving m_{13}, m_{15}, m_7 and m_5 squares. In this linking process, any square may be linked more than once in order to get the largest 2^n-square configuration forming a overall square or rectangle. In this 4-square, A and C change values while $B = 1$ and $D = 1$. So, we get BD as the result of this 4-square. Combining all the results, we get

$$F = AC\overline{D} + \overline{A}\,\overline{C} + \overline{A}D + BD.$$

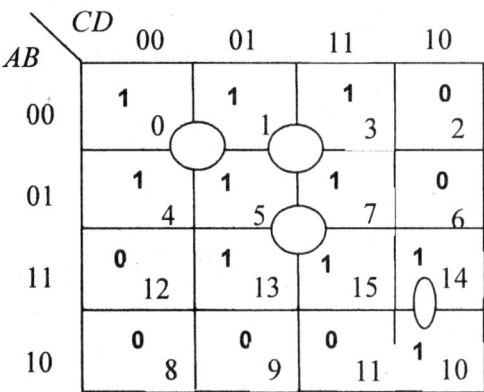

Fig. 15.3. Four variable map.

15.4 LOGIC GATES

Gates are electronic circuits meant for the realization of Boolean operations such as AND, OR, NOT, etc. A gate has one or more input and one output terminal. Depending upon the combination of inputs, a particular output result.

The truth table is a representation which shows all possible input and output values of a logic gate. The number of horizontal rows in a truth table is 2^n where n is the number of inputs of the gate. We shall discuss OR, AND, NOT, Ex-OR, NOR, Ex-NOR, and NAND gates.

15.4.1 OR GATE

This gate performs Boolean addition. The symbol of this gate is as shown in Fig. 15.4. Here, A and B are two inputs and Y is output. The truth table is as shown in Table 15.1.

$Y = A + B$

Fig. 15.4. Symbol of OR gate

A	B	$Y = A + B$
0	0	0
0	1	1
1	0	1
1	1	1

Table 15.1. Truth table of a 2-input OR gate.

$Y = 1$ if any or both of A and B are 1. The OR gate can have any number of inputs. For a 3-input OR gate, $Y = A + B + C$ Where A, B, C are inputs and Y is the output.

15.4.2 AND GATE

This gate performs Boolean multiplication. The logic symbol of a 3-input AND gate and its truth table is shown in Fig. 15.5 and Table 15.2. respectively.

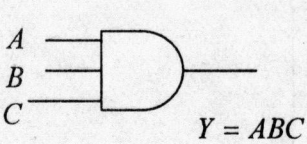

Fig. 15.5. Symbol of AND gate

A	B	C	Y = ABC
0	0	0	0
0	0	1	0
0	1	0	0
0	1	1	0
1	0	0	0
1	0	1	0
1	1	0	0
1	1	1	1

Table 15.2. Truth table of 3-input OR gate.

Here, A, B, C are 3 inputs and Y is the output. It can have any number of inputs. Here, $Y = 1$ only when all inputs are 1.

15.4.3 NOT GATE

This gate performs inversion. It has one input and one output. The logic symbol and truth table are shown in Fig. 15.6 and Table 15.3 respectively.

A ————▷o———— $Y = \overline{A}$

Fig. 15.6. Symbol of NOT gate.

A	$Y = \overline{A}$
0	1
1	0

Table 15.3. Truth table of NOT gate.

15.4.4 EX-OR GATE

This is Exclusive-OR gate. It has two inputs and one output. It gives an output $Y = 1$ when only one of the two inputs is 1, else the output is 0. It excludes the case $A = B = 1$ of the OR gate. The symbol and truth table of the Ex-OR gate is shown in Fig. 15.7 and Table 15.4 respectively.

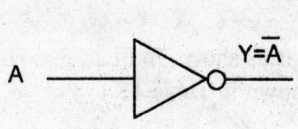

Fig. 15.7. Symbol of Ex-OR gate.

A	B	$Y = A \oplus B$
0	0	0
0	1	1
1	0	1
1	1	0

Table 15.4. Truth table of Ex-OR gate.

The symbol $\oplus$ is called Ex-OR symbol. $A \oplus B$ stands for $A\bar{B} + \bar{A}B$. Now, how can we realize an Ex-OR gate by AND, OR, and NOT gates? This is shown in Fig. 15.8.

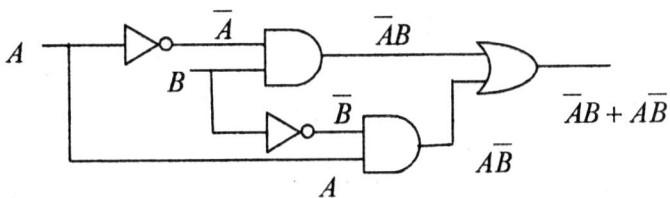

Fig. 15.8. Ex-OR gate using OR, AND, and NOT gates.

15.4.5 NOR GATE

A NOT gate following an OR gate forms a NOR gate. Logic symbol and truth table of a two input NOR gate is shown in Fig. 15.9 and Table 15.5 respectively.

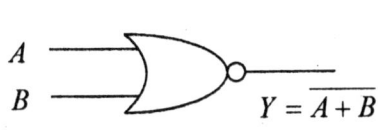

Fig. 15.9. Logic symbol of NOR gate

A	B	$Y = \overline{A + B}$
0	0	1
0	1	0
1	0	0
1	1	0

Table 15.5. Truth table of a 2-input NOR gate.

For a 3-input NOR gate, $Y = \overline{A + B + C}$, where A, B, C are inputs and Y is the output. The truth table is shown below in Table 15.6. It can have any number inputs.

A	B	C	$Y = \overline{A + B + C}$
0	0	0	1
0	0	1	0
0	1	0	0
0	1	1	0
1	0	0	0
1	0	1	0
1	1	0	0
1	1	1	0

Table 15.6. Truth table of a 3-input NOR gate.

15.4.6 NAND GATE

A NOT gate following an AND gate forms a NAND gate. It can have any number of inputs. The logic symbol and the truth table of the 2-input NAND gate is shown in Fig. 15.10 and Table 15.7 respectively.

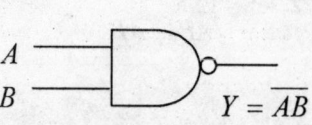

Fig. 15.10. Symbol of a 2-input NAND gate.

A	B	$Y = \overline{AB}$
0	0	1
0	1	1
1	0	1
1	1	0

Table 15.7. Truth table of a 2-input NAND gate.

For a 3-input NAND gate, the logic symbol and truth table are shown in Fig.15.11 and Table 15.8 respectively.

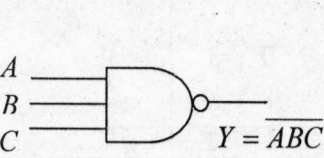

Fig. 15.11. Logic symbol of a 3-input NAND gate.

A	B	C	$Y = \overline{ABC}$
0	0	0	1
0	0	1	1
0	1	0	1
0	1	1	1
1	0	0	1
1	0	1	1
1	1	0	1
1	1	1	0

Table 15.8. Truth table of a 3-input NAND gate.

15.4.7 EX-NOR GATE

A NOT gate following an Ex-OR gate forms an Ex-NOR gate. The logic symbol and the truth table for Ex-NOR gate is shown in Fig. 15.12 and Table 15.9 respectively.

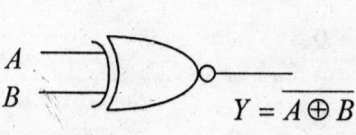

Fig. 15.12. Logic symbol of Ex-NOR gate.

A	B	$Y = \overline{A \oplus B}$
0	0	1
0	1	0
0	0	0
1	1	1

Table 15.9. Truth table of Ex-NOR gate.

Now, $Y = \overline{A \oplus B} = \overline{\overline{AB} + \overline{A}B}$

$\qquad = \overline{\overline{A}B} \cdot \overline{A\overline{B}} = \left(\overline{\overline{A}} + \overline{B}\right) \cdot \left(\overline{A} + \overline{\overline{B}}\right)$

$\qquad = \left(A + \overline{B}\right)\left(\overline{A} + B\right)$

$\qquad = A\overline{A} + AB + \overline{B}\,\overline{A} + \overline{B}B$

$\qquad = AB + \overline{A}\,\overline{B}$ (since $A\overline{A} = B\overline{B} = 0$).

15.4.8 NAND AND NOR AS UNIVERSAL GATES

The basic operations in Boolean algebra are OR, AND and NOT. Using only NAND gates we can achieve these operations. Also, using only NOR gates it is possible to realize these fundamental operations. Hence, NAND and NOR gates are called Universal gates.

15.4.9 NAND AS UNIVERSAL GATE

The OR operation can be realized as follows :

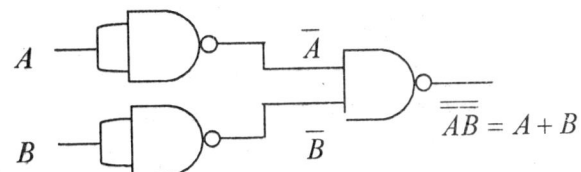

Fig. 15.13. OR operation by NAND gate.

AND operation can be realized as shown in Fig. 15.14.

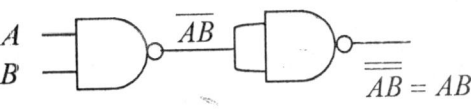

Fig. 15.14. AND operation using NAND gates.

NOT operation can be achieved by using a NAND gate as shown in Fig. 15.15.

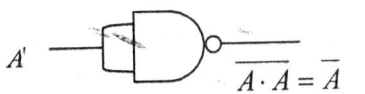

Fig. 15.15. NOT operation using a NAND gate.

15.4.10 NOR GATE AS UNIVERSAL GATE

The OR operation can be realized by using a pair of NOR gates as shown in Fig. 15.16.

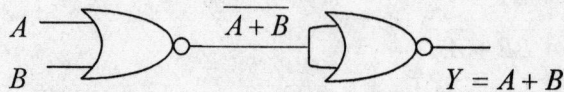

Fig. 15.16. OR function generated by NOR gates.

Note that when two inputs of a NOR gate are the same, we get inversion. That is $\overline{X + X} = \overline{X}$ and $\overline{\overline{X} + \overline{X}} = \overline{\overline{X}} = X$.

The AND operation can be achieved by NOR gates as shown in Fig. 15.17.

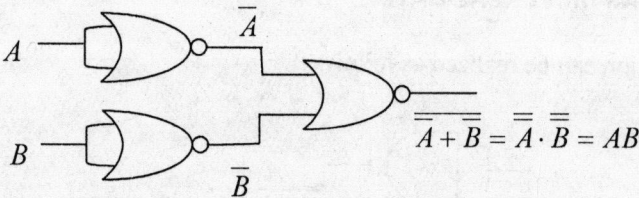

Fig. 15.17. AND operation using NOR gates.

The NOT operation can be realized as shown in Fig. 15.18.

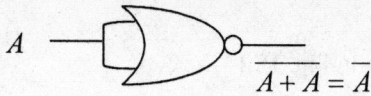

Fig. 15.18. NOT operation using NOR gates.

15.5 LOGIC SYSTEM AND GATE IMPLEMENTATION

The gates can be realized in circuits by using diodes and transistors. Depending upon the circuit realizations, the following logic families have come up:

(i)   Diode Logic (DL)
(ii)   Resistor-Transistor Logic (RTL)
(iii)   Diode-Transistor Logic (DTL)
(iv)   Transistor-Transistor Logic (TTL)
(v)   Integrated-Injection Logic ($I^2 L$)
(vi)   Emitter-coupled Logic (ECL) , and
(vii)   Complementary Metal Oxide semiconductor Logic (CMOS).

Depending upon how a bit (0 or 1) is represented in the logic hardware, logic systems can be classified as follows:

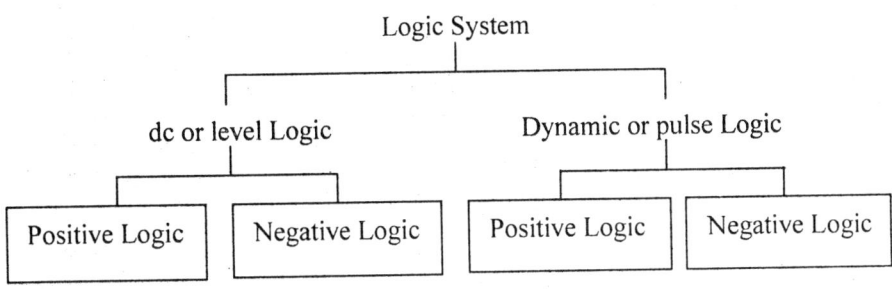

There are two digits in a digital system. These are 0 and 1. In dc or level logic system, these two digits are represented by two voltage levels. In positive level logic, the higher positive voltage is taken as level '1' while the lower positive voltage is taken as the '0' level. The absolute values of these voltage levels are not important - only the relative level, higher or lower, is important. A '1' level does not mean that the voltage is 1 volt. As an example, the collector voltage of a cut-off CE transistor equals the supply voltage. If supply voltage is $+5$ volt, level '1' is represented by $+5$ volt. When the transistor is saturated its collector voltage becomes $\simeq 0.2$ volt. So, 0.2 volt can be taken as '0' level. These levels will also have some tolerance. For example, level '1' can be effected within a range of (5 ± 1) volt, say. In the negative level logic system, the higher voltage defines the '0' level while the more negative voltage (with respective to the '0' level) defines the level '1'.

In dynamic or pulse logic, '0' bit is represented by the absence of a pulse and a '1' is represented by the presence of a pulse. In dynamic positive logic, a '1' is represented by a positive pulse (⎍) while in the dynamic negative logic system, a '1' is represented by a negative pulse (⎍).

15.5.1 OR GATE USING DIODES

In positive diode logic, the OR gate can be implemented as shown in Fig.15.19.

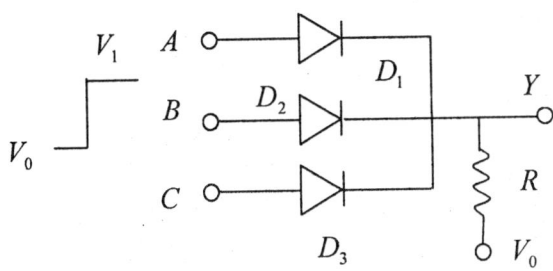

Fig. 15.19. OR gate in positive diode logic.

Here, A, B, C are 3-inputs. D_1, D_2, D_3 are 3 p-n junction diodes assumed to be identical. Let V_0 be voltage label corresponding to bit '0' and V_1 the be voltage corresponding to bit '1'.

If any one of the inputs goes high (corresponding input voltage being V_1), the corresponding diode conducts and provide a short circuit. Then, $Y = 1$ since the output voltage becomes V_1. If all the inputs go high, all the diodes provide short circuit and $Y = 1$. When all the inputs go low corresponding to input voltage V_0, all the diodes become non-conducting and the output voltage becomes the reference voltage V_0. So, $Y = 0$.

The OR gate implemented in negative diode logic is shown in Fig. 15.20.

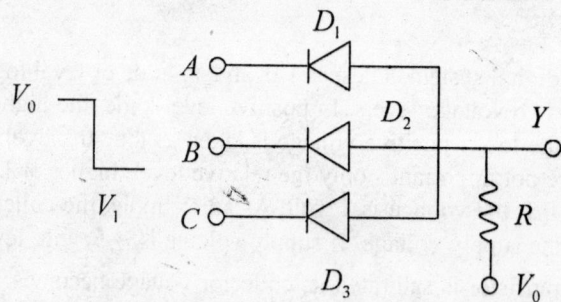

Fig. 15.20. OR gate in negative diode logic.

Here, the higher voltage V_0 stands for '0' level and the lower voltage V_1 stands for level '1'. When any or all the inputs go '1' (corresponding to voltage V_1), the corresponding diode or all the diodes become conducting. They provide short circuit and the voltage V_1 comes at the output. Then, $Y = 1$. When all the inputs becomes V_0, all the diodes become non-conducting and the reference voltage V_0 appears at the output. Now, $Y = 0$.

15.5.2 AND GATE USING DIODES

Circuit realization of the AND gate in positive diode logic is shown in Fig. 15.21.

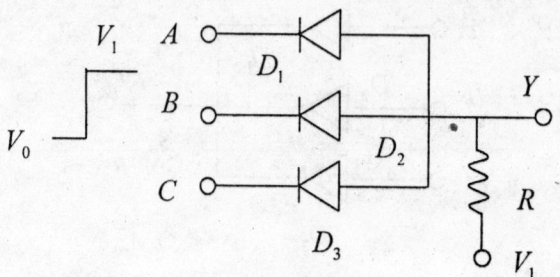

Fig. 15.21. AND gate using positive diode logic.

When any input goes low (corresponding to voltage V_0), the corresponding diode becomes conducting providing a short circuit. The output voltage becomes V_0. So, $Y = 0$. When all inputs go high (voltage V_1), all the diodes become nonconducting. Now, output voltage becomes equal to the reference voltage V_1. So, $Y = 1$ now. The circuit of the AND gate using negative diode logic is shown in Fig. 15.22.

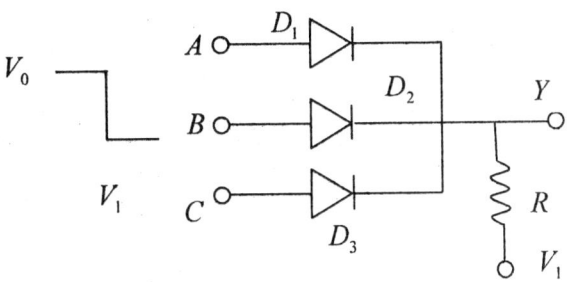

Fig. 15.22. AND gate in negative diode logic.

Here, when all inputs are '1'corresponding to voltage V_1, the diodes become nonconducting and the reference voltage V_1 appears at the output. So, $Y = 1$. When any input go '0' corresponding to voltage V_0, the corresponding diode becomes conducting providing short circuit. So, V_0 appears at the output and hence, $Y = 0$.

15.5.3 NOT GATE USING TRANSISTOR

The circuit diagram of a NOT gate in positive transistor logic is shown in Fig. 15.23. When the input voltage is V_1 corresponding to bit '1', the base current of the npn transistor becomes high enough to saturate the transistor. The collector voltage becomes very low corresponding to voltage V_0. Now, $Y = 0$. When the input voltage is V_0, the base-emitter diode is nonconducting and the supply voltage appears at the output. Now, $Y = 1$. The supply voltage is equal to V_1.

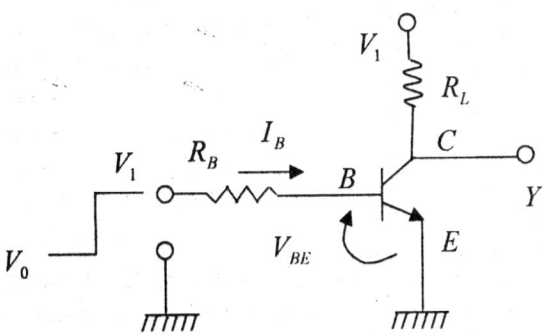

Fig. 15.23. NOT gate in positive transistor logic.

The base current $I_B = \dfrac{V_1 - V_{BE}}{R_B} = I_B\big|_{sat}$,where I_B corresponds to the saturation base current of

the transistor. Now, $I_B\big|_{sat} = \dfrac{I_C\big|_{sat}}{h_{FE}}$ and $I_C\big|_{sat} = \dfrac{V_1 - V_0}{R_L}$.

$$\therefore \qquad \frac{V_1 - V_{BE}}{R_B} = \frac{V_1 - V_0}{R_L h_{FE}} \qquad\qquad\qquad \dots (15.1)$$

Since V_{BE}, R_L, h_{FE}, V_1, V_0 are known, R_B can be found out from eqn. (15.1). Typical value of $V_1 = \pm 5$ volt and $V_0 = 0.2$ volt.

15.6 BINARY ADDER

Binary addition can be effected by using logic gates. Binary addition is the fundamental operation in binary arithmetic. Subtraction requires inversion and addition. Multiplication can be done by repeated addition. Binary division can be achieved through repeated subtraction.

For complete addition, we add two bits corresponding to a given position and we have to add the carry from the previous addition with it. So, a complete addition means addition of two bits and a carry bit, in general. This is called Full addition. Depending upon how many bits an adder adds, binary adders can be divided into two categories. These are (i) Half adder, and (ii) Full Adder.

15.6.1 BINARY HALF ADDER

Binary half adder is a digital circuit formed by logic gates whose function is to add two binary digits. It has two inputs and two outputs. Two inputs are two bits to be added. One of the two outputs is called the sum output while the other output is the carry output. Complete addition requires two half adders and hence its name is 'half adder'. The logic symbol and the truth table are shown in Fig. 15.24 and table 15.10 respectively.

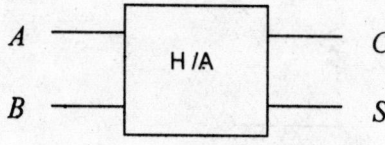

Fig. 15.24. Symbol of half adder.

Inputs		Outputs	
A	B	S	C
0	0	0	0
0	1	1	0
1	0	1	0
1	1	0	1

Table 15.10. Truth table of Half adder.

Here A and B are inputs, C is the carry output and S is the sum output. Looking at the truth table, we see that $S = 1$ only when one of the inputs is 1. This is the output of an Ex-OR gate. So, $S = A \oplus B = \overline{A}B + A\overline{B}$.

Again, $C = 1$ only when $A = B = 1$. This is the property of an AND gate. So, $C = AB$.

15.6.1.1 IMPLEMENTATIONS OF HALF ADDER

One possible implementation of the half adder is shown in Fig. 15.25.

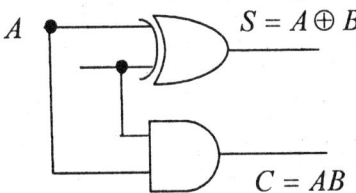

Fig. 15.25. Half adder implementation.

A second implementation of half adder in AOI (AND-OR-Invert) logic is shown in Fig. 15.26.

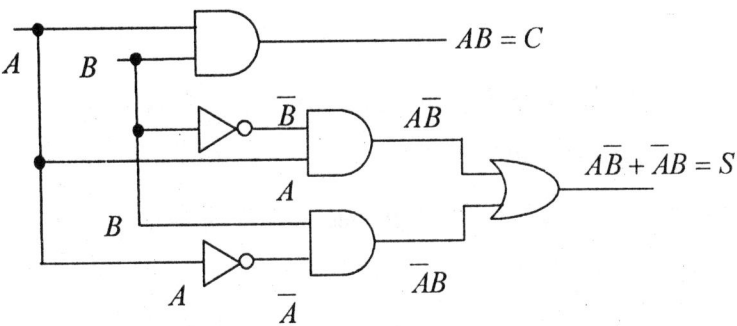

Fig. 15.26. Half adder in AOI logic.

15.6.2 BINARY FULL ADDER

It adds 3 bits and yields a sum output and a carry output. It has 3 inputs and 2 outputs. The logic symbol and truth table are shown in Fig. 15.27 and table 15.11 respectively.

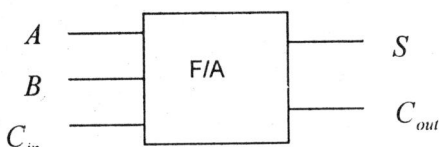

Fig. 15.27. Logic symbol of a binary full adder.

Inputs			Outputs	
A	B	C_{in}	S	C_{out}
0	0	0	0	0
0	0	1	1	0
0	1	0	1	0
0	1	1	0	1
1	0	0	1	0
1	0	1	0	1
1	1	0	0	1
1	1	0	1	1

Table 15.11. Truth table of a binary full Adder.

Boolean expressions for sum output S and carry output C are as follows:

$$S = \overline{A}\overline{B}C_{in} + \overline{A}B\overline{C}_{in} + A\overline{B}\overline{C}_{in} + ABC_{in}$$

$$C = \overline{A}BC_{in} + A\overline{B}C_{in} + AB\overline{C}_{in} + ABC_{in}$$

These expressions can be verified by putting 0 and 1 for A, B, C_{in}. Now,

$$S = C_{in}\left(\overline{A}\overline{B} + AB\right) + \overline{C}_{in}\left(\overline{A}B + A\overline{B}\right) = C_{in}\left(\overline{\overline{A}B + A\overline{B}}\right) + \overline{C}_{in}\left(\overline{A}B + A\overline{B}\right)$$

$$= C_{in}\overline{S}_1 + \overline{C}_{in}S_1 = C_{in} \oplus S_1$$

where $S_1 = \overline{A}B + A\overline{B}$ = sum output of the half adder adding A and B , and

$$C = C_{in}\left(\overline{A}B + A\overline{B}\right) + AB\left(\overline{C}_{in} + C_{in}\right) = C_{in}S_1 + AB \cdot 1 = C_{in}S_1 + C_h$$

where $C_h = AB$ = carry output of the half adder adding A and B.

15.6.2.1 IMPLEMENTATION OF FULL ADDER

An implementation using half adders is shown in Fig.15.28.

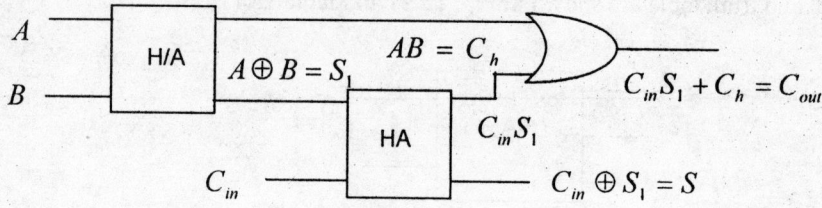

Fig. 15.28. Full adder circuit using half adders

A second implementation using AND-OR-Invert (AOI) logic is shown in Fig. 15.29.

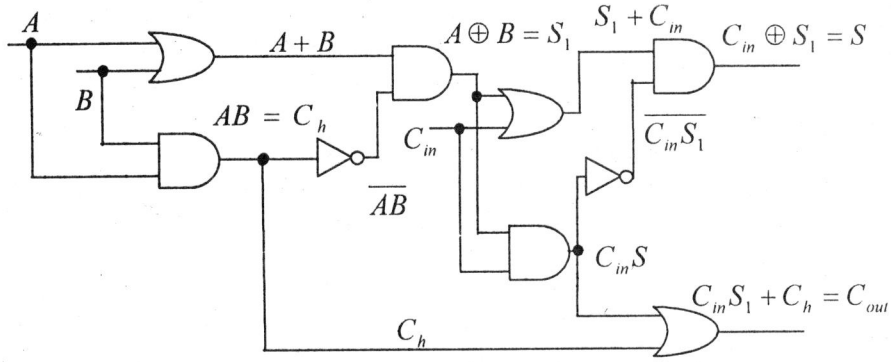

Fig. 15.29. Full adder circuit in AOI logic.

15.6.2.2 BINARY FULL ADDER OPERATION

Full adder can operated in two ways – parallel and serial. In parallel operation, all the columns of two binary numbers are added simultaneously. So, for the addition of two N-digit binary numbers are require N number of full adders. Let us consider the parallel addition of two 4 -bit binary numbers $A_3\ A_2\ A_1\ A_0$ and $B_3\ B_2\ B_1\ B_0$. A_3 and B_3 are most significant bits (MSB), and A_0 and B_0 are least significant bits (LSB). For this, we need 4 full adders as shown in Fig. 15.30.

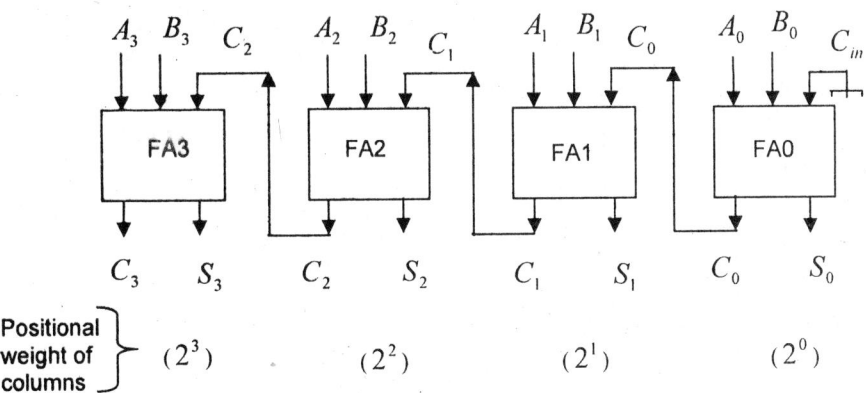

Fig. 15.30. Schematic diagram of parallel addition using full adders.

The result of addition is $C_3 S_3\ S_2\ S_1\ S_0$. The input carry of 2^0- stage is grounded since at the beginning of addition there is no carry. As an example, consider the parallel addition of two binary numbers $A = 1011\ _2$ and $B = 1001\ _2$. Now,

$$A_0 = 1, \qquad B_0 = 1 \qquad\qquad \therefore S_0 = 0 \text{ and } C_0 = 1$$
$$A_1 = 1, \qquad B_1 = 0, C_0 = 1 \qquad \therefore S_1 = 0 \text{ and } C_1 = 1$$
$$A_2 = 0, \qquad B_2 = 0, C_1 = 1 \qquad \therefore S_2 = 1 \text{ and } C_2 = 0$$
$$A_3 = 1, \qquad B_3 = 1, C_2 = 0 \qquad \therefore S_3 = 0 \text{ and } C_3 = 1$$

Thus, $1011_2 + 1001_2 = 10100_3$.

Now, we discuss the serial operation of binary full adder. Here, two binary numbers are added serially – a single column is added at a time. So, a single full adder can do the whole job. An example is shown in the following diagrams (Fig. 15.31 (a) and 15.31(b)).

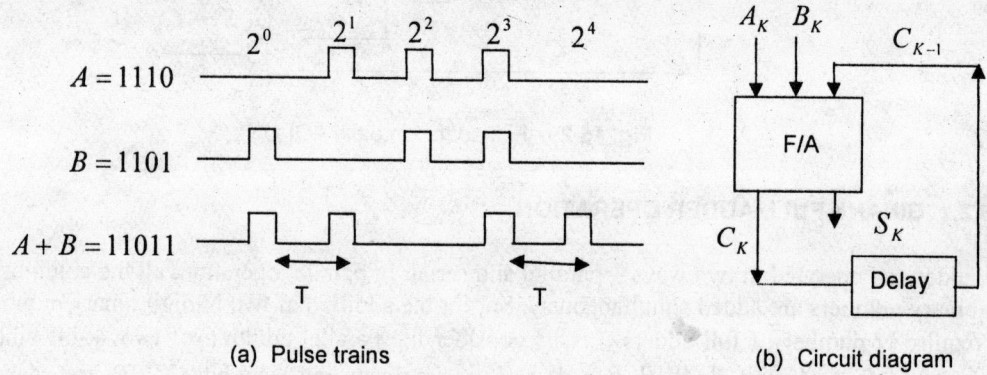

(a) Pulse trains   (b) Circuit diagram

Fig. 15.31. Serial addition of two 4-bit binary numbers.

Here, the binary numbers $A = 1110_2$ and $B = 1101_2$ are two pulse trains. Existence of a pulse means '1' and no pulse means '0'. In the addition of two bits A_K and B_K corresponding to K-th column we must add the carry from $(K-1)$th stage of addition. Now, if the carry of $(K-1)$th stage is passed through a delay circuit, which provides a delay of one pulse period, this carry will appear at the input of the full adder one period later when A_K and B_K appear at the input for addition. S_K is the sum output of the K-th stage which can be stored in a shift register. The delay can be provided by a D-flip flop.

Parallel addition is faster than serial addition. But, parallel addition has circuit complexity since it requires N full adders for the addition of two N-bit binary numbers whereas serial adder has a simple circuit since it requires only one full adder.

15.7 BINARY MAGNITUDE COMPARATOR

A binary magnitude comparator is a combinational circuit which compares the magnitude of two binary numbers. In the process of comparison, three cases can occur. If A and B are two binary numbers, then

    (i) A can be greater than B,

    (ii) A can be less then B, and

    (iii) A can be equal to B.

An Ex-OR gate can be used as the basic inequality detector. Its output is '1' only when $A \neq B$ where A and B are single bit numbers. If A and B are 2-bit numbers such as $A_1 A_0$ and $B_1 B_0$, we can detect the inequality by the following circuit (Fig.15.32).

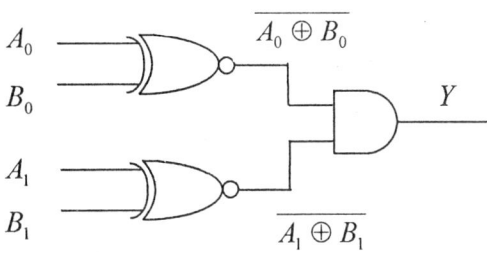

Fig. 15.32. Inequality detector for two 2-bit binary numbers.

When $A_0 = B_0$ and $A_1 = B_1$, $\overline{A_0 \oplus B_0} = A_0 B_0 + \overline{A_0} \, \overline{B_0} = 1$ and $\overline{A_1 \oplus B_1} = A_1 B_1 + \overline{A_1} \, \overline{B_1} = 1$. Hence, $Y = 1$. When $A_0 \neq B_0$ and /or $A_1 \neq B_1$, we get $Y = 0$.

15.7.1 TWO-BIT BINARY MAGNITUDE COMPARATOR

Let $A_1 A_0$ and $B_1 B_0$, be two binary numbers whose magnitude has to be compared. We define the following Boolean functions:

$$y_0 = A_0 B_0 + \overline{A_0} \, \overline{B_0}$$
$$y_1 = A_1 B_1 + \overline{A_1} \, \overline{B_1}$$

Now, when $A_0 = B_0$ we have $y_0 = 1$. Again, $y_0 = 0$ when $A_0 \neq B_0$. Similarly, when $A_1 = B_1$ we get $y_1 = 1$ and we have $y_1 = 0$ when $A_1 \neq B_1$. Then, when the two binary numbers are equal we have $y_0 = 1$ and $y_1 = 1$.

To determine whether $A > B$ or $A < B$ are define the following functions:

$$Z = A_1 \overline{B_1} + y_1 A_0 \overline{B_0} \text{ for } A > B$$

$$W = \overline{A_1} B_1 + y_1 \overline{A_0} B_0 \text{ for } A < B$$

Now, $Z = 1$ indicates $A > B$ and $W = 1$ indicates $A < B$. Suppose $A_1 = 1$ and $B_1 = 0$. Then, $y_1 = 0$. Now, $Z = 1$ so that $Z = 1$ indicates $A > B$. On the other hand, if $A_1 = 0$ and $B_1 = 1$, then $y_1 = 0$ and $W = 1$ which indicates $A < B$. Now , in this latter case, we have $Z = 0$.

For detecting the equality, we define the function $E = y_1 y_0$. If $E = 1$, we conclude $A = B$. The logic diagram of the 2-bit magnitude comparator is shown in Fig. 15.33.

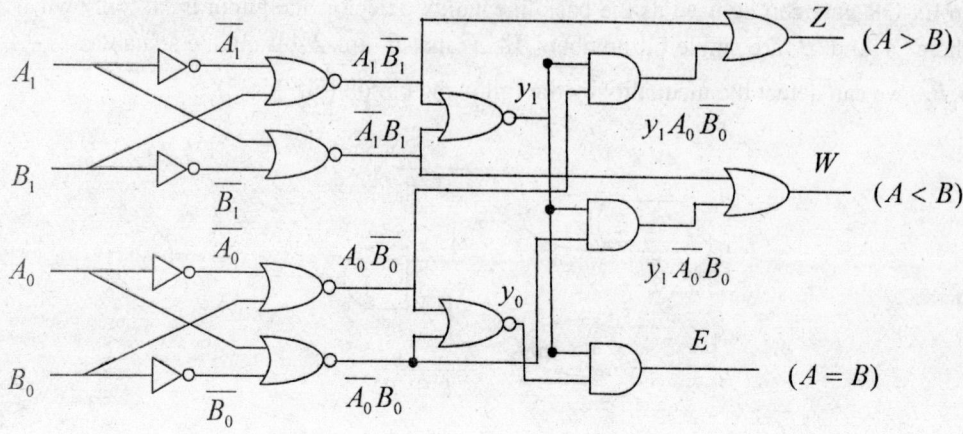

Fig. 15.33. Logic circuit diagram of a two-bit magnitude comparator.

15.7.2 FOUR-BIT BINARY MAGNITUDE COMPARATOR

Let us now compare two 4-bit binary numbers. Let $A_3 A_2 A_1 A_0$ and $B_3 B_2 B_1 B_0$ be two such numbers. For equality of A and B, all pairs of corresponding bits must be identical. This condition can be expressed by the following Boolean expression

$$y_K = A_K B_K + \overline{A_K} \overline{B_K}$$

where $K = 0, 1, 2, 3$ and y_K is the output of an Ex-NOR gate. When $A_K = B_K$, $y_K = 1$ and when $A_K \neq B_K$, $y_K = 0$. For equality of A and B, all y_K (for $K = 0, 1, 2, 3$) must be equal to '1'. This condition may be realized by the output of an AND gate having 4 inputs y_0, y_1, y_2 and y_3. So, in the comparator circuit we must generate an output $y_0 y_1 y_2 y_3$ to indicate equality of two 4-bit binary numbers.

To determine whether $A > B$ or $A < B$, we have $A \neq B$, and we first compare the pair of most significant bits A_3 and B_3. If $A_3 = B_3$, then we must compare the next significant bits A_2 and B_2. If $A_2 = B_2$, then we should compare A_1 and B_1, and so on. If in the process of comparison, we get, for example, $A_1 = 1$ and $B_1 = 0$ then we conclude $A > B$. But, if $A_1 = 0$ and $B_1 = 1$ then we conclude $A < B$.

To realize these conditions in the comparator circuit, we must generate the following functions:

$$Z = A_3 \overline{B_3} + y_3 A_2 \overline{B_2} + y_3 y_2 A_1 \overline{B_1} + y_3 y_2 y_1 A_0 \overline{B_0} \qquad \text{for } A > B$$

and $\qquad W = \overline{A_3} B_3 + y_3 \overline{A_2} B_2 + y_3 y_2 \overline{A_1} B_1 + y_3 y_2 y_1 \overline{A_0} B_0 \qquad \text{for } A < B.$

We must remember that if we find an inequality in a given higher order bits, the inequalities in lower order bits must be neglected. Otherwise, the lower order bit inequalities will lead to errorous conclusions. To explain this point, we give one example. Take 2 binary numbers

1001_2 and 0101_2. Now, $A_3 = 1$ and $B_3 = 0$. Then, $A_3 \overline{B_3} = 1$ in the condition $A > B$. Now, it is immaterial what values the other terms have in the condition $A > B$ above. This is all right. But, think of the other condition for $A < B$. W must be zero to ensure $A \not< B$. Here, $A_2 = 0$ and $B_2 = 1$. So, $\overline{A_2} B_2 = 1$. Now, unless this term in W is suppressed (i.e., made zero) by some means, $W = 1$ (in absence of the factor y_3) and which will indicate $A < B$. This is done by multiplying the term $\overline{A_2} B_2$ by proper y_K. Hence, once the inequality is detected in a given order, all lower order inequalities must be ignored. The implementation of the 4-bit binary magnitude comparator by using logic gates is shown in Fig. 15.34.

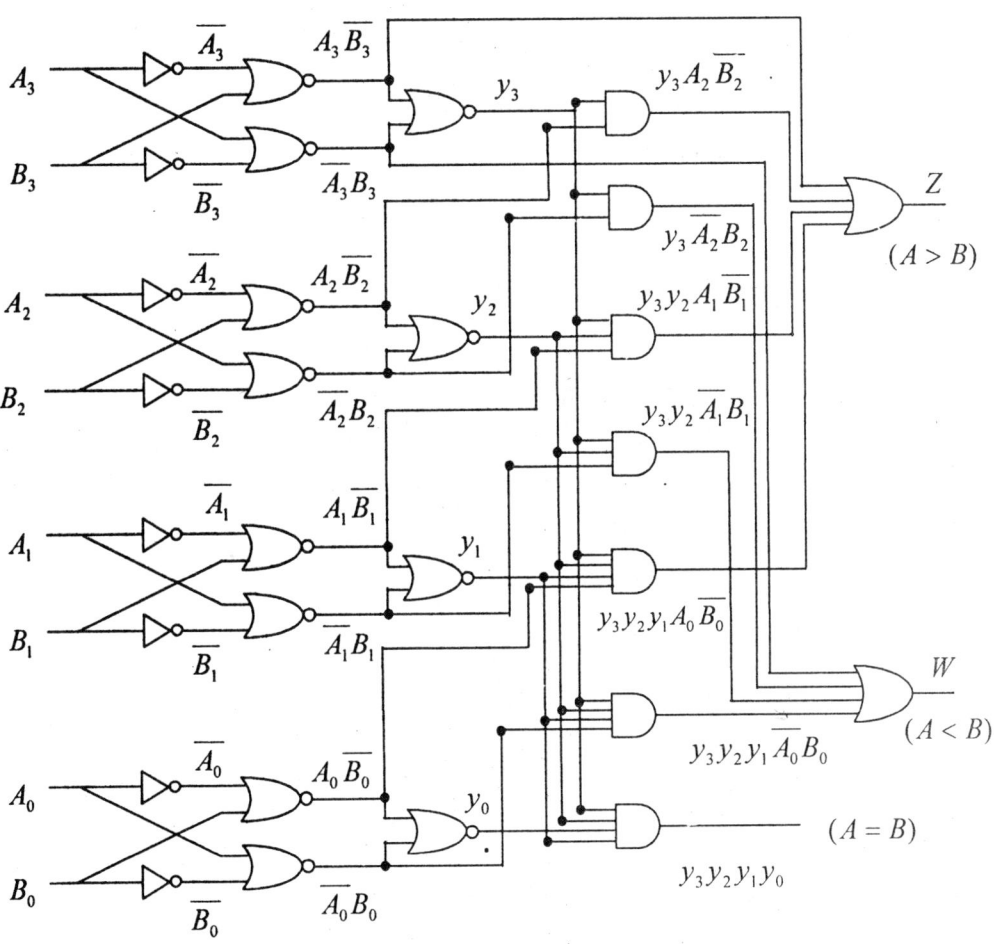

Fig. 15.34. Circuit diagram of a 4-bit binary magnitude comparator.

15.8 MULTIPLEXER

A multiplexer is a combinational logic circuit which selects a particular binary information from a large number of input lines and directs this information to a single output line. There are 2^n input lines which are selected by n number of selector bits. For $n = 1$, there are $2^1 = 2$ input lines and the corresponding circuit is called $2:1$ (or 2×1) multiplexer. The logic symbol and the logic circuit diagram of a 4×1 multiplexer is shown in Figs. 15.35 and 15.36 respectively.

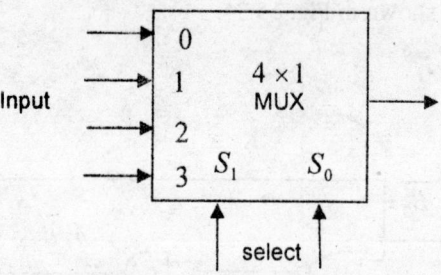

Fig. 15.35. Symbol of 4×1 multiplexer.

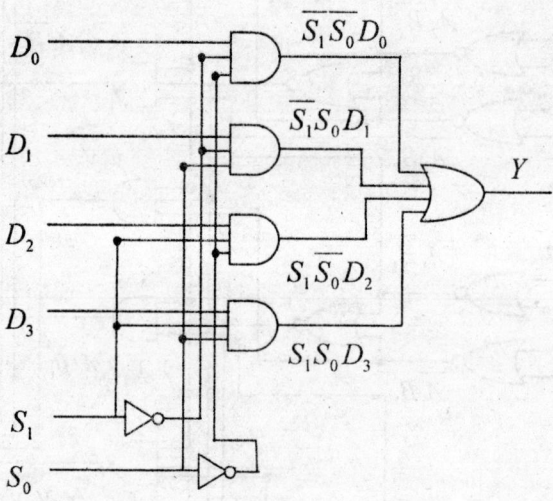

Fig. 15.36. Logic circuit diagram of a 4×1 MUX.

In Fig.15.36, D_0, D_1, D_2, D_3 are input data lines and S_1, S_0 are select input lines. The output of the circuit shown in Fig.15.36 is

$$Y = \overline{S_1}\,\overline{S_0}D_0 + \overline{S_1}S_0D_1 + S_1\overline{S_0}D_2 + S_1S_0D_3.$$

If $S_1 = 0$ and $S_0 = 0$, we get $Y = D_0$. Thus, selector inputs 0, 0 produces the input data D_0 at the output. Similarly, when the selector bits are $1,1$ the data corresponding to input line D_3 appears at the output.

An additional input, called the enable input, is added to all the AND gates in some cases. When enable input is given a particular bit (either 0 or 1), the AND gates can be made to operate as desired in a multiplexer. The other input bit supplied to the enable input will disable the multiplexer. The truth table of the 4×1 MUX is shown in table 15.12.

Select		Outputs
S_1	S_0	Y
0	0	D_0
0	1	D_1
1	0	D_2
1	1	D_3

Table 15.12. Truth table of a 4×1 multiplexer.

15.9 DEMULTIPLEXER

A demultiplexer is a combinational circuit which takes data from a single input line and directs it to one of the 2^n output lines, the transmission being controlled by a group of 'n' select bits supplied through 'n' select lines.

The input data is supplied to all of the 2^n AND gates and depending upon the binary select digits only one AND gate is enabled at a time and it makes the input data appear at a given output line. The demultiplexer is called (1×2^n) demultiplexer. The logic diagram of the (1×4) demultiplexer (DMUX) is shown in Fig. 15.37.

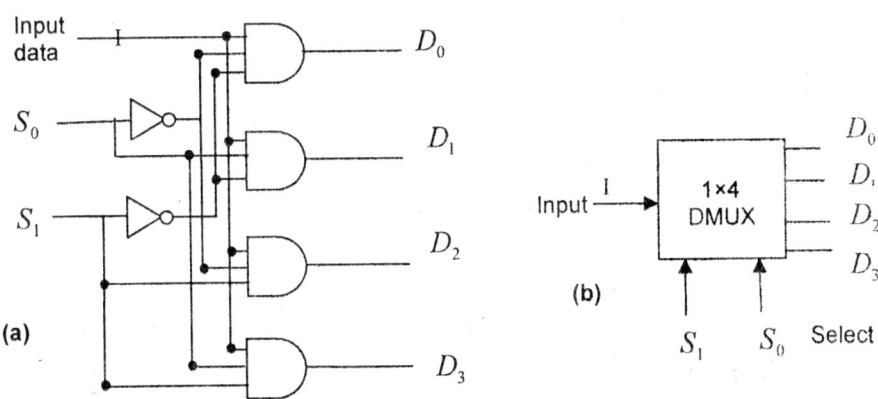

Fig. 15.37. A (1×4) demultiplexer logic diagram in (a) and its symbol in (b).

The truth table for the (1×4) demultiplexer is shown in table 15.13.

Input	Select		Outputs			
	S_1	S_0	D_0	D_1	D_2	D_3
I	0	0	I	0	0	0
I	0	1	0	I	0	0
I	1	0	0	0	I	0
I	1	1	0	0	0	I

Table 15.13. Truth table for a (1×4) demultiplexer.

SOLVED PROBLEMS

15.1 Simplify the following Boolean expressions

$$\overline{\overline{\overline{ABC + \overline{AB}}} + BC}$$

Solution

Now, $\overline{\overline{\overline{ABC + \overline{AB}}} + BC}$

$= \overline{\overline{ABC} \cdot \overline{\overline{AB}} + BC}$ [by De Morgan's theorem]

$= \overline{\overline{(A + \overline{B} + \overline{C}) \cdot (\overline{\overline{A + \overline{B}}})} + BC}$ [by De Morgan's theorem]

$= \overline{(A + \overline{B} + \overline{C})(A + B) + BC}$

$= \overline{AA + \overline{A}B + A\overline{B} + B\overline{B} + A\overline{C} + B\overline{C} + BC}$

$= \overline{AB + A\overline{B} + A\overline{C} + B(C + \overline{C})}$ [$\because \overline{A}A = 0 = \overline{B}B$]

$= \overline{AB + A\overline{B} + A\overline{C} + B}$ [$\because C + \overline{C} = 1$]

$= \overline{A\overline{B} + A\overline{C} + B(1 + A)}$

$= \overline{A\overline{B} + B + A\overline{C}}$ [$\because 1 + \overline{A} = 1$]

$= \overline{(B + \overline{B}A) + A\overline{C}}$

$= \overline{B + A + A\overline{C}}$ [$\because B + \overline{B}A = B + A$]

$= \overline{B + A(1 + \overline{C})}$

$= \overline{B + A \cdot 1}$ [$\because 1 + \overline{C} = 1$]

$$= \overline{A} \cdot \overline{B}$$

15.2 Simplify the Boolean expression

$$\overline{\left(A + \overline{BC}\right)\left(\overline{AB} + \overline{ABC}\right)}$$

Solution

Now, $\overline{\left(A + \overline{BC}\right)\left(\overline{AB} + \overline{ABC}\right)}$

$= \overline{A} \cdot \overline{\overline{BC}} \left(\overline{AB} + \overline{A} + \overline{B} + \overline{C}\right)$ [by De Morgan's theorem]

$= \overline{A} \cdot BC \left(\overline{B}(A + 1) + \overline{A} + \overline{C}\right)$ $[\because \overline{\overline{BC}} = BC]$

$= \overline{A} \cdot BC \left(\overline{B} + \overline{A} + \overline{C}\right)$ $[\because A + 1 = 1]$

$= \overline{A}BC\overline{B} + \overline{A}BC\overline{A} + \overline{A}BC\overline{C}$

$= 0 + \overline{A}BC + 0$ $[\because B\overline{B} = 0 = C\overline{C}$ and $\overline{A}\overline{A} = \overline{A}$]

$= \overline{A}BC$.

15.3 Simplify the Boolean expression

$$A + \overline{BC}\left(A + \overline{\overline{BC}}\right)$$

Solution

Now, $A + \overline{BC}\left(A + \overline{\overline{BC}}\right)$

$= A + \overline{BC}\left(A + \overline{\overline{B}} + \overline{C}\right)$ [by De Morgan's theorem]

$= A + \overline{BC}A + \overline{BC}B + \overline{BC}\overline{C}$

$= A\left(1 + \overline{BC}\right) + 0 + 0$ $[\because B\overline{B} = 0 = C\overline{C}]$

$= A \cdot 1 = A$ $[\because 1 + \overline{BC} = 1]$

15.4 Simplify the Boolean expression

$$\overline{\overline{XY} + \overline{X} + XY}$$

Solution

Now, $\overline{\overline{XY} + \overline{X} + XY}$

$= \overline{\overline{XY}} \cdot \overline{\overline{X}} \cdot \overline{XY}$ [by De Morgan's theorem]

$= (XY) \cdot \overline{XY} \cdot X$ $[\because \overline{\overline{XY}} = XY$ and $\overline{\overline{X}} = X$]

$= 0$. $[\because XY \cdot \overline{XY} = 0]$

15.5 With the help of logic diagram, design an Ex-OR gate using only four two-input NAND gates.

Solution

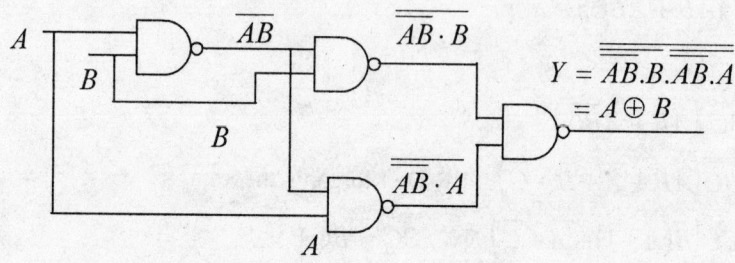

$$Y = \overline{\overline{AB}.B.\overline{AB}.A}$$
$$= A \oplus B$$

15.6 The Ex-NOR function is defined as $A \odot B = AB + \overline{AB}$.

Prove that $A \odot (B \odot C) = (A \odot B) \odot C$

Solution

Now, $B \odot C = BC + \overline{BC}$

And $A \odot B = AB + \overline{AB}$

Then, $A \odot (B \odot C) = A \odot (BC + \overline{BC})$

$= A(BC + \overline{BC}) + \overline{A}\overline{\left(BC + \overline{BC}\right)}$

$= ABC + A\overline{BC} + \overline{A}\overline{BC} \cdot \overline{\overline{BC}}$ (by De Morgan's theorem)

$= ABC + A\overline{BC} + \overline{A}\left(\overline{B} + \overline{C}\right)\left(\overline{\overline{B}} + \overline{\overline{C}}\right)$

$= ABC + A\overline{BC} + \overline{A}\left(\overline{B} + \overline{C}\right)(B + C)$ [$\because$ $\overline{\overline{B}} = B$ and $\overline{\overline{C}} = C$]

$= ABC + A\overline{BC} + \overline{A}\overline{B}C + \overline{A}B\overline{C}$ [$\because$ $B\overline{B} = 0 = C\overline{C}$]

Again, $(A \odot B) \odot C = \left(AB + \overline{AB}\right) \odot C$

$= \left(AB + \overline{AB}\right)C + \overline{\left(AB + \overline{AB}\right)}\overline{C}$

$= ABC + \overline{AB}C + \left(\overline{AB} \cdot \overline{\overline{AB}}\right)\overline{C}$ [by De Morgan's theorem]

$= ABC + \overline{AB}C + \left(\overline{A} + \overline{B}\right)\left(\overline{\overline{A}} + \overline{\overline{B}}\right)\overline{C}$ [by De Morgan's theorem]

$= ABC + \overline{AB}C + \left(\overline{A} + \overline{B}\right)(A + B)\overline{C}$ [$\because$ $\overline{\overline{A}} = A$ and $\overline{\overline{B}} = B$]

$= ABC + \overline{AB}C + \overline{A}B\overline{C} + A\overline{B}\overline{C}$

$\therefore$ $A \odot (B \odot C) = (A \odot B) \odot C$

15.7 Prove that   (i) $A \odot B = \overline{A} \oplus B$

(ii) $A \odot B = A \oplus \overline{B}$

Solution

Now, $A \odot B = AB + \overline{A}\overline{B}$

$\overline{A} \oplus B = \overline{\overline{A}}B + \overline{A} \cdot \overline{B}$

$= AB + \overline{A}\overline{B}$ [since, $\overline{\overline{A}} = A$]

$= A \odot B$

Again, $A \oplus \overline{B} = \overline{A} \cdot \overline{B} + A\overline{\overline{B}}$

$= \overline{A}\overline{B} + AB$ [since, $\overline{\overline{B}} = B$]

$= A \odot B$

15.8 Simplify the following Boolean expression:

$A\overline{B} + \overline{A}B + AB + \overline{A}\overline{B}$

Solution

Now, $A\overline{B} + \overline{A}B + AB + \overline{A}\overline{B} = \left(A\overline{B} + AB\right) + \left(\overline{A}B + \overline{A}\overline{B}\right)$

$= A\left(\overline{B} + B\right) + \overline{A}\left(B + \overline{B}\right)$

$= A + \overline{A}$ [since, $B + \overline{B} = 1$]

$= 1$ [$\because A + \overline{A} = 1$].

15.9 Use Karnaugh mapping to simplify the following Boolean function:

$F = \sum m(2,3,6,7) = \overline{A}B\overline{C} + \overline{A}BC + AB\overline{C} + ABC$.

Solution

The K-map is shown below.

A\BC	00	01	11	10
0	0 (0)	0 (1)	1 (3)	1 (2)
1	0 (4)	0 (5)	1 (7)	1 (6)

The minterms m_2, m_3, m_6, m_7 form a 4-square. So, two variables will be dropped from the result. A changes values and C takes different values over this 4-square. So, A and C will not appear in the result. Hence, $F = B$.

15.10 Reduce the following Boolean function using K-map:

$$F = \sum m\{4,6,7,15\}$$

$$= \overline{A}B\overline{C}\,\overline{D} + \overline{A}BC\overline{D} + \overline{A}BCD + ABCD$$

Solution

It is a four variable map. A, B, C and D are the four variables. The map is as shown below.

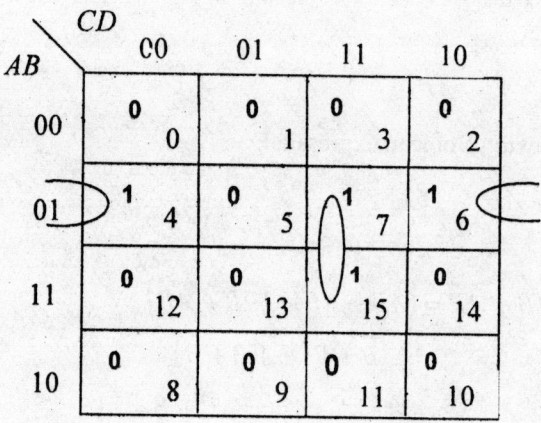

Here, folding the map vertically, m_4 and m_6, from a 2-square. So, one variable will be dropped. Since C takes different values over this 2-square, we get $\overline{A}B\overline{D}$ from this square. Then, m_7 and m_{15} from a 2-square. Here, A changes values over this 2-square. So, we get BCD from this 2-square.

$$\therefore \quad F = \overline{A}B\overline{D} + BCD.$$

15.11 Draw a truth table for the following Boolean function:

$$Y = A + B + A\overline{B}.$$

Implement using logic gate.

Solution

The truth table is as follows:

A	B	Y
0	0	0
0	1	1
1	0	1
1	1	1

This is the truth table of a two input OR gate. The logic diagram is as shown below.

$$Y = A + B$$

PROBLEMS

15.1 Prove the following Boolean identities

(i) $A + \overline{A}B = A + B$

Hint: $A = A(1 + B)$

(ii) $(A + B)(B + C)(C + A) = AB + BC + CA$

(iii) $A\overline{B} + \overline{A}B = (A + B)(\overline{A} + \overline{B})$

Hint: expand the RHS

(iv) $A \oplus B = \overline{A} \oplus \overline{B}$

Hint: RHS$= \overline{\overline{A}}\,\overline{B} + \overline{A}\,\overline{\overline{B}}$=LHS.

15.2 Implement the following Boolean functions using two-input NAND gates.

(i) $\overline{ABC} + \overline{A}B$ (ii) $\overline{\overline{ABC} \cdot \overline{ABC}}$ (iii) $AB + \overline{AB}$

15.3 Using Ex-OR gates realize an Ex-NOR gate

Hint: $A \oplus 1 = \overline{A}$. Again $\overline{A} \oplus B = \overline{A \oplus B}$. Draw an Ex-OR gate with A and 1 as inputs. Its outputs is $\overline{A}$. Draw another Ex-OR with $\overline{A}$ and B as input. Its output is Ex-NOR.

15.4 Realize the following Boolean function using Ex-OR gates:

$$F(A, B, C) = \overline{A}\,\overline{B}C + \overline{A}B\overline{C} + A\overline{B}\,\overline{C} + ABC$$

Hint:
$$F = C(\overline{A \oplus B}) + \overline{C}(\overline{A}B + A\overline{B}) = C(\overline{A \oplus B}) + \overline{C}(A \oplus B)$$
$$= C \oplus (A \oplus B) = C(\overline{A}\overline{B} + AB) + \overline{C}(\overline{A}B + A\overline{B})$$

15.5 Consider the Boolean expression $Y = (AB + C)(AB + D)$.

Realize this function Y using only NAND gates. Perform the same using only NOR gates

15.6 Simplify the following Boolean functions by K-maps:

(i) $F = \sum m(1, 3, 5, 7) = \overline{A}\overline{B}C + \overline{A}BC + A\overline{B}C + ABC$

(ii) $F = \sum m(0, 1, 4, 5, 10, 11, 14, 15)$
$$= \overline{A}\overline{B}\overline{C}\overline{D} + \overline{A}\overline{B}\overline{C}D + \overline{A}B\overline{C}\overline{D} + \overline{A}B\overline{C}D + A\overline{B}C\overline{D}$$
$$+ A\overline{B}CD + ABC\overline{D} + ABCD$$

[**Ans.** (i) C (ii) $AC + \overline{A}\overline{C}$.]

QUESTIONS

15.1 State and explain De Morgan's theorem . Explain its role in simplifying Boolean expressions.

15.2 What is Karnaugh mapping? Define the word 'Minterm'. Describe the process of drawing a Karnaugh map and explain the method of its reduction.

15.3 What do you mean by a 'Logic gate"? What are the basic logic gates? Why a NAND gate is called the building block?

15.4 Give the logic symbols, truth tables, and fuctioning of OR, AND, NOT, NAND, NOR, Ex-OR and Ex-NOR gates.

15.5 What are the logic families in the logic hardware system? What do you mean by 'positive' and 'negative' logic? What is meant by 'Level logic' and 'Dynamic logic'? Give the design of OR and AND gates in diode logic.

15.6 Give the design of a NOT gate using a transistor. Explain its operation.

15.7 Suppose you are given four 2-input NAND gates. Show the logic circuit diagram for the design of an XOR gate using these four NAND gates only.

15.8 How can you design an AND gate using OR and NOT gates?

15.9 Show using logic diagram how do you realize a half adder circuit using AOI logic.

15.10 Show using logic diagram how do you design a binary full adder using only NAND gates.

15.11 What do you mean by a binary magnitude comparator? Explain whether an Ex-OR gate can function as an inequality detector. Describe the principle of operation of a two-bit binary magnitude comparator. Explain with necessary logic diagram the functioning of a four-bit binary magnitude comparator.

15.12 What do you mean by a 'Multiplexer'? Draw the logic circuit diagram and give the truth table of a 4×1 multiplexer using basic logic gates.

15.13 What is a 'Demultiplexer'? Write down the logic symbol and the truth table of a (1×4) demultiplexer. Explain the functioning of a (1×4) demultiplexer with the help of logic circuit diagram.

16

Sequential Logic

16.1 FLIP-FLOP

Flip-flops are 1 bit memory elements. It can store 1 bit of data. Flip-flops are bistable devices whose output changes state in synchronism with a clock. Edge-triggered flip-flops change output states either on the rising edge of the pulse or on the falling edge of a clock pulse. Accordingly, it is called a positive edge-triggered flip-flop or a negative edge-triggered flip-flop. The classification of edge-triggered flip-flops is as follows.

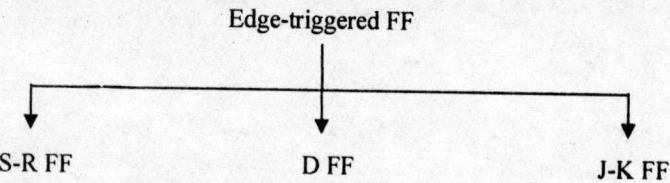

S-R means set-reset. D means delay. J-K are two inputs having no other special significance. They are consecutive letters of English alphabet.

16.1.1 S-R FLIP-FLOP

The logic symbols of positive and negative edge-triggered S-R FFs are shown in Figs. 16.1(a) and 16.1(b) respectively.

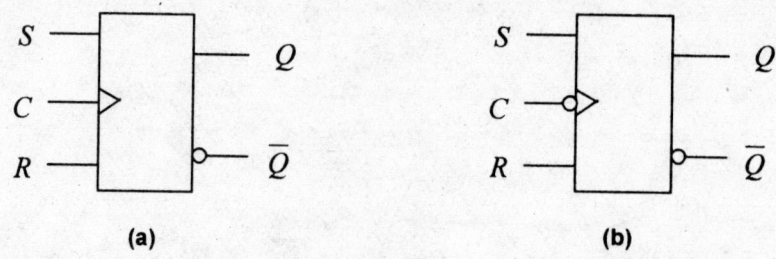

(a)   (b)

Fig. 16.1. (a) Symbol of S-R FF with positive edge triggering
(b) Symbol of S-R FF with negative edge triggering.

The triangle at the clock C indicates that the FF is edge triggere d. If there is a bubble at the C input it indicates negative edge tr iggered flip-flop. The bubble at the $\overline{Q}$ output is added to show the inversion of Q.

OPERATION OF POSITIVE EDGE-TRIGGERED S-R FF

The Set(S) and Reset (R) inputs are fed with data which are transferred to the output of the S -R FF when the clock pulse rises from a low level to a high level. When $S = 1$ (high) and $R = 0$ (low), the output $Q = 1$ and $\overline{Q} = 0$. The FF is now said to be set. When $S = 0$ (low) and $R = 1$ (high), the FF is reset and its output $Q = 0$ (low) and $\overline{Q} = 1$ (high). When $S = 0$ and $R = 0$, the FF does not change state. Its outputs remains as it were before these inputs occurred. When $S = 1$ and $R = 1$ an indeterminate condition results. The S and R inputs can change without changing the output states when the clock pulse is not in transition (low→high for positive edge-triggering). The truth table of a positive edge-triggered S-R FF is as shown below in table 16.1.

Inputs			Outputs		
S	R	C	Q	$\overline{Q}$	Remarks
1	0	⤒	1	0	Set
0	1	⤒	0	1	Reset
0	0	⤒	Q_p	$\overline{Q_p}$	Previous state
1	1	X	u	u	Indeterminate

Table 16.1. Truth table of a positive edge-triggered S-R flip-flop.

⤒ indicates rising edge of the pulse.   × means it does not matter whether positive or negative edge-triggering occurs. Q_p indicates previous output state. 'u' stands for unknown output.

The implementation of the edge-triggered S-R FF is shown below in Fig.16.2.

Let $Q = 0$ first. Also let $S = R = 0$. If the clock input $C = 0$, the outputs of steering gates G_1 and G_2 are 1. Again, since $Q = 0$, the output of gate G_4 is 1, i.e, $\overline{Q} = 1$. Now, both the inputs of gate G_3 are 1 so that $Q = 0$. Thus, there is no change of state of the FF if the clock is absent ($C = 0$). If we start with $Q = 1$, it would result in $Q = 1$ when $C = 0$. In presence of the clock ($C = 1$) and taking $S = R = 0$, if we start with $Q = 0$ or $Q = 1$ we get unchanged output.

We now consider the case, $S = 1$ and $R = 0$. When the clock input $C = 1$, the output of gate G_1 is 0. This makes the output of gate G_3 equal to 1, i.e, $Q = 1$. Since $R = 0$, the output of gate G_2 is 1. Now, both inputs of gate G_4 are 1 so that its output $\overline{Q} = 0$. The FF is now in the SET state.

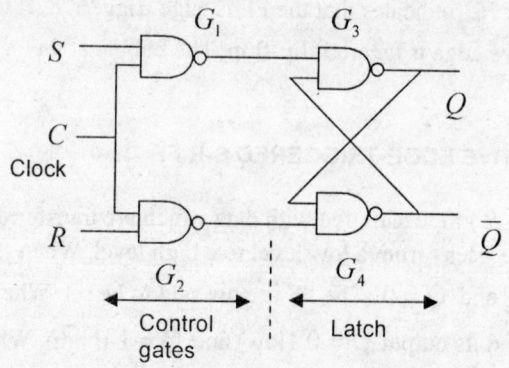

Fig. 16.2. Logic diagram of S-R flip-flop

In the next case, we take $S = 0$ and $R = 1$. Now, apply a clock pulse, $C = 1$. The output of gate G_2 is 0. So, $\overline{Q} = 1$. Then, both inputs of gate G_3 are 1. This makes $Q = 0$. The FF is now in the RESET state.

In the last case, when $S = R = 1$. In presence of $C = 1$, the outputs of gates G_1 and G_2 are both 0. This leads to $Q = \overline{Q} = 1$. But, this is logically inconsistent. This condition of the FF is indeterminate and this condition is forbidden.

16.1.2 EDGE-TRIGGERED D FLIP-FLOP

The positive edge-triggered D FF using an S-R FF and an inverter is shown in Fig. 16.3.

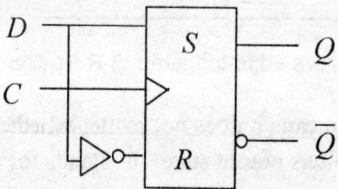

Fig. 16.3. Logic diagram of D flip-flop.

The D FF has only one input which is the D input. In addition to D input, there is a clock (C) input. When $C = 1$, $Q = 1$ when $D = 1$. The FF is set. This D input is stored at the output on the positive rising edge of the clock pulse.

When $D = 0$ and the clock pulse $C = 1$, we get $Q = 0$. The FF is reset. The FF stores a '1' when the FF is set and it stores a '0' when the FF is reset. The truth table is shown in Table 16.2 below. The D FF removes the uncertain case of $S = R = 1$ in S-R FF.

Inputs		Outputs		Remarks
D	C	Q	$\overline{Q}$	
1	⤴	1	0	Set
0	⤴	0	1	Reset

Table 16.2. Truth table of a D-flip-flop.

If the input D Corresponds to an input just after the K-th clock pulse, then this input is transferred to the output in the (K+1)th clock pulse. Thus, D FF acts as a delay device which produces a delay of one pulse period. In Fig. 16.4, W = clock pulse width, T = clock period and $W \ll T$. Bit duration is also T. Here, data bit D_K appears in the time slot T just after the clock pulse K and it is transferred to the output Q by the next clock pulse ($K + 1$).

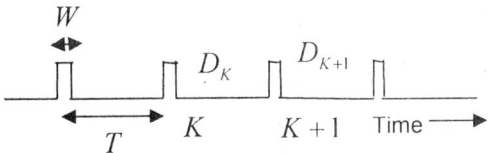

Fig. 16.4. Delay produced by D FF.

16.1.3 EDGE-TRIGGERED J-K FLIP-FLOP

The three conditions of R-S FF viz, SET, RESET and no change condition are also the 3 conditions of J-K FF. The main difference between S-R and J-K FF operation is that there is no indeterminate condition in J-K FF. The implementation of the J-K FF is shown in Fig. 16.5.

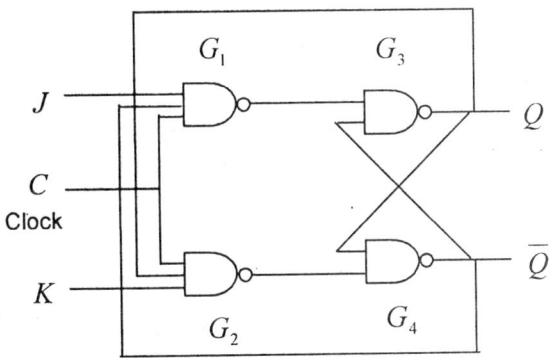

Fig. 16.5. Logic diagram of J-K flip-flop.

J and K are two inputs. If the FF is positive edge-triggered the data transfer takes place in the rising edge of the pulse. Firstly, clear the J-K FF.

Let us first assume that the FF is reset i.e., $Q = 0$. Also let $J = 1$ and $K = 0$. In the rising edge of the clock C, $C = 1$. At the input of gate G_1, we have $J = 1$, $C = 1$ and $\overline{Q} = 1$. So, output of G_1 is 0. Then, $Q = 1$. So, the JK FF is set.

Inputs			Outputs		Remark
J	K	C	Q	$\overline{Q}$	
1	0	⌐⌐	1	0	Set
0	1	⌐⌐	0	1	Reset
0	0	⌐⌐	Q_p	$\overline{Q}_p$	Unchanged
1	1	⌐⌐	$\overline{Q}_p$	Q_p	Toggle

Table 16.3. Truth table of J-K flip-flop.

In the second case, let $J = 0$ and $K = 1$. The output of NAND gate G_2 is 0, since $K = 1$, $C = 1$ and $Q = 1$. The output of NAND gate G_4 is 1, i.e, $\overline{Q} = 1$. The FF is RESET.

In the third case, if $J = K = 0$ and $C = 1$, then taking $\overline{Q} = 1$ the output of gate G_1 is 1. Then, output of gate G_3 is 0 so that $Q = 0$. Similarly, with $K = 0$, $C = 1$ and $Q = 0$, we get the output of gate G_2 equal to 1. Hence, the output of gate G_4 is 1, i.e, $\overline{Q} = 1$. Hence, in presence of the clock trigger, a no change situation will occur. The mode of operation of the JK FF with $J = K = 1$ can be analyzed by taking the FF in the RESET state with $\overline{Q} = 1$, as an example. The output of gate G_1 for $J = 1$, $C = 1$ and $\overline{Q} = 1$, is 0. The output of gate G_3 is then $Q = 1$. So, the FF is SET.

Now, consider what happens in the next triggering clock pulse. Now, $K = 1$, $C = 1$ and $Q = 1$. So, output of gate G_2 is 0. This makes the output of gate G_4 equal to 1. So, $\overline{Q} = 1$. The FF is now RESET. So, the JK FF changes its state in each triggering clock pulse at its input when $J = K = 1$. Thus, FF is said to toggle in this case.

Here, in the table Q_p means previous output. ⌐⌐ means triggering on the rising edge of the pulse. The symbol of positive edge-triggered JK FF is shown in Fig. 16.6.

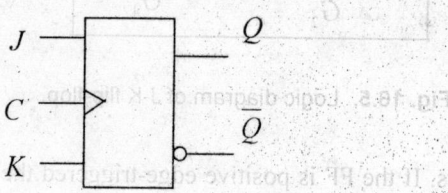

Fig. 16.6. J-K FF symbol.

16.1.4 ASYNCHRONOUS OPERATION

In S-R, D and J-K FFs input data transfer occur in synchronism with the clock pulse. The data from the inputs are transferred to the output at the triggering edge of the clock pulse. In addition to the synchronous inputs, there can be asynchronous inputs which can affect the output condition of the FF independent of the clock. These are called Preset (P_r) and Clear (C_r) inputs. They are also called direct set and direct reset inputs.

If $P_r = 1$, $C_r = 0$ and $C = 0$ then the J-K FF with active low preset and clear inputs means $Q = 1$. The FF is Set. When $P_r = 0$, $C_r = 1$ and $C = 0$, we get $\overline{Q} = 1$ with active low preset and clear inputs. The FF is Reset. The symbol of J-K FF with P_r and C_r inputs are shown below in Fig. 16.7.

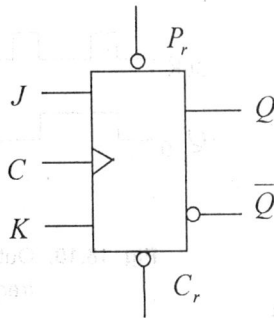

Fig. 16.7. J-K FF symbol with preset and clear inputs.

With active low P_r and C_r inputs if we make $P_r = 0$ and $C_r = 0$, we get synchronous operation of the J-K FF. This is due to the fact that $P_r = 0$ makes a '1' at the corresponding input gate G_3 and a $C_r = 0$ makes '1' at the corresponding input of gate G_4. The logic diagram of the J-K FF with P_r and C_r inputs is shown in Fig. 16.8. The condition $P_r = 1$, $C_r = 1$ makes $Q = \overline{Q} = 1$ which is an invalid operation which must be avoided.

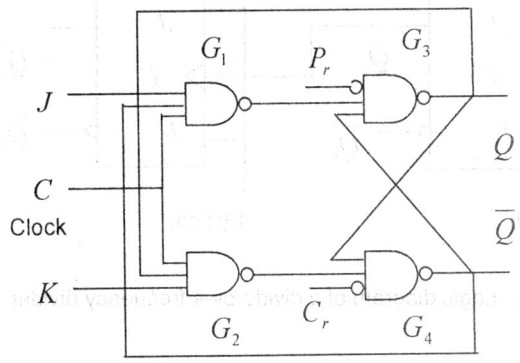

Fig. 16.8. Logic diagram of J-K FF with preset and clear inputs.

16.1.5 APPLICATIONS OF J-K FLIP-FLOPS

In this section, we describe a few applications of J-K flip-flops.

DIVISION OF FREQUENCY OF A PULSE WAVEFORM

Consider a J-K flip-flop whose J and K inputs are connected and maintained high (corresponding bit is 1). Then, this flip-flop toggles. Now apply a periodic pulse as the clock of this flip flop. At the output Q we get a square wave whose frequency is half that of the input clock pulse waveform. The flip-flop is initially reset. The corresponding logic diagram is shown in Fig 16.9, while the output waveform is shown in Fig. 16.10. The flip-flop is initially reset.

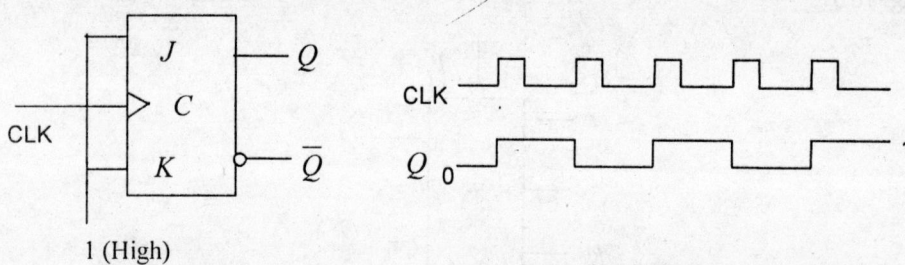

Fig. 16.9. Logic diagram for clock frequency division by 2.

Fig. 16.10. Output waveform of the clock frequency divider (divide by 2)

The flip flop is positive edge-triggered and Q changes state when the clock pulse changes from 0 to 1.

Using two J-K flip-flops drives the clock input of flip flop-2. Both the flip-flops are operated in toggle mode. Initially both the flip-flops are reset ($Q_1 = Q_2 = 0$). The logic diagram of divide-by-4 frequency divider is shown Fig. 16.11, while the output waveforms are shown in Fig. 16.12.

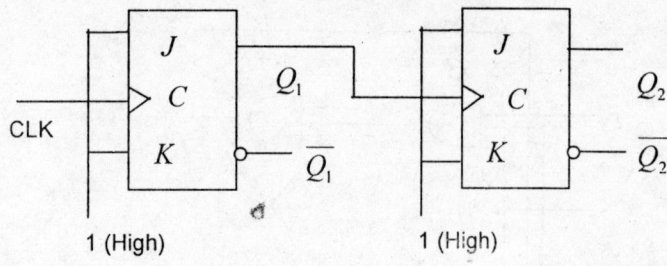

Fig. 16.11. Logic diagram of a divide-by-4 frequency divider.

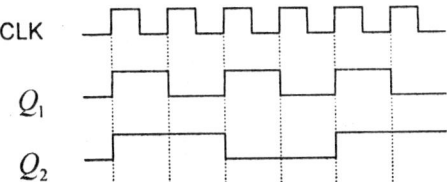

Fig. 16.12. Output waveform of the frequency divider (divide-by- 4).

16.1.6 BINARY COUNTERS USING FLIP-FLOPS

We require two J-K flip flops using negative edge-triggered mode. Both the flip-flops are reset initially. The J and K inputs are connected together and made high (level 1). Then the flip-flops toggle. The logic diagram of a 2 -bit binary counter is shown in Fig. 16.13, while the output waveform is shown in Fig. 16.14.

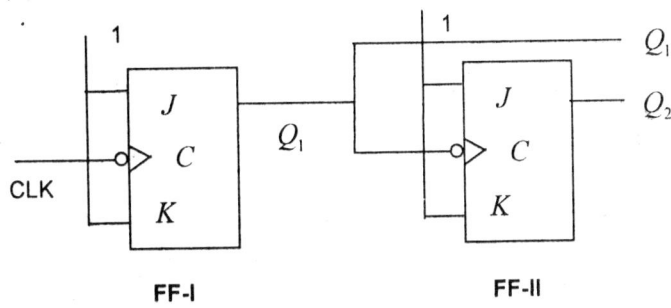

Fig. 16.13. Logic diagram of a 2 - bit binary counter.

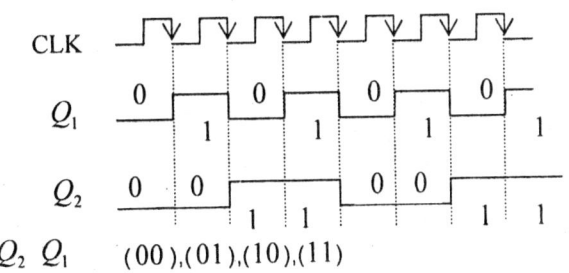

Fig. 16.14. Output waveform of a 2 -bit binary counter.

Binary sequence $(0,0)$, $(0,1)$, $(1,0)$ and $(1,1)$ is repeated after every four clock pulses. The flip-flop counts from 0 to 3.

16.2 ASYNCHRONOUS COUNTER

In this section, 2 -bit and 3 -bit binary counters are described.

16.2.1 TWO-BIT COUNTER

Here, the FFs incorporated in the counter do not change output states at the same instant of time. The clock pulses are not fed to the C inputs of all FFs other than the first FF directly. The logic diagram of the two-bit binary counter is shown in Fig. 16.15.

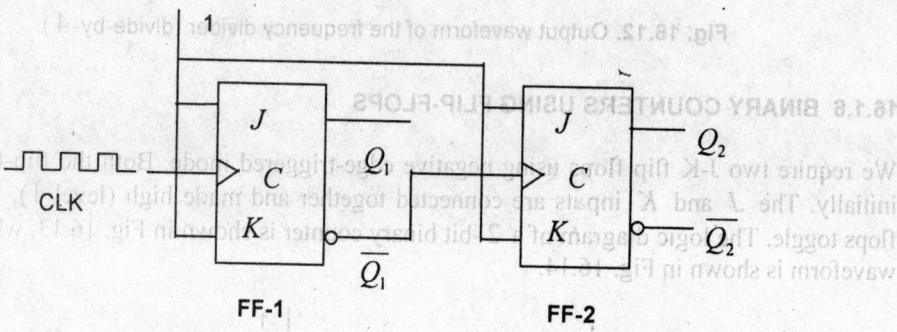

Fig. 16.15. Logic diagram of a 2-bit asynchronous binary counter.

FF1 changes output state at the positive-going edge of each clock pulse. J-K inputs are connected together and a high (1) input is given to it. FF2 is triggered by $\overline{Q}_1$ output of FF1 at the positive going edge. Due to a non-zero propagation delay through each FF, the input clock (CLK) transition and the transition of $\overline{Q}_1$ do not take place at the same time. So, FFs are not triggered at the same instant of time. This leads to asynchronous operation. $J = K = 1$ leads to toggle operation. FFs are initially RESET. The timing diagram is shown in Fig. 16.16.

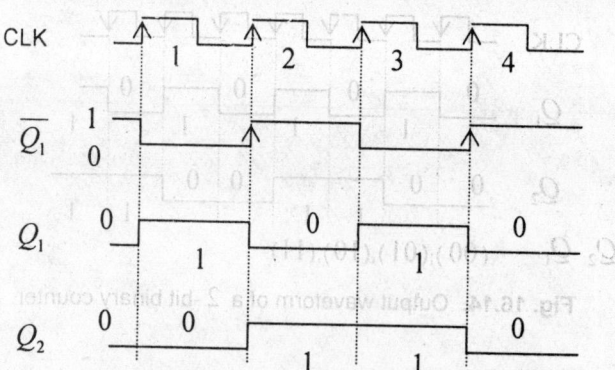

Fig. 16.16. Timing diagram of a 2-bit binary counter.

Clock Pulse	Q_2	Q_1
0	0	0
1	0	1
2	1	0
3	1	1

Table 16.4. Truth table of a 2-bit binary counter.

Q_2 is the most significant bit (MSB) and Q_1 is the least significant bit (LSB). Two bit counter exhibit 4 different states (Q_2, Q_1). It counts clock pulses from 0 to 3 in a binary sequence -- (0,0), (0,1), (1,0) and (1,1). After 3, the counter goes back to its initial state (0,0) and thus recycles. The counting process proceeds in the upward direction. i.e, from 0 to 3. That is why it is also called an up counter.

16.2.2 THREE–BIT ASYNCHRONOUS BINARY COUNTER

The logic diagram and the timing diagram of a 3-bit binary counter are given below.

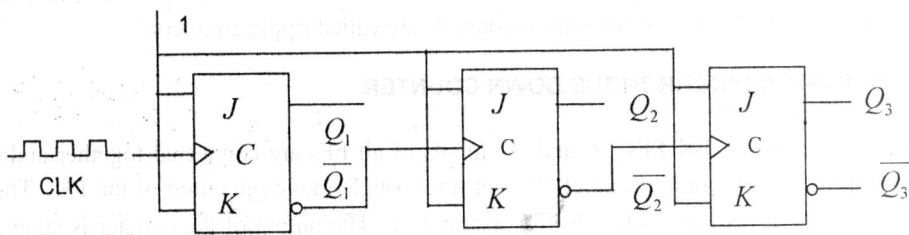

Fig. 16.17. Logic diagram of a 3-bit asynchronous binary counter.

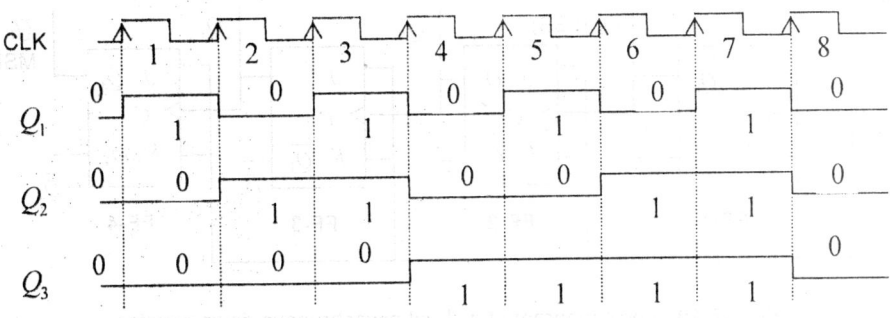

Fig. 16.18. Timing diagram of a 3-bit asynchronous binary counter.

This counter counts pulses in a binary sequence- $(0,0,0)$, $(0,0,1)$, $(0,1,0)$, $(0,1,1)$, $(1,0,0)$, $(1,0,1)$, $(1,1,0)$, $(1,1,1)$. It counts $0-7$ pulses and then recycles to initial state $(0,0,0)$. Since, the counting takes place in the upward direction it is called an up counter.

Clock pulse	Q_3	Q_2	Q_1	Remark
No pulse	0	0	0	Counter is initialized
1	0	0	1	Counting begins
2	0	1	0	
3	0	1	1	
4	1	0	0	
5	1	0	1	
6	1	1	0	Counting ends
7	1	1	1	Counter is initialized

Table 16.5. Truth table of a 3-bit binary counter.

Triggering of FF1 occurs first. FF2 is triggered by the output $\overline{Q_1}$ next and FF3 is triggered by the output $\overline{Q_2}$ last. Due to the propagation delay of the FFs, the effect of the clock pulse ripples through the series of FFs. So, asynchronous counters are called ripple counters.

16.2.3 FOUR–BIT ASYNCHRONOUS DOWN COUNTER

This counter requires 4 J-K FFs. J and K inputs of all FFs are connected together and a '1' (High) is applied to this common input. This ensures toggle mode operation of the FFs. The Q-output of one FF acts as the clock pulse for the next FF. The output of the counter is taken from the Q-outputs of all FFs. Q_4 is the most significant bit and Q_1 is the least significant bit. All FFs are initially set, i.e., $Q_i = 1$ for $i = 1, 2, 3, 4$ at time $t = 0$. We use positive edge triggering of the FFs.

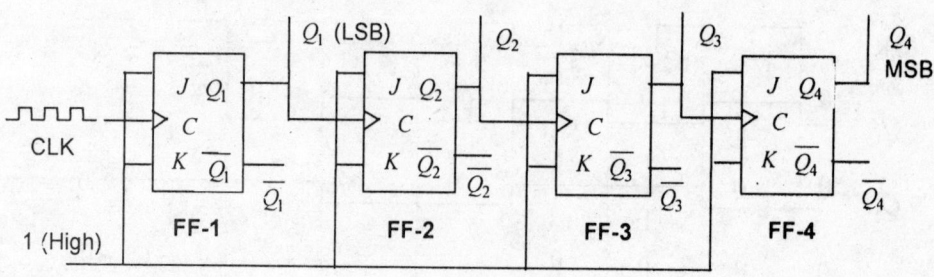

Fig. 16.19. Logic diagram of a 4-bit asynchronous down counter.

The 4-bit asynchronous down counter is shown in Fig. 16.19. The timing diagram is shown in Fig. 16.20.

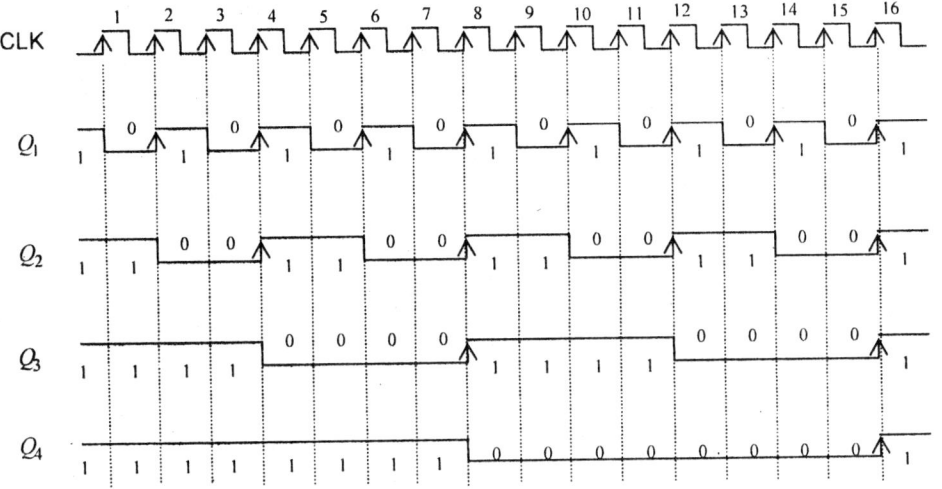

Fig. 16.20. Timing diagram of the 4-bit asynchronous down counter.

Pulse Number	Q_4	Q_3	Q_2	Q_1	Decimal equivalent
0	1	1	1	1	15
1	1	1	1	0	14
2	1	1	0	1	13
3	1	1	0	0	12
4	1	0	1	1	11
5	1	0	1	0	10
6	1	0	0	1	9
7	1	0	0	0	8
8	0	1	1	1	7
9	0	1	1	0	6
10	0	1	0	1	5
11	0	1	0	0	4
12	0	0	1	1	3
13	0	0	1	0	2
14	0	0	0	1	1
15	0	0	0	0	0
16	1	1	1	1	15

Table 16.6. Truth table of a 4-bit asynchronous down counter.

The output state of the down counter is shown in Table 16.6. The output state is represented by $(Q_4\,Q_2Q_3Q_1)$. The counter counts from 15 to 0 and then goes back to the initial state 15.

16.2.4 MODULO X COUNTER

A binary counter counts from 0 to (2^n-1) where n is the number of flip-flops in the counter. This counter has 2^n states and is called modulo 2^n counter. We can also design a counter which can count from 0 to any number $(X-1)$ and accordingly it is called modulo X counter. If we take $X=10$, the corresponding counter will count from 0 to $(10-1)=9$ and will be called a decade counter. The logic diagram of an asynchronous decade counter is shown in Fig. 16.21. To design such asynchronous modulo X counter, we have to find the value of 'n' where $2^{n-1}<X<2^n$. For $X=10$, we see that $2^3<X<2^4$. So, we have to use $n=4$ J-K flip-flops. All these J-K flip-flops must have a clear input. We consider positive edge triggering of the flip-flops. The 4-bit asynchronous counter must will be allowed to count from 0 to 9 and then it must be initialized to the zero state. To initialize the counter, a feedback circuit incorporating a NAND gate can be used. The binary equivalent of decimal 10 is 1010_2. Now, apply the bits which are '1' in the binary equivalent of X to the inputs of the NAND gate. If the output state of the counter is represented by $(Q_4\,Q_3Q_2Q_1)$ with Q_4 as MSB and Q_1 as LSB, we have $Q_4=1$ and $Q_2=1$ when $X=1010_2$. Applying Q_4 and Q_2 as the inputs of a two-input NAND gate, the output of the NAND gate is zero. With low-active clear input this zero clears all the J-K flip-flops and the output of the counter becomes (0000). When the next pulse comes to the flip-flop $Q_1=1$ and $Q_2=Q_3=Q_4=0$. This makes the output of the NAND gate equal to '1' which, in turn, makes the clear input of all flip-flops disabled and the normal operation of the counters continues. The counter is again initialized when the counter attempts to go to the state $10_{10}=1010_2$. The logic diagram of a modulo 14 counter is shown in Fig. 16.22.

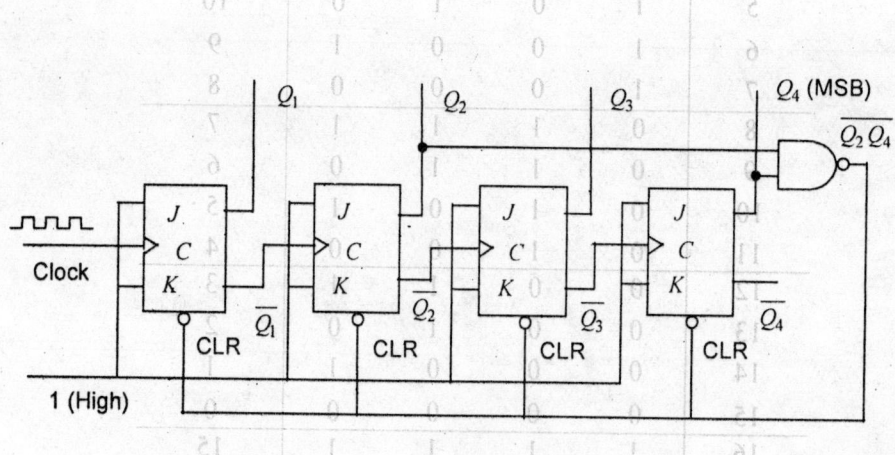

Fig. 16.21. Logic diagram of asynchronous decade counter.

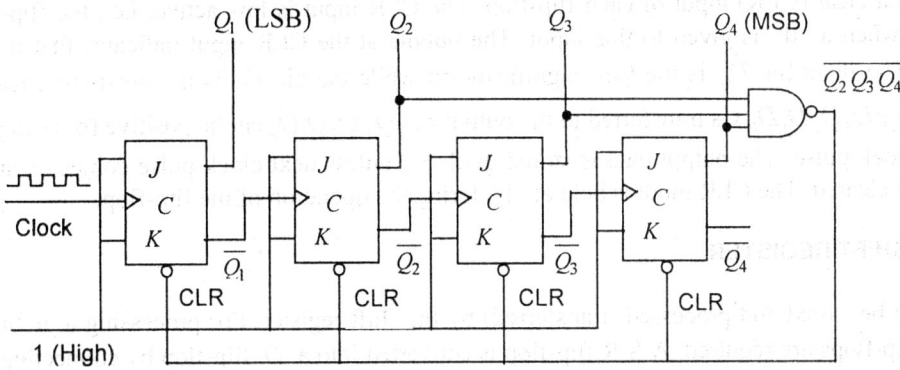

Fig. 16.22. Logic diagram of asynchronous modulo 14 counter.

16.3 REGISTER

A single flip-flop can store a single bit of data. So, n flip-flops will store n-bits of information. The device consisting of n flip-flops can store and process n-bits of data and is called a register. Registers can be divided into two classes, which are

    (i)   memory register, and

    (ii) shift register.

16.3.1 MEMORY REGISTER

The logic circuit diagram of a 4-bit memory register is shown in Fig. 16.23. There are 4 D flip-flops which are driven by a single clock. The input data are transferred to the output at the positive-going edge of the clock pulse. Each D flip-flop has a single D input.

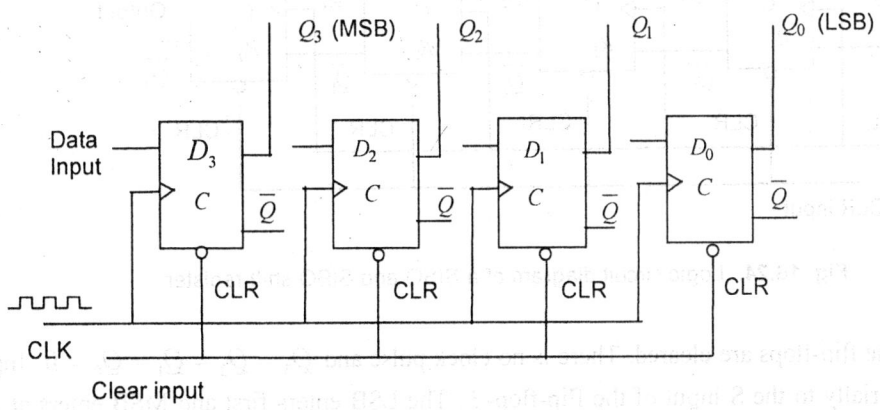

Fig. 16.23. Logic block diagram of the 4-bit memory register.

There is a clear (CLR) input of each flip-flop. The CLR input is low active, i.e., the flip-flop is cleared when a '0' is given to this input. The bubble at the CLR input indicates that it is low active. The input bit D_0 is the least significant bit while the bit D_3 is the most significant bit. The data ($D_3 D_2 D_1 D_0$) is transferred to the output as ($Q_3 Q_2 Q_1 Q_0$) at the positive (or rising) edge of the clock pulse. The output data is stored until and unless next clock pulse comes or the flip-flops are cleared. The CLR input is held at '1' during the operation of the flip-flop.

16.3.2 SHIFT REGISTER

Data can be stored and processed (transferred) by the shift register. For processing a n-bit data, n D flip-flops are required. A S-R flip-flop is converted into a D flip-flop by connecting the R input with the S-input through a NOT gate.

Depending upon the nature (serial or parallel) of input data supply and output data removal, shift registers can be classified into four groups which are

   (i)   Serial-in serial-out (SISO) shift register,
   (ii)   Serial-in parallel-out (SIPO) shift register,
   (iii)   Parallel-in serial-out (PISO) shift register, and
   (iv)   Parallel-in parallel-out (PIPO) shift register.

16.3.2.1 SISO SHIFT REGISTER

A four-bit SISO shift register circuit is shown in Fig. 16.24.

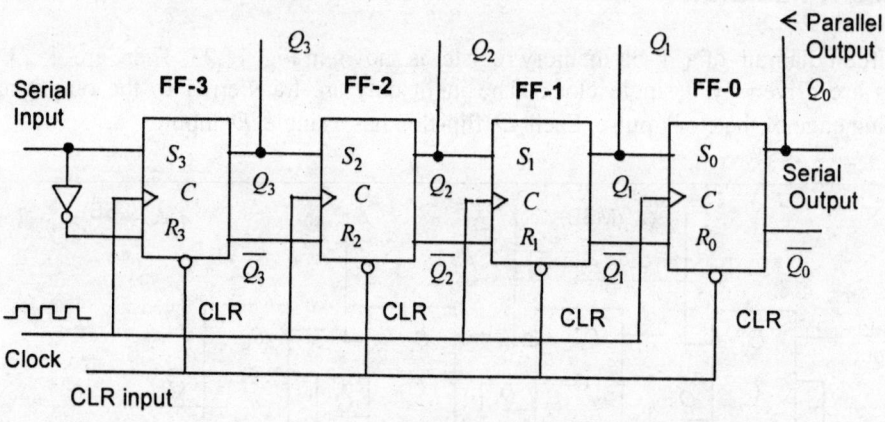

Fig. 16.24. Logic circuit diagram of a SISO and SIPO shift register.

Initially all the flip-flops are cleared. There is no clock pulse and $Q_3 = Q_2 = Q_1 = Q_0 = 0$. Input data is fed serially to the S input of the flip-flop-3. The LSB enters first and MSB enters at the end. Let us consider a 4-bit data 1010_2 which is fed serially at the S input of FF-3. The LSB is 0 here, which enters the S input of FF-3 first. At the positive going edge of the clock pulse, this data is transferred to Q_3 output. So, $Q_3 = 0$ and $\overline{Q}_3 = 1$. Now, for FF-2, $S_2 = 0$ and $R_2 = 1$. At the rising edge of the second clock pulse, this bit (0) is transferred to Q_2 output of FF-2. At the same time, the next bit '1' at the S input of FF-3 is transferred to the Q_3 output of FF-3. Thus, at the end of the second pulse, $Q_3 = 1$ and $Q_2 = 0$. At the rising edge of the third pulse, $S_1 = Q_2 = 0$ and $R_1 = \overline{Q}_2 = 1$. So, $S_1 = 0$ is transferred to Q_1 output by this clock pulse. At the same time, $S_2 = Q_3 = 1$ is transferred to Q_2 of FF-2 and $S_3 = 0$ (the third bit) is transferred to Q_3. Thus, after the third pulse, we have $Q_3 = 0$, $Q_2 = 1$ $Q_1 = 0$. Now, the last bit '1' appears at S_3 input. The fourth pulse transfers the data of FF-1, FF-2 and FF-3 to their respective outputs. Thus, at the end of fourth clock pulse we have $Q_3 = 1$ $Q_2 = 0$ $Q_1 = 1$ and $Q_0 = 0$. The output data is read from the Q_0 output. At the beginning, we get $Q_0 = 0$, the LSB. At the end of 5th clock pulse, Q_1 is transferred to Q_0 and we get $Q_0 = 1$, and $Q_2 (= 0)$ is transferred to Q_1 and $Q_3 (= 1)$ is transferred to Q_2 (making $Q_3 = 0$). At the end of the 6th clock pulse, $Q_1 = 1$ and $Q_0 = 0$ (making $Q_3 = 0$, $Q_2 = 0$). At the end of the 7th clock pulse, $Q_1 = 1$ goes to Q_0 and $Q_0 = 1$ (MSB). At the same time, $Q_3 = Q_2 = Q_1 = 0$. At the end of the 8th clock pulse, $Q_3 = Q_2 = Q_1 = Q_0 = 0$. All the FFs are cleared. The timing diagram is shown in Fig. 16.25. The truth table of the SISO register is shown in table 16.7.

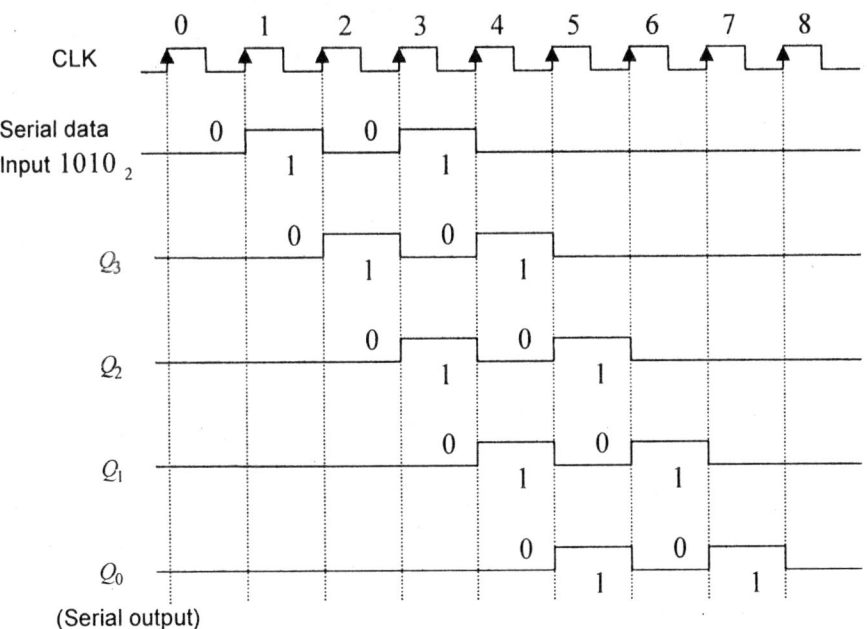

Fig. 16.25. Waveform diagram of a 4-bit SISO shift register.

Clock pulse Number	Serial data input	Q_3	Q_2	Q_1	Q_0 (Serial data output)	Remark
No pulse	0 (LSB)	0	0	0	0	Shift register is cleared
1	1	0	0	0	0	
2	0	1	0	0	0	
3	1 (MSB)	0	1	0	0	
4	0	1	0	1	0 (LSB)	Data output begins (LSB output)
5	0	0	1	0	1	
6	0	0	0	1	0	
7	0	0	0	0	1 (MSB)	Data output ends (MSB output)
8	0	0	0	0	0	Register is cleared

Table 16.7. Truth table of the 4 -bit SISO shift register.

16.3.2.2 SERIAL-IN PARALLEL-OUT (SIPO) SHIFT REGISTER

The logic circuit diagram of the SIPO shift register is shown in Fig.16.24. The register is cleared initially so that $Q_3 = Q_2 = Q_1 = Q_0 = 0$. The serial output terminal in Fig.16.24 should be deleted since the output is parallel. Let us take a binary data 1010_2. The least significant bit (LSB) is 0 and the most significant bit (MSB) is '1'. At the rising edge of the first clock pulse, the LSB (0) is transferred to the output Q_3 of the flip-flop FF-3. The next input bit 1 arrives at the input of FF-3. At the rising edge of the second clock pulse, the bit at $Q_3 (= 0)$ is transferred to the output Q_2 of FF-2. At the same time, the bit (1) at the input of FF-3 is transferred to the output Q_3. So, Q_3 becomes 1 and Q_2 becomes 0. Now, the next bit (0) appears at the input of FF-3. At the rising edge of the third clock pulse, the bit (0) at Q_2 is transferred to the output Q_1 of FF-1. Also, the bit (1) at Q_3 goes to the output Q_2 of FF-2 and the input bit (0) goes to the output Q_3 of FF-3. Thus, at the end of the third clock pulse we have $Q_3 = 0$, $Q_2 = 1$, $Q_1 = 0$. Now, the next input bit (1) appears at the input of FF-3. When the fourth clock pulse appears, $Q_1 (= 0)$ goes to Q_0, $Q_2 (= 1)$ goes to Q_1, $Q_3 (= 0)$ goes to Q_2 and the input bit (1) of FF-3 goes to its output Q_3. Thus, at the end of the fourth clock pulse, we get $Q_3 = 1$, $Q_2 = 0$, $Q_1 = 1$ and $Q_0 = 0$. Hence, we require 4 clock pulses to put a 4 -bit data into a 4 -bit SIPO shift register. The truth table is shown in table 16.8 below.

Clock pulse Number	Serial data input	Output data				Remark
		Q_3	Q_2	Q_1	Q_0	
No pulse	0 (LSB)	0	0	0	0	SIPO shift register is initially cleared
1	1	0	0	0	0	
2	0	1	0	0	0	
3	1	0	1	0	0	
4	0	1	0	1	0	Output data appears from parallel output

Table 16.8. Truth table of SIPO shift register.

16.3.2.3 PARALLEL-IN PARALLEL-OUT (PIPO) SHIFT REGISTER

The logic circuit diagram of universal shift register with both serial and parallel inputs and outputs is shown in Fig. 16.26. The circuit diagram of a PIPO shift register is obtained from Fig. 16.26 with serial input and output omitted. Here, we have to take n SR flip-flops to design a n-bit shift register. We can take J-K flip-flops instead of SR flip-flops also. The SR flip-flops are converted to D flip-flops by connecting a NOT gate between S and R inputs of the n th SR flip-flop. All the flip-flops are initially cleared by applying a '0' at the low-active clear inputs of the flip-flops.

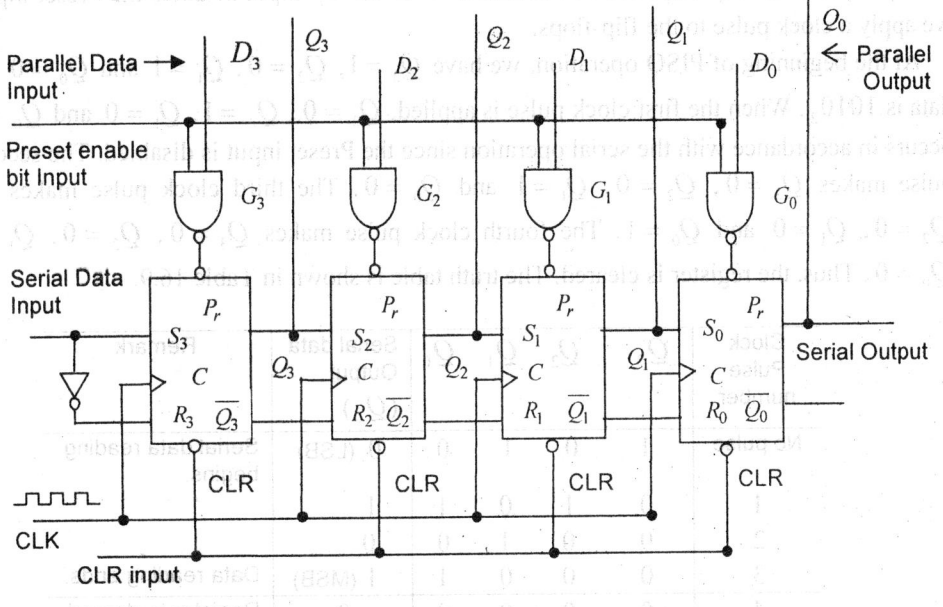

Fig. 16.26. Logic circuit diagram of a 4-bit universal shift register. The PIPO and PISO circuits are obtained by omitting the serial input terminal.

All the flip-flops are driven by the same clock. The parallel data bits D_3, D_2, D_1, D_0 is applied to the inputs of the NAND gates G_3, G_2, G_1 and G_0 respectively. The second input of the NAND gates are connected with the Preset input. The Preset input (P_r) of the flip-flops are low active in Fig. 16.26. The outputs of the NAND gates G_3, G_2, G_1 and G_0 are applied to the Preset inputs (P_r) of the flip-flops. The Preset enable bit input is made '1'. At the output of NAND gate G_K ($K = 3,2,1,0$) we get the bit $\overline{D_K \cdot 1} = \overline{D_K}$ for $K = 3,2,1,0$. Thus, if $D_K = 1$ for any K, we get a '0' at the P_r input of the flip-flop and the output of the flip-flop becomes $Q_K = 1$ since P_r is low-active input. In effect, the bit D_K is transferred to the output Q_K of the K-th flip-flop. On the other hand, when $D_K = 0$ for $K = 3,2,1,0$ we get $\overline{D_K \cdot 1} = \overline{D_K} = 1$ at the P_r input of the K-th flip-flop. Now, since P_r is low-active $P_r = 1$ disables the Preset input. Now, $Q_K = 0$ for $K = 3,2,1,0$ since the flip-flop was cleared initially. Thus, the input bit D_K is transferred to the flip-flop output Q_K. Thus, the input data is directly transferred to the the register and the output $Q_3 Q_2 Q_1 Q_0$ reads the input data. This is the function of a PIPO shift register. No action of clock is involved here.

16.3.2.4 PARALLEL-IN SERIAL-OUT (PISO) SHIFT REGISTER

The input parallel data is transferred to the shift register as mentioned in the PIPO operation. Then, make the input Preset enable bit '0'. This makes the output of all NAND gates '1'. Since the P_r input of all flip-flops are low active, a '1' at the P_r input disables the Preset input. Now, we apply a clock pulse to the flip-flops.

At the beginning of PISO operation, we have $Q_3 = 1$, $Q_2 = 0$, $Q_1 = 1$ and $Q_0 = 0$ since the data is 1010_2. When the first clock pulse is applied, $Q_3 = 0$, $Q_2 = 1$, $Q_1 = 0$ and $Q_0 = 1$. This occurs in accordance with the serial operation since the Preset input is disabled. The second clock pulse makes $Q_3 = 0$, $Q_2 = 0$, $Q_1 = 1$ and $Q_0 = 0$. The third clock pulse makes $Q_3 = 0$, $Q_2 = 0$, $Q_1 = 0$ and $Q_0 = 1$. The fourth clock pulse makes $Q_3 = 0$, $Q_2 = 0$, $Q_1 = 0$ and $Q_0 = 0$. Thus, the register is cleared. The truth table is shown in Table 16.9.

Clock Pulse number	Q_3	Q_2	Q_1	Q_0	Serial data Output (Q_0)	Remark
No pulse	1	0	1	0	0 (LSB)	Serial data reading begins.
1	0	1	0	1	1	
2	0	0	1	0	0	
3	0	0	0	1	1 (MSB)	Data reading ends.
4	0	0	0	0	0	Register is cleared.

Table 16.9. Truth table of a 4-bit PISO shift register. The input data 1010_2.

SOLVED PROBLEMS

16.1 An S-R flip-flop is driven by the following S and R inputs. This flip-flop is negative edge-triggered and initially RESET. The clock is C. Find the output (Q) waveform.

Solution

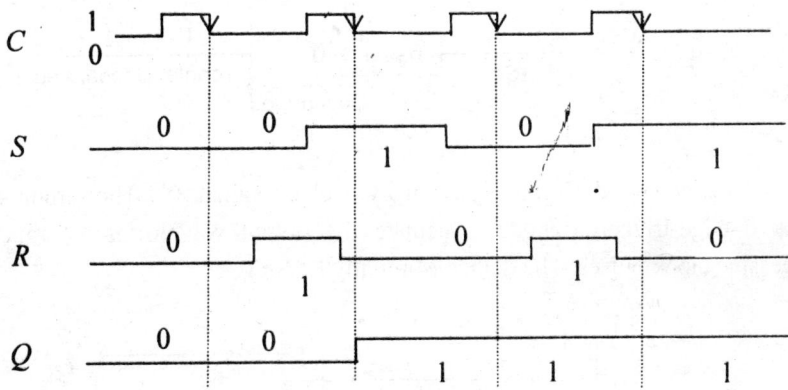

16.2 A negative edge-triggered D-flip-flop is driven by the following clock C and is initially RESET. The input D has the following waveform shown below. Find the output (Q) of this flip-flop.

Solution

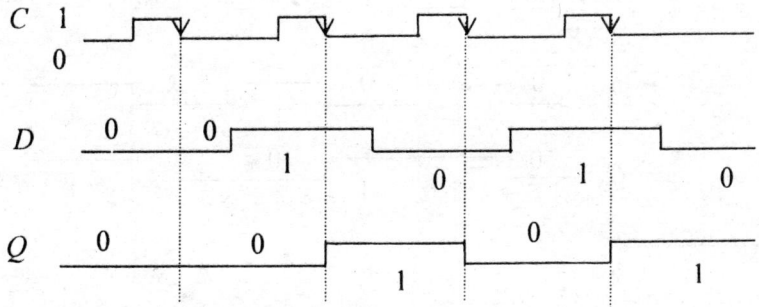

16.3 Calculate the Q output of the J-K FF with given clock C, J and K waveforms. Q is initially set.

Solution

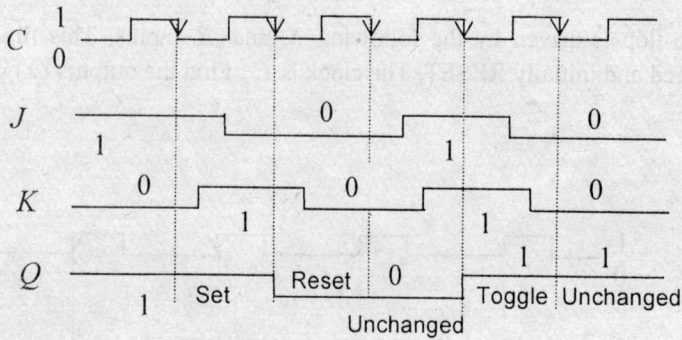

16.4 The following digital circuit is driven by a clock signal C. The circuit operates in the positive edge-triggered mode. Calculate the output waveforms Q_1 and Q_2 for five successive clock pulses. The device is initially RESET.

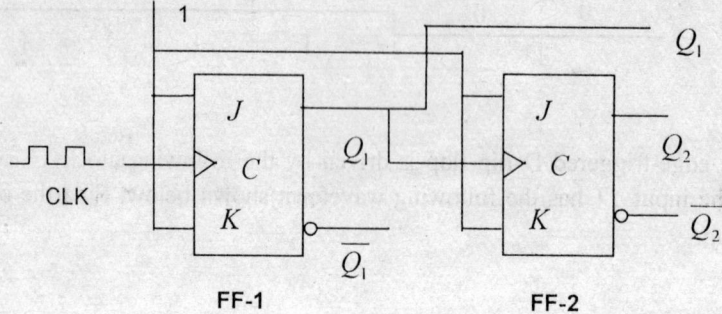

Solution

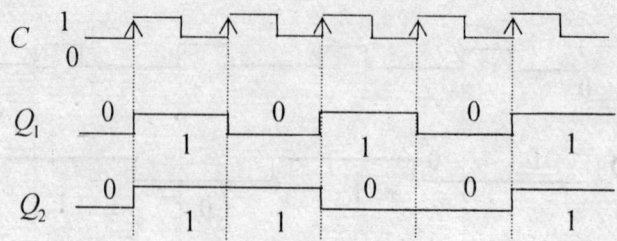

PROBLEMS

16.1 A positive edge-triggered D flip-flop is driven by the following clock C and it is initially RESET. Find the output Q of this flip-flop when the D input has the following waveform.

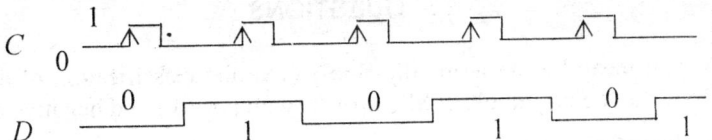

16.2 A positive edge-triggered S-R flip-flop is driven by the following clock C and it is initially RESET. Find its output waveform when S and R inputs have the following inputs.

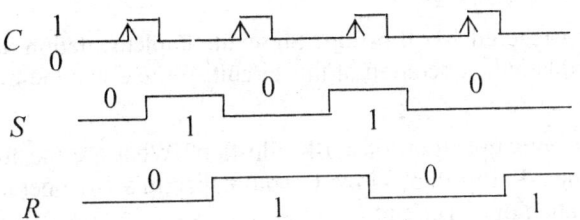

16.3 Find the Q output of the positive edge-triggered J-K flip-flop whose J and K input waveforms and clock waveform are given below. The flip-flop is initially RESET.

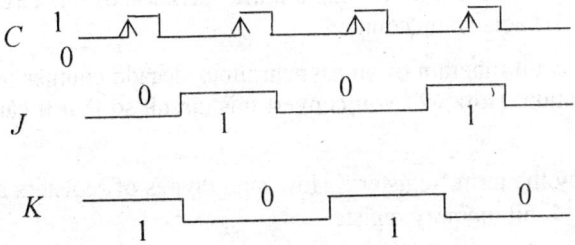

16.4 Find the pulse frequency of Q_1, Q_2 and Q_3 outputs of the following digital circuit. The J-K flip-flops are positive edge-triggered and are initially RESET. f_c is the clock frequency.

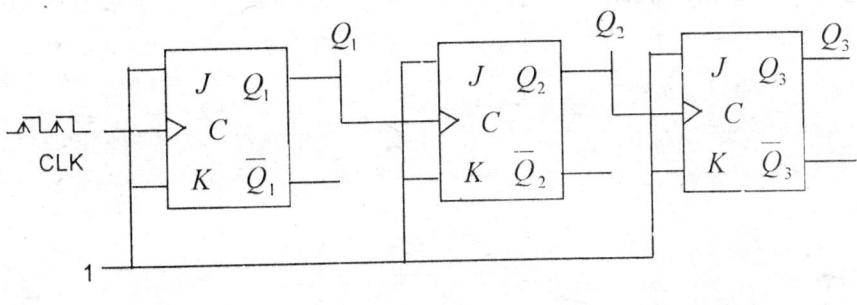

[**Ans.** $\dfrac{f_c}{2}$, $\dfrac{f_c}{4}$, $\dfrac{f_c}{8}$]

QUESTIONS

16.1 What do you mean by the term 'flip-flop'? Give the classification of edge-triggered flip-flops. Draw the logic symbols of SR flip-flop with positive and negative edge triggering.

16.2 Describe the principle of operation of an SR flip-flop. Write down the truth table of the SR flip-flop. Show using a logic diagram, the implementation of an edge-triggered flip-flop and explain the operation of this circuit.

16.3 What is a D-flip-flop? Draw the symbol of a D-flip-flop. Write down the truth table of a D-flip-flop and explain its operation.

16.4 What is an edge-triggered J-K flip-flop? Show the implementation of a J-K flip-flop using NAND gates. Explain the operation of this circuit. Write down the truth table of a J-K flip-flop.

16.5 What is asynchronous operation of a J-K flip-flop? What are the functions of preset and clear inputs of the J-K flip-flop? How do you realize this J-K operation in a logic circuit? Explain the operation of this circuit.

16.6 How do you realize a divide-by-4 frequency divider by using J-K flip flops? Describe the construction of a 2-bit binary counter using J-K flip-flops.

16.7 What do you mean by a binary asynchronous counter? Draw the logic circuit diagram of a 4-bit asynchronous down counter. Explain the operation of this circuit with the help of the timing diagram. What is an up counter?

16.8 Give the logic circuit diagram of an asynchronous decade counter using J-K flip-flops and explain its operation. How will you convert this circuit so that it can perform as a modulo 14 counter?

16.9 What is meant by the term 'register'? How many types of registers are there? Describe the functioning of a 4-bit memory register.

16.10 What is a 'shift register'? what are the different classes of the 'shift register'? Draw the logic circuit diagram of a 4-bit SISO shift register using SR flip-flops and explain its operation. Modify this circuit to perform as a 4-bit SIPO shift register.

16.11 What do you mean by the 'universal shift register'? Give the logic circuit diagram of a 4-bit universal shift register and describe its functioning. How do you obtain PIPO and PISO operations from this circuit?

Index